XIth European Conference on Animal Blood Groups and Biochemical Polymorphism

EUROPEAN SOCIETY FOR ANIMAL BLOOD GROUP RESEARCH
(E.S.A.B.R.)

XIth European Conference on Animal Blood Groups and Biochemical Polymorphism

Warsaw, July 2nd–6th, 1968

PROCEEDINGS

Springer-Science+Business Media, B.V.

Additional material to this book can be downloaded from http://extras.springer.com.

ISBN 978-94-017-5228-2 ISBN 978-94-017-5226-8 (eBook)
DOI 10.1007/978-94-017-5226-8

Originally published by PWN-Polish Scientific Publishers in 1970.
Softcover reprint of the hardcover 1st edition 1970

Organization of the Conference

The Conference was organized by European Society for Animal Blood Group Research. The Host of the Conference has been the Institute of Experimental Animal Breeding of the Polish Academy of Sciences

The Conference was held in Warsaw, July 2nd–6th, 1968

THE COMMITTEE OF THE EUROPEAN SOCIETY FOR ANIMAL BLOOD GROUP RESEARCH:

President

M. BRAEND • Norway

Secretary

J. BOUW • The Netherlands

Members

B. LARSEN • Denmark
GUNVOR LINDTRÖM • Finland
J. MATOUŠEK • Czechoslovakia
M. E. MCDERMID • Great Britain
JOLANTA GASPARSKA • Poland

THE POLISH ORGANIZING COMMITTEE:

Chairman

H. JASIOROWSKI • Warsaw

V-Chairman

JOLANTA GASPARSKA • Warsaw

Members

S. ALEXANDROWICZ • Poznań
H. BALBIERZ • Wrocław
LIDIA DOLA • Kraków
M. DUNIEC • Kraków
A. KACZMAREK • Poznań
CZESŁAWA LIPECKA • Lublin
S. WADOWSKI • Olsztyn
I. WIATROSZAK • Poznań
M. ŻURKOWSKI • Warsaw

Contents

Chapter 2: Blood Groups and Biochemical Polymorphism in Pigs

Chapter 3: Blood Groups and Biochemical Polymorphism in Poultry

Chapter 4: Blood Groups and Biochemical Polymorphism in Horse

Chapter 5: Blood Groups and Biochemical Polymorphism in Sheep and Goat

CHAPTER 6: BLOOD GROUPS AND BIOCHEMICAL POLYMORPHISM IN FISH

CHAPTER 7: BLOOD GROUPS AND HISTOCOMPATIBILITY IN MOUSE, RAT AND COYPU

CHAPTER 8: CORRELATIONS BETWEEN BLOOD GROUPS AND OTHER TRAITS

List of the Participants

AKROYD P., Unilever Research Laboratory, Colworth House, Sharnbrook, Bedford, Great Britain

ALABASTER J. S., Ministry of Agriculture, Fisheries and Food, 10 Whitehall Place, London S. W. 1, Great Britain

ALEXANDROWICZ S., Wyższa Szkoła Rolnicza, Katedra Szczegółowej Hodowli Zwierząt, Poznań-Sołacz, ul. Wołyńska 33, Poland

ANDRESEN E., Dept. of Animal Genetics, The Royal Veterinary and Agricultural College, Copenhagen, 13 Bulowsvej, Denmark

ASHTON G. C., University of Hawaii, Dept. of Genetics, Honolulu 14-96822, Hawaii, U.S.A.

BALBIERZ H., Wyższa Szkoła Rolnicza, Katedra Położnictwa, Wrocław, ul. Norwida 27, Poland

BALNER H., Radiobiological Institute TNO, Rijswijk, The Netherlands

BIEZIENKO S. T., Moskovskaya oblast, Podolskii reyon, Dubrovicy, WIZ, U.S.S.R.

BOUW J., Stichting Bloedgroepen Onderzoek, Duivendall 5, Wageningen, The Netherlands

BÖHM O., The Veterinary Institute of Slovenia, P. O. Box 257, Ljubljana, Yugoslavia

BRAEND M., Norges Veterinaerhogskole, Institut for Intermedisin, Ullevalsvn 72, Oslo, Norway

CÖP W., Stichting Bloedgroepen Onderzoek, Duivendall 5, Wageningen, The Netherlands

CSUKA J., Dept. of Genetics, University of Agriculture, Zemedelska 5, V Brno, Czechoslovakia

DĘBSKI St., Zakład Hodowli Doświadczalnej Zwierząt Polskiej Akademii Nauk, Warszawa, ul. Nowy Świat 72, Poland

DIKOV V., Institut po Biologii i Patologii na Rasmojavanete, na S.S. Jivetni, Sofia 13, Bul. Lenin 55, Bulgaria

DOBREW D., Sofia, Dragan Zankov, Nr 8, Bulgaria

DOGRUL F., Blutgruppenlabor. des Institutes für Tierzucht der Tierärztlichen Hochschule, Wien III, Linke Bahngasse 11, 1030 Wien, Austria

DOLA LIDIA, Wyższa Szkoła Rolnicza, Zakład Hodowli Bydła, Laboratorium Grup Krwi, Kraków, Al. Mickiewicza 24/28, Poland

DOSTÁL J., The Czechoslovak Academy of Sciences, Laboratory of Physiology and Genetics of Animals, Liběchov, Czechoslovakia

DORYNEK Z., Wyższa Szkoła Rolnicza, Katedra Szczegółowej Hodowli Zwierząt, Poznań-Sołacz, ul. Wołyńska 33, Poland

DROBNÁ Vlasta, Poultry Research Institute, Ivanka pri Dunaji, Czechoslovakia

DUNIEC M., Instytut Zootechniki, Zakład Immunogenetyki, Balice k/Krakowa, Poland

DUYONOW E. A., Institute of Poultry Science of the U.S.S.R. Academy of Sciences, Zagorsk 11, U.S.S.R.

EFREMOW G., Department of Physiology and Biochemistry, Faculty of Agriculture, University of Skopje, Yugoslavia

ERHARD L., Institut für Blutgruppenforschung der Tierzuchtforschung e.V. 8, München 15, Haydnstrasse, G.F.R.

ERMENKOVA LIDIA, Academy of Agricultural Sciences, Research Institute of Livestock Breeding, Sofia-Kostinbrod, Bulgaria

FESÜS L., Bloodgrouping Laboratory, University of Veterinary Medicine, VII Rottenbiller u. 50, Budapest, Hungary

FIORENTINI A., University di Milano, Facolta di Medicina Veterinaria, Cattedra di Zootecnia Generale, Via Celoria 10, Milano, Italy

GASPARSKA JOLANTA, Zakład Hodowli Doświadczalnej Zwierząt Polskiej Akademii Nauk, Warszawa, ul. Nowy Świat 72, Poland

GLASNÁK V., C. B. Station, Hradiško, okr. Praha Zapad., Czechoslovakia

GRZYBOWSKI G., Zakład Hodowli Doświadczalnej Zwierząt Polskiej Akademii Nauk, Warszawa, ul. Nowy Świat 72, Poland

HÁLA K., The Czechoslovak Academy of Sciences, Institute of Experimental Biology and Genetics, Flemingovo nam. 2, Prague 6, Czechoslovakia

HALL J. G., A.R.C. Animal Breeding Research Organization, West Mains Road, Edinburgh 9, Great Britain

HARDY J., Unilever Research Laboratory, Colworth House, Sharnbrook, Bedford, Great Britain

HASLER JUDITH, Department of Meat and Animal Science, University of Wisconsin, Madison, U.S.A.

HESS R., Institut für Künstliche Besamung V.V.B. — Tierzucht, Schönow b./Bernau, G.D.R,

HOJNÝ J., The Czechoslovak Academy of Sciences, Laboratory of Physiology and Genetics of Animals, Liběchov, Czechoslovakia

HRADECKY J., The Czechoslovak Academy of Sciences, Laboratory of Physiology and Genetics of Animals, Liběchov, Czechoslovakia

IMLAH F., University of Edinburgh, Blood Group Research Unit, Royal School of Veterinary Studies, Easter Bush Field Station, Roslin, Midlothian, Great Britain

INGRAM D. G., University of Guelph, Ontario Veterinary College, Guelph, Ontario, Canada

IVÁNYI P., The Czechoslovak Academy of Sciences, Institute of Experimental Biology and Genetics, Flemingovo nam. 2, Prague 6, Czechoslovakia

JAMIESON A., Ministry of Agriculture, Fisheries and Food Fisheries Laboratory, Lowestoft, Suffolk, Great Britain

JASIOROWSKI H., Zakład Hodowli Doświadczalnej Zwierząt Polskiej Akademii Nauk, Warszawa, ul. Nowy Świat 72, Poland

HYLDGAARD Jensen J., Dept. of Physiology, Endocrinology and Bloodgrouping, The Royal Veterinary and Agricultural College, Copenhagen V, 13 Bulowsvej, Denmark

JOVANOVIČ V., Agricultural Faculty, Novi Sad, Yugoslavia

JOVER A., Consedo Superior Investigationes Cientificas, Dept. de Zootecnia, Facultad de Veterinaria, Cordoba, Spain

JOZOV S. I., Academy of Agricultural Sciences, Research Institute of Livestock Breeding, Sofia-Kostinbrod, Bulgaria

KACZMAREK A., Wyższa Szkoła Rolnicza, Katedra Szczegółowej Hodowli Zwierząt, Poznań-Sołacz, ul. Wołyńska 33, Poland

KAMINSKI MARIE, Centre National de la Recherche Scientifique, Laboratoire d'Enzymologie, 91, Gif-sur-Yvette, France

KASZYCKA HALINA, Wyższa Szkoła Rolnicza, Zakład Hodowli Bydła, Laboratorium Grup Krwi, Kraków, Al. Mickiewicza 24/26, Poland

KIROV A., Sofia, ul. Midżur 24, Bulgaria

KLOBUKOW G. F., Moskovskaya oblast', Podolskij reyon, Dubrowitsy, WIZ, U.S.S.R.

KLOUDA P., The Czechoslovak Academy of Sciences, Institute of Experimental Biology and Genetics, Flemingovo nam. 2, Prague 6, Czechoslovakia

KOSSAKOWSKI J., Zakład Hodowli Doświadczalnej Zwierząt Polskiej Akademii Nauk, Warszawa, ul. Nowy Świat 72, Poland

KOWNACKI M., Zakład Hodowli Doświadczalnej Zwierząt Polskiej Akademii Nauk, Warszawa, Nowy Świat 72, Poland

KOVÁCS G., Bloodgrouping Laboratory, University of Veterinary Science, VII Rottenbiller u. 50, Budapest, Hungary

KŘEN V., Charles University, Faculty of General Medicine, Dept. of Biology, Albertov 4, Prague 2, Czechoslovakia

KŘENOVÁ DRAHOMIRA, Charles University, Faculty of General Medicine, Dept. of Biology, Albertov 4, Prague 2, Czechoslovakia

KRUPIŃSKA WIESŁAWA, Instytut Zootechniki, Zakład Immunogenetyki, Balice k/Krakowa, Poland

KÚBEK A., CSAV, Laboratory of Physiology and Genetics of Animals, Liběchov, Czechoslovakia

LANG B. G., University of Edinburgh, Blood Group Research Unit, Royal School of Veterinary Studies, Easter Bush Field Station, Roslin, Midlothian, Great Britain

LAZAR P., The Veterinary Institute of Slovenia, P.O. Box 257, Ljubljana, Yugoslavia

LIE HALLDIS, Norges Veterinaerogskole, Institut for Indremedisin, Ullevalsvn 72, Oslo 4, Norway

DE LIGNY WILHELMINA, Rijksinstitut voor Visserij, Onderzoek, Haringkade 1, Ijmiiden, The Netherlands

LINDSTRÖM GUNVOR, Kuriiritie 11, Tikkurila, Finland

LINHART J., C.B. Station, Hradiško, okr. Praha Zapad, Czechoslovakia

LINKE L., Zakład Hodowli Doświadczalnej Zwierząt Polskiej Akademii Nauk, Warszawa, ul. Nowy Świat 72, Poland

LIPECKA CZESŁAWA, Wyższa Szkoła Rolnicza, Katedra Szczegółowej Hodowli Zwierząt, Lublin, ul. Króla Leszczyńskiego 9, Poland

LISZKA EWA, Wyższa Szkoła Rolnicza, Katedra Hodowli Zwierząt, Kraków, Al. Mickiewicza 24/28, Poland

LÖHLE K., Institut für Geflügel und Pelztierzucht der Humboldt Universität, Berlin 104, Invalidenstrasse 42, G.D.R.

LUSH I. E., Royal Free Hospital School of Medicine, University of London, 8 Hunter Street, London WC1, Great Britain

MADEYSKA HANNA, Zakład Hodowli Doświadczalnej Zwierząt Polskiej Akademii Nauk, Warszawa, ul. Nowy Świat 72, Poland

MAJEWSKA HANNA, Zakład Hodowli Doświadczalnej Zwierząt Polskiej Akademii Nauk, Warszawa, ul. Nowy Świat 72, Poland

MAJOR, F., Institut für Tierzucht und Haustiergenetik der Universität Göttingen, Albrecht-Thaer-Weg 1, 34, Göttingen, G.F.R.

MAKAVEYEV Ts., Institute of Animal Breeding, Sofia-Kostinbrod, Bulgaria

MANSILLA RICO, Patronato de Biologia Animal, Servicio de Fisiozootecnica, Avda Puerta de Hierro, Madrid 3, Spain

MATOUŠEK J., The Czechoslovak Academy of Sciences, Laboratory of Physiology and Genetics of Animals, Liběchov, Czechoslovakia

MCDERMID M., Robert Jones and Agnes Hunt Orthopaedic Hospital, The Charles Salt Research Centre, Osvestry, Shropshire, Great Britain

MICHALAK WIESŁAWA, Zakład Hodowli Doświadczalnej Zwierząt Polskiej Akademii Nauk, Warszawa, ul. Nowy Świat 72, Poland

MICHALAK W., Zakład Hodowli Doświadczalnej Zwierząt Polskiej Akademii Nauk, Warszawa, ul. Nowy Świat 72, Poland

MOUSTGAARD J., Dept. of Physiology, Endocrinology and Bloodgrouping, The Royal Veterinary and Agricultural College, Copenhagen V, 13 Bülowsvej, Denmark

MULSOV D., Institut für Geflügel und Pelztierzucht der Humboldt Universität, Berlin 104, Invalidenstrasse 42, G.D.R.

MURAVIEV W. I., Institute of Poultry Science of the U.S.S.R. Academy of Sciences, Zagorsk 11, U.S.S.R.

NEUSSER W., Zakład Hodowli Doświadczalnej Zwierząt Polskiej Akademii Nauk, Warszawa, ul. Nowy Świat 72, Poland

NIKOŁAJCZUK Maria, Wyższa Szkoła Rolnicza, Katedra Położnictwa, Wrocław, ul. Norwida 27, Poland

OLBRYCHT FRANCISZKA, Warszawa, ul. Marszałkowska 140 m. 135, Poland

OLBRYCHT Krystyna, Zakład Hodowli Doświadczalnej Zwierząt Polskiej Akademii Nauk, Warszawa, ul. Nowy Świat 72, Poland

OKERMAN F. A., Rijksstation voor Kleinveeteelt, Burgs van Gansberghelaan 26, Lemberge-Merelbeke, Belgium

OOSTERLEE C. C., Stichting Bloedgroepen Onderzoek, Duivendaal 5, Wageningen, The Netherlands

ORMIAN M., Wyższa Szkoła Rolnicza, Katedra Szczegółowej Hodowli Zwierząt, Kraków, Al. Mickiewicza 24/28, Poland

OSTERHOFF D. R., Blood Groups Laboratory, Animal, Husbandry and Dairy Research Institute, P.O. Onderstepoort, Pretoria, South Africa

OWEN R. D., California Institute of Technology, Pasadena, California 91109, U.S.A.

PAPP M., Blood Grouping Laboratory, University of Veterinary Medicine, VII Rottenbiller u. 50, Budapest, Hungary

PAWŁOWSKI K., Wyższa Szkoła Rolnicza, Katedra Szczegółowej Hodowli Zwierząt, Kraków, Al. Mickiewicza 24/28, Poland

PAVEL ÜLO, Pälsoni 12-6, Tartu, Estonian S.S.R., U.S.S.R.

PICKUP GILL, University College of North Wales, David Jones Laboratory, College Farm, Aber Llanfairfechen Caernarvonshire, Great Britain

PIOTROWSKI J., Zakład Hodowli Doświadczalnej Zwierząt Polskiej Akademii Nauk, Warszawa, ul. Nowy Świat 72, Poland

PILZ J., Institut für Künstliche Besamung V.V.B. Tierzucht, Schönow b/Bernau, G.D.R.

PETRYK TERESA, Wyższa Szkoła Rolnicza, Katedra Szczegółowej Hodowli Zwierząt, Kraków, Al. Mickiewicza 24/28, Poland

PETROWSKÝ E., Dept. of Genetics, University of Agriculture, Zemedelska 5, V Brno, Czechoslovakia

PERRAMON A., Centre National de Recherches Zootechniques, Youy-en-Josas, France

PODLIACHOUK L., Institut Pasteur, Laboratoire d'Hemopatologie et des Groupes Sanguins, 28 Rue du Docteur Roux, Paris XVe, France

PORCZYŃSKI J., Zakład Hodowli Doświadczalnej Zwierząt Polskiej Akademii Nauk, Warszawa, ul. Nowy Świat 72, Poland

PORĘBSKA WANDA, Wyższa Szkoła Rolnicza, Katedra Szczegółowej Hodowli Zwierząt, Kraków, Al. Mickiewicza 24/28, Poland

RAPACZ J., Instytut Zootechniki, Zakład Immunogenetyki, Balice k/Krakowa, Poland

RENDEL J., Tjänste Lantbrukshögskolan, Institutionen for Husdjurstordling 75005, Uppsala 7, Sweden

RODERO A., Consedo Superior de Investigationes Cientificas, Dept. Zootecnia, Facultad de Veterinaria, Córdoba, Spain

ROGOZIŃSKA ZOFIA, Zakład Hodowli Doświadczalnej Zwierząt Polskiej Akademii Nauk, Warszawa, ul. Nowy Świat 72, Poland

RYNIEWICZ Z., Zakład Hodowli Doświadczalnej Zwierząt Polskiej Akademii Nauk, Warszawa, ul. Nowy Świat 72, Poland

Saison Ruth, University of Guelph, Ontario Veterinary College, Guelph, Ontario, Canada

Samodielkina Swietlana, Institute of Poultry Science of the U.S.S.R., Zagorsk 11, U.S.S.R.

Sandberg K., Tjänste Lantbrukshögskolan, Institutionen for Husdjursfördling 75007, Uppsala 7, Sweden

Sartore G., Istituto di Fisiologia e Chimica Biologica, Facoltà di Medicina Veterinaria, Via Nizza 52, Torino, Italy

Schleger W., Institut für Tierzucht der Tierärztlichen Hochshule, Wien III, Linke Bahngasse 11, 1030 Wien, Austria

Schmid D. O., München 22, Reitmerstrasse 25, G.F.R.

Schröffel J., Central Breeding Station, Blood Groups Dept., Liběchov, Czechoslovakia

Scott A. M., Equine Research Station, Newmarket, Suffolk, Great Britain

Siudziński S., Wyższa Sźkoła Rolnicza, Katedra Szczegółowej Hodowli Zwierząt, Poznań-Sołacz, ul. Wołyńska 33, Poland

Sirbu Z., Pasteur Institute for Veterinary Research and Biological Products, Splaiul, Independentei 105, Bucharest, Rumania

Składanowska Elżbieta, Zakład Hodowli Doświadczalnej Zwierząt Polskiej Akademii Nauk, Warszawa, ul. Nowy Świat 72, Poland

Soos P., Központi Mesterséges Termékényitesi Föállomás Vercsoport Laboratorium, Remény u. 42, Budapest XIV, Hungary

Sotillo J. L., Patronate de Biologia Animal Servicio de Fisiozootecnica, Avda Puerta de Hierro, Madrid 3, Spain

Spooner G., 3011 Bemerode, Ostarmeierstr. 12, Hannover, G.F.R.

Spooner Roger L., Agricultural Research Council, Cattle Blood Typing Service, Animal Breeding Research Organization, West Mains Road, Edinburgh 9, Great Britain

Staliński Z., Wyższa Szkoła Rolnicza, Katedra Ogólnej Hodowli Zwierząt, Kraków, Al. Mickiewicza 24/28, Poland

Štark O., Charles University, Faculty of General Medicine, Dept. of Biology, Albertov 4, Prague 2, Czechoslovakia

Stojanović Z., Agricultural Faculty, Novi Sad, Yugoslavia

Stormont C., Dept. of Veterinary Microbiology, School of Veterinary Medicine, University of California, Davis, California 95616, U.S.A.

Stratil A., CSAV, Laboratory of Physiology and Genetics of Animals, Liběchov, Czechoslovakia

Stukovsky J., Központi Mesterséges Termékényitesi Föállomás Vercsport Laboratorium, Remény u. 42, Budapest XIV, Hungary

Sukhova Nina O., Sibirskii Nauczno-Issledovatelskii Institut Ziwotnovodstva, Novosibirsk, Spartaka 7, U.S.S.R.

Suzuki Shozo, Laboratory of Animal Breeding, Tokyo University of Agriculture, Setagaya-ku, Tokyo A-1, Sakuragaoko, Japan

Szeniawska Danuta, Zakład Hodowli Doświadczalnej Zwierząt Polskiej Akademii Nauk, Warszawa, ul. Nowy Świat 72, Poland

Szmelik Ligia, Zakład Hodowli Doświadczalnej Zwierząt, Warszawa, Nowy Świat 72, Poland

Switek M., Wyższa Szkoła Rolnicza, Katedra Szczegółowej Hodowli Zwierząt, Poznań-Sołacz, ul. Wołyńska 33, Poland

Szynkiewicz Ewa, Szkoła Główna Gospodarstwa Wiejskiego, Katedra Ogólnej Hodowli Zwierząt, Brwinów k/Warszawy, Poland

Tikhonov V. N., Siberian Dept. of the U.S.S.R. Academy of Sciences, Institute of Cytology and Genetics, Novosibirsk 90, U.S.S.R.

Tucker Elisabeth, A.R.C. Institute of Animal Physiology, Babraham, Cambridge, Great Britain

TYMOWSKI J., Zakład Hodowli Doświadczalnej Zwierząt Polskiej Akademii Nauk, Warszawa, ul. Nowy Świat 72, Poland

VASENIUS L., Institutionen for Huddjurshygien Veterinärhögskolen, Helsinki, Finland

VESELSKÝ L., The Czechoslovak Academy of Sciences, Laboratory of Physiology and Genetics, Liběchov, Czechoslovakia

VYSHINSKY F. S., Institute of Poultry Science of the U.S.S.R. Academy of Sciences, Zagorsk 11, U.S.S.R.

WACHAR L. F., Estonian S.S.R. Tartu, Kreucwaldi 1, U.S.S.R.

WADOWSKA IZABELLA, Wyższa Szkoła Rolnicza, Katedra Szczegółowej Hodowli Zwierząt, Olsztyn-Kortowo, Poland

WADOWSKI S., Wyższa Szkoła Rolnicza, Katedra Szczegółowej Hodowli Zwierząt, Olsztyn-Kortowo, Poland

WAGONIS Z. C., Lithuanian S.S.R., Bajsogola, Pergales 3-1, U.S.S.R.

WIATROSZAK I., Wyższa Szkoła Rolnicza, Katedra Szczegółowej Hodowli Zwierząt, Poznań-Sołacz, ul. Wołyńska 33, Poland

WILKINS N. P., Dept. of Agriculture and Fisheries for Scotland, Marine Laboratory, P.O. Box 101, Victoria Road, Torry, Aberdeen, Great Britain

ZABŁOCKI B., Uniwersytet Łódzki, Katedra Mikrobiologii, Łódź, ul. Nowopołudniowa 12/16, Poland

ZAGULSKI T., Zakład Hodowli Doświadczalnej Zwierząt Polskiej Akademii Nauk, Warszawa, ul. Nowy Świat 72, Poland

ŻURKOWSKI M., Zakład Hodowli Doświadczalnej Zwierząt Polskiej Akademii Nauk, Warszawa, ul. Nowy Świat 72, Poland

OPENING OF THE CONFERENCE

Dr. JOLANTA GASPARSKA

Mister President, Ladies and Gentlemen,

At the Paris Conference I had the honour to invite you on behalf of our Institute of the Polish Academy of Sciences in Warsaw to take part in this Conference.

Today I have even a greater pleasure in welcoming you here and saying that the preparation of the Conference was very interesting and we enjoyed it tremendously.

We did our best to make the XIth Conference on Blood Groups a success and we hope that this week spent in Poland will remain in your memory as a useful and pleasant one.

We ask you kindly to feel here as if you were at home and if you happen to have some troubles please contact us, we are always at your disposal.

PROF. DR. S. ALEXANDROWICZ

Mister President, Ladies and Gentlemen,

In the absence of the Chairman of the Polish Academy of Sciences, Prof. Groszkowski and the Minister of Agriculture Prof. Okuniewski, who unfortunately were not able to join us today, I have the honour to welcome you all at the Conference.

We are happy that the Conference is being held in Poland. I wish you all very successful talks and hope you are going to enjoy your time in our country.

PROF. DR. H. JASIOROWSKI

Mister President, Ladies, Gentlemen, Dear Guests,

On the behalf of the Polish Organizing Committee and the Institute of Experimental Animal Breeding, which has the honour to be the host for this International Conference, I would like to welcome you in Warsaw.

We are very proud to welcome in Poland the representatives of somehow young but quickly developing and promising branch of science in the field of animal production.

Due to the small number of researchers that are reckoned among the pioneers of animal blood group research in Europe, this branch of science was at the begining based upon strong friendly international cooperation. As the result of this activity the European Society for Blood Group Research was founded.

If we can today, after 20 years from the beginning of these researches, receive over 120 scientists from many countries, if in almost all European countries a number of laboratories for animal blood groups, immunogenetics and proteins and enzymes polymorphism research were founded, it is undoubtedly due to the activity of this Society and specially of its founders.

It is the general opinion that as far as the activity, and the spirit and efficiency of international cooperation is concerned, many other branches of the animal science can learn a lot from your experience.

The Polish scientists in the field of animal production have not only many reasons to be proud, beeing the hosts for such prominent international gathering we have also many reasons to be deeply grateful to the Society and to many of its members. In Poland research work of animal blood groups has been started on the larger scale by Professor Spryszak in our institute in 1955. Dr. Spryszak has greatly contributed to the beginning of this research in Poland.

But the fact that in Poland we have now 5 animal blood group laboratories with the staff of about 30 people we owe also to our foreign friends, to some members of this Society present today. They contributed in the most important way, helping us to train the number of young scientists in the field of animal blood group research work.

11 persons working now in Poland in this field, received some long-term training abroad. Among these 7 persons from our Blood Group Laboratory.

Please let me take advantage of the opportunity to convey my warmest thanks to all foreign teachers of our research workers who kindly allowed them not only to use their laboratories but also their time and knowledge.

Let me mention those who contributed in training our staff and to whom we are really very thankful.

They are: Dr. Podliachouk from France, Dr. Braend from Norway, Dr. Bouw from The Netherlands Dr. Rendel from Sweden, Dr. Stormont and Professor Irwin from the United States of America.

Their contribution to the animal blood group research in Poland is the greatest because they helped us to build up the staff of our laboratories. Excuse my mentioning this while adressing this Conference. But the International Societies can work efficiently only in the spirit of friendly atmosphere with full understanding in spite of all the political, social, cultural and economical differences between countries and nations. The best contribution to this can be made by personal contacts.

It is our hope that your stay in Poland will contribute also to this great idea.

So you see that we have many reasons to be happy that this Conference is being held in Poland.

I would like to assure you that we will do our best so that you feel well in our country, as among the good friends. I wish you very successful meeting and discussions.

PROF. DR. M. BRAEND

Ladies and Gentlemen,

It is a great honour and pleasure for me, on behalf of European Society for Animal Blood Group Research, to express our appreciation and gratitutde to our Polish hosts for their ready willingness and their invitation to arrange the XIth European Conference on Animal Blood Groups and Biochemical Polymorphism. These our strongly felt thanks I would like, at the same time as thanking for the nice words of welcome with which we were received this morning, address to the representative for our hosts Prof. Dr. Alexandrowicz, Polish Academy of Sciences and Ministry of Agriculture and Prof. Dr. Jasiorowski, Director for the Experimental Animal Breeding.

The arrangement of our conference has grown into a very big job which puts a heavy burden on the host in the matter of expenses and work. This development is primarily caused by the rapid growth of our Society, a development which calls for some consideration and discussions. How large do we want our Society to be? We would of course not be able to control the growth of our Society entirely. But we might be able to do it to a certain extent. Thus, if we restrict ourselves to animal blood only, the Society should not grow so rapidly as if we broaden our field and include more animal fluid and tissue characteristics. We perhaps should strive against a Society of an ideal size, but to my knowledge a such one is not yet known. Most probably it should neither be too large nor to small, but some place in between.

A question related to the size of our Society is the size and scope of our proposed journal. At the end of this Conference we are going to discuss an eventual animal blood group journal. We then face the problem. How much do we want to cover with our journal. Shall we identify ourselves with animal blood only or shall we give it a much broaden scope.

The growth of our Society shows the increasing importance and interest in our research field, a fact for which we all should be happy. This development implies that many of you are participants in one of our conferences for the first time. I am happy to welcome the newcomers hoping that you will feel at home in our Society already at once and that you shall benefit from your membership. I also welcome all our old members. It is nice meeting you again. A special word of welcome I want to extend to our four guests speakers, Prof. Dr. Zabłocki, Prof. Dr. Owen, Prof. Dr.

Stormont and Dr. Balner. We are again very fortunate to have distinguished guest lecturers and are very grateful and happy that they have found time to come and speak to us. We are eagerly looking forward to listening to you.

At last but not least I have the great pleasure to extend our most sincere appreciation and thanks to our Polish colleagues who have carried through the organization of this Conference, in particular the Polish Organizing Committee represented by its chairman Prof. Dr. H. Jasiorowski. We all know the heavy work and all the time spent in planning and carrying out everything so that it should fit into the smallest detail. Our Polish hosts have done this enthusiastically and efficiently. Our Society is greatly indebted for this, but we have no other way of expressing our most sincere and deeply felt gratitude than to give our very best thanks. In doing this I want especially to mention our *ex-officio* member the vice-chairman of the Polish Organizing Committee Dr. Jolanta Gasparska. We are extremely grateful to you and your colleagues for all the good work you have done for the Society and thereby for progress in the field of animal blood groups.

Ladies and Gentlemen, I have the honour to open the XIth European Conference on Animal Blood Groups and Biochemical Polymorphism.

GENERAL REPORTS

XITH EUROPEAN CONFERENCE ON ANIMAL BLOOD GROUPS AND BIOCHEMICAL POLYMORPHISM
Warsaw, July 2nd–6th, 1968

Physicochemical and Immunological Basis of Immunoglobulin Heterogeneity

B. ZABŁOCKI

Head of the Department of Microbiology, University of Łódź, Łódź, Poland

INTRODUCTION

The history of immunoglobulins is still relatively short, their definition was introduced only 31 years ago by TISELIUS (1937), TISELIUS and KABAT (1939). Peculiar characteristics of these proteins are: (a) antibody activity (b) they are the slowest migrating electrophoretic group of serum proteins, (c) extensive heterogeneity, (d) they are formed in plasmacytes and lymphoid cells.

Their antibody function and their role in pathology, are the reasons why so much work was, and still is, devoted to their study. The immunoglobulins have been most thoroughly studied not in man only but also in a number of commonly employed laboratory animals, such as rabbits, guinea pigs, mice, horses, rats. Hitherto many excellent reviews dealing with the structure and biological properties of the immunoglobulins were published: HEREMANS (1960), FAHEY (1962), FRANKLIN (1964), COHEN and PORTER (1964), PUTNAM (1965), ZABŁOCKI (1965, 1966), NEZLIN (1966), BUCHOWICZ (1966), SKAŁBA (1967), COHEN and MILSTEIN (1967), STANWORTH and PARDOE (1967), WEIR (1967).

TABLE 1

Fractionation of proteins (methods of separation)

Methods	Remarks
I. Salt fractionation (precipitation with concentrated neutral salts), PENNELL (1960)	It is still a useful procedure, particularly when dealing with large volumes. First step to chromatograph separation
II. Precipitation with cold ethanol or other solvents under rigidly controlled conditions (COHN *et al.*, 1946, 1950)	Factors which affect solubility: pH, t°, ionic strength, dielectric constant, protein concentration, other ions. Widely employed in the commercial preparation of gamma-globulin, albumin, etc.

Methods	Remarks
III. Column chromatography:	
A. Adsorption chromatography	
1. Ion exchange resins	
2. Calcium phosphate	
3. Cellulosic ion exchangers (PETERSON and SOBER, 1956, 1960)	Anion exchangers: DEAE-, ECTEOLA-, TEAE-cellulose Cation exchangers: CM-, P-, SE-, cellulose
4. Sephadex ion exchangers	Attachement of DEAE, CM, SE groups to cross linked dextran (Sephadex)
B. Molecular-sieve chromatography (gel filtration or exclusion chromatography)	Sephadex grade: G-25, G-50, G-75, G-100, G-200 Recently also particles of polyacrylamide gel, agar, agarose
IV. Dialysis	Cellophane membranes
V. Ultrafiltration	Collodion membranes
VI. Electrophoresis	
A. Free-boundary electrophoresis (LONGSWORTH, 1959)	Classic technique of Tiselius
B. Electrophoresis convection	This method is limited to produce only two fractions in one step
C. Zone electrophoresis	
1. Density gradient electrophoresis (SVENSON, 1960)	An impressive apparatus has recently been developed for rapid automatic electrophoresis of proteins by SKEGGS and HOCHSTRASSER (1962)
2. Electrophoresis on paper strips and similar supporting structures (membranes of cellulose acetate)	One of the most widely used analytical tools
3. Continuous flow electrophoresis (WINSTEN *et al.*, 1963)	
4. Zone electrophoresis in granular media (bloc and column electrophoresis)	Materials: starch granules, cellulose powder, Sephadex, granulated agarose, Pevikon C-870, plastic powders
5. Electrophoresis in gels (WUNDERLY, 1960; SMITHIES, 1959; RAYMOND, 1962)	Electrophoresis in nonsieving gels and molecular-sieve electrophoresis (starch-gel, polyacrylamide-gel)
VII. Sedimentation	
A. Boundary ultracentrifugation in analytical rotors (WILLIAMS, 1963; TRAUTMAN, 1964)	
B. Preparative ultracentrifugation (CHARLWOOD, 1966; STANWORTH, 1967)	Zone-centrifugation is the technique of most value to the experimental immunologist

As a result of newer techniques of isolation, it is now possible to obtain these proteins and specific antibodies in a state of high purity (Table 1).

The application of a complex of modern immunological methods has contributed to our better understanding of the antigenic properties of immunoglobulins and their relationships to each other (Table 2).

The elaboration of a group of paraproteins as seen in multiple myeloma and macroglobulinemia in man, as well as by certain transplantable tumors in experimental animals, have given much insight into the structure and also the function of the immunoglobulins.

TABLE 2

Immunological methods used in protein research

Methods	Remarks
Gel-diffusion techniques for immunoprecipitation (OUDIN, 1946; OUCHTERLONY, 1948; ELEK, 1948)	Simple diffusion in one dimension (tube technique OUDIN, 1952) Simple diffusion in two dimensions (plate technique OUCHTERLONY, 1962) Double diffusion in one dimension (OAKLEY and FULTHROPE, 1953) Double diffusion in two dimensions (OUCHTERLONY, 1962)
Immunoelectrophoretic techniques (GRABAR and WILLIAMS, 1953)	Macrotechnique (GRABAR and BURTIN, 1964) Microtechnique (SCHEIDEGGER, 1955) Quantitative (BACKHAUSZ *et al.*, 1960) Antigen-antibody crossed electrophoresis (LAURELL, 1965)
Spectrofluorometric methods (PARKER, 1963, 1967)	Fluoroquenching of purified antibody (EISEN, 1964) Fluorescence polarization (HABER and BENNETT, 1962)
Equilibrium dialysis (EISEN and KARUSH, 1949; PINCKARD and WEIR, 1967)	Procedure used not only for studies on the interaction of antibody and hapten but as a tool to study changes in the antigen combining site
Immunoferritin technique (SINGER, 1959; ANDERS *et al.*, 1967)	This method was applied to a wide variety of problems (e.g. localization of antigens, cell antibody production)
Immunofluoroscence (COONS *et al.*, 1941, COONS and KAPLAN 1950)	Direct staining, indirect staining, mixed antiglobulin staining, use of two specific conjugates: one labelled with fluorescein, the other with rhodamine (NAIRN, 1964)

Methods	Remarks
Trace labelling with radioiodine (FREEMAN, 1967; CAMPBELL *et al.*, 1956)	Two iodine isotopes are available: ^{131}I, ^{125}I
Radioimmunoassay (HUNTER, 1967; BERSON *et al.*, 1956)	This technique was used firstly in endocrinology
Passive hemagglutination (MIDDLEBROOK, DUBOS, 1948; BOYDEN, 1951; PRESSMAN *et al.*, 1942; INGRAHAM, 1958; HERBERT, 1967; STAVITSKY, 1964)	Red blood cells have been found to be extremely convenient passive carriers of antigens
Passive cutaneous anaphylaxis (OVARY, 1952, 1964)	The animal of choice for PCA is the guinea pig (albino 250 g)
Plaque technique for recognition antibody producing cells (JERNE and NORDIN, 1963; INGRAHAM and BUSSARD, 1964)	This technique has considerably speeded the study of cellular events of interest to the immunologist
Method of affinity labelling (WOFSY *et al.*, 1962; METZGER *et al.*, 1963)	This method may contribute to the study of the active site of antibody

At the beginning of this review on the heterogeneity of immunoglobulins it is necessary to emphasize the fact that there is no positive method of demonstrating homogeneity. Even from a strictly logical standpoint, homogeneity can only be defined as a lack of heterogeneity. Detection of heterogeneity has practical limits linked directly to the powers of separation and detection of each method, SOBER *et al.* (1965).

Two recently introduced methods of chemical and enzymatic reducing of the immunoglobulins to smaller units (fragments) and even individual polypeptide chains had initiated a variety of studies which have succeeded in elucidating most of the structure and function of these proteins.

Enzymatic splitting of immunoglobulins

In his first experiments PORTER (1959) incubated rabbit IgG globulin with mercuri-papain in phosphate buffor pH 7.0, containing 0.01 M cysteine and 0.002 M ethylene-diamine-tetraacetate disodium salt (EDTA) to activate the enzyme. The non-dialysable digestion products were fractionated by ion exchange chromatography. Column fractionation on carboxymethyl(CM)-cellulose separated the digest into three major peaks: fragments I, II, III (Fig. 1).

Fragments I and II (now termed Fab fragments) were chemically similar, their molecular weight was 50,000–55,000. Each of these fragments had a single antigen

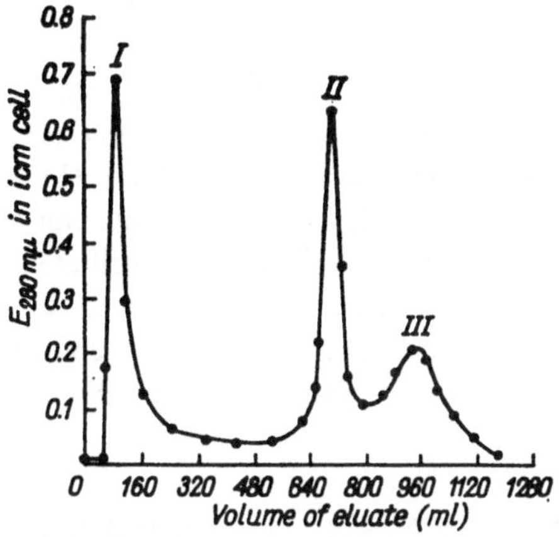

FIG. 1. Chromatography of papain digest of rabbit IgG on CM-cellulose (PORTER, 1959).

binding site (when IgG was an antibody). The fragment III (now termed Fc) had a molecular weight of 80,000, crystallized readily, did not react with antigen and demonstrated many various biological activities (antigenic specificity, complement fixing, placental crossing, skin fixation PCA, combination with rheumathoid factor to give 22S complex). It differs from the other two in antigenic properties, amino acid composition, electrophoretic mobility, sulphydryl content, carbohydrate content.

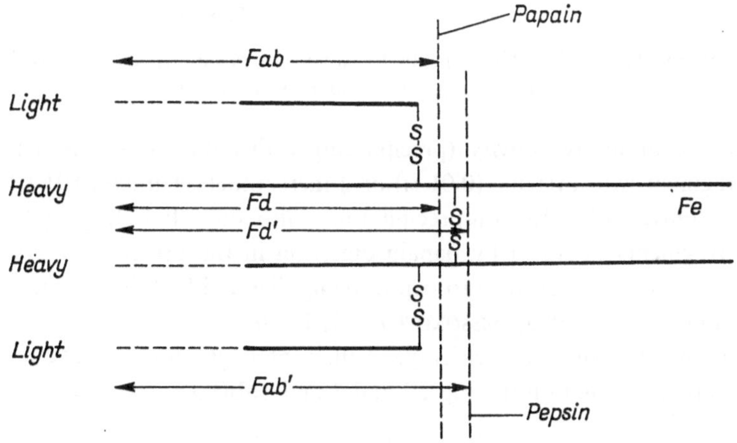

FIG. 2. Supposed sites of papain and pepsin cleavage (STANWORTH and PARDOE, 1967).

To date the most thorough studies have been performed with rabbit and human IgG. However, it seems likely that IgG globulins from a variety of mammalian species (porcine, bovine, equine, mouse, guinea pig), can readily be broken to similar fragments. The separation of the human fragments needed two-stage chromatographic procedure: firstly on CM-cellulose and secondly on DEAE-cellulose (FRANK-

LIN, 1960), while the rabbit fragments were separable by a single-stage fractionation on CM-cellulose (HSIAO and PUTNAM, 1961).

An other technique of separating the papain digests depends on zone electrophoresis in slabs of potato starch (FONGEREAU and EDELMAN, 1965) or in "Pevikon" medium-copolymer of polyvinyl acetate and chloride (MÜLLER–EBERHARD, 1960).

The cleavage of rabbit IgG molecule to fragments can also be achieved by use of other enzymes. Very interesting results (which contributed to better understanding of the mechanism of papain cleavage) were obtained with crystalline pepsin at pH 4.5 (NISSONOFF *et al.*, 1959). In addition to small peptides, the pepsin proteolysis produced a 5.25S fragment of molecular weight 106,000, now termed $F(ab')_2$, which

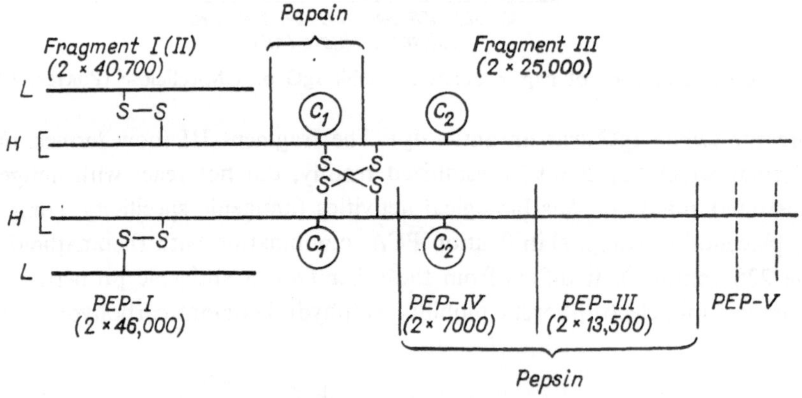

FIG. 3. Rabbit IgG relationship of peptic to papain fragments. C_1 and C_2 — carbohydrate moieties (UTSUMI and KARUSH, 1965).

preserved the antibody activity (precipitating with antigen). Subsequent reduction of this fragment with cysteine (0.01 M) or 2-mercaptoethylamine (0.01 M) at pH 8.0 gave two "univalent" 3.5S non-precipitating antibodies (Fab′ fragments) similar to the Fab fragments obtained by papain digestion in the presence of cysteine.

The Fc fragment is sensitive to pepsin being digested by this enzyme (NISSONOFF, WISSLER and LIPMAN, 1960; NISSONOFF *et al.*, 1960).

The results of comparative cleavage of the rabbit IgG molecule caused by papain or pepsin are illustrated on the Figs. 2 and 3 and Table 3.

TABLE 3

Peptic and papain digestion products of rabbit IgG globulin (STANWORTH and PARDOE, 1967)

Peptide fragment	Pep-I (5S)	Pep-II (3.3S)	Pep-III (3.3S) ⟶	Pep-III′+Pep-IV+Pep-V (2.4S)	
Coresp. papain fragment	$F(ab)_2$	Fab, Fc	Fc	?F′c (2.1S)	Peptides (< 5000)

The Fd fragment (N-terminal end of the heavy chain) located in Fab fragment is of lower molecular weight than Fd′ fragment (Pep-I 5S) after pepsin proteolysis. The sites of cleavage of rabbit IgG after pepsin or papain cleavage are different (UTSUMI and KARUSH, 1965).

Other proteolytic enzymes (trypsin, bromolain, ficin) have been used to cleave IgG molecules (HANSON and JOHANSSON, 1963; PUTNAM *et al.*, 1962), the obtained fractions 5S and 3.5S were similar to the papain digestion material.

During storage of human IgG globulins structural changes occur indicating a dimerization process affecting nearly 20% of globulin molecule.

After unfolding the IgG molecules undergo a process of proteolytic degradation similar to that achieved by papain or plasmin (CONNELL and PAINTER, 1966; JAMES *et al.*, 1963; SKVARIL, 1960).

Reduction of immunoglobulins

EDELMAN and POULIK (1961), FRANEK (1961) showed that human, bovine, equine and rabbit IgG globulins could be split into smaller units (polypeptide chains) by extensive reduction (with 0.1 M 2-mercaptoethylamine or 2-mercaptoethanol or S-sulphonation) in urea (6 M solution) which is a potent denaturating agent. These results indicated that the IgG molecule of several species had a similar number of peptide chains, contrary to results with N-terminal amino acid assay. The method applied by EDELMAN and POULIK has many disadvantages: difficulties of fractionation, insolubility of the products in usual aqueous solvents (they were soluble only in urea solution), inactivation of biological activity. FLEISCHMAN *et al.* (1961) re-examined the conditions of reduction in the absence of denaturing agent. If after reduction the ten liberated sulphhydryl groups were treated with iodoacetamide and the reduction mixture dialysed against 1 N acetic or 1 N propionic acid, dissociation into components resulted. These components could be separated either by zone electrophoresis or on a Sephadex column (Fig. 4).

The larger component A (75% yield) comprises two heavy polypeptide chains (named H chains of 50,000 mol.wt.), whilst the smaller component B (25%) comprises two light polypeptide chains (named L chains of 20,000 mol.wt.). Complete reduction of all the disulphide bonds of A or B in 6 M guanidine caused no further reduction in molecular weight. Hence, it is probable that IgG consists of only two types of peptide chains (postulated by PORTER, 1959). The molecular weights and yields are consistent with a four-chain structure (EDELMAN and BENACERRAF, 1962) and agree with the molecular weight of the whole molecule (about 150,000). Also the amino acid of the whole molecule are accounted for by the amino acids present in two A and two B chains. Further investigations (COHEN, 1963; FLEISCHMAN *et al.*, 1963; OLINS and EDELMAN, 1962) showed that Fc piece contains only heavy chains, whereas light chain is located in the Fab pieces. The N-terminal part of the heavy chains (named Fd) could be isolated from Fab piece of rabbit IgG following reduc-

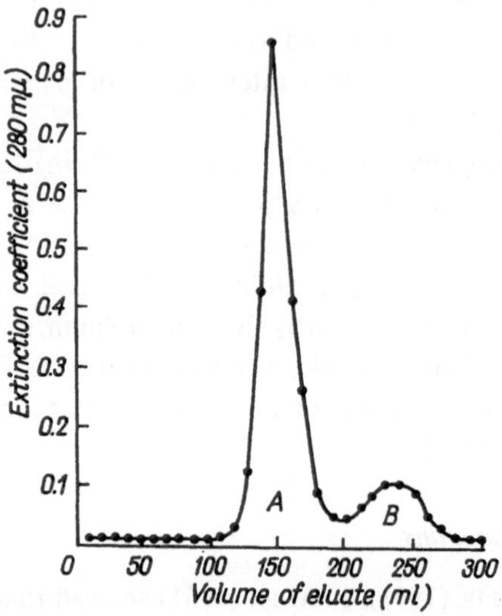

FIG. 4. Fractionation of reduced rabbit IgG on Sephadex G75 (FLEISCHMAN *et al.*, 1961).

tion and alkylation, by gel filtration on Sephadex G75 in propionic acid. A single fractionation of reduced IgG globulin on Sephadex G75 in acid fails to provide a clear cut separation of the heavy and light chains. There is a need for refractionation to avoid contamination of each chain preparation by the other (KOSHLAND *et al.*, 1966; LAMM *et al.*, 1966).

Much more complicated is the separation the $\varkappa$ and λ light chains from normal human IgG globulins. The methods of electrophoresis, chromatography, gel filtration and salt precipitation applied for this purpose were unsuccessful. COHEN (1965) isolated partially purified preparations by exploiting the fact the λ chains dissociate from reduced alkylated IgG globulin at a higher pH than $\varkappa$ chains.

MAMMALIAN IMMUNOGLOBULINS

Human immunoglobulins

The investigations of human immunoglobulins has been greatly facilitated by the ready availability of large amount of paraproteins in myeloma and macroglobulinaemic sera. The lack of similar natural sources among other species has impeded the chacterization of their various immunoglobulin constituents.

Just as electrophoresis served initially to distinguish gamma-globulins from other serum proteins, additional physicochemical and immunological procedures showed the heterogeneity of immunoglobulins. At the present time four classes of these proteins are known: IgG, IgM, IgA, IgD. Their relationship is based on structural class characteristics.

Recently ISHIZAKA *et al.* (1966) obtained data suggesting that the human reagins represent a new (fifth) class of immunoglobulins (named IgE).

The cross serologic reactivity between the four immunoglobulin classes depends on the possession of common light polypeptide chains (which are of two antigenic forms: κ or λ). The specific antigens for each immunoglobulin class are located on the heavy chains (γ, μ, α, δ) in their C-terminal portion.

The IgG globulins of normal human serum comprise about 60% of κ light chains and about 30% of λ light chains (MANNIK and KUNKEL, 1963). It is suggested that a similar distribution of light chain types occurs among the other normal immunoglobulin classes (i.e. IgA, IgM and IgD (FUDENBERG, 1965)). In contrary, the paraproteins isolated from the sera of myeloma and macroglobulinaemia patients comprise molecules with only one or other of the two light chain types. The Bence–Jones protein usually occur as light chain dimer (2×20,000 = 40,000 mol.wt.). In rare cases monomeric forms were found (BERNIER and PUTNAM, 1963).

In certain human antibodies the molecules of only one type of light chain were encountered (e.g. the molecules of cold isoagglutinins isolated from cases of haemolytic anaemia which were of the κ type), HARBOE and DEVERILL (1964).

Using monkey immune serum against normal human IgG globulin and rabbit sera against individual myeloma proteins it was possible to differentiate 4 subtypes among the heavy chains: γ2a, γ2b, γ2c, γ2d. Myeloma proteins comprise only one of the four subtypes of heavy chains (GREY and KUNKEL, 1964). Human antithyroglobulin antibodies occur in three of the different heavy chain forms: γ2a, γ2b, γ2c (LICHTER, 1964; TERRY and FAHEY, 1964).

Further antigenic differences genetically determined were discovered in human immunoglobulins. The allotypic determinants of Gm group are located on the piece Fc of human γ heavy chains. The allotypic antigenic determinants of the InV group are located on the light chains of all the human immunoglobulin classes (FRANGIONE *et al.*, 1966; FRANKLIN *et al.*, 1962; FUDENBERG *et al.*, 1966; LAWLER and COHEN, 1965; MELTZER *et al.*, 1964).

Table 4 presents the kinds of heterogeneity of human immunoglobulins.

TABLE 4

Classification of human immunoglobulins

Class	IgG	IgM	IgA	IgD	IgE
Subclass	γ1, γ2				
S	7	19, 29, 38	7, 10 etc.	7	8
Heavy chain form	γ	μ	α	δ	
Subform	γ2a, γ2b, γ2c, γ2d				
Light chain (2 types)	K, L	K, L	K, L	K, L	
Allotypes:					
(a) heavy ch.	Gm	—	—	—	
(b) light ch.	InV	InV	InV		

Mouse immunoglobulins

The investigation results of myeloma proteins in mice (C3H, BALB/c) with transplantated plasma-cell tumours indicate the class heterogeneity of mice immunoglobulins (FAHEY, 1961a, b and FAHEY *et al.*, 1964) $1/7S\gamma_1$, $2/7S\gamma_2$, $3/\gamma A$ and $4/\gamma M$. These immunoglobulins differ electrophoretically and immunologically. All four proteins possess antibody activity in accordance with the results of radioimmunoelectrophoretic analysis of mouse antiserum against haemocyanin.

In the heavy chains of $7S\gamma_2$ immunoglobulin two subtypes were discovered: γ_{2a} and γ_{2b} (WUNDERLICH *et al.*, 1964) by means of appropriate heterologous immune sera. These two globulins are of the same electrophoretic mobility and possess antibody activity.

The genetically determined isoantigen Iga-1 is located on the γ_{2a} subtype only.

There are differences in electrophoretic mobility of the fast $7S\gamma_1$ globulin of various inbred mouse lines. The differences seem to have a hereditary basis and are not accompanied by any detectable antigenic difference (COE, 1966).

Immunoglobulins of other species

Different kinds of antibodies isolated from hyperimmune sera serve as a material for investigations (HILL and CEBRA, 1964; KABAT and MAYER, 1962). The isolation of immune antibodies was performed by a two step procedure:

(1) an initial step involving precipitation with an antigen or immunoadsorption on an antigen coupled with a suitable carrier,

(2) final purification of the recovered antibody is then achieved by many physicochemical procedures (gel filtration, ion-exchange chromatography, zone electrophoresis, zone centrifugation, etc.).

The immunization of horses, cows and pigs with pneumococci or their capsular polysaccharides is a relatively simple procedure for obtaining high level of IgM antibodies. Similarly IgG antibodies are readily produced in rabbits immunized with many various antigens. However the isolation of antibodies of low concentration (e.g. IgM antibodies from rabbit) presents considerable difficulties. The same difficulties arise with the isolation of IgA antibodies (ONOUE *et al.*, 1966). Until the preparation and purification of immunoglobulins (antibodies at a very low concentration) will be solved, it is not possible to obtain conclusive answer concerning their structural relationships as was demonstrated for human or mouse immunoglobulins.

Rabbit IgG and IgM possess common antigenic determinants in the Fc fragment as well as in the Fd portion of the heavy polypeptide chain, in contrast to the four classes of human immunoglobulins where their relationships is bound with the light polypeptide chains (FEINSTEIN *et al.*, 1963).

In the sera of several animal species a fast moving form γ_1 of IgG globulin was discovered. In mice the fast moving γ_1 globulin possess specific antigenic determinants which differentiate them from the two subtypes γ_{2a} and γ_{2b}. The same two forms γ_1 and γ_2 (which differ electrophoretically and antigenically) were shown in guinea pig immune antibodies (BENACERRAF *et al.*, 1963; THORBECKE *et al.*, 1963). These forms have common antigenic determinanats in Fab fragments but specific in Fc fragments. They show also different biologic properties:

(1) the γ_1 form of guinea pig IgG elicits skin reaction of the P–K type in other animals of the same species (BENACERRAF *et al.*, 1963),

(2) whereas the γ_2 form is only capable of evoking skin reaction of the PCA type in heterologous skin (BAKER *et al.*, 1964),

(3) the γ_2 form combines with complement, mediates Arthus reaction and selectively binds to macrophages through its Fc portion (BERKEN and BENACERRAF, 1966),

(4) IgG antibodies of some species (e.g. man, rhesus monkey, dog, mouse) capable to evoke PCA reaction in guinea pig belong also to the γ_2 form of IgG (OVARY *et al.*, 1964), therefore they are not able to evoke homologous skin reaction,

(5) γ_2 forms of other animals (e.g. cow, horse, sheep) do not possess the structure (presumably within the Fc portion) necessary for binding to guinea pig tissues (OVARY, 1960),

(6) there is a tendency to think that homologous skin sensitizing antibodies found in the antisera of other species belong to the subclass γ_1 of globulins.

Natural tissue-sensitizing reagins were found in the sera of hypersensitive dogs. They were similar to human reagins showing: (1) fast electrophoretic mobility, (2) heat lability, (3) persistence at the site of passive transfer (PATTERSON *et al.*, 1962, 1964).

Some authors claimed that human reagins were of A class (ISHIZAKA *et al.*, 1964; VAERMAN *et al.*, 1964) but the lack of correlation between P–K activity and IgA globulin distribution in human allergic sera presents an evidence against such a possibility. Recently ISHIZAKA *et al.* (1966) speculated that reagin represents a new class of immunoglobulins named IgE. There is a possibility however that human isologous skin sensitizing antibodies belong to the electrophoretically fast γ_1 subclass of IgG globulin as in other animal species mentioned above and get attached passively to human skin through their Fc portion.

The separation of two forms of rabbit IgG antibodies: γ_1 electrophoretically fast and γ_2 slow was achieved by chromatography on DEAE-Sephadex A 50 (SELA and MOZES, 1966). They have different Fab fragments but common Fc fragment.

Rabbit IgA is present in colostrum in quantities much higher than in serum; it may be isolated from the whey fraction by gel filtration on Sephadex G-200. Crude estimation of the molecular weight of colostral IgA gives an approximate value of 300,000–500,000, which suggest that colostral IgA exists primarily in a polymeric form (SELL, 1967).

In the horse antisera the following classes of immunoglobulins were found: γ_1(10S), γ_2 (γ_a, γ_b, γ_c), IgM, IgA globulins (ROCKEY *et al.*, 1964). The "T" com-

TABLE 5

Immunoglobulin classification in various mammalian species (STANWORTH and PARDOE, 1967)

Mouse	γ1	γ2 γ2a, γ2b	IgM	IgA
Guinea pig	γ1	γ2	IgM	IgA
Rat	γ1 (? mast cell sensitizing antibody)	γ2 γ2a, γ2d	IgM	?IgA
Horse	γ1 (10S)	γ2 γa, γb, γc	IgM	IgA
Rabbit	γ1 (? isologous skin-sensitizing body)	γ2	IgM	IgA (6S,9S)

ponent formed in horse antisera was shown to be more closely related to horse IgG globulin than to IgA globulin (ALLEN *et al.*, 1965; SCHULTZE *et al.*, 1965).

Table 5 presents the data on the tentative classification of the immunoglobulins of various mammalian species. Similarities in structure between the corresponding classes have been inferred by their vertical alignment, but there is no direct evidence of such interspecies relationships (in most cases).

CHEMICAL STUDIES OF IMMUNOGLOBULINS AND THEIR SUBUNITS

Amino acid composition

The introduction of new techniques and especially of the automatic amino acid analysers has led to the publication of accurate analyses of immunoglobulin from several species. The results of chemical analysis of whole immunoglobulin molecules provide information of limited value, because of their great complexity. Although the amino acid composition of human immunoglobulins of the three main classes (IgG, IgM, IgA) is qualitatively identical, there are some clear quantitative differences (Table 6).

Human IgG globulins are in comparison with IgM and IgA globulins poorer in carbohydrates but richer in lysine.

Generally the immunoglobulins contain more hydroxy and dicarboxylic amino acids (particularly proline) than any other globular protein.

Table 7 presents data on the carbohydrate composition of the main classes of human immunoglobulins.

The results of comparative investigations of the amino acid composition of human, rabbit and horse immunoglobulins are summarized in Table 8.

The differences in lysine and threonine content between rabbit and human IgG globulins are evident; slight differences occur between rabbit and horse IgG globulins as well.

TABLE 6

Amino acid composition of the main classes of human immunoglobulins (HEIMBERGER *et al.*, 1964)

Amino acid	IgG (mol.wt. 160×10^3), moles/mole	IgM (mol.wt. 10^7), moles/mole	IgA (mol.wt. 160×10^3), moles/mole
Lysine	87	381	57
Histidine	27	133	23
Ammonia	146	768	202
Arginine	41	324	47
Aspartic	107	619	86
Threonine	114	716	121
Serine	178	799	146
Glutamic acid	140	825	130
Proline	100	544	104
Glycine	84	528	90
Alanine	72	493	87
1/2 (Cystine)	34	155	33
Valine	131	655	97
Methionine	10	89	10
Isoleucine	30	239	25
Leucine	102	560	110
Tyrosine	62	258	43
Phenylalanine	47	269	40
Tryptophane	33	152	28
Total	1552	8507	1479

TABLE 7

Carbohydrate compositions of the main classes of human immunoglobulins (moles/mole) (STANFORD and PARDOE, 1967)

	IgG (160,000)					IgA (160,000)	IgM (960,000)
Mannose	5.07	7.25*	9.25**	5.75	10.67 (Mannose + Galactose)	14.2	174
Galactose	3.02	3.62	4.62	3.45		14.2	87
Fucose	2.1	2.04	1.95	2.6	2.92	2.1	40
N-acetyl glucosamine	8.43	9.65	9.65	10.1	10.77	21.0	144
N-acetyl neuraminic acid	1.03	0.93	2.43	0.89	0.95	9.3	41.3

* Electrophoretically slower fraction
** Electrophoretically faster fraction
(both: Electrophoresis in polyvinyl chloride, pH 8.6.)

The sequential analyses of well defined subunits (e.g. polypeptide chains) or of fragments of particular biological interest is more important for studies of the primary immunoglobulin structure.

TABLE 8

Amino acid analysis of immunoglobulins (COHEN and PORTER, 1964)

Amino acid	Amino acid residue (g/100 g protein)				
	rabbit IgG	human IgG	human IgM	horse IgG	horse IgA(T)*
Lysine	5.76	7.06	4.91	6.77	6.50
Histidine	1.73	2.44	1.98	2.58	2.57
Arginine	4.42	4.02	4.75	3.34	3.02
Aspartic acid	8.08	7.77	6.95	7.25	7.31
Threonine	10.37	7.04	6.17	8.31	7.13
Serine	8.32	9.13	6.58	9.68	9.29
Glutamic acid	11.05	11.18	9.92	10.27	9.36
Proline	6.79	6.40	4.95	6.02	6.52
Glycine	3.98	3.37	2.91	3.68	3.53
Alanine	3.71	3.29	3.12	3.60	3.28
Valine	8.36	7.92	5.77	8.14	7.74
Methionine	1.13	0.93	1.02	0.78	0.59
Isoleucine	3.49	2.16	2.83	3.14	2.70
Leucine	6.73	7.40	6.09	6.63	6.51
Tyrosine	6.17	5.76	5.44	5.55	4.93
Phenylalanine	4.15	4.07	3.85	3.79	3.47
Cystine	2.63	2.07	2.30	2.08	1.90
Tryptophan	2.90	2.63	2.47	2.57	2.47
Carbohydrate	2.40	2.80	12.30	2.40	4.90
Total	101.6	97.4	95.70	97.80	94.80

* Now referred to as IgG(T).

In the Tables 9 and 10 the comparison of the amino acid composition of light and heavy polypeptide chains is presented.

In the Tables 11, 12 and 13 the further data on the composition of light and heavy polypeptide chains as well as of other subunits are presented.

From the data presented in the Tables the following inferences can be drawn out:

(1) there are differences between light and heavy polypeptide chains in relation to some amino acids,

(2) each light polypeptide chain possess one cysteine residue, on the other hand each heavy polypeptide chain contains 3–4 cysteine residues; these results indicate that non-reduced chains are linked by 4–5 disulphide bonds.

Terminal amino acids

The procedures for determination of the terminal amino acids are also used for demonstration of the polypeptide chain heterogeneity. The following techniques were used for this purpose:

TABLE 9

Amino acid composition of heavy chains (COHEN and PORTER, 1964)

Amino acid	Residues 50,000 g				
	rabbit IgG	human IgG	human IgM	human IgG	horse IgA(T)*
Lysine	23	29	20	26	27
Histidine	7	10	7	10	12
Arginine	16	12	18	10	10
Aspartic acid	33	33	34	30	34
Threonine	49	34	38	37	33
Serine	49	50	42	45	46
Glutamic acid	40	39	40	38	38
Proline	36	33	29	31	38
Glycine	33	28	28	28	28
Alanine	24	19	26	22	22
Valine	41	41	36	46	45
Methionine	5	4	5	4	3
Isoleucine	15	8	13	13	11
Leucine	30	31	30	28	31
Tyrosine	17	17	12	16	16
Phenylalanine	14	13	16	13	12
Half-cystine	10	7	7	3	6
S-carboxymethyl-cysteine	4	4	3(?4)	3(?4)	4
Tryptophan	8	8	7	8	7
Hexose	4.5	4	20	5	10
Hexosamine	4	4.5	12	4	7

* Now referred to as IgG(T).

(1) N-terminal amino acids. Their sequence was determined by SANGER (1945) who used 1-fluoro-2,4-dinitrobenzene as an agent reacting with the terminal amino acid, by EDMAN and SJOQUIST (1956) who treated the protein with phenylisothiocyanate, by GRAY and HARTLEY (1963) who employed a fluorescent reagent (1-dimethylaminonaphtalene-5-sulphonyl chloride known as dansyl chloride) which reacts with amino acid phenolic group and by MARGOLIASCH (1962) who defined the conditions for the use leucine aminopeptidase for sequential degradation of peptides.

An interesting outcome of N-terminal determination on rabbit polypeptide chains was the finding that the terminal amino acids: alanine, aspartic acid were confined to the light chains. On the other hand in human IgG globulins the N-terminal amino acids were located on both types of polypeptide chains, in the heavy chains they were confined to the Fd-portion.

(2) C-terminal amino acids. The consequence of this amino acids is determined by enzymatic (carboxypeptidases A and B) and chemical (hydrazinolysis) methods. C-terminal amino acids were found in the Fc fragment of rabbit heavy chain; 2 moles of glycine, 1 mole of serine, 1/2 mole of threonine and alanine per one mole of heavy chain.

TABLE 10

Amino acid composition of light chains (COHEN and PORTER, 1964)

Amino acid	Residues 20,000 g				
	rabbit IgG	human IgG	human IgM	horse IgG	horse IgA(T)*
Lysine	8	10	10	10	10
Histidine	2	3	2	2	2
Arginine	3	6	5	4	4
Aspartic acid	16	13	13	12	13
Threonine	25	15	15	17	19
Serine	19	24	24	28	32
Glutamic acid	18	20	19	16	16
Proline	10	11	11	11	12
Glycine	15	11	12	16	18
Alanine	13	12	12	12	13
Valine	18	13	13	14	14
Methionine	0.9	0.6	0.7	0.8	0.4
Isoleucine	6	5	5	10	7
Leucine	11	13	13	11	12
Tyrosine	10	8	8	6	6
Phenylalanine	5	6	5	4	4
Half-cystine	6	3	3	3	3
S-carboxymethyl-cysteine	1	1	1	1	1
Tryptophan	4	2	3	3	3
Hexose	0.3	0.25	0	0.4	0.6

* Now referred to as IgG(T).

Primary immunoglobulin structure

The primary characterization of immunoglobulins is a very difficult task because of their complexity and heterogeneity. Hence the first structural analysis were performed on the more homogeneous and available paraproteins as the urinary Bence–Jones proteins and myeloma IgG immunoglobulin.

Bence–Jones proteins comprise light chains only, which are related to normal human IgG globulin light chains. Peptide mapping and determination of amino acid composition as well as of terminal amino acids, has shown the essential differences between the light chains of type K and L (PUTNAM and EASLEY, 1965). Further investigations (BAGLIONI *et al.*, 1966; BAGLIONI and CIOLI, 1966; HILSCHMAN and CRAIG, 1965; TITANI *et al.*, 1965) on the amino acid sequences of three Bence–Jones proteins gave following results:

(1) there are structural differences between the light chains belonging to one type (K),

(2) many structural differences were found to be confined to N-terminal part, on the other hand in the C-terminal part one structural difference only was established (TITANI *et al.*, 1965),

TABLE 11

Amino acid compositions of human IgG globulin and of the heavy and light chains (CRUMPTON and WILKINSON, 1963)

Amino acid	Human IgG	Heavy chain	Light chain
Lys	7.06	7.35	6.23
His*	2.44	2.62	1.75
Arg	4.02	3.81	4.57
Asp	7.77	7.64	7.33
Thr**	7.04	6.83	7.42
Ser***	9.13	8.60	10.28
Glu	11.18	10.05	13.04
Pro	6.40	6.41	5.10
Gly	3.37	3.23	3.33
Ala	3.29	2.73	4.19
Val*	7.92	8.15	6.27
Met	0.93	1.02	0.42
Ileu	2.16	1.86	2.74
Leu	7.40	7.11	7.29
Tyr**	5.76	5.52	6.16
Phe	4.07	3.77	4.31
Cys***	2.07	1.41	1.51
Cys . CH_2 . CO_2H	0	1.15	0.69
Try	2.63	2.79	2.15
Total	94.6	92.1	94.8

* Amounts determined after 72 hr of hydrolysis only.
** Corrected to zero time for decomposition during hydrolysis.
*** Calculated as cysteic acid after oxidation and hydrolysis.

(3) such multiple structural differences are incompatible with the concept of one single point mutation (as e.g. for abnormal hemoglobins); to account for antibody variability a mechanism of chromosal rearrangement has been proposed.

The amino acid sequence studies of C-terminal peptides (heavy chains) showed close similarity of arrangement of amino acids between rabbit IgG and human myeloma IgG globulin. The amino acid sequence of one C-terminal peptide was determined (PRESS *et al.*, 1966):

NH_2 NH_2
His-Glu-Ala-Leu-His-Asp-His-Tyr-Thr-
NH_2
Glu-Lys-Ser-Leu-Ser-Leu-Ser-Pro-Gly-COOH

From these and other experiments PRESS *et al.* (1966) suggested that the Fc fragment of mammalian immunoglobulins may have a unique sequence showing little variation from one species to another, on the other hand a great variability is bound with the Fd portion of the heavy chain. The heavy chains like the light chains, within the Fc fragment of the molecule possess a stable structure at C-terminal end and a variable structure at the N-terminal end.

TABLE 12

Amino acid compositions of heavy and light chains of human immunoglobulins IgG and IgM (COHEN *et al.*, 1965)

Amido acid	Heavy chain (50,000)			Light chain (20,000)		
	IgG	IgM(1)	IgM(2)	IgG	IgM(1)	IgM(2)
Lys	29	20	21	9.7	9.6	9.5
His	9.6	7.6	8.3	2.5	2.4	2.7
Arg	12	17	18	5.8	5.3	4.6
Asp	33	33	35	13	13	14
Thr	34	38	38	15	15	15
Ser	50	42	39	24	24	24
Glu	39	40	41	20	19	19
Pro	33	29	28	11	11	12
Gly	28	28	29	11	12	13
Ala	19	26	28	12	12	12
Val	41	36	38	13	13	13
Met	3.9	5.3	5.0	0.6	0.7	0.6
Ileu	8.2	13	13	4.9	4.7	4.6
Leu	31	30	32	13	13	12
Tyr	17	12	12	7.6	8.0	6.1
Phe	13	15	16	6.0	5.3	5.3
Cys (half)	6.9	7.4	5.7	2.9	3.3	
$Cys.CH_2.CO_2H$	3.6	3.2	4.5	0.9	1.1	1.3
Try	7.5	7.6	7.8	2.3	2.7	2.8

IgM(1) was prepared from defibrinated plasma of U.K. blood donor.
IgM(2) was prepared from West African serum.

Carbohydrate content of immunoglobulins

The immunoglobulins contain carbohydrate in different quantities (glycoproteins). It seems unlikely that carbohydrate plays any part in the antibody activity (antigen combining site), because papain fragments I and II can be obtained almost free of carbohydrate, yet retaining their power to combine specifically with an antigen. The polysaccharide of immunoglobulins comprise 5 components: hexoses (D-mannose, D-galactose), L-fucose, N-acetyl-D-glucosamine, N-acetyl-neuraminic acid.

The data on carbohydrate composition of immunoglobulins are summarized in Table 14.

Extensive studies on the nature of the carbohydrate have been carried out by SMITH and others (ROSEVAR and SMITH, 1958, 1961; SMITH and JAGER, 1952). The authors extracted three glycopeptides from human IgG, their results are summarized in Table 15.

The amino acid sequence in glycopeptides isolated from human, bovine and rabbit IgG globulins was determined by the methods of SANGER and EDMAN:

Human: AspNH$_1$Tyr Glu Asp Carbohydrate,

TABLE 13

Amino acid composition of rabbit IgG globulin, the heavy and light chains and the piece Fd (CRUMPTON and WILKINSON, 1963)

Amido acid	Rabbit IgG	Heavy chain	Light chain	Piece Fd
Lys	5.76	5.79	4.99	4.22
His*	1.75	1.79	0.89	1.06
Arg	4.42	4.90	2.04	3.85
Asp	8.08	7.53	8.91	6.18
Thr**	10.37	9.79	12.20	11.98
Ser**	8.32	8.52	8.36	9.93
Glu	11.05	10.39	11.80	7.35
Pro	6.79	7.03	4.85	6.47
Gly	3.98	3.71	4.47	4.76
Ala	3.71	3.37	4.75	3.87
Val*	8.36	8.07	8.69	8.68
Met	1.13	1.21	0.33	0.77
Ileu	3.49	3.39	3.49	2.36
Leu	6.73	6.79	5.56	7.21
Tyr**	6.17	5.52	7.79	5.68
Phe	4.15	4.07	3.88	3.56
Cys***	2.63	1.91	2.68	2.00
Cys . CH_2. CO_2H	0	1.17	0.80	1.54
Try	2.90	2.91	1.71	2.98
Total	99.8	97.9	98.2	94.5

* Amounts determined after 72 hr of hydrolysis only.
** Corrected to zero time for decomposition during hydrolysis.
*** Calculated as cysteic acid after oxidation and hydrolysis.

TABLE 14

Carbohydrate composition of immunoglobulins (%) (COHEN and PORTER, 1964)

Carbohydrate	Rabbit	Human			Horse		Bovine
	IgG	IgG	IgM	IgA	IgG	IgA	IgG
Hexose	1.2	1.2	6.2	4.8	1.1	2.1	0.9
Fucose		0.3	0.7	0.2			0.2
Hexosamine	1.0	1.1	3.3	3.8	1.1	1.9	1.5
Sialic acid	0.2	0.2	2.0	1.7	0.2	0.9	0.3
Total	2.4	2.8	12.2	10.5	2.4	4.9	2.9

Bovine: Glu Glu NH_2Phe Asp Carbohydrate,
Rabbit: Glu NH_2GluNH_2 Phe Asp Carbohydrate.

In all cases investigated the carbohydrate probably combines with aspartic acid of the peptide through a stable glycosylamine bond.

The carbohydrate content in IgA and IgM is higher than in IgG.

The carbohydrate is linked mostly with heavy polypeptide chain of immuno-

TABLE 15

Composition of glycopeptides isolated from a papain digest of heat-denaturated IgG globulin (ROSEVAR and SMITH, 1961)

Residues for	Glycopeptide I 4800 g	Glycopeptide II 4200 g	Glycopeptide III 2800 g
Hexose	7.3	7.6	7.8
Glucosamine	6.0	6.0	3.1
Fucose	2.0	2.1	2.0
Sialic acid	1.0	0.6	0.2
Aspartic acid	2.1	1.8	1.8
Tyrosine	0.91	0.95	1.1
Glutamic acid	3.3	2.4	0.9
Phenylalanine	0.47	0.38	0.3
Threonine	0.27	0.08	0.1
Serine	0.45	0.22	0.2
Valine	0.24	0.30	
Alanine	0.14	0.09	
Lysine			0.4

Analytical figures are given for the anhydrous material.

globulins. Small quantities were found in the preparations of light chains (not as a contamination by heavy chain).

The chemical analysis of the papain fragments seems to indicate that carbohydrate groups are located on two sites of rabbit IgG globulins: one is located close to the C-terminal end of the papain Fab fragment, whilst the other (which contains all the fucose) derived from the Fc fragment of the heavy chain (PRESS, 1965, PRESS and PORTER, 1966; UTSUMI and KARUSH, 1965).

The antigen combining site of antibody molecules

As already mentioned papain digestion breaks the bivalent IgG rabbit antibody in Fab fragments each of them containing one antigen receptor. Therefore Fab fragments 3.5S fail to precipitate but retain their capacity to combine with antigen. The next studies have been directed to establish the role of light polypeptide chains and of Fd piece of heavy chains (located on the Fab fragments) in forming the antibody receptor.

EDELMAN *et al.* (1962) have shown that the light chains only of the antihapten (2,4-dinitrophenyl-polysine) antibody form the receptor. But the drastic methods used for the isolation of the light and heavy chains limited the value of the results obtained (FRANEK and NEZLIN, 1963). The application of a mild procedure (such as of FLEISCHMAN *et al.* (1963)) permitted the retention of antibody activity in the polypeptide chains from antihapten (anti-2,4-dinitrophenyl bovine IgG globulin) antibody (EISEN, 1964b, c; FARAH *et al.*, 1960; GROSSBERG and PRESSMAN, 1964). For the estimation of an antibody-hapten affinity the method of equilibrium dialysis was largely used (EISEN and KARUSH, 1949). The results obtained were contradictory: UTSUMI and KARUSH (1963) expressed the view that nearly all of the original hapten combining site is located in the heavy chains, no activity was found in the light chains. METZGER and SINGER (1963) suggest that the light chains may play a direct role in the combination of hapten with the heavy chains or in other terms their findings reflect the influence of hapten in maintaining the conformation of the heavy chains necessary for combination with light chains. Also LAMM *et al.* (1966) using purified polypeptide chains of guinea pig γ_2G globulin antibody have shown that the light chains were involved in the binding of hapten by the heavy chains. Specific (isolated from antihapten antibody) light chains were more effective in forming the antibody receptor than non specific light chains (isolated from non-antibody guinea pig γ_2G globulin).

The results of investigations of the iodination effects on hapten binding point to tyrosine residues playing an important role in the active sites of rabbit antibodies. The antibody activity was found to decrease rapidly with iodination, 30 iodine atoms were sufficient to destroy the binding capacity of the rabbit antibody molecule (ELLIOTT *et al.*, 1962; JOHSON *et al.*, 1960; KOSHLAND *et al.*, 1963). The amidation experiments of the lysine ε-amino groups of rabbit antibodies showed that the lysine side chains are not normally present in the active sites (WOFSY and SINGER, 1963).

For the identification of the antibody active site the technique of "affinity labelling" (METZGER *et al.*, 1963; WOFSY *et al.*, 1962) and "differential" labelling procedure (PRESSMAN and ROHOLT, 1961) were used. In the first method use is made of a bifunctional specific hapten; in the second a radioactive label is incorporated into the active site of antihapten molecules by covalent linkage to the tyrosyl residues in the presence and absence of specific hapten.

From the results of many experiments the following conclusions may be drawn:

(1) The relative roles of the light chains and Fd pieces of the heavy chains in the formation of the antigen receptor site are still far from being clear.

(2) There are suggestions that the specific antigen binding groups are located in the Fd portion of the antibody molecule, their configuration being influenced by the conformation of the neighbouring light chains.

(3) Conformational studies may throw further light on the relative contributions of the heavy and light chains to antibody activity.

(4) The elucidation of the amino acid sequence within the polypeptide chains is necessary to establish the precise role played by primary structure in the antibody specificity.

The technique of peptide mapping was used by many authors to demonstrate structural differences between IgG antibodies of rabbit immunized with different antigens. The results of these investigations can be briefly summarized as follows:

(1) The normal (non-antibody) fragments Fab(I), Fab(II), and Fc showed the same maps of peptides as the corresponded fragments isolated from rabbit antisera against Salmonella H-antigen (SEIJEN and GRUBER, 1963).

(2) On the other hand there were slight differences between the patterns of Fab(I) and Fab(II) fragments (IRRVERRE, 1965).

(3) Minor differences were demonstrated between the patterns of peptide maps obtained from rabbit antibodies against many pneumococcal polysaccharides (GITLIN and MERLER, 1961).

By the application of improved procedure: (1) highly purified antibodies, (2) subunits isolated from these antibodies, (3) using automatic amino acid analytical techniques of great precision, (4) expressing the results as molar ratios, significant differences in amino acid composition between antibodies of different specificity were demonstrated.

KOSHLAND and ENGLBERGER (1963), KOSHLAND *et al.* (1964) determined comparatively the amino acid composition of three specifically purified anti-hapten antibodies directed against the negatively charged phenyl-arsonic group, the positively charged

TABLE 16

Average amino acid recoveries from the heavy chains of three antibodies isolated from $a^1a^1b^4b^4$ rabbits (KOSHLAND *et al.*, 1966)

Amino acid	Residues per 53,000 g		
	arsonic antibody	ammonium antibody	lac antibody
Lys	25.3	25.4	25.5
His	6.73	6.87	7.17
Arg	19.7	19.5	19.6
Asp	33.1	34.5	33.5
Thr	50.4	50.7	49.9
Ser	53.1	52.4	50.3
Glu	39.3	40.0	39.2
Pro	42.9	43.1	41.9
Gly	34.6	34.9	34.5
Ala	23.0	23.1	22.7
Val	42.4	42.5	42.0
Met	5.97	5.93	6.11
Ileu	16.4	16.3	16.2
Leu	33	34	33
Tyr	17.1	17.0	14.9
Phe	15.3	15.3	15.1
CM — Cys	13.7	13.7	13.2

phenyltrimethylammonium group and the uncharged phenyl-beta-lactoside (lac) group. The results of amino acid analyses of hydrolysates of the whole anti-hapten molecules indicated that rabbit antibody against the negatively charged group possessed a higher content of arginine and iso-leucine in comparison with the antibody against the positively charged group which had a higher content of aspartic acid and leucine. These results were interpreted as evidence that antibodies possess diferences in primary structure. The results of amino acid analyses of the sub-units (heavy and light chains) isolated from the three anti-hapten antibodies indicated that the amino acid differences are located within these chains (Tables 16 and 17).

From the data presented in the Tables 16 and 17 the following conclusions can be drawn out:

(1) The differences in contents of serine and part of the tyrosine were located in the respective heavy chains.

TABLE 17

Average amino acid recoveries from the light chains of three antibodies isolated from $a^1a^1b^4b^4$ rabbits (KOSHLAND *et al.*, 1966)

Amino acid	Residues per 23,000 g		
	arsonic antibody	ammonium antibody	lac antibody
Lys	9.27	9.20	9.81
His	1.27	1.27	1.40
Arg	3.05	2.61	3.20
Asp	19.1	20.2	21.8
Thr	30.5	30.2	31.0
Ser	21.6	21.4	21.5
Glu	21.2	20.8	20.7
Pro	11.9	12.1	12.7
Gly	19.9	19.9	19.9
Ala	16.5	15.5	15.8
Val	20.8	21.7	22.5
Met	0.87	0.41	0.86
Ileu	7.47	6.79	7.31
Leu	11	11	11
Tyr	10.8	10.9	10.1
Phe	6.57	6.77	7.06
CM — Cys	7.26	7.08	7.03

(2) The differences in the aspartic acid and balance of tyrosine were located in the light chains.

(3) New differences were discovered in the content of proline, valine and alanine in the polypeptide chains not observed in whole anti-hapten molecules.

(4) The experiments confirm the role played by the heavy and light chains in the creation of antibody receptor.

(5) Light chains appear to have two regions of specificity: one of invariant amino acid composition and sequence and the other of variable amino acid composition and sequence. This suggestion is consistent with the presence of two regions (of constant and variable amino acid compositions respectively) revealed in human Bence–Jones protein.

Comparative amino acid analyses were performed for four human antibodies isolated from the serum of the same individual immunized with minimal dose of each non protein (dextran, levan, teichoic acid, blood group substance A). Some results are presented in Table 18.

From the data put in Table 18 the following conclusions may be drawn:

(1) The values for a number of amino acids (e.g. methionine, aspartic acid and glutamic acid) are very uniform throughout all the antibodies tested.

TABLE 18

Amino acid composition of antibodies and normal IgG globulin from one individual (BASSETT *et al.*, 1965)

Amino acid	Antilevan	Antidextran	Antiteichoic acid	Anti-A	IgG
Aspartic acid	122	119	122	122	123
Threonine	113	108	109	126	111
Serine	146	148	150	156	146
Glutamic acid	152	149	140	148	146
Proline	103	102	108	109	107
Glycine	102	100	119	108	107
Alanine	86.2	79.2	82.4	84.4	81.4
Valine	131	133	102	116	139
Half-cystine	37.1	39.1	34.0	31.2	35.1
Methionine	14.4	14.8	16.4	13.4	13.7
Isoleucine	33.8	31.0	33.6	33.4	34.4
Leucine	116	108	109	116	116
Tyrosine	45.3	55.9	53.6	52.4	55.9
Phenylalanine	54.9	54.1	54.5	48.4	52.0
Lysine	93.4	87.0	80.8	80.1	93.3
Histidine	28.6	25.9	29.1	28.3	30.1
Arginine	46.5	47.8	58.1	56.2	47.8
Hydroxylysine	0	3.2	3.7	0	0
Glucosamine	5.0	6.4	5.5	(17.8)	2.4

(2) Differences ranging from seven to thirty-two residues were found in the contents of glycine, valine, leucine tyrosine, arginine, lysine and threonine of the various antibodies.

(3) These differences were not correlated with the Gm types of the antibodies.

(4) Hydroxylysine was found in two of four antibodies (anti-dextran anti-teichoic acid) and in pooled IgM globulin, but not in pooled IgG globulin.

(5) The human antibodies of different specificity (from the same individual) showed more striking differences in amino acid composition than the various rabbit anti-hapten antibodies.

Not all details of the immunoglobulin heterogeneity are elucidated, although great progress was achieved in last years for better understanding of the structural complexity of these physiologically important proteins.

REFERENCES

ALLEN, P. Z., SHISINHA S. and VAUGHAN, J. H., 1965. *J. Immunol.*, **95**, 918.

ANDERS, G. A., HSU, K. C. and SEEGAL, B. C., 1967. Handbook of experimental immunology, Ed. Weir D. M., Blackwell, Oxford, 527–570.

BACKHAUSZ, R., VERES, G. and VETO, I., 1960. *Ann. Immunol. Hungariae*, **3**, 116.

BAGLIONI, C. and CIOLI, D., 1966. *J. Exp. Med.*, **124**, 307.

BAGLIONI, C., ZONTA, L. A., CIOLI, D. and CARBONARA, A., 1966. *Science*, **152**, 1517.

BAKER, A. R., BLOCH, K. J. and AUSTEN, K. F., 1964. *J. Immunol*,. **93**, 525.

BASSETT, E. W., TANENBAUM, S. W., PRYZWANSKY, K., BEISER, S. M. and KABAT, E. A., 1965. *J. Exp. Med.*, **122**, 251.

BENACERRAF, B., OVARY, Z., BLOCH, K. J. and FRANKLIN, E. C., 1963. *J. Exp. Med.*, **117**, 937.

BERKEN, A., and BENACERRAF, B., 1966. *J. Exp. Med.*, **123**, 119.

BERNIER, G. M. and PUTNAM, F. W., 1963. *Nature*, **200**, 223.

BERSON, S. A., YALOW, R. S., BAUMAN, A., ROTHSCHILD, M. A. and NEWERBY, K., 1956. *J. Clin. Invest.*, **35**, 170.

BOYDEN, S. V., 1951. *J. Exp. Med.*, **93**, 107.

BUCHOWICZ, I., 1966. *Postępy Biochemii*, **12**, 5.

CAMPBELL, R. M., CUTHBETSON, D. P., MATTHEWS, C. M. E., MCFARLANE, A. S., 1956. *Int. J. Appl. Radiation and Isotopes*, **1**, 66.

CHARLWOOD, P. A., 1966. *Brit. Med. Bull.*, **22**, 121.

COE, J. E., 1966. *Immunochemistry*, **3**, 427.

COHEN, S., 1963. *Biochem. J.*, **89**, 334.

COHEN, S., 1965. *Biochem. J.*, **95**, 136.

COHEN, S., CHAPLIN, H. and PRESS, E. M., 1965. *Biochem. J.*, **95**, 259.

COHEN, S. and MILSTEIN, C., 1967. *Nature*, **214**, 449.

COHEN, S. and PORTER, R. R., 1964. *Adv. Immunol.*, **4**, 287.

COHN, E. J. *et al.*, 1946. *J. Am. Chem. Soc.*, **68**, 459.

COHN, E. J. *et al.*, 1950. *J. Am. Chem. Soc.*, **72**, 465.

CONNELL, G. E. and PAINTER, R. H., 1966. *Canad. J. Biochem.*, **44**, 371.

COONS, A. H., CREECH, H. J. and JONES, R. N., 1941. *Proc. Soc. Exp. Biol. (N. Y.)*, **47**, 200.

COONS, A. H. and KAPLAN, M. H., 1950. *J. Exp. Med.*, **91**, 1.

CRUMPTON, M. J. and WILKINSON, J. M., 1963. *Biochem. J.*, **88**, 228.

EDELMAN, G. M. and BENACERRAF B., 1962. *Proc. Nat. Acad. Sci.* (Wash.), **48**, 1035.

EDELMAN G. M., BENACERRAF, B. and OVARY, Z., 1963. *J. Exp. Med.*, **118**, 229.

EDELMAN, G. M. and POULIK, M. D., 1961. *J. Exp. Med.*, **113**, 861.

EDMAN, P. and SJOQUIST, J., 1956. *Acta Chem. Scand.*, **10**, 1507.

EISEN, H. N., 1964a. *Methods in Medical Research*, **10**, 115.

EISEN, H. N., 1964b. *Methods in Medical Research*, **10**, 94.

EISEN, H. N., 1964c. *Methods in Medical Research*, **10**, 106.

EISEN, H. N. and KARUSH, F., 1949. *J. Am. Chem. Soc.*, **71**, 363.

ELEK, S. D., 1948. *Brit. Med. J.*, **1**, 483.

ELLIOTT, M., ENGLBERGER, F. M. and GADDONE, S. M., 1962. *J. Immunol.*, **89**, 517.

FAHEY, J. L., 1961a. *J. Exp. Med.*, **114**, 385.

FAHEY, J. L., 1961b. *J. Exp. Med.*, **114**, 389.

FAHEY, J. L., 1962. *Adv. Immunol.*, **2**, 42.

FAHEY J. L., WUNDERLICH, J. and MISHELL, R., 1964. *J. Exp. Med.*, **120**, 223.

FARAH, F. S., KERN, M. and EISEN, H. N., 1960. *J. Exp. Med.*, **112**, 1195.

FEINSTEIN, A., GELL, P. G. H. and KELUS, A. S., 1963. *Nature*, **200**, 653.

FLEISCHMAN, J. B., PAIN, R. and PORTER, R. R., 1961. *Arch. Biochem. Biophys.*, Suppl., **1**, 174

FLEISCHMAN, J. B., PORTER, R. R. and PRESS, E. M., 1963. *Biochem. J.*, **88**, 220.
FONGEREAU, M. and EDELMAN, G. M., 1965. *J. Exp. Med.*, **121**, 373.
FRANEK, F., 1961. *Biochem. Biophys. Res. Comm.*, **4**, 28.
FRANEK, F. and NEZLIN, R. S., 1963. *Folia Microbiol.* (Prague), **8**, 128.
FRANGIONE, B., FRANKLIN, E. C., FUDENBERG, H. H. and KOSHLAND, M. E., 1966. *J. Exp. Med.*, **124**, 715.
FRANKLIN, E. C., 1960. *J. Clin. Invest.*, **39**, 1933.
FRANKLIN, E. C., 1964. *Progress in Allergy*, **8**, 58.
FRANKLIN, E. C., FUDENBERG, H. H., MELTZER, M. and STANWORT, D. R., 1962. *Proc. Nat. Acad. Sci.* (Wash.), **48**, 904.
FREEMAN, T., 1967. Handbook of experimental immunology, Ed. Weir D. M., Blackwell, Oxford, 597–607.
FUDENBERG, H. H., 1965. *Ann. Rev. Microbiol.*, **19**, 301.
FUDENBERG, H. H., FEINSTEIN, D., MCGEHEE, W. and FRANKLIN, E. C., 1966. *Vox Sang.*, **11**, 45.
GITLIN, D. and MERLER, E., 1961. *J. Exp. Med.*, **114**, 207.
GRABAR, P. and BURTIN, P., 1964. Immunoelectrophoretic analysis, Elsevier, Amsterdam.
GRABAR, P. and WILLIAMS, C. A., 1953. *Biochim. Biophys. Acta*, **10**, 193.
GRAY, W. R. and HARTLEY, B. S., 1963. *Biochem. J.*, **89**, 59P.
GRÉY, H. M. and KUNKEL, H. G., 1964. *J. Exp. Med.*, **120**, 252.
GROSSBERG, A. and PRESSMAN, D., 1964. *Methods in Medical Research*, **10**, 103.
HABER, E. and BENNETT, J. C., 1962. *Proc. Nat. Acad. Sci.*, **48**, 1935.
HANSON, L. A. and JOHANSSON, B. G., 1963. *Clin. Chim. Acta*, **8**, 66.
HARBOE, M. and DEVERILL, J., 1964. *Scand. J. Haemat.*, **1**, 223.
HEIMBERGER, N., HEIDE, K., HAUPT, H. and SCHULTZE, H. E., 1964. *Clin. Chim. Acta*, **10**, 293.
HERBERT, W. J., 1967. Handbook of experimental immunology, Ed. Weir D. M., Blackwell, Oxford, 720–744.
HEREMANS, J. F., 1960. Les Globulines sériques du système gamma, Arscia, Brussels.
HILL, W. C. and CEBRA, J. J., 1964. *Biochemistry*, **4**, 2575.
HILSCHMAN, N. and CRAIG, L.C., 1965. *Proc. Nat. Acad. Sci.* (Wash,), **153**, 1403.
HSIAO, S. and PUTNAM, F. W., 1961. *J. Biol. Chem.*, **236**, 122.
HUNTER, W. M., 1967. Handbook of experimental immunology, Ed. Weir D. M., Blackwell, Oxford, 608–654.
INGRAHAM, J. S., 1958. *Proc. Soc. Exp. Biol.* (N.Y.), **99**, 452.
INGRAHAM, J. S. and BUSSARD, A., 1964. *J. Exp. Med.*, **119**, 667.
IRRVERRE, F., 1965. *Biochim. Biophys. Acta*, **111**, 551.
ISHIZAKA, K., ISHIZAKA, T. and HATHORN E. M., 1964. *Immunochemistry*, **1**, 197.
ISHIZAKA, K., ISHIZAKA, T. and HORNBROOK, M. M., 1966. *J. Immunol.*, **97**, 75.
JAMES, K., HENNEY, C. S. and STANWORTH, D. R., 1963. *Nature*, **202**, 563.
JERNE, N. K. and NORDIN, A. A., 1963. *Science*, **140**, 405.
JOHSON, A., DAY, E. D. and PRESSMAN, D., 1960. *J. Immunol.*, **84**, 213.
KABAT, E. A. and MAYER, M. M., 1962. Experimental immunochemistry, Thomas, Springfield.
KOSHLAND, M. E., ENGLBERGER, F. M. and SHAPANKA, R., 1964. *Science*, **143**, 1330.
KOSHLAND, M. E. and ENGLBERGER, F. M., 1963. *Proc. Nat. Acad. Sci.*, **50**, 61.
KOSHLAND, M. E., ENGLBERGER, F. M. and GADDONE, S. M., 1963. *J. Biol. Chem.*, **238**, 1349.
KOSHLAND, M. E., ENGLBERGER, F. M. and SHAPANKA, R., 1966. *Biochemistry*, **5**, 641.
LAMM, M. E., NUSSEZWEIG, V. and BENACERRAF, B., 1966. *Immunology*, **10**, 309.
LAURELL, C. B., 1965. *Analyt. Biochem.*, **10**, 358.
LAWLER, S. D. and COHEN, S., 1965. *Immunology*, **8**, 206.
LICHTER, E. A., 1964. *Proc. Soc. Exp. Biol. Med.*, **113**, 555.
LONGSWORTH, L. G., 1959. Electrophoresis, Ed. M. Bier, Academic Press, New York, pp. 91, 137.

MANNIK, M. and KUNKEL, H. G., 1963. *J. Exp. Med.*, **117**, 213.
MARGOLIASCH, P., 1962. *J. Biol. Chem.*, **237**, 2161.
MELTZER, M., FRANKLIN, E. C., FUDENBERG, H. H. and FRANGIONE, B., 1964. *Proc. Nat. Acad. Sci.* (Wash.), **51**, 1007.
METZGER, H. and SINGER, S. J., 1963. *Science*, **142**, 674.
METZGER, H., WOFSY, L. and SINGER, S. J., 1963. *Biochemistry*, **2**, 979.
MIDDLEBROOK, G. and DUBOS, K. J., 1948. *J. Exp. Med.*, **88**, 521.
MÜLLER-EBERHARD, H. J., 1960. *Scand. J. Clin. Labor. Invest.*, **12**, 33.
NAIRN, R. C., 1964. Fluorescent protein tracing, Livingstone, Edinburgh.
NEZLIN, R. S., 1966. Biokhimia antitiel, Izd. "Nauka", Moscow, 59–178.
NISSONOFF, A., WISSLER, F. C. and LIPMAN, L. N., 1960. *Science*, **132**, 1770.
NISSONOFF, A., WISSLER, F. C., LIPMAN, L. N. and WOERNLEY, D. L., 1960. *Arch. Biochem. Biophys.*, **89**, 230.
NISSONOFF, A., WISSLER, F. C. and WOERNLEY, D. L., 1959. *Biochem. Biophys. Res. Comm.*, **1**, 318·
OAKLEY, C. L. and FULTHROPE, A. J., 1953. *J. Path. Bact.*, **65**, 49.
OLINS, D. E. and EDELMAN, G. M., 1962. *J. Exp. Med.*, **116**, 635.
ONOUE, K., YAGI, Y. and PRESSMAN, D., 1966. *J. Exp. Med.*, **123**, 173.
OUCHTERLONY, O., 1948. *Acta Path. Microbiol. Scand.*, **52**, 186.
OUCHTERLONY, O., 1958. *Progress in Allergy*, **5**, 1.
OUCHTERLONY, O., 1962. *Progress in Allergy*, **6**, 30.
OUDIN, J., 1946. *C. R. Acad. Sci.*, **222**, 115.
OUDIN, J., 1952. *Methods in Medical Research*, **5**, 335.
OVARY, Z., 1952. *Int. Arch. Allergy Appl. Immunol.*, **3**, 293.
OVARY, Z., 1960. *Immunology*, **3**, 19.
OVARY, Z., 1964. Immunological Methods, Ed. Ackroyd J. F., Blackwell, Oxford, 259–283.
OVARY, Z., BLOCH, K. J. and BENACERRAF, B., 1964. *Proc. Soc. Exp. Biol. Med.*, **116**, 840.
PARKER, C. W., 1963. Conceptual advances in immunology and oncology, Harper, New York, 191–205.
PARKER, C. W., 1967. Handbook of experimental immunology, Ed. Weir D. M., Blackwell, Oxford, 423–462.
PATTERSON, R., PRUZANSKY, J. J. and JANIS, B., 1964. *J. Immunol.*, **93**, 51.
PATTERSON, R. and SPARK, D. B., 1962. *J. Immunol.*, **88**, 262.
PENNELL, R. B., 1960. The plasma proteins, Ed. Putnam F.W., Academic Press, New York, vol. 1, p. 9.
PETERSON, E. A. and SOBER, H. A., 1956. *J. Am. Chem. Soc.*, **78**, 751.
PETERSON, E. A. and SOBER, H. A., 1960. The plasma protein, Ed. Putnam F. W., Academic Press, New York.
PINCKARD, R. N. and WEIR, D. M., 1967. Handbook of experimental immunology, Ed. Weir D. M., Blackwell, Oxford, 493–526.
PORTER, R. R., 1959, *Biochem. J.*, **73**, 119.
PRESS, E. M., 1965. Symposium on Molecular and Cellular Basis of Antibody Formation, Prague, p. 93.
PRESS, E. M., PIGGOT, P. J. and PORTER, R. R., 1966. *Biochem. J.*, **99**, 356.
PRESS, E. M. and PORTER, R. R., 1966. Immunoglobulins. Glycoproteins, Ed. Gottschalk A. and Neuberger A., Elsevier, Amsterdam.
PRESSMAN, D., CAMPBELL, D. H. and PAULING, L., 1942. *J. Immunol.*, **44**, 101.
PRESSMAN, D. and ROHOLT, O. A., 1961. *Proc. Nat. Acad. Sci.* (Wash.), **47**, 160C.
PUTNAM, F. W., 1965. The proteins, Ed. Neurath H., vol. III, Academic Press, New York, 153–267.
PUTNAM, F. W. and EASLEY, C. W., 1965. *J. Biol. Chem.*, **240**, 1626.
PUTNAM, F. W., EASLEY, C. W. and LYNN, L. T., 1962. *Biochem. Biophys. Acta*, **58**, 279.

RAYMOND, S., 1962. *Clin. Chem.*, **8**, 455.
ROCKEY, J. H., KLINMAN, N. R. and KARUSH, F., 1964. *J. Exp. Med.*, **120**, 589.
ROSEVAR, J. W. and SMITH, E. L., 1958. *J. Am. Chem. Soc.*, **80**, 250.
ROSEVAR, J. W. and SMITH, E. L., 1961. *J. Biol. Chem.*, **236**, 425.
SANGER, F., 1945. *Biochem. J.*, **39**, 507.
SCHEIDEGGER, J. J., 1955. *Int. Arch. Allergy*, **7**, 103.
SCHULTZE, H. E., HAUPT, H., HEIDE, K., HEIMBERGER, N. and SCHWICK, H. G., 1965. *Immunochemistry*, **2**, 273.
SEIJEN, H. G. and GRUBER, M., 1963. *J. Molec. Biol.*, **7**, 209.
SELA, M. and MOZES, E., 1966. *Proc. Nat. Acad. Sci.* (Wash.), **55**, 445.
SELL, S., 1967. *Immunochemistry*, **4**, 49.
SINGER, S. J., 1959. *Nature*, **183**, 1523.
SKAŁBA, D., 1967. *Postępy Mikrobiologii*, **6**, 403.
SKEGGS, L. T., JR. and HOCHSTRASSER, H., 1962. *Ann. N.Y. Acad. Sci.*, **102**, 144.
SKVARIL, F., 1960. *Nature*, **185**, 475.
SMITH, E. L. and JAGER, B. V., 1952. *Ann. Rev. Microbiol.*, **6**, 207.
SMITHIES, O., 1959. *Adv. Protein Chem.*, **14**, 65.
SOBER, H., HARTLEY, R., CARROL, W. and PETERSON, E., 1965. The proteins, Ed. Neurath H., vol. III, Academic Press, New York, 1–97.
STANWORTH, D. R., 1967. Handbook of experimental immunology, Ed. Weir D. M., Blackwell, Oxford, 44–111.
STANWORTH, D. R. and PARDOE, G. I., 1967. Handbook of experimental immunology, Ed. Weir D. M., Blackwell, Oxford, 134–328.
STAVITSKY, A. B., 1964. Immunological methods, Ed. Acroyd J. F., Blackwell, Oxford, 363–396.
SVENSON, H., 1960. A laboratory manual of analytical methods in protein chemistry, Eds. Alexander P., Bloc R. J., Macmillan, New York, 193.
TERRY, W. D. and FAHEY, J. L., 1964. *Science*, **146**, 400.
THORBECKE, G. J., BENACERRAF, B. and OVARY, Z., 1963. *J. Immunol.*, **91**, 670.
TISELIUS, A., 1937. *Biochem. J.*, **31**, 1464.
TISELIUS, A. and KABAT, E. A., 1939. *J. Exp. Med.*, **69**, 119.
TITANI, K., WHITLEY, E., AVOGADRO, L. and PUTNAM, F. W., 1965. *Science*, **149**, 1090.
TRAUTMAN, R., 1964. Instrumental methods of experimental biology, Ed. Neuman D. N., MacMillan, New York.
UTSUMI, S. and KARUSH, F., 1963. *Fed. Proc.*, **22**, 466.
UTSUMI, S. and KARUSH, F., 1965. *Biochemistry*, **4**, 1766.
VAERMAN, J. P., EPSTEIN, W., FUDENBERG, H. H. and ISHIZAKA, K., 1964. *Nature*, **203**, 1046.
WEIR, D. M. (Ed.), 1967. Handbook of experimental immunology, Blackwell, Oxford, pp. 1245.
WILLIAMS, J. W., 1963. Ultracentrifugation in theory and experiments, Academic Press, New York.
WINSTEN, S., FRIEDMAN, H. and SCHWARTZ, E. E., 1963. *Analyt. Biochem.*, **6**, 404.
WOFSY, L., METZGER, H. and SINGER, S. J., 1962. *Biochemistry*, **1**, 1031.
WOFSY, L. and SINGER, S. J., 1963. *Biochemistry*, **2**, 104.
WUNDERLICH, J., MISHELL, R. and FAHEY, J. L., 1964. *J. Exp. Med.*, **120**, 243.
WUNDERLY, C., 1960. *J. Chromatog.*, **3**, 536.
ZABŁOCKI, B., 1965. *Postępy Mikrobiologii*, **4**, 135.
ZABŁOCKI, B., 1966. *Postępy Mikrobiologii*, **5**, 17.

DISCUSSION

J. E. RENDEL: I would like to ask Drs. Zabłocki and Owen about the nature of the antibody-antigen combining site on the antibody. There is evidence that the variable N-terminal part of the heavy chain is involved in the combining site, as this part by itself may combine with antigen.

The possible number of different combining sites would increase tremendously if the combining site would be determined by an interaction between the heavy and light chains. Is there any evidence for the occurrence of such interaction?

B. ZABŁOCKI: In the last years many works have been directed toward establishing the role in antigen binding of the light chains and of the portions of the heavy chains within the Fab fragments (Fd pieces). Special techniques were used for elucidation the nature of the active site of antibodies: highly purified antihapten antibodies, equilibrium dialysis for measurement of antibody-hapten affinity, specific or differential hapten labelling, etc. The relative roles of separated heavy or light chains are still far from clear. There are suggestions that the specific antigen-binding groups of several antibodies are heated in the Fd portions, their configuration being influenced by the conformation of the neighbouring light chains. However, with other antibodies, heavy chains were found to be completely inactive, but some antibody activity was restored on interaction with light chains (which themselves were also inactive). Conclusion: may be the importance of the heavy or the light chains in forming the active sites is not the same in different antibodies.

XITH EUROPEAN CONFERENCE ON ANIMAL BLOOD GROUPS AND BIOCHEMICAL POLYMORPHISM
Warsaw, July 2nd–6th, 1968

Genetic Aspects of the Immune Response

R. D. OWEN

California Institute of Technology, Pasadena, California, U.S.A.

The immune response probably evolved primarily for the advantages it confers on the organism in recovery from infectious disease, and specific protection against reinfection by the same pathogen. The response is characterized by recognition of substances foreign to the organism, elaboration of antibodies with combining sites complementary to parts of the foreign substances that induced antibody formation, and a specific "memory", such that later exposure to the same or a sufficiently similar foreign substance provokes a more rapid and intense antibody response than did the first exposure. Because the immune response system has evolved by selection, we can assume that its elaboration has depended on genetic variations occurring in the past. In its present form the system is highly complex, involving numerous processes of programmed cellular function, differentiation and interaction. Thus the phylogeny, the ontogeny, and the physiology of the immune response all indicate a high order of genetic dependency.

Perhaps because of its complexity, and the consequent difficulty of dissecting out unit processes for the identification of Mendelian unit control, disappointingly little progress has been made in understanding the broad scope of the biochemical or developmental genetic aspects of the immune response. Selection experiments and quantitative genetics have generally indicated that degrees of susceptibility or resistance to infectious disease are under polygenic control; rarely has a unit physiological process been identified, or associated with any particular gene. Resistance seems often to relate to secondary rather than primary processes in the immune response.

We can take the following as primary processes, not necessarily occurring in the sequence given: (a) recognition of the antigen; (b) processing of the antigen; (c) development of cellular competence to respond; (d) induction of cellular response; (e) amplification of response–cell multiplication and elaboration of intracellular machinery for synthesis and secretion of antibody; (f) regulation of response; and (g) memory and recall.

Sampling from the important secondary interactants and processes we can include: (a) complement and other cofactors involved in cytolysis or cytotoxicity;

(b) products of cascades of reactions involved in vascular damage, necrosis and other evidences of hypersensitivity; (c) interactions among diverse products of the immune machinery, producing contrary or controlling effects of one on another; (d) hormone or hormone-like factors involved in developmental and control aspects of the system, and inductive effects; and (e) transport phenomena, including *trans*-placental effects.

Even the above incomplete and broadly categorical list of primary and secondary processes involved in the immune response must impress us with the scope of our subject, and with the diversity of gene effects that must be involved in it. In attempting a relatively brief presentation of the subject, the reviewer is impressed, too, with the amount of talking and writing that has been done about it, particularly in the last 2 or 3 years. Weeks of symposium papers and discussions, recorded in thousands of pages of reports — a great list of detailed observations and measured thoughts by many people — are already at hand. Under these circumstances, it would be unwise for me to attempt a detailed review and citation of references. I shall list only a few recent sources, mainly symposia and reviews, and will not attempt to identify priority of factual discovery or idea in many instances. You can take it that little I will say is original with me. In fact, so enthusiastic has been the recent attention to this field that essentially similar ideas and observations seem often to have developed independently at two or more different sites.

It is rather remarkable that, of all the complex processes of the immune response, it is the operational one, the synthesis of specific immunoglobulin molecules, that is at present subject to the clearest genetic analysis. I will attempt first a brief summary of the state of knowledge and the significance of genetic control of immunoglobulin molecular variation. In preparing this paper, I am assuming that the preceding lecture, by Dr. Zabłocki, will already have reviewed for this audience the physical, chemical and immunological bases of immunoglobulins. We need here only to select, for brief review, some of the aspects of immunoglobulin structure most clearly relevant to matters of genetic control.

I find it somewhat easier to outline the structure and control of synthesis of immunoglobulin molecules against a background of understanding of the structure and control of synthesis of hemoglobin. As most of us know, the major adult hemoglobin of man is composed of four polypeptide chains, two alike and designated as α and the other two as β. The α- and β-chains are respectively under control of unlinked genes; mutation has occurred at each of these two loci, resulting in amino acid differences in α- and β-chains, under straightforward Mendelian control.

In addition to the major adult hemoglobin there is a predominant minor one, A2, which instead of the β-chain has a δ-chain. There are only 10 amino acid differences between the β- and δ-chains in the total of 146 for each; there is therefore very extensive homology between the two. β- and δ-chains are controlled by closely linked cistrons. In normal adult hemoglobin-synthesizing tissues, α,- β- and δ-chains are all being synthesized, producing both hemoglobins $\alpha_2\beta_2$ and $\alpha_2\delta_2$.

Heterogeneity of hemoglobin, at this first level, therefore can depend either on heterozygosity for the alleles of a cistron (such as for the sickle-cell allelic form of the β cistron, producing in the heterozygote a mixture of normal adult and sickle-cell hemoglobin), or on the operation of more than one cistron (such as the β and δ cistrons). Both of these produce mixtures; which of the two bases of heterogeneity one is dealing with in any given example is usually evident from genetic considerations. For example, hemoglobins A1 and A2 are produced by practically every normal human subject. If the mixture of A1 and A2 observed were due to heterozygosity one would have expected homozygotes to occur with predictable frequencies, having either A1 or A2 but not both. More refined genetic observations may also be applied; they need not be discussed here. Another revelation of the cistronic rather than the allelic basis for heterogeneity is when three or more chains of the same general type are present in the same individual — for example, β, δ, and γ, the chain characteristic of fetal hemoglobin. The individual has only two alleles for any given cistron; at least one of the three must therefore be dependent on a separate cistron — a conclusion that can be reached even in the absence of genetic evidence for recombination. And when, as is true with human hemoglobin, genuinely allelic variation can be demonstrated for all three of the chains, at least three cistrons must be implicated.

There is recent evidence that the γ-chain in man is commonly produced in two distinguishable forms, for which only a single amino acid difference is known, depending on cistron duplication rather than heterozygosity (SCHROEDER *et al.*, 1968). There is also good evidence, in the horse, the rabbit, the goat, and the mouse, that the α-chain cistron is duplicated.

The developmental history of hemoglobin provides additional information. Early in development, the hemoglobin produced may be a tetramer of four ε-chains, or of two α- and two ε-chains. By the end of the third month of gestation, fetal hemoglobin has succeeded these embryonic hemoglobins; it is composed of two α- and two γ-chains. Toward the end of gestation, β- and δ-chains begin to be synthesized, and to replace the γ; the synthesis of γ-chains decreases rapidly in the early postnatal period. From their amino acid sequences, ε, the two γ's, δ, and β seem clearly to be homologous, and to have derived from a common ancestral gene. There are also points of homology with the α-chains, so that phylogenetically an early ancestral gene was apparently translocated to an independent linkage group and divergence occurred by mutation in each of the separated cistrons, duplication and reduplication, and further mutational divergence. Developmentally, the tissues engaged in hemopoiesis display a programmed cistron shift, so that one kind of hemoglobin replaces another in time, but some heterogeneity of synthesis is retained all during life.

Hemoglobin studies have given some idea of why so complicated a system of protein synthesis should have evolved. First, it has been pointed out that there is a factor of safety in having at least two cistrons for a vital polypeptide chain-type, so that if one should be subject to a mutation rendering it less functional, the other is available to fill its place. Adaptability to environmental fluctuation may also be involved, because the alternative hemoglobins may function best in different internal or external environments. Second, tetrameric molecules composed of dissimilar chains have an intra-molecular regulatory mechanism available to them that may be very important for their functions. The peculiar oxygen dissociation curve of the hemoglobin molecule is thought to relate to the molecule's ability to snap into a new configuration as the two heme groups take on oxygen. Tetrameric hemoglobin molecules in which the chains are all alike behave in an "intemperate" way with regard to this function; those composed of two pairs of dissimilar chains typically show a nice molecular regulation. By translocating the gene for an ancestral chain to an independent site, and synthesizing, with mutational divergence at the two sites, two distinct but mutually adjusted chains, the principle of this kind of intra-molecular control became available to evolving organisms. Through cistron duplication, mutations can be tried out and maintained even though, at any particular step they do not work very well; cistron duplication provides a way of storing up trials until some new, emergent advantage of chain interaction can be accumulated.

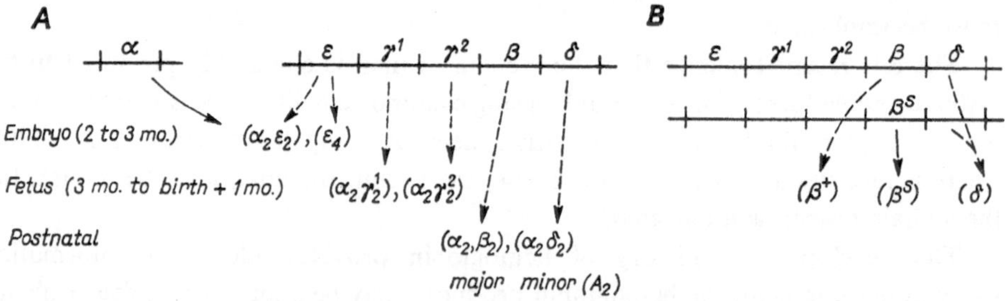

FIG. 1. A — The succession of hemoglobins in human development, based on shifts in cistron expression. B — Heterogeneity of β-related chains in human adults heterozygous for β-sicle, based on coincident expression of two alleles plus two cistrons.

The outline of this hemoglobin picture is summarized in Fig. 1. Cistrons for ε, γ^1, γ^2, β, and δ are shown as adjacent; because of their similarites, it seems likely that they are in fact rather closely related. At present, however, there is no genetic evidence for linkage of any of the cistrons listed except of β and δ.

I have given some time to the elements of the hemoglobin story, because it is becoming so relatively clear, and because a consideration of the immunoglobulins against this background may be only a little more complex, and involve only one new and important additional mechanism. The reference immunoglobulin for the following treatment is γG or IgG, the ordinary 7S antibody molecule, of molecular weight about 150,000. Like hemoglobin, it is a tetramer composed of two pairs of polypeptide chains. These types of chains are designated respectively as L and H — the L (light) type with a molecular weight of approximately 23,000, and the H (heavy) type about 50,000.

Again as in the hemoglobins, both of the types of chains are subject to both allelic and cistron variation. Taking the cistron variation first, L-chains occur in a great many species in two quite different major classes, designated as κ and λ. While κ and λ are quite different from each other, they show degrees of homology comparable to those among the β-related chains of hemoglobin, and very probably derived by duplication of a common ancestral gene. Furthermore, both the κ and the λ classes are heterogeneous, occurring in several subclasses. Thus, an assortment of L-chains are indicated as available for use by the immunoglobulin synthesizing machinery of the typical vertebrate organism.

The H-chains, too, display cistron diversity. The γ-class, characterizing γG immunoglobulin molecules, is present in at least four different subclasses, known as γG1, γG2, γG3, and γG4. A normal person produces all four of these γG subgroups, with a ratio in normal human serum of approximately 65 : 23 : 8 : 4% respectively. This kind of γ-chain variation cannot, therefore, depend on allelic alternatives, but reflects cistron variation. These cistrons appear to be closely linked, in the following sequence: γG4–γG2–γG3–γG1. Other H-chains characterize other classes of immunoglobulins: α-chains for IgA (γA), the peculiar antibody secreted at tissue surfaces; and μ-chains for IgM (γM), the very large molecular weight class of antibody composed of regular multimeric associations of (L2 H2) units. In the mouse, a heavy chain linkage group has been identified which contains the structural cistrons designated as G($\gamma 2^a$ or Ig-1), H($\gamma 2^b$ or Ig-3), and A(γA or Ig-2). The following table, from POTTER and LIEBERMAN (Cold Spring Harbor Symposia on Quantitative Biology (1967), p. 188), gives concordances for various nomenclatures applied to mouse immunoglobulins:

Polypeptide chains	Light		Heavy				
1	κ	λ	μ	α	φ	γ	η
2			μ	α	γ1	$\gamma 2^a$	$\gamma 2^b$
3			μ	α	γ1	$\gamma 2^a$	$\gamma 2^b$
Genes							
1	K	L	M	A	F	G	H
2			γM	γA	γ1	$\gamma 2^a$	$\gamma 2^b$
3				Ig-2		Ig-1	Ig-3
Antigenic determinants							
1*				A^n		G^n	H^n
3**				Ig-2.n		Ig-1.n	Ig-3.n

1 = POTTER and LIEBERMAN (1967).
2 = FAHEY, WUNDERLICH and MISHELL (1964a, b).
3 = HERZENBERG, WARNER, and HERZENBERG (1965).

* We use consecutive numbers, irrespective of chain; thus a determinant that is common to two chains would have the same number, e.g. G^8H^8.

** n = number of the specificity or determinant. HERZENBERG *et al.* (1965) use consecutive numbers for each chain type; thus Ig-1.8 and Ig-2.8 are different.

Given a choice of six different L-chain cistrons, and six different H-chain cistrons, and with heterozygosity for each cistron, a typical cell would have built into it the potentiality for 144 different tetrameric molecules, each molecule with two like L-chains and two like H-chains. This is a crude approximation, of course, but in view of the fact that our information is clearly incomplete, it may well represent a minimal estimate of potential for variation at this level.

Much greater diversity becomes evident when the detailed amino acid sequences of individual chains are taken into account. Based mainly on work with myeloma proteins, it has for some time been evident that the aminoterminal halves of the L-chains within any given class (such as λ) are highly variable. There are stretches of sequence within this "variable half" that are closely conserved and homogeneous, but the majority of the amino acids within that region may be varied (Fig. 2).

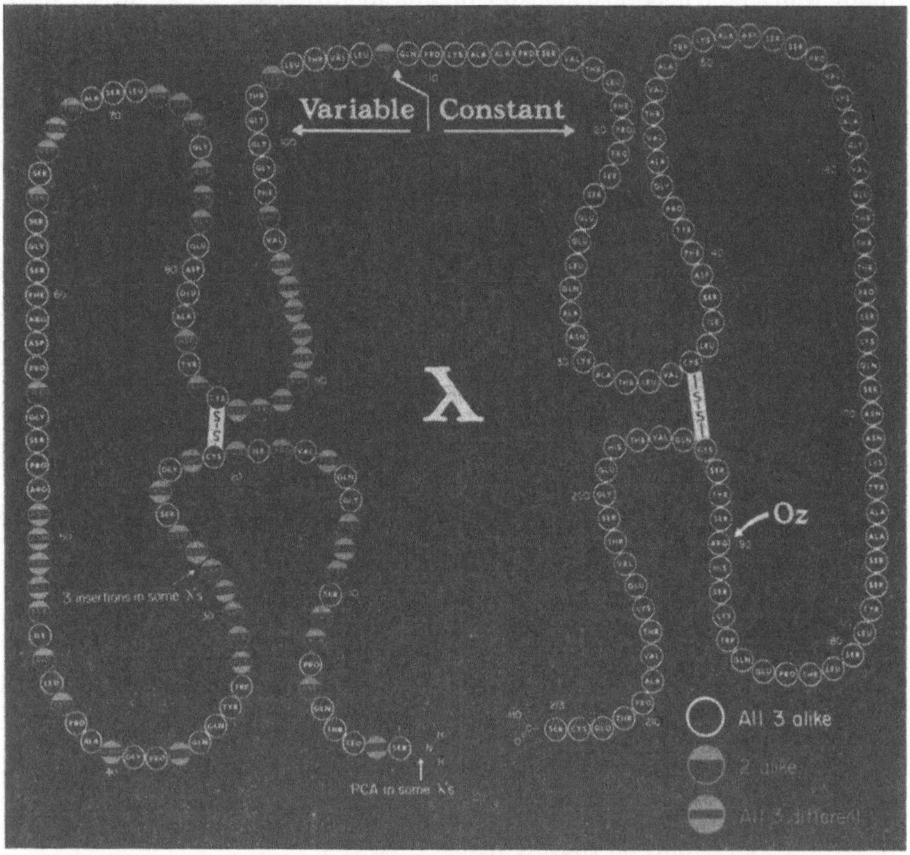

FIG. 2. Amino acid sequence of the human λ Bence-Jones protein Sh. Circles for amino acids distinguish those that are identical in λSh, λBo, and λHa, from those that are alike in any two of the three, and from those that differ in all three. (From PUTNAM, TITANI, WINKLER and SHINODA, 1967. Cold Spring Harbor Symposia on Quantitative Biology XXXII: Antibodies, p. 29.)

It is probably this latter type of variation that is most relevant to the great diversity of antibody specificities the organism can produce. The way this order of diversity is generated is still being argued, almost entirely on the basis of amino acid sequence data rather than genetic evidence. There are those who have believed that the variation must all be represented in the fertilized egg, and reflect the long course of mutation, selection, genetic transmission, and evolution. While thoughtful and persuasive sequence arguments have supported this viewpoint, and it has by no means been disproved, advocates of this kind of explanation must deal with difficult genetic questions. For example, what selection pressures may be adduced to explain the maintenance of so large a diversity of genetic potential to react with the unforeseen, protecting it against randomization by recurrent mutation, against drift and loss? And how may so very large a degree of genetic similarity-within-diversity be maintained without compelling meiotic difficulty — unequal pairing, inversion, and all the kinds of chromosomal disturbance that may lead to disruption of the complicated library of relevant genetic information through the processes of sexual reproduction?

Others have maintained that the main sources of this diversity must be somatic rather than germinal. For the most part, until recently, this kind of explanation has centered on chromosomal mechanisms; given a degree of duplication and divergence something like that already established, somatic recombination, inversion, and other chromosomal phenomena could lead to the generation of large amount of somatic diversity. Sequence data indicate that at least three distinguishable germinal cistrons or part-cistrons would have to be involved in such processes, because more than two base-differences must be present at particular amino acid specifying sites to account for the amino acid differences that have already been noted.

The generation of diversity by chromosomal events of the sorts indicated would have to be rather closely specified, in order to maintain the regularity in length and the constancy of sub-regions observed within the variable parts of the chains. Another sort of explanation of the generation of somatic diversity has been gene hypermutation, leading to assorted amino acid substitutions following base changes in limited regions of the relevant DNA. It is difficult to specify, however, how such hypermutation may be confined so regularly to particular regions in the DNA strand. Most recently, Dr. Melvin COHN of the Salk Institute has developed an approach based on the assumption that normal gene-mutational processes, at normal rates, coupled with somatic selection for diversity, might be adequate to explain the phenomena of the variable regions. His paper was delivered at a symposium I did not attend, and which is not yet published (COHN, 1967). He did, however, generously make a draft of his paper available to me. In my judgment, you will find this paper well worth studying if you are interested in the subject in depth. COHN reminds us that there is an enormous turnover of cells in the mammalian thymus; only about one percent of the cells produced there ever leave it. The not-unreasonable mutation rate of 10^{-4} per cistron controlling a variable region per cell generation, in a total cell population of 10^7, with a generation time of one day, would provide for the production of a great deal of variability, one mutation at a time. If the normal non-mutated chains do not fit with each other and, under the conditions of the thymus, cells bearing these chains are eliminated, strong local conditions may be visualized for the selection of mutants; only cells with chains that fit for critical segments of the variable regions might leave the thymus to seed the lymphatic tissues. There, on contact with an appropriate antigen, these cells may be subjected to a second selective step in the elaboration of an immune response — a specific selection, based on antigens. In my very quick and inadequate description here, the theory may seem excessively speculative. Personally, though, I find it more satisfying than any of the other speculations that have gone before it in this provocative field.

To provide an outline summary for this aspect of the genetics of the immune response we may refer to a model of the IgG half-molecule by SINGER and THORPE (1968) (Fig. 3). We can take the L-chain on the left as a κ-chain; the top half of it, shown as a loop, is the N-terminal and variable half of the chain. In any given myeloma protein the amino acid sequence in this region is constant, but among immunoglobulin molecules produced by an individual and containing the κ-chain there are many amino acid differences in this region. At positions 23 and 88, however, there is always a cysteine, providing for an intrachain disulfide bridge that is a constant structural feature of the molecule. The amino acids closely associated with this structural region are relatively constant, and show a degree of conservatism in evolution. At position 34 there is an asparagine or aspartic acid, to which, for some κ-chains, a rather complex carbohydrate is attached. The attached sugar is N-acetylglucosamine, in a type of carbohydrate that includes at least five additional glucosamines, four mannoses, two galactoses, two galactose-sialic acid groups, and two fucoses (Cold Spring Harbor Symposia on Quantitative Biology (1967), p. 256). Incidentally, we might take occasion here to divert to the carbohydrate attached to position 150 from the C-terminal end of the heavy chain — also attached to an asx, and very similar in composition to the carbohydrate noted for the L-chain. Carbohydrate on the γ-chain is a common characteristic of the molecule; there is evidence that the first sugar groups are attached while the chain is still associated with ribosomes, and that additional sugar attachments grow in the cell and in the process of secretion from the cell. The completed carbohydrate chain, in fact, seems to be necessary for secretion of the completed antibody molecule from the cell. The carbohydrate in the H-chain is associated with a part of the molecule that is important in such processes as complement fixation and transport across the placenta; it is also a closely conserved structure in evolution, together with the associated amino acids in that part of the polypeptide (HOWELL *et al.*, 1967). The genetic control of the synthesis and addition of the sugars and carbohydrates adds another dimension for genetic consideration, comparable in many ways to the problems of glycoproteins and blood groups.

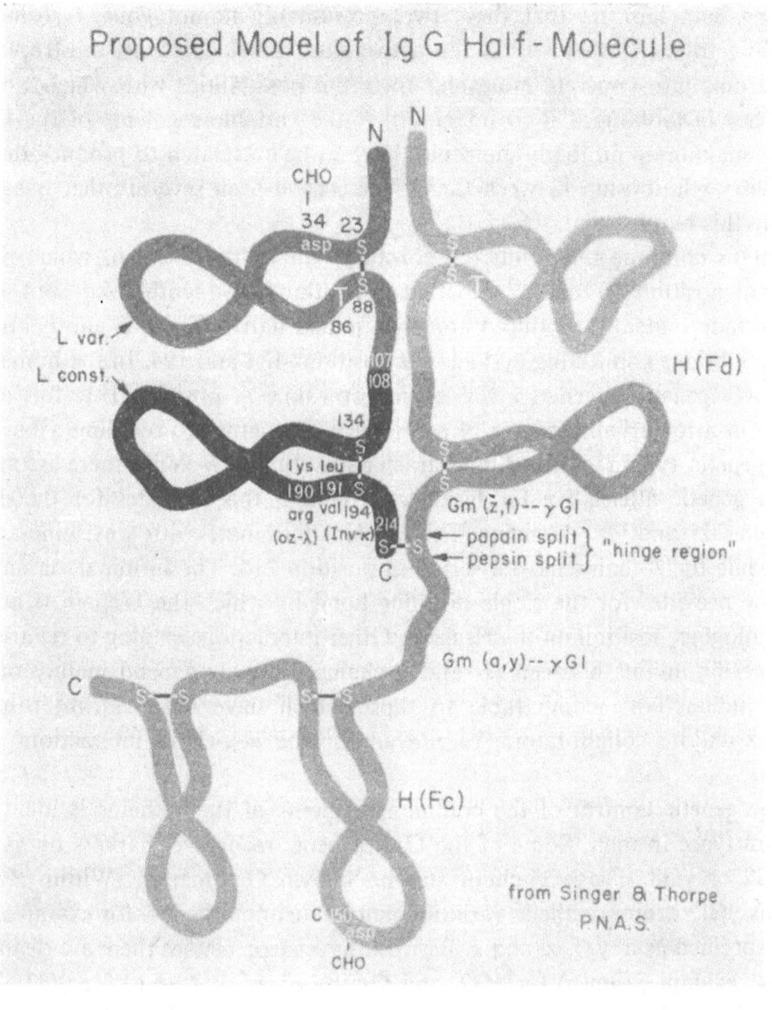

FIG. 3. Based on a figure in "On the location and structure of the active sites of antibody molecules", by SINGER, S. J. and THORPE, N. O., 1968. Proc. Nat. Acad. Sci. (U.S.A.) (in press). (See text.)

At position 86 on the L-chain there is a tyrosine, which SINGER and THORPE have shown, by affinity labeling technique, to be an integral part of the combining sites of antibody molecules directed against 2,4-dinitrophenyl hapten in rabbits. There is also a tyrosine in the variable part of the heavy chain, which participates in the active sites of antibody against this hapten. This is the first conclusive demonstration that the variable parts of both the H- and the L-chains are involved in the active site of the antibody molecule. It further suggests that the L-chains of a number of species have an evolutionarily conserved tyrosine residue at position 86, and that H-chains also have the homologous, conserved tyrosine residue. The affinity labeling data indicate that these two, presumably homologous, tyrosine residues are close to each other within the active sites directed against dinitrophenyl. The model puts into concrete molecular form the possibilities with which COHN dealt; it is easy in looking at it to imagine that the variable segments of the H- and L-chains making up antibody molecules have to be correlated to produce the required symmetry relationship between them. There have been several other bases for suggesting this requirement.

Let us continue now with the constant half of the L-chain, which appears to begin at position 108 in a chain that is continuously synthesized, not developed as two independently produced and then joined halves. There is another intra-chain disulfide bridge connecting cysteines at positions 134 and 194. In the human κ-chain there is a genetic alternative for leucine or valine at position 191; this alternative results in a remarkable degree of serological distinctness, providing a basis for immunogenetic typing of the Inv characteristic. In the λ-chain there is, at position 190, a genetic alternative for lysine or arginine; this provides for the distinction between Oz^+ and Oz^- λ-chains. The κ-chains terminate with a cysteine at position 214, while the λ-chains have a serine in position 215. The terminal or subterminal cysteine provides for the single disulfide bond by which the L-chain is attached to a homologous position on the H-chain. Other interactions leading to relatively stable and specific unions between H- and L-chains probably depend mainly on van der Waals interactions, comparable to those which have been shown, primarily by PERUTZ and his collaborators, to prevail for the α–β-chain interactions in hemoglobin.

The genetic control of the common segments of the γ-chains is identified with the Gm types in man. Some of the Gm reagents recognize markers on γGl; others on γG2 or γG3. The γG4-chain has no known Gm marker. Within a sub-class, such as γGl, ordinary allelic variation can be distinguished — for example, for Gm types specified as a, y, f, z, and x. Between the cistron classes there are distinct markers; for example, Gm(n) for γG2, and Gm(b, b^3, b^4, s, t, c, g) for γG3. The rules for distinguishing allelic from cistron variations, outlined for hemoglobin earlier in this paper, have helped greatly in making sense of the complexity of the Gm types.

The principles of the molecular genetics of immunoglobulins are therefore, in many respects, very similar to those for hemoglobin. Superimposed on the hemoglobin-determining kind of system, with translocation, duplication and reduplication, and mutational divergence among both allelic and reduplicated cistrons, is the special problem posed by the variable regions of the immunoglobulin chains. For molecules like hemoglobin, or for the constant parts of the immunoglobulin chains, we can well imagine that the somatic generation of mutants can either go unselected, and therefore result in contamination of the major molecular product only with undetectably low levels of diverse mutant forms, or we can conceive of somatic selection against diversity to reduce contaminating mutant forms to even lower concentrations. In the immunoglobulin system, which functions through diversity, somatic selection for diversity in the variable parts of the chains may very well provide the major new attribute of the genetics of this class of molecules.

SINGER and THORPE have chosen for considered reasons to show the half-molecule structure as a series of loops. First, of course, are the disulfide bonds making intra-chain and interchain ties. In addition, there is evidence from amino acid sequence homology studies that phylogenetically there was probably one ancestral gene producing a polypeptide comparable to one of the loops shown. This was duplicated to provide for variable and constant halves of the L-chain, and reduplicated *in situ* to provide for diverse classes and sub-classes of L-chains. It was also translocated, duplicated, reduplicated and duplicated again to provide for the long constant fragment of the H-chain, its variable region to be correlated with the corresponding fragment of the L-chain, and the diversity of H-chains that characterize sub-classes and major classes of the immunoglobulins. Through all this complex evolutionary process there was a good deal of conservatism, imposed by the necessity of maintaining a reasonably constant ultimate structure for the tetrameric molecule, appropriate regions for chain interaction, and molecular functions that require interaction with cofactors external to this molecular system itself — for example, complement fixation and cell-binding affinities for some of the classes of immunoglobulins. Not all, however, fix complement or bind to cells; the distinctions in molecular composition or structure associated with the alternative of doing so or not doing so may very well provide critical information on the nature of these important secondary immunologic processes.

It should be emphasized that the Singer–Thorpe model is of the half-molecule. In any given complete IgG molecule there is a symmetrical second half, identical in details that include the composition of the variable and constant halves of the L-chains and the variable and constant fragments of the H-chains.

It will be noted that the various Gm factors characteristic of a given cistron, such as that for γGl, are distributed along the linear structure of the chain; some of the factors are found in the so-called Fc fraction, a constant fraction split off by papain, and others extend into the part of the H-chain left in association with the L-chain after such enzymatic treatment. Some of these factors undoubtedly reflect a colinear relationship between the controlling, DNA its cistronic messenger RNA, and the synthesized polypeptide chains. Other Gm specificities may reflect the tertiary structure of the molecule, depend on its integrity, and deviate from colinearity in the sense noted. As sequence information becomes available on the γ-chains, and is associated with genetic data, in which intracistronic recombination has already been recognized, we may expect this approach to immunogenetics to contribute importantly to our understanding of the relationships between nucleic acids and the antigenic variations they control.

The expression of the genetic information also offers many points of similarity between the hemoglobins and the immunoglobulins. There are, however, two important points of difference, involving the phenomena of allele exclusion and cistron suppression in the cellular expression of the nucleic information for immunoglobulins. Perhaps because some functions of the intact IgG molecule depend on its containing two identical valence sites, a regular control process appears to have developed such that only one of the members of an allelic pair is expressed in the cell's recognition of antigen and its immunoglobulin synthesis. The other is excluded. Furthermore, with a variety of cistrons available, for expression cells appear to "turn on" only one L-chain-controlling and one H-chain-controlling cistron at a time. Perhaps the best evidence for the latter assertion is that in a myeloma the product is uniform; only one complex of Gm types, for example, characterizing one of the γG sub-classes, is produced by a given myeloma, even after prolonged periods of multiplication and many transplant generations. It is still not fully clear, however, whether under normal circumstances in the life history of a given cell a cistron-shift may not be possible, resulting in a succession of IgM (μ-chain) and IgG (e.g. γGl-chain) synthesis. In the hemoglobins, it does seem clear that two homologous cistrons, such as those for β and γ, can work in the same cell, probably at the same time. For the hemoglobins, there is no allelic exclusion either; β^{s} and the normal β alleles are both expressed in individual cells in a sickle cell heterozygote.

Normally, allele exclusion appears to operate nearly at random; some cells express one allele and the other its alternative. Under experimental conditions, however, it has proved possible to set the organism in such a way that most or nearly all of the cell population expresses only one of a pair of heterozygous alleles. This is achieved in rabbits for prolonged periods in the life of the individual, by the passive transfer of antibody from a female rabbit immunized against the immunoglobulin type her progeny will inherit from their sire, or through passive experimental immunization of the young themselves in the early postnatal period. In the mouse, passive suppression of allotype is much less effective and of briefer duration than in the rabbit. The Herzenbergs have pointed out that this difference may relate mainly to the types of variation available for suppression by passively administered antibody. In rabbits, the L-chain antigens with which the suppressing antisera react are present on most of the immunoglobulins, and comprise more than 90% of the total immunoglobulins. If suppression depends on the elimination of stem cells expressing the activity of particular chromosomes, most of the cells would be affected, and when the suppressing antiserum disappeared only a few would be left to shift and repopulate, "leaving a more or less permanently allotype-suppressed rabbit as a result" (Cold Spring Harbor Symposia on Quantitative Biology (1967), p. 186). In mice, on the other hand, the antigens with which the suppressing antisera react reflect only the chromosomal region coding for the heavy chain that is characteristic of γG_{2a} globulin. Various other cistrons capable of expression will be unaffected, leaving a generous population of unaffected cells to contribute quick recovery of the treated animals. In the rabbit, through especially clever experimental design, it has been demonstrated that even homozygotes can be rather fully suppressed. Under these circumstances, new classes of immunoglobulins are revealed (Cold Spring Harbor Symposia on Quantitative Biology (1967), p. 312). It seems possible that these include λ-chain molecules; λ-chains are known from chemical studies to comprise only about 10% of the L-chains of ordinary rabbit sera.

It is tempting to speculate that the principles of the molecular basis of allele exclusion and cistron suppression occurring in the normal development of competent cells may be the same as that for experimental allotype suppression — that in the latter case passively administered antibody complementary with the combining site exposed on the cell surface induces suppression, while in the normal cell the product of one allele or cistron that happens to appear first, and to achieve dominance, represses all the others or is rather directly involved in their repression. Such a scheme might suggest that, while it is the combining sites of the treating antiglobulin that are responsible for its direct and specific attack on the relevant cells, it may be the common regions of these molecules that are involved in allotype suppression, as a part of the normal machinery for the control of antibody synthesis. Such a line of thought even suggests, in hazy outline, an interpretation of the basis of specific immunologic tolerance which contrasts with the present, largely accepted, interpretation in terms of the developmental elimination of "forbidden clones." In other words, tolerance may depend on direction of clonal differentiation, rather than on clonal elimination. Other phenomena of current interest, such as the desensitization of Rh-sensitized women by treatment with anti-Rh sera, and the induction of prolonged tolerance achieved in the presence of anti-lymphocyte sera, with a regenerating lymphocyte population developing as tolerant to tissues present at the time of regeneration, may also be fitted into such a scheme.

When rabbit peripheral lymphocytes are subjected in culture to anti-allotype sera, many are stimulated to transform into immature blast cells. It has recently been demonstrated (SELL, 1968) that antisera to one L-chain allotype stimulate 77–82% of the lymphocytes of the corresponding homozygote to transform. Only about half that frequency of transformation is observed in heterozygote cultures, stimulated with antiserum specific for the product of either allele. This provides a direct demonstration of allele exclusion for lymphocytes in culture, comparable to a similar conclusion reached in earlier studies on allotypic expression by immunoglobulin-producing plasma cells (Cold Spring Harbor Symposia on Quantitative Biology (1967)).

It suggests that the action of an anti-allotype serum is not to eliminate directly the relevant cell-type, but to promote its transformation (LEVEY and MEDAWAR, 1966). Note, however, that the anti-allotype sera react with the constant portions of the immunoglobulin molecules, while specific antigen stimulation to cell division and immunoglobulin synthesis relate to specificity regions on the same molecules. The nature and degree of association between the phenomena of allele exclusion, allotype suppression, stimulation to blast formation, action of anti-lymphocyte sera, and specific antigen triggering and induction invite further thought and work.

In our consideration of the molecular genetics of the immunoglobulins to this point we have subtly invaded other primary aspects of the immune response — the cellular commitment that must precede induction of the response, and some of the aspects of amplification through cell multiplication and elaboration of the intracellular machinery for synthesis and secretion of the primary molecular products. I shall set aside here, for lack of time and knowledge on my part of information closely relevant to genetic aspects of the system, further consideration of the control, memory, and recall aspects of the response. Something should be said, though, about the early events in the sequence of the response — the processing and recognition of antigens.

At least in the primary response, and probably in the secondary as well, most if not all antigens appear to be picked up and processed by macrophages. There is at present no good reason to believe that the ingestion of foreign materials by macrophages depends on a specific recognition phenomenon. Neither is it clear what macrophage processing entails; it is generally assumed to be digestion of the antigen to fragments, but as has been particularly pointed out by SELA (Cold Spring Harbor Symposium on Quantitative Biology (1967)), some antigenic determinants of proteins are properties of the three-dimensional structure of the molecule. Ordinary digestion, as proteolysis, would be expected to destroy this structure. It appears, therefore, that recognition for the induction of antibody synthesis precedes complete dissolution of the antigen by proteolysis.

There have been repeated reports that macrophage RNA, perhaps in association with fragments of the antigen, is an effective stimulus to the reactive cells. For the most part, the relevant experiments could be adequatelly explained without postulating specificity of the macrophage RNA, at least as an informational macromolecule (macrophage RHA+antigen fragments = superantigen, the RNA acting essentially as an adjuvant to immunization). My colleagues, CAMPBELL and GARVEY, have for several years been recording the storage of antigen fragments in the form of RNA complexes, in immunized animals. In studies by ADLER *et al.* (1966), observations have been recorded that, at least at face value, seem to make it likely that for the elaboration of IgM, the first immunoglobulin to be formed in culture, donor macrophage RNA serves as an informational macromolecule. RNA from peritoneal exudate cells which had been exposed to bacteriophage was added to cultures of lymph node fragments from allotypically distinct normal rabbits. The IgM anti-phage antibody, but not the later-formed IgG, bore the light-chain marker of the RNA donor, rather than that of the lymph node cells. It seems important that further studies be conducted of this still-mysterious early phase in the immune response, and of the role of macrophages and cell interaction in it.

At least one clearly segregating genetic unit has been recognized as being involved in the recognition or processing stage of the response. Inbred strain 2 guinea pigs respond to hapten conjugates of poly-L-lysine, while guinea pigs of inbred strain 13 respond only to some polymers containing lysine, and then, typically, only weakly or variably (LEMELIN *et al.*, 1968; MAURER and PINCHUK, 1968). Hybrids between the two strains, immunized with 2-4-dinitrophenyl-poly-L-lysine (DNP/PLL), are responders, and segregating generations, most-studied in the non-inbred Hartley strain, give results compatible with the conclusion that a single gene is involved, with the responder allele dominant. In the segregating population, only the individual guinea pigs that can give responses to DNP/PLL conjugates can also react to a linear copolymer of L-glutamic acid and L-lysine (GL), which is not cross-reactive with DNP/PLL (LAMELIN *et al.*, 1968). This suggests that the same genetic factor which affects the immune response to one specific determinant also affects the response to another. DNP/PLL, which fails to elicit an immune response when it is injected by itself into non-responder guinea pigs, evokes a good specific antibody when it is injected in complex with a foreign protein. In other words, the genetic distinction involves, not an antibody structural gene specific for the determinant, but instead a genetically controlled process earlier in the immune response, concerned with the recognition or processing of the carrier molecule rather than with the specificity-determining haptenic sites.

Poly-L-lysine as a backbone, with side chains of poly-DL-alanine, (A–L) is non-antigenic to inbred strains of mice, but when short sequences of tyrosine and glutamic acid are added to the tips of the side chains, (T, G)-A–L is a good antigen, inducing antibody mainly specific for tyrosine, glutamic acid and alanine (McDEVITT and TYAN, 1968). Histidine (H, G)-A–L, or phenylalanine, (P, G)-A–L, can be substituted for tyrosine to give varied and effective antigenic determinants.

C3H mice respond only slightly to (T,G)-A–L; C57 black mice produce about ten times as much antibody as do C3H to this antigen. The distinction between good and poor response appears to be discontinuous, and to be determined by a single genetic alternative, with the good response dominant. When the immunizing antigen has histidine instead of tyrosine, the degree of response is opposite: C3H mice respond well, C57 poorly. Both strains respond well when the immunizing antigen has phenylalanine instead of tyrosine or histidine.

Spleen cells from (C3HXC57B1/6)F_1 mice, transferred into lethally irradiated C3H recipients, respond according to the donor cell type; the genetic difference therefore must be expressed in the donor cell system itself, and not in some more general aspect of the host environment. Most provocatively, this response appears to be closely linked to the H-2 complex, the major histocompatibility region in the IXth linkage group. The association indeed tempts speculation, but I believe it best at this point for you to do your own thinking about it.

Studies of the responses of one systematically random-bred and six inbred strains of rats to synthetic polypeptides containing glutamic acid and lysine, with or without tyrosine, have revealed fluctuations within the random-bred strain which may be related to unidentified environmental factors or to genetic variation among sub-samples from a variable population. Inbred strains show repeatable interstrain differences, which the authors (GILL and GERSHOFF, 1968) tend to ascribe to polygenic differences concerned with responses to various structural parameters of the antigens. Hybrids between a good and a poor responder strain were good responders. In the absence of data on F_2 or backcross generations, it is impossible to assess the detailed genetic nature of these differences. The studies are further complicated by the existence of thresholds for paralysis of response, which appear to vary among the strains of rats studied.

Among the interactants and processes that may be considered secondary to the specific immune response machinery itself, but which nevertheless are often of compelling importance in immunology, a few examples of defined genetic variation may be cited. Simply-inherited deficiencies are known for several of the components of the complement system. Tendencies to display various kinds and degrees of atopic hypersensitivity have been known for many years to be familial; genetic variation may affect, for example, the amount or site of histamine release, or susceptibility to histamine damage, or the kinds and amounts of antibodies with complement-fixing or cell-binding or damaging properties. Secondary interactions among the diverse products of the immune machinery may lead to the stimulation of higher levels of response or reaction in some instances, or the suppression of response in others. In hormonal and inductive systems, genes are known to control both the elaboration and nature of the active substances and the competence of target cells to react. Gross aberrations in the structure of relevant organs — the scaffolding provided by lymph nodes and spleen for lymphoid cell differentiation, or the virtual absence of a thymus, for example — undoubtedly often reflect genetic variation, and have in some instances been shown to be associated with definable genetic units*. Immunologic interactions between mammalian mother and fetus, or oviparous female and progeny, depend upon transport, by one route or another, of immunologically active molecules or cells from one generation to the next; genetic variations in the structure of the placenta, in the regulation of relevant transport phenomena, and in characteristics of the molecules themselves essential to their being transported, are indicated in some instances, and undoubtedly exist in many. Genetic segregation obviously affects these inter-generation interactions; it may involve histocompatibility genes and their effects.

Taken altogether, the genetic aspects of the immune response impress one these days more with the potential than with the present state of the field. As the components of the immune machinery become more clearly defined, and as competent and systematic studies are undertaken with modern genetic methodology, we can

* Spontaneous autoimmune diseases often have important genetic components. See, for example, COLE, R. K., KITE, J. H., JR. and WITEBSKY, E., 1968. Hereditary autoimmune thyroiditis in the owl, *Science*, **160**, 1357–1358; and FUDENBERG, H. H., 1967. Genetically determined abnormalities in antigen-antibody interaction, Proceedings of the 3rd International Congress of Human Genetics, Johns Hopkins Press, Baltimore, pp. 233–246.

confidently expect very highly significant contributions to come from this field in the future. These contributions will be of value in both directions: on the one hand, the immunologic system provides excellent working material for obtaining increased insight into genetics and how it works; and on the other, the application of genetic technology is sure to expand and intensify greatly our perception of immunology.

REFERENCES*

A Discussion on the Chemistry and Biology of Immunoglobulins. Organized by PORTER, R. R., 1966. *Proc. Roy. Soc.*, **B166**, 113–243, 11 papers and discussions.

ADLER, F. L., FISHMAN, M. and DRAY, S., 1966. Antibody formation initiated *in vitro*. III. Antibody formation and allotypic specificity directed by RNA from peritoneal exudate cells, *J. Immunol.*, **97**, 554–558.

COHN, M., 1967. The molecular biology of expectation. Rutgers Symposium on Nucleic Acids in Immunology. (In press).

Cold Spring Harbor Symposia on Quantitative Biology, 1967. **XXXII**, Antibodies. Published by the Cold Spring Harbor Laboratory of Quantitative Biology, Cold Spring Harbor, Long Island, New York, 68 papers, pp. 619

FLEISCHMAN, J. B., 1966. Immunoglobulins, *Annual Review of Biochemistry*, **35**, 835–872.

Genetics of the Immune Response. Report of a World Health Organization Scientific Group on the Genetics of the Immune Response, E. S. LENNOX, Chairman. To be published by World Health Organization.

GILL, T. J. III and GERSHOFF, S. N., 1968. Genetic control of the antibody response in the rat to structurally different synthetic polypeptide antigens, *J. Immunol.* (In press).

GREENWALT, T. J. (ed.), 1967. Advances in immunogenetics, J. B. Lippincott, Philadelphia, 7 papers, including 3 on immunoglobulins.

HOLLAENDER, A. (ed.), 1966. Symposium on differentiation and growth of hemoglobin and immunoglobulin synthesizing cells, *J. of Cellular Physiol.*, Suppl. 1, **67**, pp. 215.

HOWELL, J. W., HOOD, L. and SANDERS, B. G., 1967. Comparative analysis of the IgG heavy chain carbohydrate peptide, *J. Mol. Biol.*, **30**, 555–558.

KABAT, E. A., 1968. Structural concepts in immunology and immunochemistry, Holt, Rinehart and Winston, Inc., New York. A concise, up-to-date teaching textbook, pp. 310.

KILLANDER, J. (ed.), 1967. Gamma globulins: structure and control of biosynthesis, Proceedings of 3rd Nobel Symposium, Interscience Publishers, John Wiley & Sons, Inc., New York, 48 papers and discussions, pp. 643.

LAMELIN, J.-P., PAUL, W. E. and BENACERRAF, B., 1968. The immune response of random-bred Hartley strain guinea pigs to 2,4-dinitrophenyl conjugates of a copolymer of L-glutamic acid and L-lysine, *J. Immunol.*, **100**, 1058–1061.

LENNOX, E. S. and COHN, M., 1967. Immunoglobulins, *Annual Review of Biochemistry*, **36**, 365–406.

LEVEY, R. H. and MEDAWAR, P. B., 1966. Nature and mode of action of antilymphocytic antiserum, *Proc. Nat. Acad. Sci. (U.S.A.)*, **56**, 1130–1137.

LOCKE, M. (ed.), 1967. Control mechanisms in developmental processes, 26th Symposium of the Society for Developmental Biology, Academic Press, New York. Note especially the reviews by

* *Note.* Several of these references are to volumes containing numerous individual papers relevant to the subject of this lecture. No consistent effort is made here, or in the text, to identify these sources more precisely; rather, this list is to be taken as a general guide to reading and to further references.

I. M. LONDON *et al.*, on erythroid cells and hemoglobin; and R. AUERBACH, on development of immunocompetent cells.

MAURER, P. and PINCHUCK, P., 1968. Immune response of strain 13 guinea pigs to lysine-containing polymers of amino acids, *J. Immunol.*, **100**, 1141–1142.

MCDEVITT, M. D. and TYAN, M. L., 1968. Genetic control of the antibody response by spleen cells and linkage to the major histocompatibility (H-2) locus, *J. Exp. Med.* (July).

RICH, A. and DAVIDSON, N. (eds.), 1968. Structural chemistry and molecular biology: a volume dedicated to Linus Pauling by his students, colleagues, and friends, W. H. Freeman & Co., San Francisco.

SCHROEDER, W. A., HUISMAN, T. H. J., SHELTON, J. R., SHELTON, J. B., KLEIHAUER, E. F., DOZY, A. M. and ROBBERSON, B., 1968. Evidence for multiple structural genes for the γ chain of human fetal hemoglobin, *Proc. Nat. Acad. Sci. (U.S.A.)* (In press).

SELL, S., 1968. Zygosity of lymphocyte donor and maximum percent blast transformation induced by antiallotype sera, *Federation Proceedings*, **27**, 277.

SINGER, S. J. and THORPE, N. O., 1968. On the location and structure of the active sites of antibody molecules. *Proc. Nat. Acad. Sci. (U.S.A.)* (In press).

SMITH, R. T., MIESCHER, P. A. and GOOD, R. A. (eds.), 1966. Phylogeny of immunity, University of Florida Press, Gainesville.

SMITH, R. T., GOOD, R.A. and MIESCHER, P. A. (eds.), 1967. Ontogeny of immunity, University of Florida Press, Gainesville.

WOLSTENHOLME, G. E. W. and O'CONNOR, M. (eds.), 1967. Antilymphocytic serum, Ciba Foundation Study Group No. 29, Little, Brown & Co., Boston. 6 discussion sections, pp. 165.

DISCUSSION

B. ZABŁOCKI: It is generally accepted that phylogenetically and ontogenetically the IgM antibodies appear as the first immune response before IgG antibodies are produced. After the beginning (in later stages of immunization) of the IgG production the synthesis of IgM antibodies is put to a stop. Would you be kind, Professor Owen, to express your opinion from genetic point of view on the mechanism of this phenomenon.

R. D. OWEN: First I should say that it is not clear that IgM antibody production regularly precedes IgG; part of the common opinion that it does may depend on the greater sensitivity to IgM of test system that have been employed. Setting this reservation aside, however, from a genetic point of view I suppose the primary process involved in a switch from IgM to IgG antibody production would be the activation of a gene controlling one of the gamma polypeptide chains, in place of a gene controlling μ-chain production. How this comes about, when it does, is almost completely obscure at present. It is not even known for sure whether the switch occurs in the life history of a single antigen-sensitive cell lineage, or whether it involves the mobilization of a new set of cells, as a similar switch appears commonly to do in the normal developmental conversion from γ–β-chain production in hemoglobin synthesis. The biophysical and biochemical mechanics of the switch process are similarly obscure; there is some evidence that when γ-chain production begins there is some kind of feedback inhibition of μ-chain synthesis, at least for the organism as a whole, but this is still in the category more of an informed speculation than an established experimental fact. The most provocative riddle in this situation seems to me to be how antibodies of comparable specificity continue to be produced by the organism under the conditions of this shift. On the surface, the situation would seem to suggest that the same variable regions are being coded for two successive kinds of antibodies. However, such amino acid sequence information as is known to me does not seem to support that relatively simple interpretation.

B. ZABŁOCKI: Professor Owen, I would like to have your opinion on the possible mechanisms of the immunosuppression of antibody formation caused by introducing specific antibodies before the application of the homologous antigen. A feedback mechanism was proposed for the explanation of this phenomenon?

R. D. OWEN: I referred to the possibility of feedback inhibition, with particular regard to effects of IgG synthesis on pre-established IgM synthesis, in response to your earlier question. In more general terms, I am reminded by your question of the interaction between Rh sensitization and ABO type of mother and child. Currently, the effects of preformed alpha and beta antibodies on the effectiveness of an Rh stimulus (e.g. the distinction between the effective 0 Rh stimulus in an 0 Rh-negative mother, and the ineffective A Rh stimulus in 0 Rh-negative mothers), is almost universally interpreted as a matter of disposition of the stimulating antigen. It is assumed that the alpha antibody of 0 mothers sequesters A Rh-positive cellular material in such a way that does not come to the attention of the immune machinery of the mother. It still seems possible, however, that the suppression of the anti-Rh response may be at least in part dependent on more profound effects, perhaps falling in the categories of allele or cistron exclusion as discussed in my paper. The most interesting and, probably, clinically important extensions of these observations relate to the apparently effective desensitization of Rh-negative women by the injection of anti-Rh antibody, quite apart from any ABO relationship. Again, the effects of this treatment are generally accepted as relating to disposition of the antigen, but could conceivably also reflect more basic cellular phenomena.

R. L. SPOONER: If a group of inbred animals or identical twins are injected with one antigen is there any variation in the specificity of the antibody produced in these animals with the same genetic make up?

R. D. OWEN: A great heterogeneity of antibodies is produced even within highly inbred lines or identical twins. Even a single individual gives a heterogeneous response. So we might well expect similar variation even in a genetically homogeneous population of separate individuals. The work on genetic aspects of the immune response, which is reviewed in my manuscript but had to be treated rather hastily in this oral presentation, seems to me only to indicate that there are definable genetic limits on the character of immune responses, affecting many different phases of this complex process. There remains, however, a great diversity of potential within any given genotype. Possibly Professor Stormont may have some further comment, based upon his experience, in this context.

J. RENDEL: Dr. Spooner evidently raised a very important question. I would also like to comment on it. There will be a paper by Sellei and myself in *Journal of Genetical Research* dealing with this question, it will come out very soon, where we immunized a number of monozygous twins against the very same antigen and there was very high degree of similarity of the antibody response in most of the monozygous twin-pairs. However there were a few exceptions, one a very marked one. And we made a lot of tests, biological tests, and we could not detect any signs of dizygosity? And we thought that this difference in antibody response may be due to some mutation in the antibody forming cells. So, I think there is room for diverse antibody response within monozygous twins and in general they respond very similarly.

C. STORMONT: I have one point to add to what Dr. Owen had to say and it concerns the animals with previous immunological experience. Even if you were working with identical twins or with highly inbred lines of animals — no two animals could have had exactly the same previous immunological experience, and it is that which helps determine any new immunological experience.

R. D. OWEN: The heterogeneity within a highly inbred line can't all be allelic variations resulting from residual heterozygosity, first because these mice should be almost completely homozygous, and second because we know that different cistrons, often adjacent on the genetic map, can be mobilized in the response of a single individual, just as occurs in the succession of hemoglobin types. The most provocative recent development in the field, to me, is the evidence that responses to certain copolymers in mice are closely linked to, or may even be an integral part of, the H-2 complex. It makes one wonder whether this important histocompatibility and blood-type complex may be engaged in a vital aspect of immune responses — perhaps the generation of heterogeneous and variable regions of immunoglobulin molecules.

B. Zabłocki: Hemoglobulin system is a very convenient material for genetic investigations. Much more complicated is the immunoglobulin system where the variability of the N-terminal section of the polypeptide chains (which is just a half of the whole chain) is very great; variation has been observed of forty amino acid sites.

In the last years we are more and more convinced that the antibody specificity depends on the amino acid sequence in the antigen combining site. At the same time we have very little data how the conformational properties of antibody molecule influence the antibody specificity. In this case we need the help from biophysicist and new research techniques.

From the experiments of Fishman *et al.* two principal conclusions were drawn out: (1) two lines of cells (macrophages and limphocytes) are taking part in the process of antibody formation, (2) the RNA isolated from the macrophages contains the antigen information. The first conclusion is accepted (but not by all authors). The second conclusion is not supported, the RNA seems to be a stimulator (without specificity) to the antigen activity as an immunogen.

XITH EUROPEAN CONFERENCE ON ANIMAL BLOOD GROUPS AND BIOCHEMICAL POLYMORPHISM
Warsaw, July 2nd–6th, 1968

Isozyme Polymorphism of the Blood

C. J. STORMONT
Serology Laboratory, University of California, Davis, California, U.S.A.

INTRODUCTION

Having been invited at a rather late date to stand in for Professor Harry HARRIS on the subject of isozymes, my first reaction was to suggest the title "Isozymes and Their Possible Role in Evolution". However, in the course of assembling literature and thoughts, it occurred to me that a review of the literature in isozymes might be more appropriate and rewarding. Helping me to make that decision was the fact that SHAW's timely review (1965) is now well antidated. BECKMAN's more recent review (1966) is restricted to man whereas LUSH's monograph (1966) is restricted to vertebrates other than man.

In assembling literature for my review, I had included articles on isozyme polymorphism in whole organisms such as Drosophila and in tissues other than blood. However, by the time I took pencil in hand, it became clear that I would do well if I could cover the subject of isozyme polymorphism of the blood in the time available.

Two technological events opened the door to this new field: (1) SMITHIES' development (1955) of zone electrophoresis in starch gels, a method which so many of you have employed successfully in exploring protein polymorphism involving the "carrier" proteins such as hemoglobin and transferrin, and (2) the application of the methods of histochemical staining (HUNTER and MARKERT, 1957; and MARKERT and MØLLER, 1959) to reveal zones of enzymatic activity directly in the gels. This combination of gel electrophoresis and histochemical staining is referred to as the "zymogram" display. We are also indebted to MARKERT and coworkers for the term "isozyme" which they used to denote multiple molecular forms of the same enzyme, a cogent example being the multiple molecular forms of lactate dehydrogenase (MARKERT and MØLLER, 1959).

Some idea of the rapidity of progress being made in the use of zymogram methodology in exploring for enzyme polymorphism is evidenced by the papers of this present Conference in contrast with those of the two previous Conferences (Paris, 1966; Prague, 1964). Twenty of the 118 or so reports listed in the program of this

Conference are concerned in part or in whole with isozymes, whereas there were less than half that number in the two previous Conferences. At this rate of acceleration, one might predict that over half the papers of the 1970 Conference will be devoted to isozymes.

Why, may we ask, has there been this rather sudden switch, as it were, from the classical approach to the study of blood groups? The most obvious answer would seem to be that the rewards, per increment of effort, are much greater. With appropriate gel techniques and staining procedures it is a relatively easy task to expose protein and enzyme polymorphism in virtually any vertebrate species. It is also a relatively easy task to expose blood group polymorphism, that is, given an adequate battery of blood-typing reagents, but there are many of us here who well know the countless hours of work it takes to develop those blood-typing reagents.

The real meaning of this difference became apparent to me when M. BRAEND from Norway joined our staff as a visiting investigator in 1961. He introduced us to the technique of zone electrophoresis in starch gels and, as one of the by-products, discovered transferrin polymorphism in horses. J. RENDEL from Sweden came to us in the same capacity in 1963. He introduced us to the zymogram methodology and went on to establish the relationship between plasma alkaline phosphatase zones and the R–O blood groups of sheep. G. SARTORE from Italy came in 1963 and returned again in 1964 to develop the zymogram display for esterases and in the process discovered the carbonic anhydrase polymorphism in cattle. By then we were fully convinced of the utility of gel methods, and made a partial but not complete change over to the methods they introduced. Most rewarding has been the zymogram method for esterases. Using that method, with the usual modifications, consisting mainly in changes in the buffer systems, we have, for example, detected at least four polymorphic systems of esterases in rabbits alone (GRUNDER *et al.*, 1965; SCHIFF and STORMONT, 1968; STORMONT and SUZUKI, 1968).

But there are more compelling reasons for the change, and this I believe, can be best illustrated by the events that took place when Elizabeth M. TUCKER from Britain came to our laboratory as a visiting investigator in 1965. Among other things, Dr. TUCKER called our attention to a report by LEE (1964) concerned with genetic differences in sheep with respect to their ability to hydrolyse certain di-(2-chloroethyl) aryl phosphates, such as the anthelmintic Haloxon. LEE spoke of those sheep whose plasma contained an esterase that would rapidly hydrolyse Haloxon as "Halon high", whereas those sheep whose plasma showed little or no hydrolytic activity for Haloxon were designated "Halon low". He also provided convincing evidence that the Halon high character is inherited, in contrast with Halon low, as a simple Mendelian dominant, thereby providing the first evidence of a pharmacogenetic difference in sheep.

At about that same time, we had, by means of the zymogram method for esterases, exposed an A or arylesterase which was present in the plasma of some but not all sheep. I will not bother you here with the details of the experiments which identified this A esterase with the esterase present in the plasma of Halon high sheep because those details are recorded in reports by TUCKER *et al.* (1967, 1968).

The point I wish to make here is that we had for the first time enlisted the serious interests of persons outside our immediate discipline. For example, Dr. N. F. BAKER, a parasitologist and member of the Faculty of this School, became keenly interested in what we were doing when we convinced him, with his assistance that it is only the Halon low or A-esterase-negative sheep which develop neurotoxic symptoms when given more than the recommended anthelmintic dose of Haloxon. This recessive gene, and perhaps others like it, would explain the serious side-effects, often referred to by the profession as "drug idiosyncracies" which occur in some but not all animals when being treated with drugs. The opportunity is here to discover other enzyme differences that are certainly involved in these idiosyncratic reactions. I have no doubt that with our present knowledge of esterase isozymes in rabbits we could selectively breed a number of experimental strains that would differ markedly in their responses to drugs. But this idea is apparently so new that we are having difficulties trying to sell it.

It can be predicted that zymogram methodology will be used extensively in studies of diseases (both inborn and of an infectious nature), in examining evolutionary affinities and taxonomic relationships among animals and plants, in ontogenetic studies, in tracing enzyme changes in the various clones of cells in culture, in exploring for evidence of hybridization of cells in culture and, eventually, the breeding of more efficient and healthy livestock.

There is, however, one other good reason for the change over to gel methodology as so aptly remarked at the Paris Conference by Dr. HALL. "Gels are fun!".

However, let us not completely forget the classical approach and other methods such as immunoelectrophoresis, because there are correlations and linkage relationships to be disclosed between the markers detected with those techniques and the markers detected by the gel techniques. The synchrony between all these methods will become even more exciting when effective methods are developed for releasing unaltered, the enzymes and perhaps antigen-carrier-lipid-phosphate molecules that are involved in the synthesis of the red cell membrane. The assembly mechanism for red cell antigens with carbohydrate specificities is not expected to be much different from that being worked out for the assembly of the somatic antigens of the Salmonella group of organisms (see ROBBINS *et al.*, 1967; NIKAIDO, 1967). After all, the code is universal.

The balance of this report is concerned with the listing, along with comments, of many if not all isozyme polymorphism that have been revealed to date in studies of the zymogram patterns of red cells and blood plasma.

RED CELL ISOZYMES

Acid phosphatases (AcP)

The phosphatases or phosphomonoesterases are widely distributed in tissues and the elements of blood. We distinguish "acid" phosphatases (AcP), with pH optimums around 5 to 6, from alkaline phosphatases, with pH optimums around 7 to 8, and adjust our buffers accordingly.

HOPKINSON *et al.* (1963) developed an effective method for the resolution of AcP isozymes in human red cells. In staining the gels, they used the sodium salt of phenolphthalein diphosphate as substrate and ammonia as the developing agent. The phenolphthalein released by the hydrolytic activity of AcP develops a deep pink color in the alkaline environment. This stain, of course, is transient.

HOPKINSON *et al.* observed five AcP phenotypes involving three AcP zones (isozymes), namely F (for fast), I (for intermediate) and S (for slow).

At least six AcP alleles are now known in man (see SUTTON, 1967), and some of the alleles are rather common.

In a footnote added in proof, HOPKINSON *et al.* pointed out that improved separations of the AcP zones are obtained by halving the concentration of the gel buffer. This cogent point reminded me of what happened when I inadvertently misread one of our graduate student's recipes for an esterase gel buffer. This error n reading (or writing, which he would admit) amounted to a 10-fold dilution of the buffer, a mere difference of one decimal place. But this error led to a much shorter running time (about 2½ hr as contrasted with 12 hr) and to the resolution of numerous zones of esterase that could not be seen at all when using the "recommended" technique. We need not apologize for such errors because they have led to many important discoveries and innovations such as the discontinuous system of buffers. That error, with the incorporation of a few minor modifications, has led to the development of a most effective method for resolving zones of esterase in rabbit red cells. The method is described by SCHIFF and STORMONT in the January 1968 issue of the *Immunogenetics Letter*.

I am not aware of any published reports on RBC–AcP polymorphisms in species other than man. We shall, however, be hearing more about AcP phenotypes in the session scheduled for Friday morning.

Adenylate kinase (AK)

Kinase is the general term for any enzyme that transfers a phosphate group to another substrate.

FILDES and HARRIS (1966) developed a zymogram method for the display of human RBC–AK (red-blood-cell adenylate kinase) isozymes. They detected three phenotypes controlled by a pair of codominant autosomal alleles. More recently, CARSON and GOWER (1967) added two alleles to that series.

I am not aware of any published reports on RBC–AK polymorphisms in species other than man.

Carbonic anhydrase (CA)

This important enzyme functions not only as a hydratase but it also has esterase activity for certain substrates such as the α and β forms of naphthyl acetate.

Using a zymogram method for esterases, TASHIAN and colleagues (see TASHIAN, 1965) have disclosed a number of isozymes in RBC-esterases of man and other primates. Of particular interest are the zones which were eventually classified as carbonic anhydrases by means of inhibition tests with acetazolamide, a specific inhibitor of CA. Four alleles code for the CA I zones of man that are identified by the symbols CA Ia, CA Ib, CA Ic and CA Id. The allele that codes for CA Ia is the common or "normal" allele. The others are rare.

Two alleles are known to code for two CA phenotypes in cattle (*Bos taurus*) and three CA alleles are recognized in American buffalo (*Bison bison*) (SARTORE *et al.*, 1968). Similarly, two alleles code for the two CA zones recognized in domestic sheep (TUCKER *et al.*, 1967). As in primates, the CA locus of the Bovidae is autosomal and the alleles act as codominants. Similarly, these CA zones of the Bovidae are specifically inhibited by acetazolamide and, because of the abundance of CA in red cells, they are readily visualized following staining with the ordinary protein dyes such as nigrosin and buffalo black.

The allele which codes for the fast CA zone in both cattle and sheep is designated CA^F, and in each species it is the rarer of the two. In a survey (SARTORE *et al.*, 1968) if nine breeds of cattle in the United States, allele CA^F ranged in frequency from a low of 0.01 in Aberdeen Angus to a high of 0.41 in Jerseys. In a much more limited survey with respect to number of animals studied, TUCKER *et al.* (1967) observed CA^F in only 3 of 9 breeds of sheep. The bison survey was limited to 51 animals all from one national herd numbering about 150 animals. The frequency of the three bison alleles, designated CA^1, CA^2, and CA^3 was, respectively, 0.35, 0.60 and 0.05. This is probably representative of the species because all American buffalo now living are descendants of a few animals that survived the slaughter of the 19th century. In that connection, it may be of some interest to point out that this is the most extensive polymorphism so far revealed in studies of that species.

None of the CA alleles in the three species of Bovidae can be homologous with the possible exception of CA^S of cattle and CA^3 of bison. These two alleles code for CA isozymes that occupy the same or nearly the same electrophoretic position.

Human RBC–CA has a molecular weight of about 30,000 and is composed of a single polypeptide chain. In peptide and amino acid analysis, and in studies of kinetic and other properties, TASHIAN *et al.* (1966), SHOWS (1967) and TASHIAN and YU (1966) have already accumulated a body of information about CA Ia, CA Ic and CA Id of man. The difference between CA Ia and CA Ic seems to be the substitution of a single amino acid residue, namely, arginine for glycine in CA Ic. Similarly, the difference between CA Ia and CA Id appears to be the substitution of a single lysine residue for threonine in CA Id.

There were no significant differences between CA Ia and CA Ic in their carboxylesterase and hydratase activities. In contrast, there were numerous differences between CA Ia and CA Id including heat stability. CA Id is much more temperature sensitive than CA Ia. In the presence of lead, and using β-naphthyl acetate as substrate, the hydrolytic activity of the CA Id variant showed a five-fold increase while that of CA Ia was slightly depressed. Repeating the same experiment but with another substrate (*p*-nitrophenyl acetate) the hydrolytic activity of CA Ia for that substrate was reduced about 1/4 that of normal whereas the hydrolytic activity of CA Id was reduced about 1/2. It would be interesting to speculate on what the effect would be of substituting allele CA I^d for CA I^a in a person of genotype CA I^a CA I^d.

Catalase

SHAW (1965) listed one report of a rare electrophoretic variant of human RBC-catalase. I know of no other reports.

It will be interesting to see whether the kinetic forms of canine RBC-catalase described by ALLISON *et al.* (1957) can be differentiated in zymograms.

Esterases

Except for acetylcholinesterase, the precise metabolic function of the numerous esterases is unknown; however, recent evidence (SKARNES *et al.*, 1968) indicates that some esterases are involved in the degradation and inactivation of endotoxins.

Polymorphic systems of RBC-esterases have been detected by the zymogram method in the mouse (PELZER, 1965), Peromyscus, or deer mice, (RANDERSON, 1965) and in the rabbit (GRUNDER *et al.*, 1965; SCHIFF and STORMONT, 1968), and I suspect that before this Conference is over we will be hearing about RBC-esterase polymorphism in at least one other species.

Glucose-6-phosphate dehydrogenase (G6PD)

G6PD is a key enzyme in the degradation of glucose through the pentose monophosphate pathway. Molecular variants of G6PD are responsible for the disease in man known as favism and for pharmacogenetic differences in the response of persons to such drugs as primiquine. Some of the variant forms of human RBC-G6PD differ in their electrophoretic mobilities and others only in their metabolic activities. At least 20 genes are involved (see SUTTON, 1967). They are all located on the X-chromosome and presumably constitute a single allelic series.

As a result of the confusing nomenclature for human G6PD genotypes and phenotypes, the World Health Organization convened a committee in 1966 to resolve the problem. The report of that committee is reprinted in the November 1967 issue of *the American Journal of Human Genetics.*

G6PD variants have played an important role in the confirmation of the LYON hypothesis which, by the time you hear me speak these words, will have been reviewed by E. ANDRESEN in the common session on Tuesday morning.

The locus for RBC–G6PD in the genus *Equus* and the genus *Lepus* is also located on the X-chromosome, but in the absence of recognizable intraspecific variants this required a rather ingenous procedure. TRUJILLO *et al.* (1965) and MATHAL *et al.* (1966) showed that the zone of G6PD in the species *E. asinus* migrates ahead of that of the species *E. caballus.* In a study of the hybrids (mules and hinnies) the aforementioned investigators showed that the male hybrids exhibited only a single zone of G6PD which corresponded in mobility with the G6PD of their female parents, whereas the female hybrids exhibited the G6PD zones of both parents. Precisely the same results were obtained by OHNO *et al.* (1965) in hybridizing hares of the species *Lepus europeaus* and *L. timidus.* The results of these experiments were in complete agreement with expectation based on sex-linkage.

Lactate dehydrogenase (LDH)

This important enzyme in the glycolytic cycle occurs in five isozymic forms (MARKERT and MØLLER, 1959), although all of these isozymic forms are not evident in all tissues. Also, additional forms, the so-called "X" components, occur in testicular tissue as will be discussed elsewhere in this Conference.

Each of the five regular isozymic forms is a tetramer composed of combinations of A and B subunits (MARKERT, 1963). As reviewed by VESELL (1965), the two subunits are coded for by two different structural genes that are genetically independent, as evidenced in the studies of SHAW and BARTO (1963) on the LDH of Peromyscus, and the studies of BOYER *et al.* (1963) and NANCE *et al.* (1963) on human LDH. At least seven mutants, some involving the A cistron and some involving the B cistron are known in man. RBC–G6PD electrophoretic variants also occur in baboons, as reported by SYNER and GOODMAN (1966).

A multiplicity of electrophoretic forms of LDH has been detected in studies of the tissues of a variety of marine and terrestial animals, and elsewhere during this Conference we will hear more about those variants as well as the evidence for a third subunit, namely C, detected by VALENTA *et al.* (1967) in studies of pig spermatozoa.

Peptidases

LEWIS and HARRIS have recently reported (1967) an effective method of displaying zones of peptidase on zymograms. Their procedure depends upon the following sequence of reactions:

$$(1)\ \text{peptide}+H_2O \xrightarrow{\text{peptidase}} \text{L-amino-acid},$$
$$(2)\ \text{L-amino-acid}+O_2 \xrightarrow{\text{L-amino-acid oxidase}} \text{keto-acid}+NH_2+H_2O_2,$$
$$(3)\ H_2O_2+\text{O-dianisidine} \xrightarrow{\text{peroxidase}} \text{oxidized dianisidine}.$$

The complete reaction sequence results in the appearance of dark brown zones at the sites of peptidase activity. They used rattlesnake venom as the source of L-amino-acid oxidase, that of *Crotalus adamantus* being particularly active with dipeptides and tripeptides containing leucine, isoleucine, phenylalanine, tyrosine, methionine and tryptophan.

Using 12 dipeptides and 4 tripeptides, they were able to demonstrate five separate systems (A, B, C, D and E) of human RBC-peptidases. They presented evidence that at least four alleles code for the variant forms of peptidases in each of the systems A and B. They were also aware of electrophoretic variants in the remaining systems but will report on those at a later date.

Obviously, from those initial results this would appear to be the most effective method devised to date for the exploration of isozyme polymorphism, and that is why I have taken this much time to call your attention to it.

Hexokinase

HOLMES *et al.* (1967) have developed a zymogram method for the display of zones of human RBC-hexokinase. Two zones, designated I and III, were detected in normal adults. Zone II, which appeared to be completely associated with fetal hemoglobin, made its appearance in adults with hereditary persistence of fetal Hb.

I believe it is likely that the intermediate zone results from complexing but I will have more to say on that when we come to the subject of plasma alkaline phosphatases.

The technique described by HOLMES *et al.* may prove to be most useful in studies of such inigmas as Hb–C (also N) of sheep, a hemoglobin which makes its appearance in anemic sheep of Hb types A and AB but apparently never in sheep of type B (see BEALE *et al.*, 1966; and BRAEND and EFREMOV, 1965).

Phosphogluconate dehydrogenase (PGD or 6PGD)

According to PARR (1966) and others, at least 5 alleles code for the variant electrophoretic forms of PGD in human red cells. One of the alleles namely PGD^0, is a null or silent allele. All heterozygotes, excepting those involving the null allele, exhibit a hybrid zone, thereby indicating that each zone is a dimer.

Two PGD alleles are known in rats (PARR, 1966) and here, also, the heterozygote exhibits a hybrid zone.

THULINE *et al.* (1967) presented evidence for three RBC–PGD phenotypes in cats, the heterozygote exhibiting a hybrid zone. Apparently this is the first example of enzyme polymorphism in cats.

Phosphoglucomutase (PGM)

Two loci coding for subunits of RBC–PGM are known in man. As a leading reference to the studies of human PGM isozymes, attention is called to the paper by BREWER *et al.* (1967). At least seven alleles are known to code for one of the subunits and 2 for the other (see SUTTON, 1967).

Phosphoglucose isomerase (PGI)

This enzyme is another of the series of enzymes which participates in the glycolytic cycle. It is known to occur in two electrophoretic forms controlled by codominant autosomal alleles (CARTER and PARR, 1967). The heterozygote exhibits a hybrid zone, thereby indicating that PGI isozymes are dimers.

There are two additional RBC-enzymes of man, namely, glutathione reductase and malate dehydrogenase, for which electrophoretic variants are known (see SUTTON, 1967). However, like PGI, no electrophoretic variants of those enzymes in animals have as yet been recorded. We turn now to the plasma isozymes.

PLASMA ISOZYMES

Blood plasma is a repertory for enzymes that have leaked, so to speak, from virtually all tissues including the blood cells. It is known that the amount of certain plasma enzymes, like alkaline phosphatase, is influenced by numerous things including diseases.

Alkaline phosphatase (AlP)

The so-called "non-specific" AlP of blood plasma is traceable to many tissues including intestinal mucosal cells (LANGMAN *et al.*, 1968).

As reviewed by ROBINSON and GOLDSMITH (1967) two different zones of AlP may be seen in zymograms of human plasma. This is also true for sheep (RENDEL and STORMONT, 1964), cattle (GAHNE, 1967a,b), chickens (LAW and MUNRO, 1965; WILCOX, 1966) and perhaps many other vertebrate species. Although the picture in cattle seems to be similar to that in man and sheep, I shall, for the purpose of this discussion, limit most of my remarks to the picture in man and sheep. In each of those species, the A zone migrates ahead of the B zone, and seems to be present in all individuals. The B zone is not. However, when it is present in sheep, it is almost invariably associated with the presence of another substance in the plasma namely, the soluble 0 blood group substance. This association has been studied in detail by RENDEL *et al.* (1964). Similarly the presence of the B zone in human plasma is highly correlated but not absolutely correlated with soluble H and B substances in the plasma. The ramifications of this association and many other aspects of the problem have been thoroughly reviewed by ROBINSON *et al.* (*loc. cit.*).

Numerous hypotheses have been advanced to explain the relationship between the B zone and the soluble blood group substances of specificity 0 (or H) and B. ARFORS, BECKMAN and LUNDIN, as cited by ROBINSON *et al.*, have proposed, in essence, that the B zone in man represents a complex of AlP with soluble blood group substance of H and/or B specificity. I am inclined to accept that hypothesis even though critical proof is still wanting. I find it most difficult to accept the hypothesis that the B zone represents an isozyme coded for by a gene or genes different from the gene or genes that codes for AlP in the A zone. To accept the notion that the B zone is a different isozyme, one must also be prepared to accept the conclusion that its controlling gene (or genes) is recessive to a gene (or genes) for its absence.

A similar situation is encountered in chickens excepting that no demonstrable blood group substance is involved. Furthermore, the fast and slow zones are mutually exclusive. The mystery of the recessive inheritance of the slow zone was solved by LAW (1967). He showed that when plasma containing the fast zone was treated with neuraminidase, the mobility of that zone is converted to the mobility of the slow zone. On the other hand, when plasma containing the slow zone was treated with neuraminidase, there was no effect on the mobility of the slow zone. Thus, the fast zone can be attributed to a gene which, when present, brings about the attachment of sialic acid to AlP. He also showed that this same gene will account for the two mutually exclusive forms of plasma leucine aminopeptidase, which, tncidentally, were absolutely correlated with the two electrophoretic phenotypes of AlP.

Amylase

Here I refer to the persons involved in the presentation of six papers on amylase phenotypes in other sessions of this Conference.

Ceruloplasmin

This copper-containing enzyme possesses oxidase activity and is classifiable as a laccase. Its precise metabolic function is, however, still unknown.

Insofar as I am aware, the first report on ceruloplasmin polymorphism in animals other than man was that of IMLAH (1964). He described three multi-banded phenotypes in zymograms of pig serum and showed that the phenotypes are controlled by a pair of codominant autosomal alleles.

SHREFFLER *et al.* (1967) presented evidence for three alleles controlling multiple molecular forms of ceruloplasmin in man. We shall hear about ceruloplasmin polymorphism in cattle during the Friday morning session of this Conference.

Esterases

I have already alluded to the zone of A esterase in the plasma of certain sheep and to the identity of that zone with the Halon high phenotype described by LEE (1964). In a further study of the esterase activity of sheep plasma using a variety of synthetic ester substrates, LEE (1966) was able to distinguish six phenotypes thereby indicating that the Es locus supports at least three alleles (Es^A, Es^B, and Es^C). It is only the allele Es^A which gives rise to the zone of esterase described by TUCKER *et al.* (1967), and it is only that esterase which protects sheep against the neurotoxic effects of overdoses of Haloxon.

GAHNE (1966) described a system of equine plasma esterases involving at least four alleles, including one allele (Es^0) which acts as a silent member of the series, that is, with respect to the substrates he employed. It is of more than passing interest to point out that in the same report GAHNE described a new system of equine prealbumins involving at least four alleles, and increased the number of known equine transferrin alleles from 6 to 7.

SPRAGUE (1967) presented evidence for plasma esterase polymorphism in three species of tuna (bluefin, bigeye and yellowfin). He proposed that four codominant autosomal alleles would be necessary to explain the five phenotypes which he observed in a sample of 70 southern bluefin tuna, and that four alleles would also be necessary to explain the four phenotypes observed in bigeye tuna. Only two phenotypes were seen in the yellowfin sample. One of the major problems was accounting for certain specimens that lacked any demonstrable esterase zones in the zymogram displays. As he pointed out, this could be due to deterioration or to a silent allele. When he scoured the samples with respect to staining intensity it was evident that there was a preponderance of females among the strong reactors.

In their report on the electrophoretic forms of esterase in sheep plasma, TUCKER *et al.* (1967) described a "quick tube test" for the instantaneous diagnosis of the A esterase. When a drop of plasma or serum from an A-positive donor is added to two drops of the staining solution, a deep brown coloration develops almost immediately. With plasma from A-negative sheep the reaction is much slower (about 30 seconds) and the color change is to dark green.

Using the quick tube test in screening serum samples from other species, STORMONT and SUZUKI (1968) were able to distinguish two phenotypes in rabbits and went on to show that the esterase responsible for the fast reaction is readily demonstrable in starch gel. However, this intensely staining zone of esterase made its appearance almost midway in a series of zones involved in an esterase polymorphism which may or may not be independent of the zone with this intense activity.

Playing a hunch that the intensely staining esterase in rabbit plasma might be the product of the atropinesterase gene (As) identified by SAWIN and GLICK in 1943, the next step was to try the simple physiological test for the identification of rabbits with the As gene. In this test, a few drops of a 2% solution of atropine is added directly to the eye. This causes immediate dilation of the pupil by relaxing the pupil reflex. However, rabbits with the As allele begin to recover this reflex in a manner of 15 min to 1 hr, whereas those without the As allele do not begin to recover the reflex until upwards of 12 hr. To be certain, it was only the rabbits identified as having the intensely staining zone of esterase that began to recover their pupil reflexes in a period of 15 min to 1 hr. In essence, we had discovered two more ways to test for the presence of the As allele, namely, the quick tube test and the zymo gram display for esterases.

SEMEONOFF and ROBERTSON (1968) presented evidence for an interesting but genetically somewhat puzzling esterase polymorphism in the plasma of field voles (*Microtus agrestis L.*). They designated the system E_1 and described animals of four phenotypes: (1) those with the fast zone of E_1 having "standard" esterase activity, 2) those with the fast zone having "double" the standard activity, (3) those with the slow form of E_1 (all having standard esterase activity), and (4) those having no demonstrable E_1. Like the aforementioned alkaline phosphatase polymorphism o- chickens (LAW, 1967), animals with both the slow and fast forms of E_1 were never encountered. Although the authors advanced an explanation based on four alleles, the genetic and population data are not convincing. It seems likely to me that a dominant gene at another locus, analogous to that detected by LAW (1967), may account for the two electrophoretic forms of E_1.

In another session of this Conference we shall be hearing about plasma esterase polymorphism in chickens, and with that remark I conclude this review.

REFERENCES

ALLISON, A. C., AP REES, W. and BURN, G. P., 1957. Genetically-controlled differences in catalase activity of dog erythrocytes, *Nature*, **180**, 649–650.

BEALE, D., LEHMANN, H., DRURY, A. and TUCKER, E. M., 1966. Haemoglobins of sheep, *Nature*, **209**, 1099–1102.

BECKMAN, L., 1966. Isozyme variations in man. Monographs in human genetics, Vol. 1, S. Karger, Basel and New York.

BOYER, S. H., FAINER, D. C. and WATSON-WILLIAMS, E. J., 1963. Lactate dehydrogenase variant from human blood: evidence for molecular subunits, *Science*, **141**, 642–643.

BRAEND, M. and EFREMOV, G., 1965. Haemoglobin N of sheep, *Nature*, **205**, 186–187.

BREWER, G. J., BOWBEER, D. R. and TASHIAN, R. E., 1967. The electrophoretic forms of phosphoglucomutase, adenylate kinase, and acid phosphatase in the American negro, *Acta genet.*, **17**, 97–103.

CARSON, P. E. and GOWER, M. K., 1967. Population, family and biochemical investigations of human adenylate kinase polymorphism, *Nature*, **214**, 1156–1158.

CARTER, N. D. and PARR, C. W., 1967. Isoenzymes of phosphoglucose isomerase in mice, *Nature*, **216**, 511.

FILDES, R. A. and HARRIS, H., 1966. Genetically determined variation of adenylate kinase in man, *Nature*, **209**. 261–263.

GAHNE, B., 1966. Studies on the inheritance of electrophoretic forms of transferrins, albumins, prealbumins and plasma esterases of horses, *Genetics*, **53**, 681-694.

GAHNE, B., 1967a. Inherited high alkaline phosphatase activity in cattle serum, *Hereditas*, **57**, 83–99.

GAHNE, B., 1967b. Alkaline phosphatase isoenzymes in serum, seminal plasma and tissues of cattle, *Hereditas*, **57**, 100–114.

GRUNDER, A. A., SARTORE, G. and STORMONT, C., 1965. Genetic variation in the red cell esterases of rabbits, *Genetics*, **52**, 1345–1353.

HOLMES, E. W., MALONE, J. L., WINEGRAD, A. I. and OSKI, F. A., 1967. Hexokinase isoenzymes in human erythrocytes: association of type II with fetal hemoglobin, *Science*, **156**, 646–648.

HOPKINSON, D. A., SPENCER, N. and HARRIS, H., 1963. Red cell acid phosphatase variants: a new human polymorphism, *Nature*, **199**, 969–971.

HUNTER, R. L. and MARKERT, C. L., 1957. Histochemical demonstration of enzymes separated by zone electrophoresis in starch gels, *Science*, **125**, 1294–1295.

IMLAH, P., 1964. Inherited variants in serum ceruloplasmins of the pig, *Nature*, **203**, 658–659.

LANGMAN, M. J. S., CONSTANTINOPOULOS, A. and BOUCHIER, I. A. D., 1968. ABO blood groups, secretor status, and intestinal mucosal concentrations of alkaline phosphatase, *Nature*, **217**, 863–865.

LAW, G. R. J., 1967. Alkaline phosphatase and leucine aminopeptidase association in plasma of the chicken, *Science*, **156**, 1106–1107.

LAW, G. R. J. and MUNRO, S. S., 1965. Inheritance of two alkaline phosphatase variants in fowl plasma, *Science*, **149**, 1518.

LEE, R. M., 1964. Di-(2-chloroethyl) aryl phosphates, a study of their reaction with B-esterases, and of the genetic control of their hydrolysis in sheep, *Biochem. Pharmacol.*, **13**, 1551–1568.

LEE, R. M., 1966. Genetic control of the hydrolysis of aromatic esters by sheep plasma A-esterase, *Genet. Res. Camb.*, **7**, 373–382.

LEWIS, W. H. P. and HARRIS, H., 1967. Human red cell peptidases, *Nature*, **215**, 351–355.

LUSH, I. E., 1966. The biochemical genetics of vertebrates except man, North-Holland Publishing Co., Amsterdam, pp. 118.

MARKERT, C. L., 1963. Lactate dehydrogenase isozymes: dissociation and recombination of subunits, *Science*, **140**, 1329–1330.

MARKERT, C. L. and MØLLER, F., 1959. Multiple forms of enzymes: tissue, ontogenetic and species specific patterns, *Proc. Nat. Acad. Sci.*, **45**, 753–763.

MATHAL, C. K., OHNO, S. and BEUTLER, E., 1966. Sex linkage of glucose-6-phosphate dehydrogenase gene in Equidae, *Nature*, **210**, 115–116.

NANCE, W. E., CLAFLIN, A. and SMITHIES, O., 1963. Lactic dehydrogenase: genetic control in man, *Science*, **142**, 1075–1077.

NIKAIDO, H., 1967. Bacterial cell wall-deep layers. The specificity of cell surfaces, Prentice Hall, Inc., Englewood Cliffs, New Jersey, pp. 3–30.

OHNO, S., POOLE, J. and GUSTAVSSON, I., 1965. Sex-linkage of erythrocyte glucose-6-phosphate dehydrogenase in two species of wild hares, *Science*, **150**, 1737–1738.

PARR, C. W., 1966. Erythrocyte phosphogluconate dehydrogenase polymorphism, *Nature*, **210**, 487–489.

PELZER, C. F., 1965. Genetic control of erythrocyte esterase forms in *Mus musculus*, *Genetics*, **52**, 819–828.

RANDERSON, S., 1965. Erythrocyte esterase forms controlled by multiple alleles in the deer mouse, *Genetics*, **52**, 999–1005.

RENDEL, J. and STORMONT, C., 1964. Variants of ovine alkaline phosphatases and their association with the R–O blood groups, *Proc. Soc. Exptl. Biol. Med.*, **115**, 853–856.

RENDEL, J., AALUND, O., FREEDLAND, R. A. and MØLLER, F., 1964. The relationship between the alkaline phosphatase polymorphism and blood group 0 in sheep, *Genetics*, **50**, 973–986.

ROBBINS, P. W., DRAY, D., DANKERT, M. and WRIGHT, A., 1967. Direction of chain growth in polysaccharide synthesis, *Science*, **158**, 1536–1542.

ROBINSON, J. C. and GOLDSMITH, L. A., 1967. Genetically determined variants of serum alkaline phosphatase: a review, *Vox Sang.*, **13**, 289–307.

SARTORE, G., STORMONT, C., MORRIS, B. G. and GRÜNDER, A. A., 1968. Electrophoretic forms of esterases in red cells of cattle and bison, *Genetics*. (In press).

SAWIN, P. B. and GLICK, D., 1943. Atropinesterase, a genetically determined enzyme in the rabbit, *Proc. Nat. Acad. Sci.*, **29**, 55–59.

SCHIFF, R. and STORMONT, C., 1968. The genetic control of rabbit red cell esterase variation, Proc. 12th Int. Cong. Genet., Tokyo. (In press).

SEMEONOFF, R. and ROBERTSON, F. W., 1968. A biochemical and ecological study of plasma esterase polymorphism in natural populations of the field vole, *Microtus agrestis L*, *Biochem. Genet.*, **1**, 205–227.

SHAW, C. R., 1965. Electrophoretic variation in enzymes, *Science*, **149**, 936–943.

SHAW, C. R. and BARTO, E., 1965. Autosomally determined polymorphism of glucose-6-phosphate dehydrogenase in Peromyscus, *Science*, **148**, 1099–1100.

SHOWS, T. B., 1967. The amino acid substitution and some chemical properties of a variant human erythrocyte carbonic anhydrase: carbonic anhydrase Id Michigan, *Biochem. Genet.*, **1**, 171–195.

SHREFFLER, D. C., BREWER, G. J., GALL, J. C. and HONEYMAN, M. S., 1967. Electrophoretic variation in human serum ceruloplasmin: a new genetic polymorphism, *Biochem. Genet.*, **1**, 101–115

SKARNES, R., RUTENBERG, S. and FINE, J., 1968. Fractionation of an esterase from calf spleen implicated in the detoxification of bacterial endotoxin, *Proc. Soc. Exp. Biol. Med.*, **128**, 75–80.

SMITHIES, O., 1955. Zone electrophoresis in starch gels: group variations in the serum proteins of normal adults, *Biochem. J.*, **61**, 629–641.

STORMONT, C. and SUZUKI, Y., 1968. Electrophoretic variation in plasma esterases of rabbits, Proc. 12th Int. Cong. Genet., Tokyo. (In press).

SUTTON, H. E., 1967. Human genetics. *Ann. Rev. Genet.*, **1**, 1–36.

SYNER, F. M. and GOODMAN, M., 1966. Polymorphism of lactate dehydrogenase in Gelada baboons, *Science*, **151**, 206–208.

TASHIAN, R. E., 1965. Genetic variation and evolution of the carboxylic esterases and carbonic anhydrases of primate erythrocytes, *Am. J. Hum. Genet.*, **17**, 257–272.

TASHIAN, R. E. and YU, Y. L., 1966. Alterations accompanying the mutation of a human erythrocyte carbonic anhydrase, Third Int. Cong. Hum. Genet., p. 96 (abstract).

TASHIAN, R. E., RIGGS, S .K. and YU, Y. L., 1966. Characterization of a mutant human erythrocyte carbonic anhydrase lc Guam, *Arch. Biochem. Biophys.*, **117**, 320–327.

THULINE, H. C., MORROW, A. C., NORBY, D. E. and MOTULSKY, A. G., 1967. Autosomal phosphogluconic dehydrogenase polymorphism in the cat (*Felis catus L.*), *Science*, **157**, 431–432.

TRUJILLO, J. M., WALDEN, B., O'NEIL, P. and ANSTALL, H. B., 1965. Sex-linkage of glucose-6-phosphate dehydrogenase in the horse and donkey, *Science*, **148**, 1603–1604.

TUCKER, E. M., SUZUKI, Y. and STORMONT, C., 1967. Three new phenotypic systems in the blood of sheep, *Vox Sang.*, **13**, 246–262.

TUCKER, E. M., BAKER, N. F. and STORMONT, C., 1968. The toxicity of the anthelmintic Haloxon in relation to esterase activity in sheep, *Amer. J. Vet. Res.* (In press).

VALENTA, M., HYLDGAARD-JENSEN, J. and MOUSTGAARD, J., 1967. Three lactate dehydrogenase isoenzyme systems in pig spermatozoa and the polymorphism of subunits controlled by a third locus C, *Nature*, **216**, 506–507.

VESELL, E. S., 1965. Polymorphism of human lactate dehydrogenase isozymes, *Science*, **148**, 1103–1105.

WILCOX, F. H., 1966. A recessively inherited electrophoretic variant of alkaline phosphatase in chicken serum, *Genetics*, **53**, 799–805.

DISCUSSION

B. ZABŁOCKI: The method of zone electrophoresis in starch gel was successfully used for investigations of the light and heavy chains of IgG immunoglobulins. This medium is not appropriated for the eventual investigations of enzymes splitting carbohydrates. In this case the "Perikon" (copolymer of polyvinyl acetate) may be useful.

XITH EUROPEAN CONFERENCE ON ANIMAL BLOOD GROUPS AND BIOCHEMICAL POLYMORPHISM
Warsaw, July 2nd–6th, 1968

Leukocyte Antigens of Man and Subhuman Primates*

H. BALNER**
Radiobiological Institute TNO, Rijswijk Z. H., The Netherlands

INTRODUCTION

Human leukocyte antigens have now been studied for some 10 years. During that period a relatively small number of investigators have defined the major antigens, proven that most of them belong to one genetic system (comparable to the H-2 locus of mice), and shown that they are of importance for histocompatibility (Histocompatibility Testing, 1965 and 1967). This last circumstance is probably the reason for the tremendous interest in leukocyte typing throughout the world: at present, matching for leukocyte antigens is the only practical method of histocompatibility testing for non-related human individuals. It should be emphasized in this context that, unlike in rodents, leukocyte or tissue antigens of primates are not recognizable on red blood cells and that the classical red cell blood groups are, except for the strong antigens of the human ABO system, probably not important for histocompatibility.

Soon after the first human leukocyte antigens had been defined with isoantisera, parallel studies to determine the leukocyte antigens also for other primate species were started. The reasons for this research effort were two-fold: firstly, the sharing of tissue antigens between the various primate species was of great phylogenetic interest and, secondly, apes and monkeys had become particularly attractive subjects for preclinical transplantation research; methods of histocompatibility testing for these species were therefore needed. While leukocyte typing in subhuman primates has not reached the degree of refinement presently available for man, rapid progress is being made, particularly for the antigens of chimpanzees and Rhesus monkeys.

Chimpanzees have indeed been shown to share many leukocyte antigens with man (SHULMAN and HILLER, 1967; METZGAR *et al.*, in press; BALNER *et al.*, 1967) and several of the chimpanzee's "own" leukocyte antigens are presently being defined serolo-

* Part of this work has been performed under contract with Euratom (European Atomic Energy Community), Contracts Nr. 029-63-1 BIAN and 062-66-1 BIAN.

** Member of Euratom.

gically; within one or two years we should know to what extent the main leukocyte antigen system of the chimpanzee is identical with that of man.

As for the lower primates, intensive efforts have been made to identify the leukocyte antigens of Rhesus monkeys (BALNER *et al.*, 1965b, 1966), the primate species most widely used in medical research. In fact, certain crucial early experiments establishing the importance of leukocyte antigens for histocompatibility have been carried out in Rhesus monkeys (BALNER *et al.*, 1965a). More recently, leukocyte typing of Rhesus monkeys has reached the stage where the major system of their leukocyte antigens can be broadly outlined (BALNER and BEKKUM, 1967).

Leukocyte typing has lately been taken up also in a number of other species such as the rabbit (DÉMANT, 1967) and the dog (EPSTEIN *et al.*, 1968) but in neither has it reached the stage just described for man and monkeys, namely the recognition of the majority of the antigens of the main histocompatibility locus. In mice and rats, finally, where histocompatibility typing has classically been based on red cell typing, methods are gradually changing; many investigators now prefer to define transplantation antigens on nucleated cells rather than on red cells, the method of choice for the last 20 years.

The present paper deals exclusively with the leukocyte antigens of man and the subhuman primates. We shall describe the methods by which leukocyte antigens are recognized serologically, how the necessary iso-antisera were obtained in the various species and what criteria are presently being used for the serological definition of an antigen (degree of specificity of sera, etc.). Also, several methods will be described that have been used, both in man and monkeys to prove the relevance of leukocyte antigens for histocompatibility.

Data for frequencies of leukocyte antigens and associations between the various antigens based on population studies are available for humans as well as monkeys. Genetic associations of antigens however are far more reliably deduced from family studies; for man such studies have recently led to an insight into the actual antigenic composition of the "alleles" responsible for the inheritance of those particular traits. These same family studies also revealed that several of the defined human leukocyte antigens segregate independently of the antigens of the main system (HL-A) i.e. belong to a different "locus".

Finally, attention will be paid to the sharing of antigens between man and the subhuman primates. Several authors, using various techniques, have proven such sharing (SHULMAN *et al.*, 1965; METZGAR and ZMIJEWSKI, 1966; BALNER *et al.*, 1967b). As already stated, striking similarities for leukocyte antigens were found between man and chimpanzees; as a consequence, chimpanzees are now being used for the production of serological reagents (iso-antisera) useful for human tissue typing (BALNER *et al.*, 1967b, a). For the same reason, chimpanzees are being employed in areas of transplantation research where the use of human volunteers would be the only alternative (BALNER *et al.*, 1968). As for the lower monkeys, there also seems to be a degree of similarity between the spectrum of leukocyte antigens in

Rhesus monkeys on the one hand and the apes and man on the other (BALNER and BEKKUM, 1967); the experiments on which these assumptions are based will be described.

IDENTIFICATION OF LEUKOCYTE ANTIGENS

Since hetero-antisera do not easily distinguish individual-specific antigens within a species, iso-antisera are generally used for the recognition of leukocyte antigens. For humans, there are sera from multiparous women (containing iso-antibodies against paternal fetal antigens), from poly-transfused patients or from individuals deliberately immunized with allogeneic skin or leukocytes. In non-human primates the last category comprises the majority of reagents being used for typing purposes.

Iso-antibodies against leukocytes can be demonstrated by a multitude of serological tests. Initially, leukoagglutination was the method of choice (DAUSSET, 1958; ROOD and LEEUWEN, 1963; PAYNE, 1957; AMOS and PEACOCKE, 1964), but complement fixation and a few other methods are also employed (SHULMAN *et al.*, 1962; COLOMBANI *et al.*, 1964); in recent years several variations of the cytotoxicity test have become more popular and are now preferred by many investigators (TERASAKI, 1964; KISSMEYER-NIELSEN and KJERBYE, 1967; ROGENTINE, 1967). The advantages and disadvantages of the various methods have been critically reviewed (ROOD f*et al.*, 1966).

To define antigens of any kind, the serological reagents must have a rather high degree of specificity. Most of the available anti-leukocyte sera however, be it human, chimpanzee or Rhesus iso-antisera, are by no means monospecific and thus not particularly suited for grouping work. The difficult task of selecting those reagents that, because of their degree of specificity would be useful for typing purposes, was greatly facilitated by VAN ROOD's computer approach (ROOD and LEEUWEN, 1963; Histocompatibility Testing 1967; CEPPELLINI, 1968). This method, employing a 2×2 association analysis of positive and negative reactions against leukocytes, is based on the assumption that iso-antisera, no matter how they are produced, will be of "interest" if they show similar or identical reactivity patterns when tested with a large panel of leukocyte samples. The reasons for this being that the antigen or closely linked antigens recognized by such randomly produced sera of identical reactivity pattern are likely to be "strong" (and important for matching) and that such sera will probably contain not more than one or two different antibodies. That these assumptions were correct has since been proven for leukocyte antigens of several species; in fact most of the known leukocyte antigens of man and monkeys have been defined by sera selected by this computer approach.

It is evident that the ultimate analysis of a reagent's specificity will depend on meticulous absorption studies (WALFORD *et al.*, 1967; BATCHELOR and SANDERSON, 1967). However, recent developments have shown that even sera of limited specificity (oligo-specific sera) can be quite useful typing reagents; if sufficiently specific on an "operational" basis, they can still help to define serological phenotypes that are inherited and segregate as mono-factorial traits (CEPPELLINI *et al.*, 1967).

Once a sufficient number of antigens can be serologically defined and the phenotype of a large number of individuals has been determined, the question of associations between these antigens and their mode of inheritance arises. A certain amount of information regarding their genetic relationship can be obtained from the positive or negative associations observed in population studies (PAYNE *et al.*, 1964; ROOD *et al.*, 1965; DAUSSET *et al.*, 1965). Far more reliable data however are obtained from transmissional analysis, which implies studying segregation and linkage of the various recognizable factors in large families. Under favourable circumstances, such analysis permits the determination of the antigenic components of the "alleles" responsible for the inheritance of these traits and even an assignment of the phenotypes to one or more genetic systems of loci (ROOD *et al.*, 1967; CEPPELLINI *et al.*, 1967; DAUSSET *et al.*, 1967). For man, such family studies have recently led to the conclusion that the majority of the recognizable leukocyte antigens belong to one system or locus, comparable to the H-2 locus of the mouse (see further below).

RELEVANCE OF LEUKOCYTE ANTIGENS FOR HISTOCOMPATIBILITY

As mentioned in the introduction, the universal interest in leukocyte typing is largely due to the now generally recognized importance of leukocyte antigens for transplantation immunology. Just how far leukocyte typing has been perfectionated to serve as a histocompatibility test for random populations (prospective matching of host/donor combinations) will be discussed separately for the various primate species. In this section only a brief summary is given of the experiments that led to the realization that transplantation antigens of primates can be identified serologically on leukocytes.

While evidence that transplantation antigens are carried by most nucleated cells, including the leukocytes, had been available for quite some time (MEDAWAR, 1946), the question remained whether the antisera capable of identifying leukocyte antigens are indeed recognizing the major transplantation antigens of a species (i.e. the principal tissue antigens responsible for rejection of allografts).

The simplest type of experiment capable of answering that question, was successfully carried out in 1964 in Rhesus monkeys (BALNER *et al.*, 1965a)*. *In vitro* reactivity against leukocytes of an immunized individual's serum was compared with rejection times of skin allografts from donors whose leukocytes reacted positively or negatively with the recipient's serum. As shown in Fig. 1 grafts from donors whose leukocytes reacted positively with the sera were rapidly rejected, which proves that in the majority of cases these antisera (produced after grafting allogeneic skin)

* This type of experiment had previously been tried on human volunteers by ROOD *et al.* (1964); the outcome was negative but that may have been due to the long interval between immunization and test grafting.

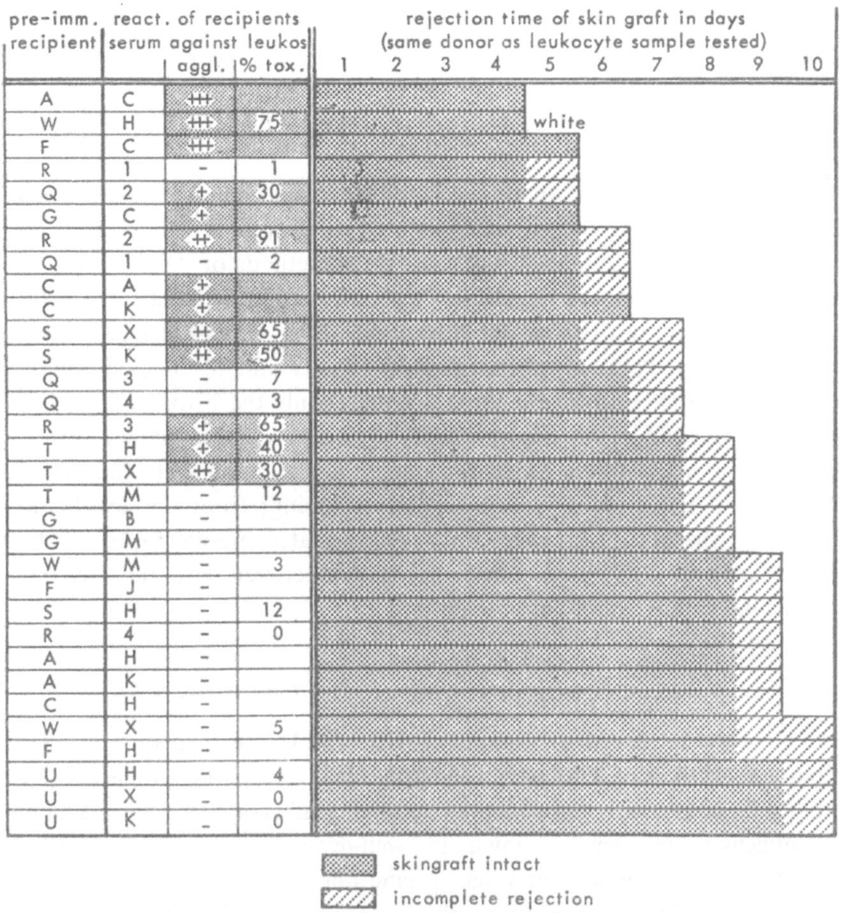
pre-imm. recipient
react. of recipients serum against leukos
aggl.
% tox.
rejection time of skin graft in days (same donor as leukocyte sample tested)
1 2 3 4 5 6 7 8 9 10
A C +++
W H +++ 75
F C +++
R 1 - 1
Q 2 + 30
G C +
R 2 ++ 91
Q 1 - 2
C A +
C K +
S X ++ 65
S K ++ 50
Q 3 - 7
Q 4 - 3
R 3 + 65
T H + 40
T X ++ 30
T M - 12
G B -
G M -
W M - 3
F J -
S H - 12
R 4 - 0
A H -
A K -
C H -
W X - 5
F H -
U H - 4
U X - 0
U K - 0
white
skingraft intact
incomplete rejection

FIG. 1. Comparison serum activity and skin graft rejection time. Recipients A, C, S, T, U and W were preimmunized with skin grafts only, F G received skin grafts and blood intravenously, Q and R, skin grafts leukocytes intradermally. From BALNER *et al.* (1966), pp. 541–557.

recognized transplantation antigens on leukocytes. Shortly thereafter ROOD *et al.* (1965) as well as DAUSSET and co-workers (1965) were capable of proving, with a more refined variation of this test, that several of the defined human leukocyte antigens are transplantation antigens. This type of test (shortened rejection time of grafted skin from donors carrying a particular antigen against which the recipient had been specifically immunized) was also useful for proving that the human red cell antigens A_1 and B are strong histocompatibility antigens (DAUSSET and RAPAPORT, 1966).

Does this type of evidence prove that leukocyte typing will be useful for prospective matching of host/donor combinations? Not quite, since the procedure may lend undue importance to the antigen used for pre-immunization. Many attempts have been made to correlate prospective or retrospective matching for the known leukocyte antigens to survival time of allogeneic skin grafts or the success of organ or bone marrow transplants in non-immunized individuals. The extent to which such a correlation can be found in the various primate species is described in the next sections.

Remains the question whether other histocompatibility tests are available for primates and how their (predictive) matching value compares with the present state of leukocyte typing. In the last few years three methods of "direct" matching have emerged as potential histocompatibility tests for man and animals, the MLC (mixed lymphocyte culture test (BACH and HIRSCHORN, 1964; BAIN and LOWENSTEIN, 1964), the NLT (normal lymphocyte transfer test (BRENT and MEDAWAR, 1963) and the Third Man Test (MATHÉ *et al.*, 1961). Of these, only the MLC test has been maintained as a method of histocompatibility testing. BACH has recently evaluated his unidirectional MLC test as a histocompatibility test for man and compared its potential with the present state of leukocyte typing; from this excellent work it appears that the MLC test, even in its perfectionated present form, can be a most welcome refinement of histocompatibility testing between close relatives but not yet for matching within a random population.

PRESENT STATE OF LEUKOCYTE TYPING IN MAN

Even now that the data obtained at the highly informative Turin Histocompatibility Workshop of 1967 have been "digested" and partly published (Histocompatibility Testing 1967; CEPPELLINI, 1968) the number of well-defined human leukocyte antigens is not exactly known. If "well-defined" means antigens that are recognized by several mono- or oligo-specific antisera, preferably in different laboratories and by different techniques, the main groups of investigators (VAN ROOD, DAUSSET, PAYNE, BODMER, CEPPELLINI and TERASAKI) presently recognize between 20 and 30 different antigens or specificities. Many of these specificities as recognized by these and several other groups of investigators are identical although the designations used by them may still be different. Table 1, taken from a recent review by ROOD and EERNISSE (1968), depicts the state of affairs and gives the reader an impression of the intricate situation with regard to the present nomenclature.

In anticipation of a uniformly accepted nomenclature which is now being worked out by a World Health Organization committee, we shall use the established nomenclature of VAN ROOD and collaborators. Although DAUSSET had described th- first human leukocyte antigen "Mac" in 1958 (synonyms 8a, La2, etc.), VAN ROOD established the first "system" of leukocyte antigens, the "Four" system, comprising the antigens 4a and 4b which behave as genetic alternatives and now form the "backbone" of the major system of human leukocyte antigens (presently called the HL–A system).

In the years preceding the Turin Histocompatibility Workshop of 1967, "allelic" relationships between a number of the (then) recognized antigens had been established; apart from VAN ROOD's "Four" and "Five" groups, the LA antigens of PAYNE and BODMER, for instance, were shown by 2×2 association analysis in population studies as well as by transmissional analysis (family studies) to form an allelic system (PAYNE *et al.*, 1964).

The next step was obviously to investigate the interrelationships between the various recognized "allelic systems" of antigens (4a/4b, 5a/5b, the LA-antigens, etc.). Mainly on the basis of family studies, VAN ROOD *et al.* suggested in 1965 already, that several of his allelic systems of antigens were genetically linked and probably part of one system (chromosomal locus) comparable to the H-2 locus of mice, while other antigens (5a and 5b) segregated independently and thus belonged to a different locus or system (ROOD *et al.*, 1965). DAUSSET, also in 1965, came to a similar conclusion (DAUSSET *et al.*, 1965); his reasoning however was based on associations observed in the distribution of the antigens in the population. The merits and disadvantages of both approaches have been recently outlined in a paper by CEPPELLINI *et al.* (1967) in which he concludes that a final answer regarding the genetic associations between the various recognized antigens could only be provided by large scale family studies. The mentioned Turin Workshop was more or less the consequence of such reasoning. There, the leukocytes of 11 large families (22 parents and 55 children) were typed by 16 teams, the participants using their own reagents and techniques. The results unambiguously showed that the majority of the recognizable antigens segregate as part of one genetic system (the HL-A locus or chromosomal region) while a few antigens, such as 5a and 5b, do not. Another conclusion was that the defined leukocyte antigens segregate independently of the ABO red cell antigens (CEPPELLINI, 1968).

Meticulous analysis of the Turin data have also led to the recognition of the HL-A genotypes of parents and children. For a detailed description of the methods used to determine the actual antigenic factors carried by the HL-A alleles, the number of different alleles so far identified, their frequencies in the population and certain conclusions and predictions based on these data, the reader is referred to the proceedings of the Turin Workshop (Histocompatibility Testing 1967) as well as recent publications by ROOD and EERNISSE (1968), DAUSSET *et al.* (1967) and CEPPELLINI *et al.* (1967).

A theoretical and practical point of major importance with regard to human leukocyte antigens is obviously their relevance for histocompatibility. As pointed out in a previous section, the two principal approaches are skin grafting on preimmunized individuals and evaluation of the influence of leukocyte typing and matching on the survival time of allografts in unimmunized individuals. The latter approach comprises mainly retrospective studies and only a few prospective ones. The earliest available significant data are those of ROOD *et al.* (1965) and DAUSSET *et al.* (1965) in which it was shown that the HL-A antigens 4a, 4b, 7c, 7d and 8a (Mac) are transplantation antigens; these were skin grafting experiments on recip-

TABLE 1

Van Rood	Dausset	Payne	Terasaki	Ceppellini	Bodmer	Engelfriet	Kissmeyer	Lalezari	Lawler	Walford	Zmijewski	Amos	Batchelor	Balner	Amiel	Shulman	Freq. of anti-gens (%)
				2													90
4a					4a		4a1		4a					*4a*			66
	3	4a		3			{ 4a* *4a1*				D5	16					69
	4		*4*	*11*	LA4		{ *T12* 4a2					{ 17 18					44
				4													64
	{ *5** *5*			5		*7*					L2	{ 12 45				F1	14
		4c	6		4c												26
4b				*6*					4b		E1			*4b*			90
6a																	90
6b					4d	6											55

TABLE 1 (ctnd.)

Aggl. EDTA	Tox. C'fix	Aggl. Def.	Tox.	Tox.	Tox.	Tox.	Tox. C'fix	Aggl. EDTA	Aggl. EDTA	Tox.	Aggl. Def.	Tox.	Tox.	Tox.	Aggl. EDTA	C'fix	
7a							T19									*C1*	49
7b																	74
7c	*10*		*5**			*3*	*4b*	6b		{ *C65* *C66*		{ *2* *36* 38					37
7d	*8*			*7* *19*	*7d*	*1*				*7d*		41	*2*				23
LA1	*11*	*LA1*	*1*	*8*	*LA1*	*4*	*LA1*			LA1			1				34
8a	2	LA2	*2*	14	*LA2*	2	*LA2*		8a	*LA2*		42	5			B1	61
	1			9								*1*					
LA3	*12*	LA3		*10*	*LA3*		*LA3*			LA3		4				HILL	25
				12						*HUN*			*3*				28
5a																	34
5b								*5b*									97
				1				*NA1*	*5b*								56

Note. Most of these data were collected during the Torino Workshop 1967 (Histocompatibility Testing 1967). At the top of the Table the name of the principal investigator is given. At the bottom of the Table is indicated which technique(s) was (were) used by the investigator. If an antigen is in italics it implies that the sera recognizing this antigen and the other italicized antigens reacted identically with all 96 leukocyte samples of the panel. Asterisks indicate subgroups. From ROOD and EERNISSE (1968).

ients previously immunized against those particular antigens. A year later both AMOS *et al.* (1966) and VAN ROOD's and CEPPELLINI's group (ROOD *et al.*, 1966) showed that skin grafts exchanged between non-immunized sibs survived longer if matched for the known HL-A antigens; a similar coerrelation could not be demonstrated for skin grafts exchanged between unrelated individuals or between parents and sibs.

As for studies of host/donor combinations in kidney transplantation TERASAKI *et al.* (1966, 1967) as well as other investigators (DAUSSET *et al.*, 1967; PAYNE *et al.*, 1967), did a number of mainly retrospective studies to establish a correlation between the fate of a graft and the degree of compatibility for leukocyte antigens. The results suggested that compatibility for leukocyte antigens was of importance if patient survival and graft histology were used as parameters. However, some of these data were obtained from rather unhomogenous patient material and the typing, especially in the earlier years, was performed with batteries of antisera of sometimes rather poorly defined specificity. In more recent studies, ROOD *et al.* (1967) avoided these sources of error by strictly dividing the material into groups with homogeneous genetic background (sib–sib, parent–sib and random host–donor combinations) while the accuracy of typing (number of antigens determined, specificity of sera, etc.) had greatly improved compared with the earlier work. From this material it appeared that identity for the known HL-A antigens not only improved graft expectancy when a sib, but also when a parent was the kidney donor. The implications of the finding that "identity" for the recognizable HL-A antigens must be distinguished from mere "compatibility" (i.e. the donor not possessing antigens the recipient lacks) are elaborately discussed in a review by ROOD and EERNISSE (1968).

LEUKOCYTE ANTIGENS OF CHIMPANZEES

Because of their known phylogenetic closeness to man chimpanzees have been used as donors for hetero- or xeno-transplantation to man long before serological typing of their tissue antigens had begun. The relatively long functioning of kidneys transplanted from randomly chosen chimpanzees to man were indeed the first empirical evidence that chimpanzee tissues must have important antigens in common with man (REEMTSMA *et al.*, 1964). The sharing of certain red cell antigens between non-human primates and man had previously been demonstrated by a number of investigators but the vast literature dealing with that subject will not be discussed in the context of the present paper; for details the reader is referred to a number of excellent monographs and reviews (MOOR-JANKOWSKI and WIENER, 1968; WIENER and MOOR-JANKOWSKI, 1967; SCHMITT, 1968). The role of red cell antigens of non-human primates for tissue-typing (histocompatibility) was not covered by those authors; in fact, experimental data available about that subject are still very limited (MURPHY *et al.*, 1965; BALNER *et al.*, 1966, 1967b).

The first serological data proving the sharing of leukocyte antigens between chimpanzees and man comes from SHULMAN *et al.* (1965) and from METZGAR's group (METZGAR and ZMIJEWSKI, 1966) who showed that human iso-antisera reacted with some but not all chimpanzee leukocytes. These authors had also been able to demonstrate that chimpanzee sera obtained after immunization with chimpanzee or human material reacted with a proportion of human leukocyte samples (SHULMAN *et al.*, 1964; METZGAR *et al.*, 1964). The techniques used were mainly leukocyte agglutination or complement fixation.

Actual typing of chimpanzees for defined human leukocyte antigens was first taken up by our own group in 1966 (BALNER *et al.*, 1967b). Using leukocyte agglutination, cytotoxicity and absorption techniques, 7 of 12 defined human leukocyte antigens could be demonstrated, first on a small panel of 9 chimpanzees. Recently 13 more chimpanzees were typed in the same fashion and the combined results for 22 leukocyte samples are depicted in Table 2. Other investigators confirmed the presence of these same and a number of other defined human leukocyte antigens on primate cells; they used a mixed agglutination technique on cells cultured *in vitro* (METZGAR *et al.*, in press).

In our first publication on this subject a number of tentative conclusions were drawn with regard to the similarities and differences in the distribution of human

TABLE 2

Human leukocyte antigen on chimpanzee leukocytes

	4a	4b	6a	6b	7a	7b	7c	7d	8a	9a	5a	5b
Debbie	+											+
Fred	+											+
Henkie	+											+
Ufford	+		+			+						+
Anita	+		+	+								+
Mario	+		+	+		+						+
Regina	+		+	+		+						+
Tineke	+			+		+	+					+
Abe	+						+					+
Elvis	+	+	+									+
Katie	+	+	+			+						+
Nico	+	+	+	+		+						+
Quarles	+	+	+	+		+						+
Isac	+	+				+						+
Simon	+	+				+	+					+
Diana	+	+				+	+					+
Brigitte		+				+	+					+
Jan		+	+			+						+
Pietje		+	+			+						+
Claudia		+	+	+	+	+						+
Dorus		+	+		+	+	+					+
Gerrit		+	+				+					+

leukocyte antigens in chimpanzees and in man (Balner *et al.*, 1967b). Most of those rather speculative conclusions still hold true, some have to be revised. It seems wisest however to postpone further comparisons and conclusions until the remainder of our chimpanzee colony (50 individuals) has been typed and a statistically more sound evaluation of frequencies, correlations, etc. can be made. One thing however seems likely, namely that the two major antigenic specificities of the human HL-A system, antigens 4a and 4b, are also amongst the main determinants of chimpanzee leukocytes and seem to behave in an allelic fashion (contrasting distribution) as they do in man. Whether the chimpanzee's 4a and 4b antigens are identical with or only similar to these specificities in man remains an open, much discussed question. It should be remembered that even for man the exact role of 4a and 4b in the spectrum of leukocyte antigens is not quite clear; some investigators in fact regard them as the main antigenic determinants of the HL-A locus, with most of the other identifiable antigens of this locus possibly being so called "inclusions" which together determine the over-all phenotype of a leukocyte sample.

To identify the chimpanzee's "won" leukocyte antigens, a large number of iso-antisera are presently being produced in our laboratory and tested against all available chimpanzee leukocyte samples, using the cytotoxicity test and absorption techniques. With the computer approach described above, a number of chimpanzee leukocyte antigens are likely to be identified soon; those antigens, when sufficiently defined, may or may not turn out to be carried by human leukocytes. Whatever the result, some further light will then be shed on the phylogeny of primate tissue antigens.

Apart from these aspects, leukocyte typing in chimpanzees has a number of not insignificant practical consequences. The availability of leukocyte-typed chimpanzees has already led to a program of "planned" immunizations in our laboratory with the intent of producing antisera of a particular specificity. It was a rather pleasant confirmation of prior theoretical speculations that, for instance, a 4a-negative chimpanzee immunized with skin and leukocytes from a 4a-positive chimpanzee produced an antiserum of distinct anti-4a reactivity in chimpanzees but also when reacted with a large number of human leukocyte samples! Similarly it has been possible to produce an anti-4b serum usable both for human and chimpanzee leukocyte typing (Balner *et al.*, 1967b; Ceppellini, 1968). It is clear that the chimpanzee iso-antisera presently being produced serve a dual purpose: the identification of chimpanzee iso-antigens (as outlined above) and the planned production of reagents usable for human tissue typing.

A few other potential benefits to be derived from tissue typing in chimpanzees should also be mentioned. The development of more potent immunosuppressive agents as well as promising methods to induce specific tolerance with soluble or semi-soluble antigens have renewed the interest in clinical heterografting. Some of the basic pre-clinical work in these areas will no doubt be done with tissue-typed chimpanzees, although it is not certain whether these rather exotic and rare animals will be the species of choice to provide the material for future tissue and organ "banks". On the other hand, some of the still unsolved problems of the role HL-A antigens in histocompatibility, the feasibility of inducing tolerance to them and the role of the weaker HL-A leukocyte antigens in transplantation are research areas in which chimpanzees can certainly play a role in the future. At present, chimpanzees are already indispensable experimental animals: because of the discussed similarities with regard to tissue antigens, chimpanzees are used to evaluate the efficacy and toxicity of antihuman antilymphocyte sera (BALNER, EYSVOOGEL and CLETON, 1968; BALNER *et al.*, 1968); the clinical application of these highly promising immunosuppressive agents is now regarded as hazardous and of doubtful value unless prior *in vivo* testing in chimpanzees has shown a serum's usefulness. Finally, chimpanzees may also be helpful in certain areas of cancer research, particularly in tumour immunology as has recently been pointed out by ROOD *et al.* (in press).

LEUKOCYTE ANTIGENS OF RHESUS MONKEYS

The reasons for our special interest in leukocyte typing of monkeys have already been briefly outlined in the introduction. In short, the Rhesus monkey figures as the optimal experimental animal for our pre-clinical transplantation studies so that a reliable histocompatibility test was needed; when the direct histocompatibility tests (MLC, NLT and Third Man Test, see before) had shown disappointing results (BALNER, 1966), a maximum effort was made to perfectionate leukocyte typing to a point where it might be useful for matching purposes. At the same time it seemed worthwhile to study the leukocyte antigens in yet another primate species for other reasons: for instance, to establish whether in Rhesus monkeys the principal antigens are also genetically governed by one chromosomal locus. Finally it was of phylogenetic interest to determine to what extent the major leukocyte or tissue antigens of man, chimpanzees and lower monkeys are phenotypically related.

For details of the early work on leukocyte typing in Rhesus monkeys the reader is referred to a number of our previous publications (BALNER *et al.*, 1965b, 1967; BALNER and DERSJANT, 1965). Suffice it to say that the iso-antisera used for the identification of antigens were selected by VAN ROOD's computer method as described for human leukocyte typing (ROOD and LEEUWEN, 1963). Since Rhesus iso-antisera have a tendency to be mono-specific even less frequently than their human counterparts, a special method of immunizations and absorptions was developed to increase the specificity of oligo-specific reagents (BALNER *et al.*, 1966). The first antigens or antigenic groups determined in this fashion were called 1a and 1b because of their seemingly allelic relationship (negative association of reactivity patterns in population studies). Once these two specificities were sufficiently defined, the production of new iso-antisera (to obtain antibodies against "new" antigens) was based on prior matching host/donor combinations for antigens 1a or 1b. This approach was indeed successful; of some 30 antisera prepared in this fashion the computer selected 5 groups of antisera (3 or 4 for each) which had rather similar reactivity patterns and could be expected to recognize new antigens (BALNER, 1967). Figure 2, taken from that same reference, shows the cytotoxicity reactivity of those groups of sera against a panel of 78 Rhesus leukocyte samples. The specificities thus recognized were tentatively called "antigens" 3–7. Since that time many new iso-antisera have been

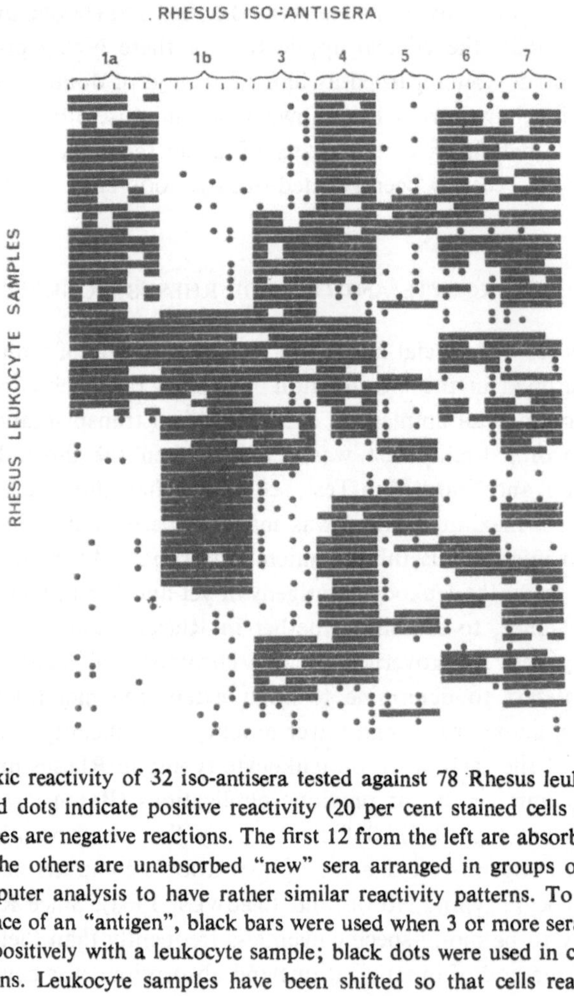

FIG. 2. Cytotoxic reactivity of 32 iso-antisera tested against 78 Rhesus leukocyte samples. Black areas and dots indicate positive reactivity (20 per cent stained cells in cytotoxicity test), open spaces are negative reactions. The first 12 from the left are absorbed anti-1a and anti-1b sera; the others are unabsorbed "new" sera arranged in groups of four reagents shown by computer analysis to have rather similar reactivity patterns. To accentuate the probable presence of an "antigen", black bars were used when 3 or more sera of a particular group reacted positively with a leukocyte sample; black dots were used in cases of isolated positive reactions. Leukocyte samples have been shifted so that cells reacting positively with anti-1a sera and negatively with anti-1b sera are at the top. From BALNER (1967).

prepared, some again on the basis of previous typing results (for instance monkeys typed as 5-negative immunized with skin grafts and leukocytes from monkeys typed as 5-positive, etc.), others on a random basis. The gratifying results of this programme was the following: when 38 new reagents were tested against about 100 Rhesus leukocyte samples the reactivity patterns were mostly "as could have been expected" on the basis of the planned immunizations, while the randomly produced sera often "fell into place" (that is, their reactivity pattern against a panel of test leukocytes was similar to that of sera of an already classified specificity). Only 3 sera showed "new" specificities: the reactivity pattern of one having a highly significant negative correlation with that of the sera recognizing antigen 6; two others were similar to each other but were not strongly correlated with the other groups of sera. The probably allelic alternative for antigen 6 is now called antigen 8 (although only one serum of that kind is available), while the specificity recognized by the other two sera is provisionally called antigen 9. Although not suited for a review paper, these results have been described in some detail because they allow the cautious prediction that, as in man, the majority of the "major" leukocyte antigens might now be recognized also in Rhesus monkeys. The gene frequencies for the various antigens, their correlations in population studies as well as their mode of inheritance in Rhesus families will be published elsewhere. It is noteworthy that these family studies have already provided the valuable information that what we call antigens 1a, 1b etc., although defined by reagents that are not strictly monospecific, are inherited and segregate as so called "monofactorial traits" (CEPPELLINI *et al.*, 1967).

The relevance of antigens 1a and 1b* for histocompatibility was shown in 1965 with skin grafting experiments using pre-immunized donors (BALNER *et al.*, 1965a). Surprisingly, identity of host and donor for antigens 1a and 1b seemed to prolong skin graft survival also on immunized individuals and to mitigate the graft vs. host reaction in irradiated monkeys treated with allogeneic bone marrow (BALNER, DERSJANT *et al.*, 1967; BALNER, BEKKUM *et al.*, 1967). A reasonable explanation for the suggestive evidence that matching merely for those two antigens was of influence also in transplantation experiments on non-immunized individuals, could be the following: presuming that in Rhesus monkeys (as in man and mice) two alleles of one single locus are determining the strong histocompatibility antigens, selecting individuals identical for "antigens" 1a or 1b may have increased the chance of dealing with individuals that have one or possibly even two alleles of identical or near-identical antigenic composition in common. If that is so, then antigens 1a and 1b may well be the Rhesus monkey's counterparts of the human 4a and 4b antigens. A detailed analysis of the actual antigenic composition of the "alleles" of Rhesus monkeys will only be possible when the data of family studies (that are now in progress) become available.

Remains the question to what extent the major leukocyte antigens of Rhesus monkeys are similar to or even identical with those of the higher primate species. The similarities of antigenic determinants on leukocytes (and by implication of tissue antigens) between man and chimpanzees have already been discussed. For species as closely related as man and chimpanzees conventional serological *in vitro*

* In future publications antigen 1b will, more logically, be called antigen 2.

tests can be used (leukocyte agglutination or cytotoxicity tests performed with iso-antisera of one species on the other species' cells). If cells from phylogenetically more remote species have to be tested with human iso-antisera, for instance, inhibition tests and certain other techniques are also required.

In trying to prove the sharing of antigens between anthropoids and lower monkey species we have mainly used absorption techniques to supplement the conventional cytotoxicity test; so far we think to have demonstrated the following similarities or cross reactivities (BALNER, DERSJANT *et al.*, 1967; BALNER, 1967): (1) the "Rhesus" antigens 1a and 1b may well be related to the human 4a and 4b antigens of VAN ROOD. This was shown by absorption of human sera of anti-4a and anti-4b specificity with respectively 1a and 1b Rhesus leukocytes. (2) The Rhesus 1b antigen was present on 9 out of 9 chimpanzee leukocyte samples tested, the 1a antigen on none. Other Rhesus specificities (antigens 3–7) were not demonstrable on chimpanzee cells. (3) The Rhesus antigen 1a and 1b as well as several of the other specificities (3–7) showed polymorphism also in Speciosa monkeys (*Macaca speciosa*), but none of these specificities was demonstrable on the leukocytes of a small panel of Baboons (*Papio cynocephalus*).

Finally, using the mixed agglutination technique on short-term cultures of skin cells, METZGAR *et al.* (in press) have recently been able to prove the presence of many leukocyte antigens on cells of apes but also on those of macaques and African Green monkeys.

In conclusion it can be said that, as expected, more antigens are shared amongst the higher primates than amongst anthropoids on the one hand and monkeys on the other. As a consequence, for certain aspects of pre-clinical transplantation research (in particular in experiments in which sub-human primates substitute for human volunteers) replacement of the valuable and rare apes by more readily available and less expensive lower monkeys may not always be possible.

CONCLUSIONS

(1) The identification of human leukocyte antigens has rapidly progressed in recent years; between 20 and 30 antigens can now be defined serologically. The inheritance of the majority of these factors is controlled by one complex genetic system. This locus, called HL-A, is in more than one respect similar to the H-2 system of the mouse.

Several of the human leukocyte antigens have been shown to be transplantation antigens and the relevance of serological leukocyte matching for histocompatibility in man has been proven.

(2) The chimpanzee is known to be phylogenetically close to man. Several of the human leukocyte antigens have indeed been shown to be carried by chimpanzee leukocytes and chimpanzee iso-antisera obtained after carefully planned immuni-

zations can be used also for human tissue typing. A programme for the serological identification of the chimpanzee's "own" leukocyte antigens is in progress.

(3) The leukocyte antigens of Rhesus monkeys have been the subject of intensive research for several years. An increasing number of antigens can now be defined serologically and for some of these the relevance for histocompatibility has been proven. As in man, the majority of the leukocyte antigens of Rhesus monkeys seem to be controlled genetically by one single locus.

The possibility to match Rhesus monkeys for the major leukocyte antigens has made this primate species a particularly attractive model to study certain aspects of clinical bone marrow and organ transplantation.

(4) The sharing of leukocyte or tissue antigens between monkeys, apes and man has been studied by a number of investigators. The consequences of the observed antigenic similarities for the understanding of the phylogenetic relationship between these species as well as for certain aspects of transplantation research are briefly discussed in the present paper.

ACKNOWLEDGEMENTS

The data on leukocyte antigens of sub-human primates are mainly based on research performed at the Radiobiological Institute TNO. Of my collaborators, Mr. H. DERSJANT and Prof. Dr. D. W. VAN BEKKUM should be mentioned as having contributed substantially to the results described. Furthermore, the help of the maintenance personnel of the primate colony at the Radiobiological Institute is gratefully acknowledged.

Dr. J. J. VAN ROOD's share in this entire research program should be specifically mentioned; his pioneering work in the field of human leukocyte antigens provided the necessary inspiration to continue parallel studies in sub-human primates. Most of the described experiments were in fact planned and executed in close cooperation with him and one of his collaborators, Miss A. VAN LEEUWEN.

REFERENCES

AMOS, D. B., HATTLER, B. G., HUTCHIN, P., MCCLOSKEY, R. and ZMIJEWSKI, C. M., 1966. *The Lancet,* **1**, 300.

AMOS, D. B. and PEACOCKE, N., 1964. Histocompatibility Testing, p. 161, Nat. Acad. Sci., Washington D. C., 1965, publication 1229.

BACH, F. H., Progress in Medical Genetics. (In press).

BACH, F. and HIRSCHHORN, K., 1964. *Science,* **143**, 813.

BAIN, B. and LOWENSTEIN, L., 1964. *Science,* **145**, 1315.

BALNER, H., 1966. *Vox. Sang.*, **11**, 306.

BALNER, H., 1967. Proceedings of a Conference of Experimental Medicine and Surgery in Primates, N. Y. Acad. Sci. (In press).

BALNER, H. and BEKKUM, D. W. VAN., 1967. Proceedings of a Symposium on the Use of Non-human Primates in Medical Research, Lyon. (In press).

BALNER, H., BEKKUM, D. W. VAN, VRIES, M. J. DE, DERSJANT, H. and PUTTEN, L. M. VAN, 1967. Advance in Transplantation, Proc. of the First Int. Congr. of The Transplantation Society, Paris, p. 449, Munksgaard, Copenhagen, 1968.

BALNER, H. and DERSJANT, H., 1965. Histocompatibility Testing (Series Haematologica 11), Report of a Conference and Workshop, Leiden, 1965, p. 103, Munksgaard, Copenhagen.

BALNER, H., DERSJANT, H. and BEKKUM, D. W. VAN, 1968. Proceedings of a Symposium on Immunological and Clinical Problems of Organ Transplantation, Bonn. (In press).

BALNER, H., DERSJANT, H., LEEUWEN, A. VAN and ROOD, J. J. VAN, 1967. Histocompatibility Testing, Report of a Conference and Workshop, Torino and Saint-Vincent, 1967, p. 267, Munksgaard, Copenhagen.

BALNER, H., DERSJANT, H. and ROOD, J. J. VAN, 1965a. *Transplantation*, **3**, 230.

BALNER, H., DERSJANT, H, and ROOD, J. J. VAN, 1965b. *Transplantation*, **3**, 402.

BALNER, H., DERSJANT, H. and ROOD, J. J. VAN, 1966. *Ann. N. Y. Acad. Sci.*, **129**, 541.

BALNER, H., EYSVOOGEL, V. P. and CLETON, F. J., 1968. *The Lancet*, **1**, 19.

BALNER, H., LEEUWEN, A. VAN, DERSJANT, H. and ROOD, J. J. VAN, 1967a. Histocompatibility Testing, Report of a Conference and Workshop, Torino and Saint-Vincent, 1967, p. 257, Munksgaard, Copenhagen.

BALNER, H., LEEUWEN, A. VAN, DERSJANT, H. and ROOD, J. J. VAN, 1967b. *Transplantation*, **5**, 624.

BATCHELOR, J. R. and SANDERSON, A, R., 1967. Histocompatibility Testing, Report of a Conference and Workshop, Torino and Saint-Vincent, 1967, p 139, Munksgaard, Copenhagen.

BRENT, L. and MEDAWAR, P. B., 1963. *Brit. Med. J.*, **2**, 269.

CEPPELLINI, R., 1967. Advance in Transplantation, Proc. of the First Int. Congr. of The Transplantation Society, Paris, p. 195, Munksgaard, Copenhagen, 1968.

CEPPELLINI, R., CURTONI, E. S., MATTIUZ, P. L., MIGGIANO, V., SCUDELLER, G. and SERRA, A., 1967. Histocompatibility Testing, Report of a Conference and Workshop, Torino and Saint-Vincent, p. 149, Munksgaard, Copenhagen.

COLOMBANI, J., COLOMBANI, M. and DAUSSET, J., 1964. Histocompatibility Testing, p. 163, Nat. Acad Sci., Washington D. C., 1965, publication 1229.

DAUSSET, J., 1958. *Acta Haematol.*, **20**, 156.

DAUSSET, J., IVÁNYI, P., COLOMBANI, J. and FEINGOLD, N., 1967. Advance in Transplantation, Proc. of the First Int. Congr. of The Transplantation Society, Paris, p. 231, Munksgaard, Copenhagen, 1968.

DAUSSET, J., IVÁNYI, P., COLOMBANI, J., FEINGOLD, N. and LEGRAND, L., 1967. Histocompatibility Testing, Report of a Conference and Workshop, Torino and Saint-Vincent, 1967, p. 189, Munksgaard, Copenhagen.

DAUSSET, J., IVÁNYI, P. and IVÁNYI, D., 1965. Histocompatibility Testing (Series Haematologica 11), Report of a Conference and Workshop, Leiden, 1965, p. 51, Munksgaard, Copenhagen.

DAUSSET, J. and RAPAPORT, F. T., 1966. *Ann. N. Y. Acad. Sci.*, **129**, 408.

DAUSSET, J., RAPAPORT, F. T., IVÁNYI, P. and COLOMBANI, J., 1965. Histocompatibility Testing (Series Haematologica 11), Report of a Conference and Workshop, Leiden, 1965, 1965, p. 63, Munksgaard, Copenhagen.

DAUSSET, J., RAPAPORT, F. T. and LEGRAND, L., 1967. Advance in Transplantation, Proc. of the First Int. Congr. of The Transplantation Society, Paris, p. 749, Munksgaard, Copenhagen, 1968.

DÉMANT, P., 1967. Advance in Transplantation, Proc. of the First Int. Congr. of The Transplantation Society, Paris, p. 325, Munksgaard, Copenhagen, 1968.

EPSTEIN, R. B., STORB, R., RADGE, H. and THOMAS, E. D., 1968. *Transplantation*, **6**, 45.

Histocompatibility Testing 1965 (Series Haematologica 11), Report of a Conference and Workshop, Leiden, 1965, Munksgaard, Copenhagen.

Histocompatibility Testing 1967. Report of a Conference and Workshop, Torino and Saint-Vincent, 1967, Munksgaard, Copenhagen.

KISSMEYER-NIELSEN, F. and KJERBYE, K. E., 1967. Histocompatibility Testing, Report of a Conference and Workshop, Torino and Saint-Vincent, p. 81, Munksgaard, Copenhagen.

MATHÉ, G., AMIEL, J. L. and NIEMETZ, J., 1961. *Rev. Fr. Et. Clin. Biol.*, **6**, 684.

MEDAWAR, P. B., 1946. *Brit. J. Exp. Pathol.*, **27**, 15.

METZGAR, R. S., SEIGLER, H. F. and ZMIJEWSKI, C. M., Proceedings of the National Symposium on the Immunological Aspects of Polymorphism, Quebec, University of Laval. (In press).

METZGAR, R. S. and ZMIJEWSKI, C. M., 1966. *Transplantation*, **4**, 84.

METZGAR, R. S. ZMIJEWSKI, C. M. and AMOS, D. B., 1964. Histocompatibility Testing, p. 45, Nat. Acad. Sci., Washington D.C., 1965, publication 1229.

MOOR-JANKOWSKI, J. and WIENER, A. S., 1968. Primates in medicine, Vol. I, p. 49, Karger, Basel/ /New York.

MURPHY, G. P., MELBY, E. C., WELDON, C. S., MIRRAND, E. A. and HUSER, H. J., 1965. *Invest. Urol.*, **3**, 244.

PAYNE, R., 1957. *Arch. Intern. Med.*, **99**, 587.

PAYNE, R., PERKINS, H. A. and NAJARIAN, J. S., 1967. Histocompatibility Testing, Report of a Conference and Workshop, Torino and Saint-Vincent, p. 237, Munksgaard, Copenhagen.

PAYNE, R., TRIPP, M., WEIGLE, J., BODMER, W. and BODMER, J., 1964. *Cold Sp. Harb. Symp. Quant. Biol.*, **29**, 284.

REEMTSMA, K., MCCRACKEN, B., SCHLEGEL, J., PEARL, M., PEARL, C., DEWITT, C., SMITH, P., HEWITT, R., FLINNER, R. and CREECH, O., 1964. *Ann. Surgery*, **160**, 384.

ROGENTINE, G. N. JR., 1967. Histocompatibility Testing, Report of a Conference and Workshop, Torino and Saint-Vincent, 1967, p. 371, Munksgaard, Copenhagen.

ROOD, J. J. VAN and EERNISSE, J. G., 1968. *Seminars in Hematology*, **5**, 187.

ROOD, J. J. VAN and LEEUWEN, A. VAN, 1963. *J. Clin. Invest.*, **42**, 1382.

ROOD, J. J. VAN, LEEUWEN, A. VAN, BRUNING, J. W. and EERNISSE, J. G., 1966. *Ann. N. Y. Acad. Sci.* **129**, 446.

ROOD, J. J. VAN, LEEUWEN, A. VAN, BRUNING, J. W. and PORTER, K. A., 1967. Advance in Transplantation, Proc. of the First Int. Congr. of The Transplantation Society, Paris, p. 213, Munksgaard, Copenhagen, 1968.

ROOD, J. J. VAN, LEEUWEN, A. VAN, EERNISSE, J. G., FREDERIKS, E. and BOSCH, L. J., 1964. *Ann. N. Y. Acad. Sci.*, **120**, 285.

ROOD, J. J. VAN, LEEUWEN, A. VAN, SCHIPPERS, A. and BALNER, H., Cancer research. (In press).

ROOD, J. J. VAN, LEEUWEN, A. VAN, SCHIPPERS, A. M. J., PEARCE, R., BLANKENSTEIN, M. VAN and VOLKERS, W., 1967. Histocompatibility Testing, Report of a Conference and Workshop, Torino and Saint-Vincent, 1967, p. 203, Munksgaard, Copenhagen.

ROOD, J. J. VAN, LEEUWEN, A. VAN, SCHIPPERS, A. M. J., VOOYS, W. H., FREDERIKS, E., BALNER, H. and EERNISSE, J. G., 1965. Histocompatibility Testing (Series Haematologica 11), Report of a Conference and Workshop, Leiden, 1965, p. 37, Munksgaard, Copenhagen.

SCHMITT, J., 1968. Immunobiologische Untersuchungen bei Primaten, Karger, Basel, New York.

SHULMAN, N. R. and HILLER, M. C., 1967. Proceedings of a Conference on Experimental Medicine and Surgery in Primates, N.Y. Acad. Sci. (In press).

SHULMAN, N. R., MARDER, V. J., ALEDORT, L. M. and HILLER, M. C., 1962. Proc. 9th Congr. Intern. Soc. Blood Transf., Mexico City, p. 439, Karger, Basel, New York.

SHULMAN, N. R., MARDER, V. J., HILLER, M. C. and COLLIER, E. M., 1964. *Progr. Haematol.*, **4**, 222.

SHULMAN, N. R., MOOR-JANKOWSKI, J. and HILLER, M. C., 1965. Histocompatibility Testing (Series Haematologica 11), Report of a Conference and Workshop, Leiden, 1965, p. 113, Munksgaard, Copenhagen.

TERASAKI, P. I., 1964. Histocompatibility Testing, p. 171, Nat. Acad. Sci., Washington D. C., 1965, publication 1229.

TERASAKI, P. I., VREDEVOE, D. L. and MICKEY, M. R., 1967. *Transplantation*, **5**, 1057.

TERASAKI, P. I., VREDEVOE, D. L., PORTER, K. A., MICKEY, M. R., MARCHIORO, T. L., FARIS, T. D., HERRMANN, T. J. and STARZL, T. E., 1966. *Transplantation*, **4**, 688.

WALFORD, R. L., SHANBROM, E., TROUP, G. M., ZELLER, E. and ACKERMAN, B., 1967. Histocompatibility Testing, Report of a Conference and Workshop, Torino and Saint-Vincent, 1967, p. 221, Munksgaard, Copenhagen.

WIENER, A. S. and MOOR-JANKOWSKI, J., 1966. Proceedings Hartford Conference on Blood Groups and Blood Transfusion, Vol. 1, p. 75, Better Bellevue Assoc., New York, 1967.

DISCUSSION

R. D. OWEN: You referred to van Rood's study in which a husband and his unrelated wife had identical HL-A alleles. I understand that more extensive typing has so far failed to reveal truly identical alleles for three locus, among unrelated individuals (using the terms "alleles" and "locus" in a broad sense) suggesting a remarkably extensive allelic diversity at this locus. I wonder if you would care to comment further about the extent of this diversity.

H. BALNER: When van Rood reports individuals with identical HL-A alleles he is referring to identity *only* for the HL-A antigens recognizable with his panel of antisera. If many other reagents had been used it is quite likely that differences would have been observed.

P. IVÁNYI: 1. What is the frequency of antibody production in chimpanzees and Rhesus in random alloimmunization? 2. Which serological method did you use in your work on Rhesus monkey? 3. It is interesting that antigen 8^{A} (MAC) was not present in chimpanzee. Could you detect other HL-A antigens which behave as alternatives for MAC (probably produced by a chromosomal sub-region) in the chimpanzee?

H. BALNER: Thank you very much for that question. This gives me a chance to say things I didn't have time for during the lecture. Your first question was about the frequency of iso-antibodies occurring in immunized monkeys and chimpanzees. The answer is: virtually 100%. If you immunize them strongly you are bound to get an antibody. If you ask how many of those antibodies are sufficiently specific to be of use for typing purposes, I would say that it should be somewhere around 60% for monkeys as well as chimpanzees. As for the serological methods for monkeys, we started out with agglutination techniques but we later adopted the cytotoxicity test which is more reliable. We have done a number of experiments using complement fixation but the incidence of positive reactions was very low.

Now, let us discuss the HL-A antigens that we did not find in chimpanzees. This intrigued us as much as it does you. Why do not chimpanzees not have the 8A antigen, for instance? There is obviously a chance that we may still find it. It was not found in the 22 chimpanzees typed so far. This does not mean that we shall not find it in the next 30 chimps that are waiting to be typed. If that antigen is indeed never found, that would be extremely interesting, philogenetically.

B. ZABŁOCKI: Some surgeons are using a very similar test for histocompatibility, i.e. the introduction to the recipient introcutaneously a leukocyte suspension of the donors.

The first investigations of the histocompatibility antigens were made with many different inbred mouse lines (Snell). The mouse H_2 antigen (the strongest) has some common determinants with erythrocytes. I fully agree with you that leukocytes antigens (but not erythrocytes) are of great importance in the field of homotransplantation for finding the best donors.

H. Balner: If you take the recipients' lymphocytes and inject them intradermally into the potential donors, you get an *in vivo* picture of how the host would probably react against organs of the various donors. A weak intradermal reaction means that there are probably only minor histocompatibility differences, a strong reaction means that we would probably not choose that particular donor. However, most investigators do not regard the NLT-test as a practical and reliable histocompatibility test. It has too many drawbacks and is not sensitive enough. One has to measure that size of erythema and induration in the skin. For one thing, people differ very much in their cutaneous reactions to the injection of cells. This method has been more or less abandoned as a dependable histocompatibility test, although it seems that some people still use it as an additional test.

THE SHORT REPORTS REFERRING TO ALL ANIMALS

XITH EUROPEAN CONFERENCE ON ANIMAL BLOOD GROUPS AND BIOCHEMICAL POLYMORPHISM
Warsaw, July 2nd–6th, 1968

A Contribution to the Definition of Monospecific Antileukocyte Sera

P. KLOUDA, P. DÉMANT and P. IVÁNYI
Institute of Experimental Biology and Genetics, Czechoslovak Academy of Sciences, Prague, Czechoslovakia

The majority of antileukocyte sera used for leukocyte antigen typing are polyvalent. Since leukocyte antigens are complexes of antigenic factors*, a polyvalent serum reacting solely with the antigenic factors (a.f.) of one antigen is called a relatively monospecific or a pure serum (IVÁNYI and DAUSSET, 1966). The number of antibodies present in a serum can be reduced by means of absorptions with leukocytes of suitably chosen individuals, which react positively with the serum. The serum absorbed with leukocytes of a positively reacting individual chosen at random is retested with the leukocytes of the donor. If the absorbed serum does not react with the leukocytes of the donor, the antibody population has not been refined, since the leukocytes used for absorption possessed antigens (a.f.) for all which antibodies present in the serum. If, after absorption, the serum reacts positively with the leukocytes of the donor, it has been refined, since the leukocytes used for absorption did not possess all antigens (a.f.) against all antibodies present in the serum. A serum whose polyvalency cannot be proven by as many as thirty absorptions is called an operationally monospecific or a defined serum (WALFORD and TROUP, 1967; WALFORD *et al.*, 1967). The number of absorptions needed in order to divide the serum gives a rough idea of the association of antigens (a.f.) against which the antibodies in the serum react. If the association of antigens (or a.f.) in a population is low, there is a relatively high probability of finding an individual possessing only some of the antigens (or a.f.) against which the serum reacts, out of a small number of individuals, whose leukocytes are used for absorptions. The absorption of the serum by leukocytes of such an individual will result in the refinement of the serum. On the other hand if the association of antigens (or a.f.) against which the antibodies in the serum present react is high, the majority of randomly chosen positive-

* In this paper the distinction between antigen and antigenic factor is operational. The term antigen is used for a complex of antigenic factors associated with each other more closely than with other antigenic factors in a given population. The term antigenic factor is used for the designation of units of which the complex is composed. It is preferred to the terms "determinant, component, partigen, subfactor" etc. because it is neutral and thus serves better as operational term.

ly reacting individuals, whose cells are used for absorptions, possess all antigens (or a.f.) against which the serum reacts. Absorptions with leukocytes of individuals possessing all antigens (or a.f.) do not lead to refinement of the serum.

The association of antigens (or a.f.) in a population or that of antibodies in a serum can be expressed in different ways: by means of the χ^2 or the coefficient of correlation or by means of percentage (ROOD and LEEUWEN, 1965). In this paper the association of antibodies in a serum is expressed by the coefficient of correlation $r = \sqrt{\frac{\chi^2}{N}}$, which the authors find to be the best way of estimating the association (FEINGOLD, 1966; DÉMANT *et al.*, 1966).

METHODS AND RESULTS

In a series of absorptions with a serum, this serum will most probably, at first be refined into antibodies with the lowest associations. The more absorptions that are carried out, the higher will be the probability of refining this serum into antibodies with higher associations. If the association is expressed by the coefficient of correlation, it is possible to express mathematically the dependence of the increasing number of absorptions on the refinement of the serum into antibodies with increasing associations and thus to calculate this relationship. The value of the coefficient of correlation depends on the reaction frequency of the serum as shown by the frequency of the members of the panel, who react positively with the serum, and by the association of the antigens (or a.f.) with which the antibodies in the serum react. The lowest value of the coefficient of correlation at a given reaction frequency of the serum and a given part of the panel, reacting positively with both antibodies in the serum occurs if the two antibodies have the same reaction frequency. Given the same reaction frequency of the serum and the same part of panel reacting with both antibodies present in the serum the value of the coefficient of correlation is the highest in the case of an inclusion of one antibody within the other. Inclusion is operationally defined as the relationship of two antigens (or a.f.) or antibodies, where the included antigen (a.f.) or antibody occurs only in the presence of the other, the including antigen (a.f.) or antibody.

The present paper calculates the lower limits of the interval in which the coefficients of correlation of antibodies in a serum of a known reaction frequency, which was not refined by a certain number of absorptions, fluctuate. The upper limit always equals 1. The lower limits of the interval were calculated for two levels of probability (0.95 and 0.99 respectively), and for five reaction frequencies of a serum. The resulting values are shown in Figs. 1 and 2. The horizontal ordinate gives values for the coefficients of correlation, the vertical ordinate gives the number of absorptions. For every serum with a given reaction frequency two curves are drawn, which determine the lower limits of interval, in which the coefficients of correlation of antibodies in the serum fluctuate. The right curve, called the inclusion curve, determines

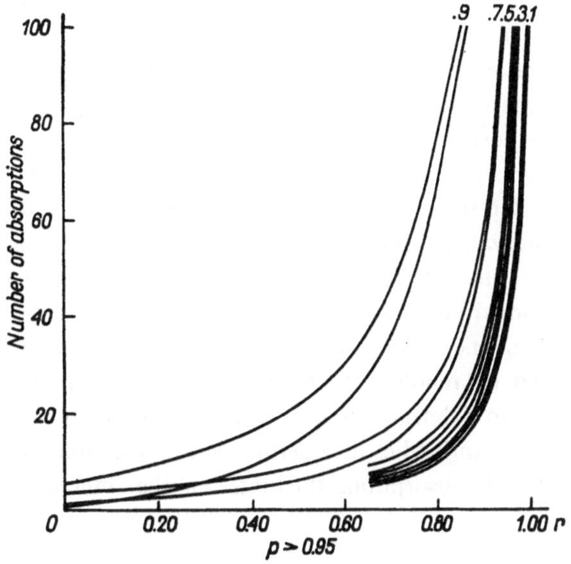

FIG. 1. Absorption curves for the 95% probability level.

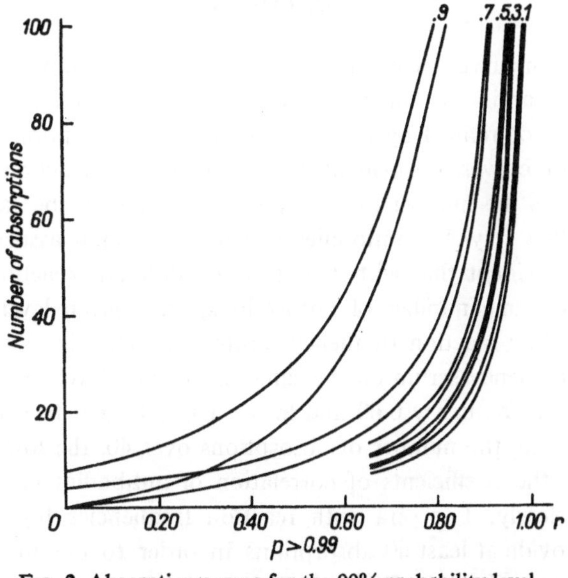

FIG. 2. Absorption curves for the 99% probability level.

the lower limits of the interval in which the coefficients of correlation of antibodies present in the serum fluctuate in the case when one antigen (or a.f.) determined by the first antibody is an inclusion into the antigen (or a.f.) determined by the other antibody. The left curve determines the lower limits of the interval in which the coefficients of correlation of the two antibodies present in the serum fluctuate when both antibodies have the same reaction frequency.

The lower limits of the interval in which the coefficients of correlation fluctuate were calculated for from one to a hundred absorptions. For graphical reasons the curves for the reaction frequency of sera equal to 0.1, 0.3 and 0.5 are shown for values of coefficients of correlation higher than 0.65 only. The absorption curves were calculated for positive values of coefficients of correlation only.

If a certain number of absorptions is carried out with a serum with a known reaction frequency and the serum is not refined, it is possible to detect on the absorption curves the interval, in which the coefficients of correlation of two anitbodies in a serum fluctuate, at a given level of probability, if at least two antibodies are present in the serum. For example 30 absorptions were carried out with a serum with a reaction frequency 0.70. This number of absorptions did not lead to the serum being refined by renoval of an antibody population. From the absorption curves it is possible to determine that at a probability level of 95% the coefficient of correlation of two antibodies in a serum with a reaction frequency 0.70 which was not refined by 30 absorptions fluctuates in the interval between 0.82 (0.83) to 1.00.

DISCUSSION

The absorption curves enable us to ascertain the interval, in which the coefficients of correlation of antibodies in the serum of a known reaction frequency fluctuate, if this serum is not refined by a certain number of absorptions. From the course of the absorption curves it is evident, that the higher the reaction frequency of a serum, the higher is the number of absorptions that have to be carried out to refine the serum so that it yields antibodies which are in high associations. From this fact it becomes evident that with two sera of different reaction frequencies it is possible, by the same number of absorptions, on a given level of probability to refine from the low reaction frequency serum antibodies in associations, which in high reaction frequency serum on the same probability level cannot be refined. For sera with reaction frequency 0.60 and lower it is sufficient to carry out 40 absorptions. By increasing the number of absorptions over 40, the lower limits of the interval in which the coefficients of correlation of antibodies in such sera fluctuate increases very slowly. For sera with reaction frequencies higher than 0.60 it is necessary to provide at least 60 absorptions in order to obtain sera designated as effectively monospecific sera. Effectively monospecific sera are those sera, in which a further increase of the number of absorptions contributes only slightly to the increase of the lower limit of the interval in which the coefficients of correlation of antibodies which the serum was not refined fluctuate.

The absorption curves are calculated for the situation, where the absorptions are carried out on a sample of the population in which the reaction frequency of the serum was ascertained. When the absorptions are carried out on a different population than that for which reaction frequency of the serum was ascertained, the division of the serum caused by the different distribution of antigens (or a.f.) in different populations (DAUSSET *et al.*, 1965; DAUSSET *et al.*, 1967) may result in a refinement of the serum with a much lower or higher number of absorptions than the absorptions than the curves indicate.

The absorption curves indicate the interval in which the coefficients of correlation of not refined antibodies in a serum of a given reaction frequency fluctuate. It is therefore necessary to interpret the results of absorptions evaluated by means of absorption curves in the sense that if a serum was not divided by a given number of absorptions, the association of the not divided antibodies in a serum is not lower than the value indicated by the absorption curves at a given probability level. However the absorption curves cannot be interpreted as indicating, that for to refuse a serum of two antibodies with a given coefficient of correlation and a given reaction frequency it is necessary to provide the number of absorptions indicated by the absorption curves at a given probability level.

The absorption curves show the impossibility of obtaining a monospecific serum even by a relatively high number of absorptions. If two antigens (or a.f.) occur in a population so often together, that their coefficient of correlation reaches values higher than 0.90 then even by a very high number of absorption there may be no refinement of the serum containing antibodies against these antigens (or a.f.).

The coefficient of correlation of antibodies still present in a serum, of arbitrary reaction frequency, subjected to a given number of absorption has an interval of fluctuation which can be determined immediately by reference to absorption curves.

CONCLUSIONS

The number of antibodies in a polyvalent serum can be reduced by means of absorptions with leukocytes of suitably selected individuals and thus defined or operationally monospecific sera can be obtained. The results of absorptions can be evaluated by means of the absorption curves presented in this paper. The absorption curves indicate for a serum with an arbitrary chosen reaction frequency and with which a certain number of absorptions was carried out, the interval, within which fluctuate the coefficients of correlation of antibodies still present in the serum at a given probability level will not fluctuate.

From the course of absorption curves it is evident, that for sera with the reaction frequency equal or lower than 0.60 it is sufficient to carry out 40 absorptions. For sera with reaction frequency higher than 0.60 it is sufficient to provide 60 absorptions to obtain effectively monospecific sera. Effectively monospecific sera are such sera, in which a further increase of the number of absorptions contributes only slightly to the increase of the lower limit of the interval in which the coefficients of correlation of the antibodies, still present in the serum at a given probability level, fluctuate.

By means of absorption curves it is possible to establish immediately the interval in which the coefficients of correlation of antibodies still present in a serum of an arbitrary reaction frequency, absorbed a given number of times fluctuate.

REFERENCES

DAUSSET, J., IVÁNYI, P. and IVÁNYI, D., 1965. Tissue alloantigens in humans: Identification of a complex system (Hu-1). In: Histocompatibility Testing, Munksgaard, Copenhagen, 51.

DAUSSET, J., IVÁNYI, P., COLOMBANI, J., FEINGOLD, N., and LEGRAND, L., 1967. The Hu-1 system. In: Histocompatibility Testing, Munksgaard, Copenhagen, p. 189.

DÉMANT, P., IVÁNYI, P. and IVAŠKA, M., 1966. The values of the correlation coefficient of antigens controlled by allelic genes in a population at equilibrium, *Folia Biol.* (Praha), **12**, 437.

FEINGOLD, N., 1966. Utilisation du coefficient de corrélation dans l'étude des groupes leucocytaires, *Rev. Franç. et Clin. Biol.*, **11**, 1036.

IVÁNYI, P. and DAUSSET, J., 1966. Allo-antigens and antigenic factors of human leucocytes. A hypothesis, *Vox Sang.*, **11**, 326.

ROOD, J. J. VAN and LEEUWEN, A. VAN, 1965. Defined leukocyte antigenic groups in men. In: Histocompatibility Testing, Nat. Acad. Sci. Publ. 1229, Washington, p. 21.

WALFORD, R. L. and TROUP, G. M., 1967. Monospecific lymphocytotoxic antisera, an absorption study, *Vox Sang.*, **12**, 173.

WALFORD, R. L., SHANBROM, E., TROUP, G. M., ZELLER, E. and ACKERMAN, B., 1967. Lymphocyte grouping with defined antisera. In: Histocompatibility Testing, Munksgaard, Copenhagen, p. 221.

XITH EUROPEAN CONFERENCE ON ANIMAL BLOOD GROUPS AND BIOCHEMICAL POLYMORPHISM
Warsaw, July 2nd–6th, 1968

A New Technique for the Permanent Recording of Electrophoretic Results

D. R. OSTERHOFF

Department of Zootechnology, Faculty of Veterinary Science, University of Pretoria, Onderstepoort South Africa

SUMMARY. A new Dalcomatic automatic copier from the firm "Nederlandsche Fotografische Industrie N.V." was installed in our laboratory and has been used for the production of copies of all kinds of gels by a true photographic process. The Dalcomatic proved to be very sensitive as it even gave full tonal separation of black from red in cases of gels showing results of serum proteins and enzymes simultaneously. Technical data of the apparatus and details regarding exposure time and paper types are given.

INTRODUCTION

In almost all immunogenetical laboratories difficulties have been encountered with regard to the storage of starch gels for later reference. Various methods have been described for the preparation of transparent gel strips and preservation of thin-layer and large thick-layer starch gels (DE LIGNY, 1968). The final aim of all attempts are fast and reliable reproductions of the migration results in the different kinds of gel.

In this short note a new approach to the solution of keeping permanent records of starch gels will be outlined.

METHODS

Most of the gel formers consisting of plastic trays have the following internal dimensions 13.5×20.0×0.6 cm. After electrophoresis the gel is split along its lenght in a horizontal plane by means of a very fine nylon fishing line. The two gel slices are separated and transferred to a dish for staining with the cut surface uppermost. The staining-dye employed in most of the work is a saturated solution of Amido-Black 10 B in solvent of composition: methanol–distilled water–glacial acetic acid (50 : 50 : 10 v/v). Three to four changes of the same solvent are usually adequate for destaining. In the fifth bath the gels are left overnight, and the solvent is once more changed the next morning.

The gels are then wrapped in thin polythene sheets to avoid contamination of the glass platform of the machine with the solvent mentioned. One gel is placed on the glass platform with a sheet of "Dalcopy" paper on top of it with the sensitized side face down. The lid is pressed firmly down to locking position and the timer operating device is started by pushbutton.

Depending on the type of protein separation either negative or positive prints can be made, the latter being done by repeating the process with the same printing paper.

The processing unit is part of the machine and special Dalcopy developer and stabilizer are used. Dalcopy paper must at all times be fed into the automatic processing unit with the sensitized side face down.

The working space of the machine is 55×25 cm and four gels can be copied at the same time.

RESULTS AND DISCUSSION

Gels with haemoglobin and serum protein separation of several domesticated species have been investigated and it was found that all gels could be used for instant printing the only exception being the amylase gels.

Exposure times for starch gel prints — "Dalcopy"

Serum fraction	Negative, sec	Positive, sec	Paper
Bovine haemoglobin	30	5–15	Dalcopy Gloss soft grade (1)
" transferrin	20	15	"
" albumin	20	20	"
Equine haemoglobin	30	5–15	"
" transferrin	15	20	"
" albumin	20	20	"
" esterase	1.5	8	Dalcopy Gloss normal (2)
" pre-albumin	20	25	Dalcopy Gloss soft grade (1)
" transferrin	20	25	"
Ovine haemoglobin	30	5–15	"
" transferrin	10	25	"

The establishment of the best exposure times was based on gels with the normal thickness of 3 mm. A decrease in thickness would possibly give better prints but very thin gels are usually difficult to handle. Prints of gels rendered transparent give clearer readings, but for routine analyses the preparation of transparent gels is not necessary.

The optimal exposure times have been worked out by trial and error for both negative and positive prints and are given in the Table for different serum fractions together with the type of paper giving the best prints.

Especially valuable gels, e.g. with samples from international comparison tests or with newly detected types can be fixed permanently with "Hypo" photographing permanent fixing solution. All other prints have been stapled on the working sheets without extra fixing and after filing for a period of more than a year no fading whatsoever could be observed. It is not necessary to include an example of the working sheet in this note because these or similar sheets are used in most laboratories. Copies made by the Dalcopy machine have actually been accepted by printers for publishing.

ACKNOWLEDGEMENTS

The author wishes to thank Mr. Johnstone from the Firm Holtung & van Maasdyk, Distributors of Dalcopy, Johannesburg for his assistance in the copying of gels.

REFERENCE

De Ligny, W., 1968. A routine method for the preservation of starch gels, *Immunogenetics Letter*, 5, 128–129.

XITH EUROPEAN CONFERENCE ON ANIMAL BLOOD GROUPS AND BIOCHEMICAL POLYMORPHISM
Warsaw, July 2nd–6th, 1968

Allotypes (Serum Antigens) in Farm Animals

J. RAPACZ and Judith HASLER

University of Wisconsin, Department of Genetics and Department of Meat and Animal Science, Madison, Wisconsin, U.S.A.

SUMMARY. This paper presents our attempts to detect allotypes (serum antigens) in cattle, pigs, mink and sheep. It proved possible to demonstrate that isoimmunization can result in production of antibodies. Thus serum proteins of the species studied are sufficiently different in structure to induce serum protein isoimmunization. At least eleven antigens were discovered in cattle, representing variety of proteins: three macroglobulins, two high density lipoproteins, one albumin, one intra-erythrocyte non-hemoglobin protein. In mink five allotypes were found which represent α-, β- and γ-globulins. Five isoprecipitins were produced in sheep. A large number of allotypes were detected in pigs by using isoimmunosera. Within this group the most interesting are two low density lipoprotein allotypes (LDLpp-1 and LDLpp-2) which are determined by allelic genes.

INTRODUCTION

The existence of many polymorphic forms of molecules, first described for erythrocyte antigens, was later recognized in serum proteins. Genetically controlled variations have been found within many functionally groups of proteins through the application of electrophoretic methods.

The possibility, that serum proteins differed in antigenic specificity, was long recognized as potentially useful for the study of gene action, but however contributed little until recent years, due to difficulties of distinguishing single antigenic differences in proteins.

Two technical advances, immunodiffusion and immunoelectrophoresis opened the way to examination of these differences. In these methods antigen-antobody gel precipitation reactions are essential to demonstrate differences. This presentation will describe our attempts to detected allotypes (serum antigens) in cattle, pigs, mink and sheep.

MATERIAL AND METHODS

Preparation of antigen. An ammonium sulphate fraction and an alum precipitate of serum protein and immunoglobin coated on to red cell were used with complete adjuvant as inocula to produce iso- and heteroprecipitins in cattle and rabbits against cattle serum antigens. Isologous whole serum and immunoglobulin coated red cells with complete adjuvant were used in pigs, mink and sheep to produce isoprecipitins.

Cattle. In the first group, where the ammonium sulphate fraction was used, six cows were immunized. The second group was composed of four cows and alum precipitate of serum was used for injections. Five cows were injected with immunoglobulins coated on red cells. Each recipient received from five to nine intramuscular injections at 7 to 25 days intervals.

The antisera were tested against individual serum by a double diffusion test on microscope slides, using the technic described by RAPACZ *et al.* (1968). A modification of the immunoelectrophoresis procedure described by HIRSCHFELD (1960) was used.

RESULTS

Cattle. Table 1 presents allotypes which were found in cattle. A short characteristic of allotypes presented in this Table shows that in cattle 11 different antigens were recognized. They belong to different groups of proteins; three macroglobulins, two lipoproteins, two gamma-globulins, two immunoglobulins, one albumin and one intra-erythrocyte non-hemoglobin protein. Study with details were carried out on three of them: Mc-1, macroglobulin, RAPACZ *et al.* (1968), Ec-1 (intra-erythrocyte non-hemoglobin antigen) and Lpc-1 (high density α-lipoprotein). Further investigation on the other allotypes is on the way.

Pigs. Table 2 presents allotypes which were found in pigs. Three series of isoimmunization, using 22 pigs, resulted in production of 12 antisera with isoprecipitins, which differentiate at least 11 antigens.

Within this group the most interesting are two β-lipoprotein allotypes (LDLpp-1 and LDLpp-2) which are determined by allelic genes (paper in preparation). These alleles as demonstrated by gel precipitate test in agar, control synthesis of partly common antigens, located on two independent molecules.

Mink. In mink five allotypes were found, using isoprecipitins, obtained from antisera, produced in six series of immunization. In two series of heteroimmunization where rabbits and ferrets were used, all eight antisera completely lost activity when fractionated by absorption. The two strongest isoprecipitins were obtained, when young females, called "misses", i.e. animals which had lost their kits, were immunized with serum of their males. When immunoelectrophoresis was applied, both of them were found in the area of α-globulins.

Sheep. In attempt to produce isoprecipitins in sheep, two series of injections were carried out. In first whole serum with complete adjuvant was used. In second series immunoglobulins coated on red cells were injected.

TABLE 1

Allotypes (serum antigens) in cattle

Serum No.	Source of antibody	Tentative nomenclature	Type of serum antigen	Strength of reaction	Frequency
		(1) Mc-1 (166-1)	Macroglobulin (Mg) (Immunoglobulin IgM?)	Med. or weak	3–76%
166 (11-15-65)			MW 200,000 19S	Weak	>60%
	Isoimmune	(2)— (166-2)	MW 200,000 19S		
166 (8-19-66)		(3)— (166-3)	MW 200,000 19S	Very weak	<40%
		(4) HD Lpc-1 (306-1)	High density lipoprotein, α-Lp 1063-1210 MW < 200,000	Med. or weak	<20%
306 (12-22-65)	Isoimmune	(5)— (306-2)	Lipoprotein	Very weak	?%
		(6) EC-1 (306-3)	Intra erythrocyte protein (non-hemoglobulin) MW about 80,000	Med.	15–25%
802 (12-8-65) 779 (4-5-65)	Isoimmune	(7) 9	α-Globulin	Med.	<1%
604 G-6 (3-17-67)	Heteroimmune	(8) 4	Variable γ-globulin (IgA?) present in serum and milk at the same time	Med.	Variable%
766 (11...66)	Naturally occurring	(9) 5	Albumin	Med.	<1%
98A 106S 045	Naturally occurring	(10) 6	Immunoglobulin?	Weak	20%
Lidka serum produced in Poland	Isoimmune	(11) Igc-1	Immunoglobulin (IgG) MW 160,000 7S	Med.-weak	15%

In this experiment 12 sheep were used five of which gave positive response. As shown in Table 3, two antigens are carried by immunoglobulins and three detected by precipitation with whole serum are unidentified.

DISCUSSION

The purpose of this investigation was to initiate study of allotypes in farm animals. After many attempts by different laboratories, it was rather surprising, to detect serum antigens existing in so many forms.

TABLE 2

Immunization against whole serum and immunoglobulins coated on erythrocytes in pigs

Series	Donor	Recipient	Injected material	Results
I	Pooled sera of 5 pigs (Chester)	H-1-6	Pooled sera with complete adjuvant	Anti-α-globulin
	Pooled sera of 5 pigs (Poland China)	C-2-10	”	Very weak antibodies
	Pooled sera of 5 pigs (Hampshire)	C-16-4	”	Two antibodies against lipoproteins anti-α-globulin
	Pooled sera of 2 pigs (Chester and Poland China)	C-11-2	”	Precipitins against 3 unidentified antigens
II	C-22-4	C-26 4	Pooled sera with complete adjuvant	Precipitins against 3 unidentified antigens
	C-26-3	H-19-6	”	Preciptins against an unidentified antigen
	H-16-4	C-23-8	”	”
	P-29-4	H-16-6	”	”
III	C-26-4	C-22-4	Immunoglobulins with complete adjuvant	Precipitins against immunoglobulins
	C-23-8	H-16-4	”	”
	H-16-6	P-29-4	”	”

TABLE 3

Immunization against whole serum and immunoglobulins coated on erythrocytes in sheep

Series	Donor	Recipient	Injected material	Results
I	Shropshire 581	Hampshire 719	Serum with complete adjuvant	A precipitin against 1 unidentified antigen
	Targhee 4-573	Southdown 7-988	”	”
	Hampshire 720	Southdown 7-1014	”	”
II	Shropshire 298	Southdown 7-1013	Immunoglobulin with complet adjuvant	A precipitin against immunoglobulins
	Hampshire 945	Targhee 4-564	”	”
			”	”

The allotypes found in cattle suggest that each functionally different group of serum proteins is determined by many genes which can produce altered antigenic specificity. It is evident from complete study on Mc-1, Ec-1 in cattle, Lpp-1 and Lpp-2 in pigs and the α-globulin antigens in mink, that allotypes are inherited in simple Mendelian fashion.

The existence of many molecules the presence of which was difficult to demonstrate previously, could be very important in explanation of their function. Since these allotypes belong to different groups of proteins, it will be possible to study their association with enzymes and hormones, or other important forms of macromolecules and explain the relation between the genes and the interaction of these substances. The best example of this assumption are two lipoproteins (Lpp-1 and Lpp-2) in pigs, where the completed study has revealed how two allelic genes regulate the synthesis of macromolecules. It is interesting to note, that product of alleles, which is in part identical, is not produced on one molecule and in heterozygote pigs does not contribute to the synthesis of the same lipoprotein molecule.

The results contrast with present knowledge about A and B antigens of the human blood groups, determined by allelic genes which occur on the same mucopolysaccharide molecules. The discovery of the new Ec-1 system (intra-cellular non-hemoglobin protein) is as interesting as unexpected. It is the best example the many unknown genetic controlled macromolecules which are waiting to be discovered.

REFERENCES

RAPACZ, J., KORDA, N. and STONE, H. W., 1968. Serum antigens of cattle I. Immunogenetics of a macroglobulin allotype, *Genetics*, **58**, 387–389.

HIRSCHFELD, J., 1960. Immunoelectrophoresis procedure and application to the study of group-specific variations in sera, *Science Tools*, **7**, 18–25.

DISCUSSION

D. R. OSTERHOFF: Have you done any studies with the allotypes in twins. And can you identify I'm sure, that you identify monozygous twins.

J. RAPACZ: Two allotypes: Mc-1 and Lpc-1 were included in our study. Tests of these antigens in twins of various breed indicated that they are useful genetic markers for diagnosing zygosity. Serum of each of 25 pairs of twins judged monozygotic by blood type, hemoglobin and transferrin type, had identical reactions with anti-Mc-1. In contrast ten of 38 pairs of dizygotic twins from which 37 pairs were chimeras gave discordant reactions with anti-Mc-1. These data argue strongly against chimerism of the Mc-1 factor. Frequency of Lpc-1 was low in studies twins but also showed concordance with other tests for zygosity with exception of one pair where other tests did not exclude monozygosity and test for Lpc-1 allotype showed presence in one, and absence in the other co-twin.

L. PODLIACHOUK: Are the antibodies which cause the hemolytic disease of type I_gG or I_gM?

J. RAPACZ: During my first visit in the U.S.A. I was not able to check the type of hemagglutinins present in normal sera of the females which lost offspring or did not litter. Especially it would be interesting to analyse these two sera Nos. 1 and 90. During my second visit we found for the first time hemolysins in a serum of one mink (T-520). We separated this serum using Sephadex G-200. Hemolysins were recovered from the second peak what would indicate that antibodies were of the 7S type. We did not proceed those studies further.

CHAPTER 1

BLOOD GROUPS AND BIOCHEMICAL POLYMORPHISM IN CATTLE

XITH EUROPEAN CONFERENCE ON ANIMAL BLOOD GROUPS AND BIOCHEMICAL POLYMORPHISM
Warsaw, July 2nd–6th, 1968

Structure of Loci Controlling Complex Blood Group Systems in Cattle

J. BOUW and A. FIORENTINI

Stichting Bloedgroepen Onderzoek, Landbouwhogeschool, Wageningen, The Netherlands
Istituto di Zootecnia Generale, Universita di Milano, Milano, Italy

SUMMARY. In this report a series of cases of irregular transmission of blood groups of the system B in cattle will be presented. The irregularities are considered to be a result of crossing-over within the chromosomal region which controls this blood group system. On the basis of a series of cases of crossing-over a tentative map is proposed in which a linear order is presented for the genes or gene complexes which control the blood groups of this system.

INTRODUCTION

In their studies on the inheritance of blood group factors in cattle STORMONT *et al.* (1945) observed that some factors were regularly transmitted in combination with each other. The combinations of factors which are inherited as genetic units are called blood groups or phenogroups.

In the first phases of the blood grouping work in cattle no definite answer was given to the question whether these phenogroups were controlled by non-separable genetic units or by a series of closely linked genes, each one of which controlled a single factor.

STORMONT (1955), BOUW *et al.* (1964) and several others reported a number of cases of abnormal transmission of blood groups of the B system. BOUW *et al.* proposed that most of these irregularities could be explained on the assumption of linked genes in which each gene corresponded to a part of the complex blood group.

Following these initial observations of the blood groups of the B system of 30,000 Dutch cattle were compared with those of their parents. In these comparisons a modification in the structure of the B locus could definitely be established in 57 cases. Out of this material a series of cases has been selected to establish the linear order of units of the chromosomal region which controls the blood group system.

MATERIAL AND METHODS

In the period 1960–1967 the parentage of over 30,000 Dutch cattle was controlled by blood grouping tests. By comparing the blood groups of the B system in parents and offspring it was clearly evident that the complex phenogroups were usually transmitted as genetic units.

The phenogroups were found to be transmitted as such through several generations of animals. Phenogroups which had occurred in intensively used breeding animals were found to have a relatively high frequency. Examples of such distributions of phenogroups were presented by BOUW (1958). After the reported observations on irregular transmission an analysis was made of the tested material. In this analysis it turned out that a relatively large number of phenogroups of the B system as such could not be traced back to the ancestors of the tested animals. Out of this material 57 cases could be extracted in which it was possible to point out in which individual a modification had occurred. Examples giving evidence for these modifications were presented by BOUW *et al.* (1964).

An analysis of the 57 cases showed that several cases were in fact more or less identical or showed a close similarity. Those cases which were considered to be essential for a study of the genetic control of the blood groups of the B system are presented in Table 1.

TABLE 1

Recombinations of blood groups of the B system in cattle

Genotype parent	Recombinated type	Genotype parent	Recombinated type
1a $BGKO_xY_2A'O'/I'$	$BGKO_xA'O'$	10 $BO_1/OQJ'K'O'$	OQK'
1b $BGKO_xY_2A'O'/I'$	Y_2I'	11 $BO_1Y_2D'I'/O$	O_1I'
2 $BGKO_xY_2A'O'/PI'$	$BGKO_xY_2A'I'O'$	12 BO_1Y_2D'/E'_3H_8	$Y_2D'E'_3H_8$
3 $BGKO_xY_2A'O'/I_2$	$BGKO_xY_2O'$	13 $GE'_1/Y_1D'G'I'$	$D'G'I'$
4 $BGKO_xY_2A'O'/E'_3H_8$	$BGKO_xA'E'_3O'H_8$	14 $GY_2E'_1/O_1QD'$	O_1D'
5 $BGKO_xY_2A'O'/$ $/O_xY_2D'E'_1F'O'$	$BGKO_xY_2A'D'O'$	15 $GY_2E'_1/E'_3H_8$	$Y_2E'_1$
6 $BO_3Y_2A'E'_3G'P'/$ $/E'_3H_8$	$Y_2E'_3G'$	16 $GY_2E'_1/I'$	GY_2
7 $BO_3Y_2A'E'_3G'P'$ $/I_1QE'_1I'$	$O_3QA'E'_1$	17 $O_xY_2D'E'_1F'O'/I_2$	$I_2Y_2D'E'_1$
8 $I_1QE'_1I'/I_2$	I_2I'	18 $BGKO_xY_2A'O'/$ $/GO_xY_2A'D'E'_3F'$	O_xY_2A'
9 $I_1OJ'K'O'/I_2$	$I_2OJ'K'O'$	19 $GO_xE'_3F'O'/$	GO_xO'

In this Table the antigenic factors are written in alphabetical order as this is usually done in blood grouping work in cattle. In the genotypes of the parents the segregation which is normally expected is indicated with a solidus between the two phenogroups. The recombinations which were found to be transmitted as such to the descendents are written to the right of these genotypes. To this material two cases have been added which were found in other laboratories. Case No. 10 was

TABLE 2

Crossing-over in the bloodgroup system B of cattle

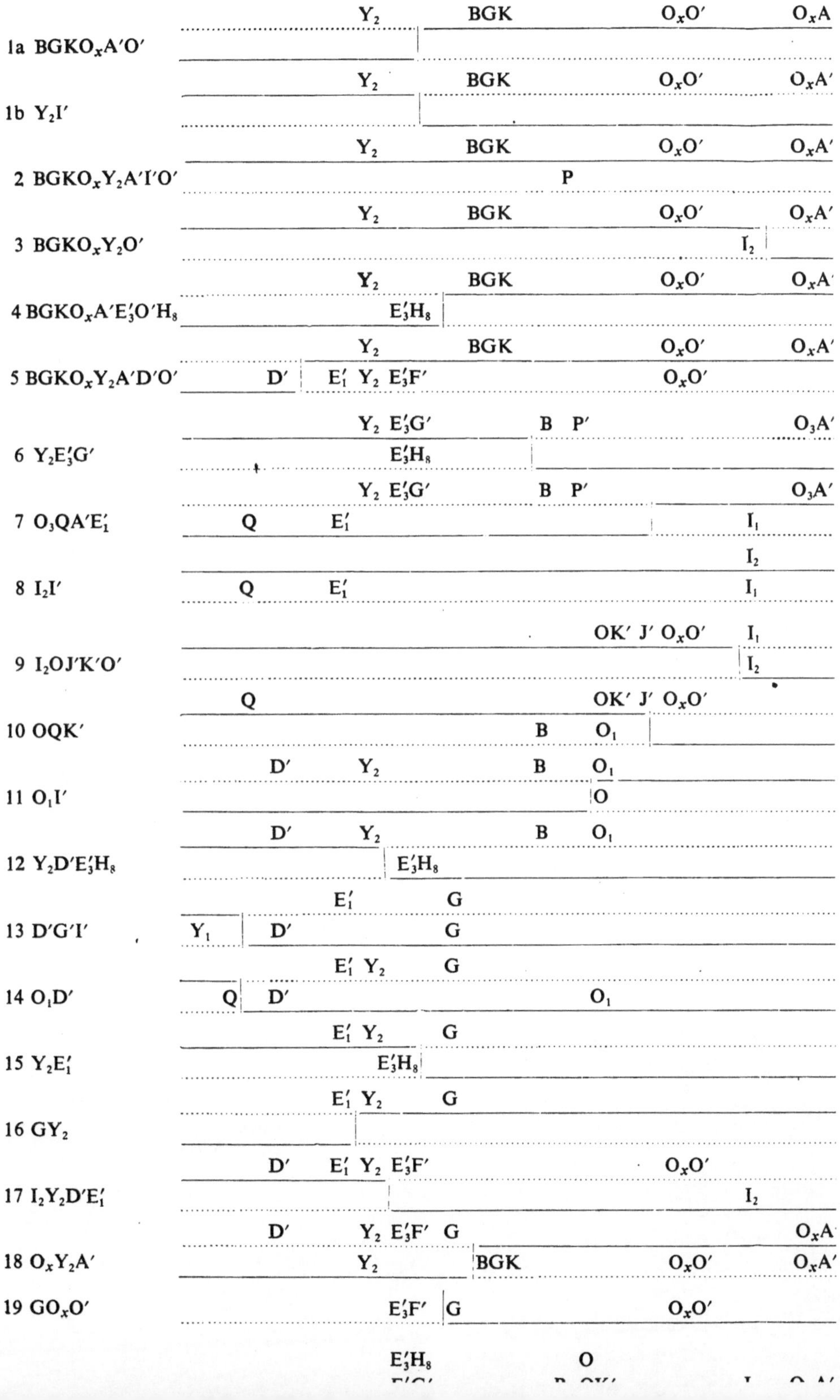

reported by MOUSTGAARD and NEIMANN-SØRENSEN (1962), case No. 19 was established in cooperation with ERHARD (1967) in German Fleckvieh.

The cases 1a and 1b have been included to show that cases which may look very different when the newly established phenogroups are considered can in fact be considered as complementary to each other (see also Table 2).

RESULTS AND CONCLUSIONS

The recombinations of Table 1 are presented again at the left-hand side of Table 2. In the central part of this Table the factors are written on 2 parallel lines which can be considered as the respective parts of the chromosomes of each case. In this second Table the order of the factors on the lines is based on the conclusions which are written at the right-hand side of the figures.

The complex of factors originating from one parent is written above the topline of each figure, the complex from the other parent above the lower line. The factors above the solid line were transmitted to the descendant.

On the basis of the conclusions at the right-hand side of the Table a line has been drawn at the bottom of Table 2 which can be considered as a map for this region of the chromosome. The marks ▬ and · on this line can be considered as indicators for the controlling elements of the antigenic substances. On this bottom line the 16 places indicated with ▬ have been established definitely on the basis of the available data. The position for the factors B, P′ and J′ has been indicated with · . For P′ and J′ this means that a position one space more to the left or the right is still possible. For B it was not yet possible to decide whether the symbol should be written either left or right or above BGK.

For 8 antigenic factors which are also known to belong to the blood group system B no conclusive data are available as yet, partly because these factors are rare in Dutch cattle and partly because the reagents for some of these factors have been used for only a small part of the material.

From the available material it has become evident that the frequency of observed breakages within the detectable parts of the complexes depends upon:

(1) The frequency of the group in the tested material.

(2) The distance between the extreme ends of the detectable factors.

In the total material of 30,000 parents–offspring combinations the complex $Y_2BGKO_xA'O'$ was found in a frequency of 10%, while for this complex 11 cases of breakage were observed. This means a frequency of breakages of 0.3%. In the same material the group $Y_1D'G'I'$ had a frequency of 0.5%, while 7 cases of breakage were found. The frequency of breakage within $Y_1D'G'I'$ can thus be calculated as 4.6%.

The available data and material were considered to be insufficient to start accurate calculations about the percentages of breakage between the various parts of the chromosome.

DISCUSSION

All cases of irregular transmission of blood groups which have been presented here can be explained very well on the basis of crossing-over in the chromosomal region which controls the B system. This conclusion does not, however, exclude other explanations for irregular transmission of blood groups. In the bottom line of the Table some factors or complexes of factors are written above each other. It is quite possible that a breakage between some of these structures will be found in the future. On the other hand the differences in serological specificity of those structures which are written above each other seem to be rather slight. These differences can also be explained by assuming slight changes within the controlling genes. On the bottom line various substances indicated by E′, O and Y are found at more than one place. FIORENTINI and BOUW (1970) will explain at this Conference how reagents by which these substances are detected can react with 2 or more different substances, while they can be independent from each other from a genetical point of view. None of the antigenic substances which have been indicated more precisely by the use of a subscript has been written more than once on the bottom line. This does not mean that the reagents detecting these factors are absolutely specific.

Further work in this field may reveal that the reagents for these factors also detect different antigenic factors controlled by different parts of the chromosome. This possibility must most certainly be taken into consideration when animals belonging to other breeds of cattle are tested with these reagents. Further investigations should make it possible to improve the precision of our knowledge of the relevant positions and the frequencies of recombinations of the structures in this region of the bovine chromosome.

ACKNOWLEDGEMENTS

The authors are indebted to Mr. C. BUYS who very critically and accurately collected all cases of irregular transmission which he observed in Dutch cattle.

REFERENCES

BOUW, J., 1958. Blood group studies in Dutch cattle breeds, Thesis, Utrecht. Veenman en Zn., Wageningen, pp. 84.

BOUW, J., NASRAT, G. E. and BUYS, C., 1964. The inheritance of blood groups in the blood group system B in cattle, *Genetica*, **35**, 47–58.

ERHARD, L., 1967. Personal communication.

FIORENTINI, A. and BOUW, J., 1970. Specificities of antibodies detecting antigenic substances on cattle red blood cells, XIth European Conference on Animal Blood Groups and Biochemical Polymorphism, pp, 117–122, The Hague–Warszawa.

MOUSTGAARD, J. and NEIMANN-SØRENSEN, A., 1962. Possible interallelic crossing-over in the bovine B blood group system, *Immunogenet. Letter*, **2**, 62–64.

STORMONT, C., 1955. Linked genes, pseudo alleles and blood groups, *Amer. Nat.*, **89**, 105–116.
STORMONT, C., IRWIN, M. R. and OWEN, R. D., 1945. A probable allelic series of genes affecting cellular antigens in cattle (Abstract), *Genetics*, **30**, 25–26.

DISCUSSION

J. RENDEL: What is the frequency of B-system recombinants?

J. BOUW: This is largely dependent upon the completeness of the battery of blood-typing reagents and the breed or breeds of cattle under study. In my own material, I previously suggested (1964) a frequency of 1 : 500 but believe now this should be halved (1 : 250).

A. JAMIESON: Now that we have learned about many examples of genetic crossing-over within the cattle **B** locus, is it necessary to reconsider the present application of genetic information at this locus in cattle parentage tests? Should the use of this locus be suspended or modified without delay?

J. BOUW: The observations on crossing-over can most certainly be considered as a warning for the laboratories using blood groups for parentage control. For those who are aware of the possibilities of crossing over such data are only stimulating the interest in studying and comparing pheno groups in parents and offspring. For those who wish to keep the matters more simply we can say that it has never been observed in cattle that a blood group factor not present in one of the parents was found in an offspring as a result of crossing-over.

XITH EUROPEAN CONFERENCE ON ANIMAL BLOOD GROUPS AND BIOCHEMICAL POLYMORPHISM
Warsaw, July 2nd–6th, 1968

A Probable Crossing-over Between Two B-alleles of Cattle Blood Groups

J. SELLEI and J. RENDEL

Department of Animal Breeding, Agricultural College, Uppsala 7, Sweden

During the past decades several mutation and recombination-like events have been recorded by studies of the inheritance of the B phenogroups in cattle (STORMONT, 1963; BOUW *et al.*, 1964; LARSEN, 1966).

Lately a rather informative case was detected in our laboratory at a routine test for parentage control. One young bull (5) was found to have a combination of B factors which had never before been observed in Swedish cattle.

The B types of No. 5 and its parents are given in Fig. 1.

♂ (77)	$O_2QJ'K'O'/BO_3YA'E'_3$	♂ (5)	$BO_1YD'E'_3J'K'O'$
♀ (22)	$BO_1YD'/BO_3YA'E'_3$		

FIG. 1.

The genotype of the sire (77) was confirmed by pedigree studies and by testing his offspring along with their dams. The mother (22) had previously produced one BO_1YD' offspring in a mating to a sire with the B locus genotype $-/-$. Her genotype was therefore almost certainly $BO_1YD'/BO_3YA'E'_3$. The three alleles of the parents are very common in the Swedish Red and White Cattle (SRB) (RENDEL, 1958) to which these animals belong. All the available breeding data were controlled in the herd and at the A. I. centre and there was no indication of illegitimacy. So far not a single animal has been encountered with the bull's (5) exceptional phenotype among the more than 15,000 animals of the SRB breed tested in our laboratory. In case of illegitimacy one of the parents would therefore have had to carry an extremely rare B-allele.

To exclude technical errors new samples of each animal were tested. Repeat samples were also sent to the Oslo laboratory for checking our results. (We take this occasion to thank Dr. BRAEND for his kind cooperation).

When analysing this parentage it can be concluded that the antigens: B, O_1, Y and D' must have been transmitted by the cow (22). The remaining antigens: E'_3, J', K', O' most probably were inherited from the sire (77). The E'_3 antigen might of course alsó have come from the dam in which case the J', K', O' antigens would have need to come from the sire (77). However this explanation of the transmission of the E_3' antigen requires simultaneous changes in the B loci of both the sire and the dam. As this seems extremely improbable so this possibility will be disregarded.

It seems more reasonable that one crossing-over happened during the process of spermatogenesis in the sire. In this case the following assumptions must be made for a satisfying explanation for the observed phenotype (Fig. 2).

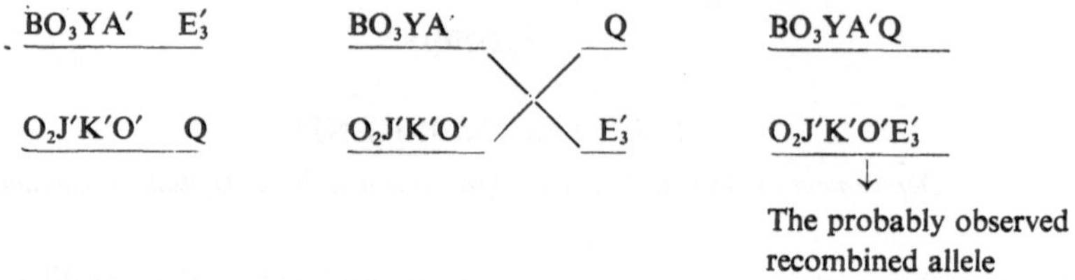

FIG. 2. The probable crossing-over during the process of spermatogenesis in the sire.

(1) The DNA sequences controlling the Q and E_3' antigens have the same positions in the compound B-locus.

(2) The Q and E_3' "sequences" must lay terminally as contrasted to the "BO_3YA'" and "$O_2J'K'O'$" sequences.

REFERENCES

STORMONT, C., 1963. Mammalian immunogenetics, Proc. 11th Intern. Congr. Genet., **3**, 715-722.

BOUW, J., NASRAT, G. E. and BUYS, C., 1964. The inheritance of blood groups in the blood group system B in cattle, *Genetica*, **35**, 47–58.

RENDEL, J., 1958. Studies of cattle blood groups. IV. The frequency of blood group genes in Swedish cattle breeds, with special reference to breed structure, *Acta Agr. Scand.*, **VIII**, 3.

LARSEN, B., 1965. Irregular inheritance of a phenogroup in the B system of cattle, *Immunogenetics letter*, **4** (1), 10–11.

XITH EUROPEAN CONFERENCE ON ANIMAL BLOOD GROUPS AND BIOCHEMICAL POLYMORPHISM
Warsaw, July 2nd–6th, 1968

Specificities of Antibodies Detecting Antigenic Substances on Cattle Red Blood Cells

A. FIORENTINI and J. BOUW

Istituto di Zootecnia Generale, Universita di Milano, Milano, Italy
Stichting Bloedgroepen Onderzoek, Landbouwhogeschool, Wageningen, The Netherlands

SUMMARY. In this report an analysis is presented of the specificities of antibodies which are in use in studies of blood groups of the system B in cattle. This analysis is based upon observations which are made in studies of the genetic control and the serologic reactivity of the antigens which are detected with these antibodies. On basis of the results of these analyses the authors will demonstrate that several populations of antibodies which are used as reagents are reacting specifically with 2 or more distinctly different antigenic factors.

INTRODUCTION

Already from the experiences of LANDSTEINER and VAN DE SCHEER (1936–1945) is known that specificities of antibodies are not always absolute. EISEN *et al.* (1964) demonstrated that even the antibodies which are produced against single hapten species can be mixtures of antibody molecules with different specificities. STORMONT (1955) and OWEN (1958) suggested that also for the blood group systems of cattle the relation between genes antigens and antibodies can be differing from the one-to-one-to-one correspondence. STONE and MILLER (1961) and GROSCLAUDE (1965) demonstrated that certain reagents for the S system reacted with 2 or 3 different antigenic factors, while absorptions with each of these factors exhausted all reactivity for cattle red blood cells.

During their investigations on the map for the chromosomal region which controls the blood group system B in cattle, BOUW and FIORENTINI (1970) were faced with the question whether also for this system one specificity of antibodies could react with different antigenic properties. In this report a series of data will be presented which demonstrate that some reagents for the blood group system B react with 2 or more substances which are differing in their genetic control and their serologic activities.

MATERIAL AND METHODS

The studies on specificities of reagents for the blood group system B are based upon:

(1) The analysis of the composition of the complex blood groups (phenogroups) of the blood group system B. This analysis was based upon the results of tests and genetic information of around 50,000 Dutch and 8000 Italian Brown Swiss cattle.

(2) The analysis of a series of cases of crossing-over in the chromosomal region which controls the blood group system B. In total 77 cases of crossing-over were used. Ten of these cases were brought together from the literature by STORMONT (1964), 10 cases were observed in Italian Brown Swiss and 57 in Dutch cattle.

(3) The analysis of the results of over 200 iso-immunizations with cattle red blood cells. The techniques and results of these immunizations have been reported by FIORENTINI and GALIZZI (1968).

(4) The analysis of series of iso-immune sera for cattle red blood cells on basis of absorptions of these sera. The techniques for these absorptions were first described by FERGUSON *et al.* (1942).

RESULTS AND CONCLUSIONS

From an analysis of the well established phenogroups of the B system in Italian Brown Swiss and Italian and Dutch Fresian cattle, it appears that some antigenic factors condition the presence of others. All antigenic factors, the conditioning as well as the conditioned, are detected by their reactivity with iso-immune antibodies, which are sufficiently homogenous to be considered as reagents. In Table 1 a series of such conditioning and conditioned factors is presented.

TABLE 1

Conditioning and conditioned factors of the B system

Conditioning factors	Conditioned factors	Conditioning factors	Conditioned factors	Conditioning factors	Conditioned factors
K	B–G	F′	E_3'	O′	O_x
I_1	I_2	$M_7(H_8)$	E_3'	Y′	Y_2
O_1	O–O_3–O_x	I′	H_4	G″	E_3'
A′	(O_3)–O_x	J_1'	O′–O_x	$M_6(H_{12})$	E_2'–E_3'–F′
E_1'	E_2'–E_3'	K′	O–O_3–O_x	M_{13}	J_2'–O_3–O_x

These conditioning and conditioned factors can form simple phenogroups by themselves like BGK–O_xA′–O_xO′–I′H_4, or participate in various combinations to the constitution of complex phenogroups like BGKO_xO′H_4–BGKO_xA′O′H_4–G$O_x$$E_3'$F′O′$H_4$.

From Table 1 it is evident that the presence of some factors can be conditioned

TABLE 2

Reaction patterns for O specificities

Phenogroups	Conditioning factors	Conditioned factors	Reagents						
			O_1	O	O_3	O_x	K′	O′	A′
GO_1–BO_1–$O_1T_1Y_2E'_3F'H_4$	O_1	O–O_3–O_x	+	+	+	+	—	—	—
$OJ'_1K'O'$–OK'	K′	O–O_3–O_x	—	+	+	+	+	—	—
O_3A'	A′	O_3–O	—	—	+	+	—	—	+
O_xO'–$O_xO'H_4$	O′	O_x	—	—	—	+	—	+	—
O_xA'	A′	O_x	—	—	—	+	—	—	+

TABLE 3

Reaction patterns for E′ specificities

Phenogroups	Conditioning factors	Conditioned factors	Reagents						
			E'_1	E'_2	E'_3	E'_1	F′	M_7	G″
$GY_2E'_1$–$I_1E'_1$–$Y_2D'E'_1$	E'_1	E'_2–E'_3	+	+	+	+	—	—	—
$O_xE'_3F'O'$–$OT_1E'_3F'K'$	F′	E'_3	—	—	+	—	+	—	—
E'_3M_7G''–$I_1E'_3G'G''H_4M_7$	M_7	G″–E'_3	—	—	+	—	—	+	+
E'_3G''–$E'_3G'G''$	G″	E'_3	—	—	+	—	—	—	+

by various different antigenic properties. The majority of these "conditioned factors" can be brought together into two groups, the O and the E′ specificities.

In Tables 2 and 3 the reaction patterns for the specificity groups O and E′ of Table 1 are demonstrated for a number of phenogroups.

These two Tables are demonstrating clearly that the reaction patterns for reagents like O_3 or O_x and E'_2 or E'_3 are conditioned by a large variety of different antigenic substances in the investigated breeds of cattle.

From observations of crossing-over cases in the blood group system B it has become evident that the loss or gain of reactivity for certain antigenic factors is usually accompanied by the loss or gain of reactivity for others BOUW and FIORENTINI (1970) will demonstrate that the conditioning substances O_1–K′–A′ and O′ of the O specificity group and E'_1–F′ and $M_7(H_8)$ of the E′ specificity are controlled by different parts of the chromosome. Their data also demonstrate that the loss or gain of these conditioning substances is usually accompanied by the loss or gain of factors which are according to Table 1 — conditioned by these substances.

In Tables 4 and 5 a demonstration is given of the possibilities to produce reagents for the factor E'_3, of which factor is known that its presence can be conditioned by several different conditioning substances (see Table 1). The antisera presented in these Tables are produced by using different types of donors, either separately or in combinations. All presented antisera were produced in recipients which lacked reactivity for reagents of the E′ specificity group. In low dilutions the four

TABLE 4

Pattern of reactions of E′ antisera before absorption in different dilutions

Donors	$E_1'\ M_6\ F'\ M_7$			E_1'			M_6			F'		
dil.	$\frac{1}{4}$	$\frac{1}{16}$	$\frac{1}{64}$	$\frac{1}{4}$	$\frac{1}{16}$	$\frac{1}{64}$	$\frac{1}{4}$	$\frac{1}{16}$	$\frac{1}{64}$	$\frac{1}{4}$	$\frac{1}{16}$	$\frac{1}{64}$
Test-cells												
E_1'	+	+	+	+	+	+	+	+	—	+	+	—
M_6	+	+	+	+	+	±	+	+	+	+	±	—
F'	+	+	+	+	+	—	+	±	—	+	+	+
M_7	+	+	+	+	±	—	+	—	—	+	—	—

TABLE 5

Pattern of reactions of E′ antisera after absorption in low dilution

Donors	$E_1'\ M_6\ F'\ M_7$				E_1'				M_6				F'			
Abs.-cells Test-cells	E_1'	M_6	F'	M_7	E_1'	M_6	E'	M_7	E_1'	M_6	F'	M_7	E_1'	M_6	F'	M_7
E_1'	—	+	+	+	—	—	±	±	—	—	—	±	—	—	—	—
M_6	+	—	+	+	—	—	—	—	—	—	±	+	—	—	—	—
F'	+	+	—	+	—	—	—	—	—	—	—	—	—	—	—	±
M_7	+	+	+	—	—	—	—	—	—	—	—	—	—	—	—	—

presented antisera demonstrate the same pattern of reactions when tested with a variety of antigenic substances. All substances belonging to the E′ specificity were found to react with these four antisera, while no reactivity was found with cells lacking E′ specificity.

The following remarks can furthermore be made by these Tables:

(1) The reagents show the same pattern of reactions in low dilutions, while almost only the donor's cells react in higher dilutions.

(2) The antisera produced with a system of "multiple donor's" (E_1'–M_6–F'–M_7) cannot be exhausted by absorption with a single antigen.

TABLE 6

Pattern of reactions of an E_3' reagent before and after absorption

B groups donor: $BO_3Y_2A'E_3'G'P'Q''G''H_4M_8/E_3'G''M_8$
B groups recipient: $BGKO_xY_2A'O'H_4/I'H_4$

Absorbed with	—	$GY_2E_1'Q'$	$E_3'G''M_7$	$E_3'G'G''$	$O_1T_1E_3'F'K'$	$BGKO_xE_2'F'O'M_6$
Tested with:						
$GY_2E_1'Q'$	+	—	—	—	—	—
$E_3'G''M_7$	+	—	—	—	—	—
$E_3'G'G''$	+	—	—	—	—	—
$O_1T_1E_3'F'K'$	+	—	—	—	—	—
$BGKO_xE_2'F'O'M_6$	+	—	—	—	—	—
Donor	+	—	—	—	—	—

(3) The antisera produced with a "single donor" are completely exhausted after absorption with the donor's type and partly after absorption with other E′ types.

In Table 6 it has been demonstrated that it is possible to produce antisera which react with all E′ types, while all types of this specificity group are able to take out all reactivity.

The data of this Table could thus be used to demonstrate that anti-E_3' is a single specificity of antibodies which reacts with one antigenic factor — E_3'. The data of Tables 1–3 are demonstrating, however, that this "reagent" is detecting various antigenic substances which are differing in their genetic control and in their serologic activity. The data of Tables 4–5 are revealing that apparently simple and similar reagents can cover a variety of differences in specificities.

DISCUSSION

From the presented data it is first of all evident that a discussion about specificities of antibodies for cattle red blood cells is directly connected with the serologic activity of the substances which are detected with these antibodies. Furthermore it is evident that any discussion on the nature of cattle red cell antigens cannot disregard the fact that until now these antigens are defined only on basis of their reactivity with these antibodies. STORMONT (1964) pointed out the similarity of the serological behaviour of the antigenic structures of the system B of cattle and sheep and the antigen 0 of Salmonella. Today we know that as well in the 0 antigen of Salmonella (ROBBINS and UCHIDA, 1962; LUDERITZ *et al.*, 1966), as in the ABO- Lewis-MN- and probably also the Rh blood groups in man, the serologic specificities reside in the carbohydrate part of the antigen molecules (WATKINS, 1966). Most probably also the antigenic specificities of cattle red cells have to be ascribed to the polysaccharides. From our data we can only say that several populations of antibodies, sufficiently homogenous to be considered as reagents do react with various antigenic factors which are different in their antigenic specificity and which are controlled by different parts of the chromosome. The logical consequence of this is that all these different antigenic factors have a common "reacting site". This means that they have to be closely related in their biochemical structure, in which the "immunodominant" group (LUDERITZ *et al.*, 1966) is only partially responsible for the total specificity. Since we do not have any exact information on the biochemical composition and the physical dimensions of the "immunodominant units" in comparison to the total antigenic sites which determine the antigenic specificities of cattle red cells, any further conclusion on this direction should be merely academical. The authors are of the opinion that further studies in inhibition tests and on the biochemical basis of the antigenic structures will offer sufficient possibilities to elucidate these questions.

REFERENCES

BOUW, J. and FIORENTINI, A., 1970. Structure of loci controlling complex blood group systems in cattle. XIth European Conference on Animal Blood Groups and Biochemical Polymorphism, pp. 109–113, The Hague–Warszawa.

EISEN, H. N., SIMMS, E. S., LITTLE, JR. and STEINER, L. A., 1964. Affinities of anti-2,4-dinitro phenyl (DNP) antibodies induced by ε-41-mono-DNP-ribonuclease, *Fed. Proc.*, **23**, 559.

FERGUSON, L. C., STORMONT, C. and IRWIN, M. R., 1942. On additional antigens in the erythrocytes of cattle. *J. Immun.*, **44**, 147–164.

FIORENTINI, A. and GALIZZI, S., 1968. La produzione dei reagenti iso-immuni anti erythrocitari. Nota 1° Influenza dell'età e dello stato fisiologico, *Atti Soc. Ital. Sc. Veter.* (In press).

GROSCLAUDE, F., 1965. Studies on the blood group system in French Cattle Breeds, Proc. 9th Eur. An. Blood Group, 79–85.

LANDSTEINER, K. and SCHEER, J. VAN DE, 1936. On cross-reactions of immune sera to azoproteins, *J. Exp. Med.*, **63**, 325–339.

LANDSTEINER, K., 1945. The specificity of serological reactions, Harward Univ. Press, Cambridge, Mass., **14**, 310.

LUDERITZ, O., STAMB, A. M. and WESTPHAL, O., 1966. Immunochemistry of 0 and R antigens of Salmonella and related Enterobacteriaceae, *Bacteriology Rev.*, **30**, 192–195.

OWEN, R. D., 1958. Immunogenetics, Proc. Int. Congr. Genetic 10th, Montreal, I, 364–374.

ROBBINS, P, W. and UCHIDA, I., 1962. Determinants of specificity in Salmonella: changes in antigenic structure mediated by bacteriophage, *Fed. Proc.*, **21**, 702–710.

STONE, W. H. and MILLER, W. J., 1961. Naturally occurring iso-antibodies of the S blood group system in cattle, *J. Immunology*, **86**, 165–169.

STORMONT, C., 1955. Linked genes, Pseudo-alleles and blood groups, *The An. Nat.*, **845**, 105–115.

STORMONT, C., 1964. Mammalian immunogenetics. Proc. 11th Int. Cong. of Genetics, 715–722.

WATKINS, W. M., 1966. Blood group substances, *Science*, **152**, 172–180.

XITH EUROPEAN CONFERENCE ON ANIMAL BLOOD GROUPS AND BIOCHEMICAL POLYMORPHISM
Warsaw, July 2nd–6th, 1968

Antibody Titre and the Level of Gamma Globulin in Immune Sera in Cattle

Wiesława MICHALAK, Krystyna TOMASZEWSKA-GUSZKIEWICZ and M. ŻURKOWSKI

Institute of Experimental Animal Breeding, Polish Academy of Sciences, Warsaw, Poland

SUMMARY. This investigation was carried out on 92 immunizations. The distribution of proteins in the serum was determined by paper electrophoresis.

The results which were obtained in our investigations indicate: (1) that a possible interdependence exists between the level of gamma globulin in serum at the start of immunization, on the one hand, and the antibody titre obtained after immunization, on the other; (2) that there is not, however, any distinct interdependence between an increased level of gamma globulin during the course of immunization, on the one hand, and the antibody titre in serum on the other.

INTRODUCTION

In producing immune sera in cattle we know well of two primary problems which in distinct ways make the stimulation of antibodies difficult. A number of antibodies exist that have proved very difficult to stimulate. The other problem is the fact that certain animals do not respond to immunization.

The research which we have carried out has been aimed at finding out (1) whether and to what extent the level of gamma globulin in serum exerts an influence on the titre of antibody that is obtained as a result of immunization; (2) if an increased titre of antibody in the course of immunization, has a visible effect on the level of gamma globulin in serum.

MATERIALS AND METHODS

The sera examined were obtained from cattle of the Black and White Lowland breed. The pairs selected for immunization were chosen in an attempt to obtain antibodies against all the antigens present in the cattle population being examined.

Immunization was carried out using 50% suspension of blood cells in physiological salt solution and giving 4–5 injections at 7 day intervals injecting each time 20 ml of the suspension of blood cells.

The following is the total number of immunizations carried out: 49 immunizations, 36 reimmunizations and 7 second reimmunizations.

There was a time interval of two months between successive immunization courses. The level of gamma globulin in the serum was determined by the use of paper electrophoresis. Separation of proteins in the serum was achieved by use of a veronal buffer pH = 8.6 of ionic strength = 0.1. The supporting medium used was Whatman's No. 1 paper size: 3.5×35 cm. Sample insertion was 15 cm from cathodic end. Staining was with 0.01% bromophenol blue in 96% ethyl alcohol for 10 min. Destaining was carried out with a 2% solution of acetic acid. The time of electrophoretic separation was 18 hr.

The level of gamma globulin in each particular test sample was determined from the curve obtained by photometric scan of the stained electrophoretogram. The results are given as percentages.

RESULTS

Figure 1 shows an interdependence occurring between the level of gamma globulin in the serum expressed as percentage before immunization on one hand, and the antibody titre obtained after immunization on the other.

Out of 49 injected cows 6 animals did not produce any antibodies. In these animals the level of gamma globulin varied within the range of 39% to 49%. In cattle that had produced antibodies of a titre of 1 : 16 the level of gamma globulin was 34% to 57%. There were two animals, however, in which the antibody titre (1 : 512), and they were marked by a low level of gamma globulin (35% and 36%). The lowest level of gamma globulin was found in a cow, which had antibodies of a titre of 1 : 128.

The data is summarized in Table 1.

TABLE 1

Gamma globulin level (in %) before immunization and antibodies titre

Gamma globulin level in %	Antibodies titre				χ^2
	⩾ 1 : 16		< 1 : 16		
	number	%	number	%	
Below or equal 36	1	14	6	86	5.2
From 36 up to 44	10	40	15	60	3.0
More than 44	13	77	4	23	6.1

Amongst animals showing a low level of gamma globulin (⩾ 36%) there were 6 cows that had antibody titres above 1 : 16. There was the only one animal showing a low level of gamma globulin and a low level of antibody too. A different response was given by animals which before immunization were marked by a high level of

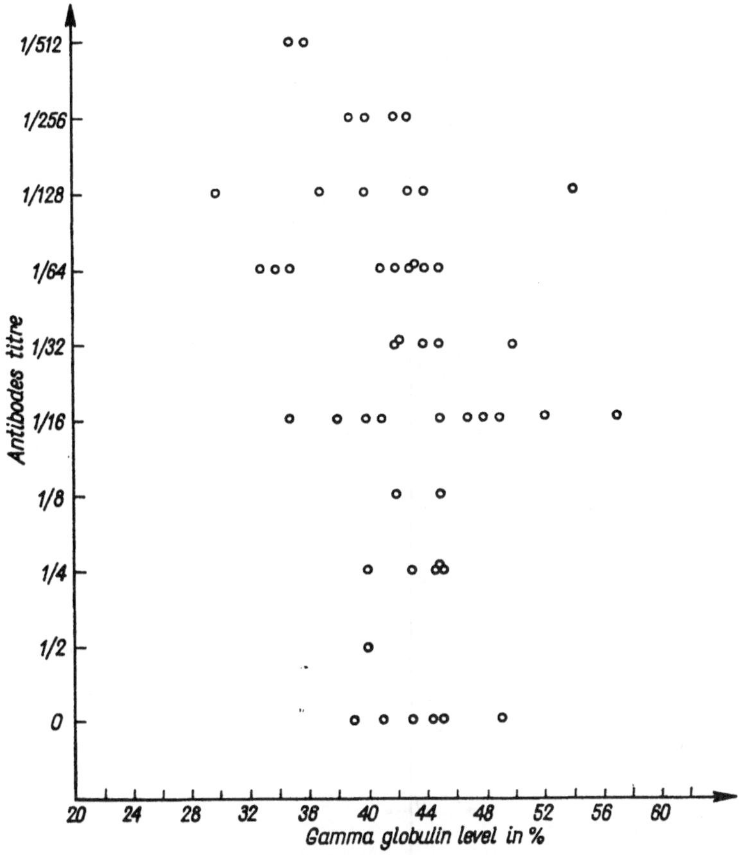

FIG. 1. Gamma globulin level in serum before immunization and antibodies titre after immunization.

gamma globulin. The majority (77%) of these recipients produced antibodies of a low titre and there were 4 cows only (23%) that produced antibodies of a titre above 1 : 16. In both cases these differences were statistically significant.

Figure 2 shows the differences between the level of gamma globulin before immunization and that after immunization depending on the antibody titre.

Out of the total number of 86 immunizations we found in 19 cases (22%) a decreased level of gamma globulin in serum in the course of immunization; in 6 cases (7%) no variation was found in the level of gamma globulin, and in 61 cases (71%) an increased level of gamma globulin was found.

Apart from two cases where a very high antibody titre was obtained (1 : 1024) and an increased level of gamma globulin of 13% and 14% was found — no differences in level of gamma globulin dependent on antibody titre were found. The increase in level of gamma globulin in serum of cows with antibody titres of 1 : 16 was of the same range as that in cows with antibody titres of 1 : 256.

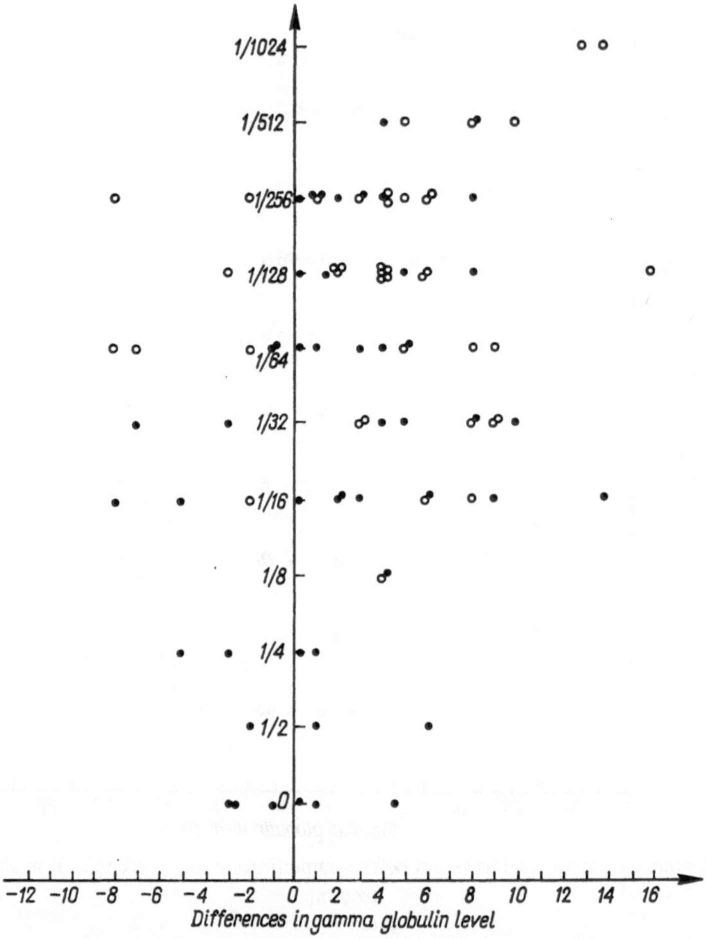

FIG. 2. Differences in gamma globulin during immunization and antibodies titre: ● immunization, ○ reimmunization.

No differences were observed in the level of gamma globulin in the immunized and in reimmunized animals.

The results obtained are shown in Fig. 2, and summarized in Table 2.

In examining cases of decreased levels of gamma globulin we observed a certain consistency in that alongside an increase in antibody titre there occurred a decreasing number of animals showing a decline of gamma globulin levels in the course of immunization.

In the group of animals marked by low antibody titre (less than 1 : 16) the difference between the number of animals showing a decreased or an increased level of gamma globulin in the course of immunization is on the border of significance ($\chi^2 = 3.7$), whereas in the group of animals of the high antibody titre (above 1 : 128) the difference is highly significant.

TABLE 2

Difference in the level of gamma globulin found out before and after immunization and antibodies titre

Antibodies titre	Difference in the level of gamma globulin				χ^2
	—		+		
	number	%	number	%	
Below or equal 1 : 16	9	47.4	15	24.2	3.7
In between 1 : 32–1 : 64	7	36.8	16	25.8	0.9
More than 1 : 128	3	15.8	31	50.0	6.8
Total	19	100.0	62	100.0	

DISCUSSION

The results which we obtained in our investigations have indicated a possible interdependence between the level of gamma globulin in serum before the start of immunization on one hand, and the antibody titre obtained as a result of immunization on the other. The animals marked by a low level of gamma globulin appear to be more likely to produce antibodies of relatively high titre. The results summarized in Table 1 constitute evidence supporting this thesis. Out of 7 recipients marked by a low level of gamma globulin ($\geqslant 36\%$), there were 6 cows that showed subsequent high antibody titre (3 animals with titres of 1 : 64, 1 cow with a titre of 1 : 128, and 2 animals with a titre of 1 : 512). But out of 17 recipients showing a high level of gamma globulin ($< 44\%$) there were 4 cows only that subsequently showed high antibody titre (2 animals with a titre of 1 : 32, 1 cow with a titre of 1 : 64, and 1 animal with a titre of 1 : 128). The differences are statistically significant ($\chi^2 = 5.2$ and $\chi^2 = 6.2$).

Further evidence of interdependence may be seen in the fact that the antibodies of the highest titre after the first immunization appeared in the cows showing a low level of gamma globulin (titre 1 : 512, level of gamma globulin 35 and 36%, and titre 1 : 128, level of gamma globulin 30%).

The hypothesis which has been presented and discussed above calls for a further investigation of more extensive material for its verification.

We have not, found any distinct interdependence between an increased level of gamma globulin during the course of immunization on one hand, and the antibody titre of serum obtained on the other.

XITH EUROPEAN CONFERENCE ON ANIMAL BLOOD GROUPS AND BIOCHEMICAL POLYMORPHISM
Warsaw, July 2nd–6th, 1968

Investigation on the Production of Immune Antisera in Cattle: Three New Reagents Anti-I_0, -G_0, -R_0

M. DUNIEC, J. RAPACZ and J. WĘGRZYN

Zootechnical Institute, Immunogenetics Laboratory, Balice n/Cracow, Poland

SUMMARY. In an effort to produce reagents for red blood cell antigens in cattle, three reagents anti-I_2, anti-G_2, and anti-R_2 were produced which showed interesting patterns of reaction. These reagents, when diluted, gave much weaker reactions with red blood cells, possessing the I_1, G_1 and R_1 antigens.

After absorption with any of the weakly reacting red cells, the reagents still had antibodies against the I_2, G_2, R_2 antigens. We called these reagents anti-I_0, -G_0, -R_0.

Using the reagent anti-I_0, 5600 cattle were tested. They represent four breeds: Black and White, Polish Red, Danish Red and Simental cattle.

It was observed that the reagent anti-I_0 gave reactions in all cases (872) where the antigen I_2 occurred without I_1.

There were 490 individuals with I_1 antigen, but only 26 of them gave reaction with anti-I_0. When the genotypes were established for this group of 26 individuals it appeared that each of them had in one phenogroup the blood factor I_1 and in the other the I_0 antigen.

The results demonstrate that reagent anti-I_0 identifies the product of the gene which we propose to call B^{I_0}, since antigen I_2 is produced by both genes.

INTRODUCTION

The study of cattle blood groups has yielded extensive and useful data for practical breeding in addition to that of value for theoretical use in research on inheritance. The extent of the work is limited first of all by the number of reagents. Numerous papers deal with procedure for production of the reagents and almost every researcher in field of immunogenetics is familiar with this work. In this routine work unpredicted results appear from time to time, which permit better understanding of the serological and genetical relation between genes and antigens.

RESULTS AND COMMENTS

In the course of work on the fractionation and testing of the immune sera, the Immunogenetic Laboratory of the Zootechnical Institute in Balice has produced 60 reagents. Within this group there were three; anti-I_2, anti-G_2 and anti-R_2 which gave stronger reaction with appropriate subtype antigens: I_2, G_2 and R_2 when used in high dilution, and weaker reaction with the I_1, G_1 and R_1 antigens. When careful absorptions of the I_2 reagent with the I_1 the G_2 with the G_1 and the R_2 with the R_1 red cells were performed, the reagents lost reactivity for all red cells with the I_1, G_1 and R_1 antigens, but antibodies still remained for subtype antigens: I_2, G_2 and R_2. The reagents with these patterns of reaction we named anti-I_0, -G_0 and -R_0. In order to establish the relationship between I_1 and I_2 antigens and to identify the phenogroups in which they occur, 5600 blood samples of Black and White, Polish Red, Danish Red, and Simental cattle were tested using the following reagents: A_1, A_2, B, G_1, G_2, K, I_1, I_2, I_0, O_1, O_x, P, Q, T, Y_1, Y_2, A′, B′, G′, D′, I′, J′, K′, O′, Y′, E'_1, E'_2, K′, O′, Y′, E'_1, E'_2, G''_1 G''_2, P′, Ba5, C_1, C_2, R, W, X_1, X_2, L′, F, V, J_1, J_2, L, M, S, H′, U_1, U_2, U′, U″, Z.

Within the tested individuals there were 19 different phenogroups in which antigen I_1 or I_0 occurred. Table 1 presents these data. It was observed that anti-I_0 reagent gave reactions in all cases (892) where the antigen I_2 but not I_1 was present. There were 492 cases of the I_1 antigen, but only 26 of these gave reaction with anti-I_0. When the genotypes were established for this group of 26 individuals it appeared that each of them had in one phenogroup the blood factor I_1 and in the other the I_0 antigen. Genotypes for these individuals are given in Table 2. This can be ex-

TABLE 1
Phenogroups of the B system in which I_1 and I_0 antigens were encountered

No.	Phenogroups	No. of animals
1	BI_1	3
2	BI_1Q	50
3	BI_1QT_1P'	4
4	G_1I_1	2
5	I_1	30
6	$I_1''_xQA'K'E'_1$	46
7	$I_1O_xJ'K'O'$	89
8	$I_1QI'A'$	34
9	$I_1QI'E'_1$	2
10	I_1Y_1I'	2
11	I_1Y_2I'	46
12	$I_1G'G''$	139
13	$I_1I'E'_1$	2
14	I_1E_1G''	41
15	I_0*	851
16	$I_0O_xJ'K'O'$	2
17	$I_0D'E'_1G''$	15
18	$I_0E'_1G''$	3
19	I_0I'	1

* I_0 = the former I_2.

TABLE 2
Genotypes in the B system of cattle in which R. C. reacted with anti-I_1 and anti-I_0 reagents

No.	Genotype	No. of animals
1	$I_1O_xJ'K'O'/I_0$	9
2	I_1/I_0	6
3	BI_1/I_0	1
4	$I_1G'G''/I_0$	6
5	$I_1O_xQA'K'E'_1/I_0$	1
6	$I_1G'G''/I_0D'E'_1G''$	3

plained graphically by using a diagram which RAPACZ *et al.* (1968) worked out for the A blood group system in mink. The presented model explains the mutual dependence, serological and genetical relationship between these three antigens: I_1, I_2 and I_0 determined by two genes B^{I_1} and B^{I_0}. This presentation does not include data on anti-G_2 and anti-R_2 which are analogous to the anti-I_0 reagent, and study of them is being carried out. The results demonstrate that reagent anti-I_0 identifies the product of the gene which we propose to call B^{I_0}, since antigen I_2 is produced by both genes.

Three blood group antigens I_1, I_2, I_0 which are detected by three reagents anti -I_1,-I_2,-I_0. *Two genes B^{I_1}, B^{I_0} which determine three antigens I_1, I_2, I_0.* *Patterns of reactions*

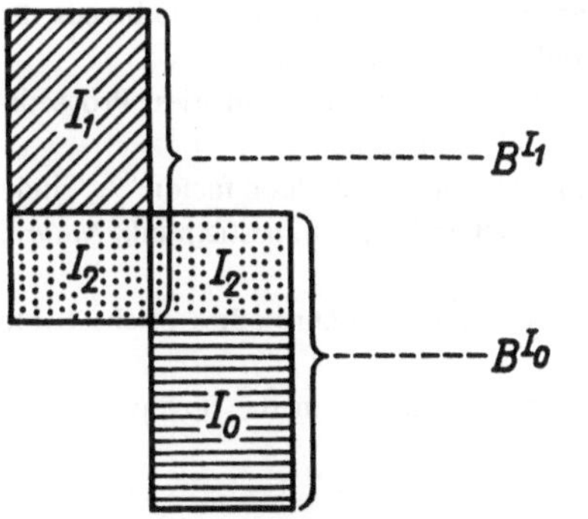

Phenotype	Genotype	Reagents Anti I_1	Anti I_2	Anti I_0
I_1	$B^{I_1}../B^{I_1}..$	+	+	−
	$B^{I_1}../-$	+	+	−
	$B^{I_1}../B^{I_0}..$	+	+	+
I_0	$B^{I_0}../B^{I_0}..$	−	+	+
	$B^{I_0}../-$	−	+	+
−	−/−	−	−	−

Genetical and serological relation between three antigens I_1, I_2, I_0 in the B system in cattle.

The relationship between these three antigens is the same as RAPACZ *et al.* (1968) demonstrated in mink between following antigens: B, B_2 and C, and A, A_2 and C. A similar relationship between U_1 and U′ in the S system in cattle was described by STORMONT (1950) and confirmed by STONE and MILLER (1962). Later MILLOT (1965) observed association in the B system between G′ and G″, however, he did not explain the relationship between genes, which determined these factors. At the same time GROSCLAUDE (1965) described reagents, which react with two or three antigens in the SU system and since he could not fractionate them, he described the reagents as possessing a secondary antibody. The data presented in his work shows, that similar relationships exist between antigens in the B system as these in the SU system, a fact previously unrecognized. It seems that the described cases call for a new explanation for a previously unrecognized relationship between the genes, which determine blood groups.

Aside from the theoretical aspects, this work also has a practical application; using the I_0 reagent it was possible to establish genotypes for I_1 and I_0 in all 490 individuals, which had the I_1 antigen. It was especially important to use this reagent to test the 26 individuals which had in one phenogroup the I_1 and in the other the I_0 antigen. For the first time it was possible to demonstrate presence of subtype I_2 when the I_1 antigen was present. It would be interesting to find a phenogroup in which all three antigens I_1, I_2 and I_0 occur. This would support the hypothesis that the B system in cattle is composed of subunits with different numbers of antigens, and that the great variety of phenogroups is created by crossing over. An example of an allele with two homologous units of genes would be the first case in a blood group system in mammals. A recent study by RAPACZ *et al.* (1968) carried out in the U.S.A. suggests that such a case of homologous but unequal crossing was detected.

REFERENCES

GROSCLAUDE, F., 1965. Studies on the S blood-group system in French cattle breeds, Proceedings of the 9th European Animal Blood Group Conference, 79–85, Prague.

MILLOT, P., 1965. Bovine isohaemolysins seeming to have several specificities, Proceedings of the 9th European Animal Blood Group Conference, 75–78, Prague.

RAPACZ, J., HASLER, J. and SHACKELFORD, R. M., 1968. Additional blood factor "A_2" independently determined by two allelic genes: A^a and A^c of A system in the domestic mink, *Nature*, *Genetica Polonica*. (In press).

STONE, W. H. and MILLER, W. J., 1962. Naturally occurring iso-antibodies of the S blood group system in cattle, *J. Immunol.*, **86**, 165–169.

STORMONT, C., 1950. Additional gene-controlled antigenic factors in the Bovine erythrocyte, *Genetics*, **35**, 76–94.

DISCUSSION

G. LINDSTRÖM: Have you made cross-absorptions with I_1, I_2 and I_0 (?) cells for determination the new reagent?

J. RAPACZ: Yes, we did absorbed our reagent anti-I_0 with all phenogroups where I_1 occurred as well as we absorbed by red cells with I_2 antigens.

R. SPOONER: Who said globulins had antibody activity — I asked what was reference to this statement as I had never heard such statement before and did not believe it.

J. RAPACZ: Yes, there are more and more evidence that γ-globulins have antibody specificities.

I can not recall at this moment papers but I believe you may find it in "Chromatography of the antibody globulins on anion-exchange cellulose columns" by H. C. Goodman; in "γ-globulin and Antibodies" by R. R. Porter, in "The structure of Antibodies" by R. R. Porter where he put it "Although the complexity of *immunoglobulin gamma (and of the other classes of immunoglobulin)* has presented investigators with a most difficult puzzle however considerable progress has now been made in solving much of it" and in "Physical Heterogenity of Bovine γ-globulins ...", by F. A. Murphy *et al.*, where we read — "In fractions 21–27 another protein was encountered in association with the γ- G-globulin. This was identified as transferrin by its colour ...".

XITH EUROPEAN CONFERENCE ON ANIMAL BLOOD GROUPS AND BIOCHEMICAL POLYMORPHISM
Warsaw, July 2nd–6th, 1896

F^f — A New Allele in the Bovine FV Blood Group System

D. R. OSTERHOFF and N. POLITZER

Department of Zootechnology, Faculty of Veterinary Science, University of Pretoria, Onderstepoort, South Africa

SUMMARY. The FV system has so far been regarded as a closed system, but immunological and genetical evidence is brought forward to show that in Zebu-type cattle this system has characteristics similar to the A–B–0 system in man.

Blood group studies on indigenous cattle led to the discovery of a new allele in the bovine FV blood group system. In accordance with the international nomenclature the allele was named F^f, the small letter indicating that the allele is responsible for the absence of antigenic factors F and V in the FV system.

INTRODUCTION

The FV system was originally described by STORMONT (1952) as a two allele closed system. Three phenotypes are present, the heterozygotes are expressed jointly by both alleles and may be counted directly. Therefore, this system is usually used for tests on the existence of genetic equilibrium in the breeds investigated. If the gene frequencies calculated from the direct results in the FV system are in close agreement with the frequencies calculated according to Hardy–Weinberg's formula, a population is regarded as being in genetic equilibrium.

RENDEL (1958) was able to subdivide V into V_1 and V_2, and in 1962 STORMONT described the following subtypes of F and V, namely F_1 (the conventional F) and F_2, V_1, V_2 (the conventional V) and V_3. The reactions of cells of the different pheno-

TABLE 1

Reaction of possible phenogroups with the respective reagents in the FV system (STORMONT, 1962)

Phenogroups	Reagents				
	F_1	F_2	V_1	V_2	V_3
F_1	+	+	·	·	·
F_2	·	+	·	·	±
V_1	·	·	+	+	+
V_2	·	+	·	+	+

types with the various reagents, according to this "revised" system, are shown in Table 1.

STORMONT stated that from the 10 diploid combinations of the four phenogroups eight distinctive patterns of reactions were provided. The diploid combinations F_2/V_2 and V_2/V_2 yield the same reactions, and also the combinations F_2/V_1 and V_1/V_2 and can thus only be differentiated by means of genetic tests.

In the indigenous breeds in South Africa a number of FV-negative reacting animals were found and many attempts failed to prove that all these animals possessed the F_2 antigen.

METHODS AND MATERIAL

The usual blood typing and absorption techniques were used. A total of 7333 blood samples of 17 different breeds of cattle was collected during the period 1960–1967 in three provinces of South Africa. Furthermore, 331 Boran blood samples from the East African Research Organization, Kikuyu, Kenya were obtained during 1960 and could be included in this study.

RESULTS AND DISCUSSION

Table 2 gives the gene frequencies of F^F and F^V in all breeds investigated.

The results of the tests of genetical equilibrium in the FV system are presented in Table 3. The FV negative animals were regarded as positives, added to the respective classes according to the observed percentage in Table 2.

Even after making up the number of FV negative animals in the *Bos indicus* breeds by classifying them as positives according to the observed percentages the observed data were not in agreement with those expected on the basis of Hardy-Weinberg's law. In the Drakensberger and Bonsmara breeds belonging also to the Zebu-crosses, no statistically significant deviation of the expected from the observed values was found, but in the Afrikaner, Boran and Nguni these differences were highly significant. The difference was also highly significant in the Sussex breed. In this case, however, it is due to the fact that in classes with numbers of 5 or smaller somewhat unrealistic high values of χ^2 are obtained. Corrections for small numbers should actually be applied in these cases, but this would include complicated calculations from which little extra information would be obtained.

Before further work was initiated series of tests were performed to ensure that these FV negative samples were truly negative. The technique of determining the F and V factors in the haemolytic test was definitely trustworthy. The South African reagents for these two factors have never shown any deviation in all international comparison tests; thus, the deviation from Hardy's law must be explained in another way.

TABLE 2
Gene frequency at the FV locus in South African cattle breeds

Breed	Observed percentage			Calculated frequencies	
	F/F	F/V	V/V	F	V
Aberdeen Angus	52	42	6	0.73	0.27
Afrikaner*	73	13	14	0.80	0.20
Ayrshire	74	25	1	0.86	0.14
Beef Shorthorn	87	13	0	0.94	0.06
Bonsmara*	82	15	2	0.90	0.10
Boran*	76	19	5	0.85	0.15
Brown Swiss	64	32	4	0.80	0.20
Dairy Shorthorn	56	39	5	0.76	0.24
Drakensberger*	91	9	0	0.96	0.04
Friesian	66	30	4	0.81	0.19
Guernsey	87	13	0	0.94	0.06
Hereford	86	15	0	0.93	0.07
Jersey	46	42	12	0.67	0.33
Nguni*	76	18	6	0.85	0.15
Red Poll	63	31	6	0.78	0.22
Simmentaler	66	30	4	0.81	0.19
South Devon	85	15	0	0.92	0.08
Sussex	95	4	1	0.97	0.03

* In these breeds a number of animals did not react with the classical F and V reagents, they were regarded as "not tested" and excluded from the calculations: Afrikaner 239, Bonsmara 9, Boran 2, Drakensberger 13, Nguni 28.

The first investigation included a test of all possible F_2 reagents available in the laboratories in the world to ensure that possible subtypes of F and V could be excluded. One F_2 reagent was obtained from California, one from Czechoslovakia and two from Denmark. Requests were also made to obtain this reagent from France and Poland, unfortunately without success.

A Norwegian V_2 reference reagent was also sent to South Africa by Dr. BRAEND from Oslo. A typical section of these tests is shown in Table 4.

It is obvious that a number of our "negative" cells reacted with all four F_2 reagents obtained especially with the Danish F_2-reagent. But even if the tests stood overnight and some more trace reactions could be seen, a number of definitely negative reactions remained. In Table 4 three animals — 1233, 6348 and 6747 — reacted with the overseas F_2 and the South African V, which is identical to the Norwegian V_2 reagent. This is in accordance with STORMONT's revised FV system as outlined in Table 1.

The reactions outlined in Table 4, however, prove that the Californian reagent used in STORMONT's studies gave many more negative reactions than the Danish F_2 reagent and that the "revised" system did not "cover" all possibilities in this system. If, furthermore, the "revised" system is to be accepted, it would mean that

TABLE 3

Test of genetical equilibrium for the FV blood group system

Breed	Observed			Expected			χ^2
	F^F/F^F	F^F/F^V	F^V/F^V	F^F/F^F	F^F/F^V	F^V/F^V	
Aberdeen Angus	163	132	19	167.3	123.8	22.9	1.318
Afrikaner	569	102	109	499.2	249.6	31.2	116.443***
Ayrshire	376	127	5	375.7	122.4	9.9	2.598
Beef Shorthorn	129	19	0	130.8	16.7	0.5	0.342
Bonsmara	115	21	3	112.6	25.0	1.4	2.519
Boran	251	63	17	239.1	84.5	7.4	18.516***
Brown Swiss	231	116	14	231.0	115.6	14.4	0.012
Dairy Shorthorn	54	38	5	56.0	35.4	5.6	0.334
Drakensberger	202	20	0	204.9	17.0	0.4	0.962
Friesian	592	296	36	588.5	276.1	32.4	0.204
Guernsey	122	18	0	123.7	15.8	0.5	0.829
Hereford	207	35	0	209.3	31.5	1.2	1.614
Jersey	562	513	147	548.6	540.3	133.1	3.158
Nguni	517	122	35	487.0	171.8	15.2	42.076***
Red Poll	233	115	22	225.1	127.0	17.9	2.350
Simmentaler	73	33	5	72.8	34.2	4.0	0.293
South Devon	296	52	0	294.5	51.3	2.2	2.218
Sussex	381	16	4	377.3	23.3	0.4	35.767***

The degree of significance is given as $*P < 0.05$, $**P < 0.01$ and $***P < 0.001$.

TABLE 4

The haemolytic tests with all reagents used in the FV system

Sample No.	F S.A. Standard	F_2 California	F_2 Czecho-slovakia	F_2 Denmark	F_2+R_1 Denmark	V S.A. Standard	V_2 Norway
G 29	0/0*	0/0	0/0	0/1	0/0	0/0	0/0
1233	0/0	0/1	0/1	3/3	1/3	3/4	4/4
1245	4/4	4/4	tr/3	2/3	1/3	4/4	4/4
1256	0/0	2/2	0/0	2/3	tr/1	0/0	0/0
1261	0/0	2/2	0/0	2/2	1/2	0/0	0/0
1262	4/4	4/4	4/4	2/3	3.4	0/0	0/0
6348	0/0	0/0	0/0	2/3	tr/1	3/4	3/4
6747	0/0	0/tr	0/1	2/2	3/3	3/4	3/4
7626	0/0	2/2	0/0	4/4	2/2	0/0	0/0
7717	4/4	4/4	4/4	3/3	2/3	4/4	4/4
7836	0/0	0/0	0/0	0/0	0/0	3/4	4/4
8710	4/4	4/4	4/4	3/4	3/3	0/0	0/0
2447	0/0	0/0	0/0	3/4	2/3	0/0	0/0
2449	0/0	0/0	0/0	2/3	2/2	0/0	0/0
2450	0/0	0/0	0/0	0/0	0/0	0/0	0/0

* Both readings — after 1.5 and after 3 hr — are given.

TABLE 5
Results of absorption tests with different F_2 reagents

Sample No.	F (SA) 1261*	F_2(Cal.) 1256	(F_2(Cal.) 1261	F_2(Cz.) 1233	F_2(Dk.) 1233	F_2(Dk.) 1256	F_2(Dk.) 1261	F_2+R_1(Dk.) 1261
G 29	0**	0	0	0	0	0	0	0
1233	0	0	0	0	0	0	0	0
1245	4	4	0	4	3	0	0	3
1256	0	0	0	0	0	0	0	0
1261	0	0	0	0	0	0	0	0
1262	4	4	0	0	3	0	0	4
6348	0	0	0	0	0	0	0	0
6747	0	0	0	0	0	0	0	0
7626	0	0	0	0	0	0	0	0
7717	4	4	0	4	3	0	0	3
7836	0	0	0	0	0	0	0	0
8710	4	4	0	4	4	0	0	3
2447	0	0	0	0	0	0	0	0
2449	0	0	0	0	0	0	0	0
2450	0	0	0	0	0	0	0	0

* Absorption cells.
** Only final reading given.

a large number of animals classified as V/V not only in South Africa but in all other laboratories in the world, should be classified as F_2/V or F_2/V_2. In all laboratories the genetic equilibrium of breeds was established with the aid of the FV system (see also Table 3) and the two allele, three phenogroup system is still used in present studies by various authors.

The next step was a series of absorptions to test the validity of the results obtained with the different F_2 reagents. Table 5 gives the results of several of these absorption tests.

The absorption results show that practically all antibodies against non-F cells were taken out, while the F cells still gave strong reactions. Cells of animal 1256 and 1261 could take out all F antibodies from the Danish F_2 reagent, but not from the Danish F_2+R_1 reagent, while only cells of 1261 were able to remove these antibodies from the Californian F_2 reagent. On the other hand the same cells were not in the position to weaken the South African F standard reagent. The results were inconclusive, but it was clear that all three F_2 reagents used were different.

In addition to the above tests 30 samples of negative cells were sent to Dr. J. RENDEL, Uppsala to throw some more light on these differences. Dr. RENDEL came to the following conclusions: "There is no evidence that your cells contain V. The weak reactions obtained with one of our F reagents (U355(U382)) have probably no connection with F. The antibodies against F were unaffected by absorption with the weakly reacting non-F cells".

Enough evidence was now available to postulate a new allele in the FV system, the allele F^{f} which in addition to the known alleles completes the FV system. The best way to test this postulation is a control of family data. For simplicity's sake only Afrikaner cattle were included. Since the allele F^{f} is similar to the allele L^{1} (= 0) in the AB0 blood group system in man, the FV-negative cells are designated "0" in Table 6, for the sake of simplicity. The F^{F_2} and F^{V_2} alleles were disregarded.

TABLE 6

The distribution of FV blood types among offspring 408 matings

Mating types	No. of matings	Blood types of offspring			
		F	V	0	FV
F × F	127	108	0	19	0
V × V	1	0	1	0	0
0 × 0	62	0	0	62	0
FV × FV	0	0	0	0	0
F × V	27	8	3	2	14
F × 0	164	101	0	63	0
F × FV	5	2	0	0	3
V × 0	16	0	13	3	0
V × FV	1	0	1	0	0
0 × FV	5	3	2	0	0

So far no offspring have been found that do not fit the proposed theory of three alleles. A few matings were included where one of the parents possesses both F and V. If these animals ever had FV negative offspring a second locus and epistasis would have to be postulated. But with the evidence available the three allele hypothesis for *Bos indicus* cattle breeds seems to be appropriate. The gene frequencies in these breeds are calculated as follows:

p = frequency of F^{F}

q = frequency of F^{V}

r = frequency of F^{f}, and

$\bar{F}$ = frequency of phenotype F

$\bar{V}$ = frequency of phenotype V

$\bar{0}$ = frequency of phenotype 0, and

$$r = \sqrt{\bar{0}}$$

$$p = 1 - \sqrt{\bar{V} + \bar{0}}$$

$$q = 1 - \sqrt{\bar{F} + \bar{0}}$$

The first gene frequency calculation in the FV system was performed on 431 animals from the Afrikaner herds from three Government farms at Potchefstroom, Glen and Pretoria.

This material was later on enlarged to 780 animals, both calculations being

TABLE 7

The proportions of four phenotypes and frequencies of three alleles in the FV system in three materials of Afrikaner cattle

Phenotypes	Number	Percentages observed	Percentages expected	Gene frequency	
Material I					
F	174	40.37	46.11		
V	43	9.98	14.53	F^F	0.2969
FV	44	10.21	6.33	F^V	0.1066
0	170	39.44	39.44	F^f	0.6280
Material II					
F	444	56.95	56.69	F^F	0.3811
V	60	7.69	7.53	F^V	0.0643
FV	37	4.75	4.90	F^f	0.5533
0	239	30.61	30.61		
Material III					
F	601	57.73	58.25		
V	65	6.24	6.58	F^F	0.3864
FV	48	4.61	4.32	F^V	0.0559
0	327	31.41	31.41	F^f	0.5605

shown in Table 7. The same Table shows frequency calculations on another completely independent Afrikaner sample of 1041 animals.

The sum of the allele frequency of F^F (p), F^V (q) and F^f (r) is near unity in all three materials, confirming the theory of triple alleles. A test of significance of the discrepancy between unity and the calculated sum of p, q and r may be necessary and BERNSTEIN has shown that the said discrepancy is

$$\bar{0} \text{ disp.} = \frac{1}{\sqrt{N}} \sqrt{\frac{pq}{2(1-p)(1-q)}}$$

where N is the total number of individuals tested. The discrepancy in the three materials was 0.0076, 0.0047 and 0.0036 respectively, all being not significant statistically.

The three allele hypothesis for the *Bos indicus* breeds should therefore be accepted as correct.

The practical application of these findings is twofold. Firstly, the FV system may only be used in *Bos indicus* breeds for calculations of genetical equilibrii, if the additional allele is included, and secondly the FV system does not allow many parentage exclusions in these breeds. It was recently shown on a large material of 1165 parentage cases that with the aid of the FV system 16% of cases could be excluded in Friesians, 12% in Jerseys and only 4% in Afrikaner cattle (OSTERHOFF, unpublished).

ACKNOWLEDGEMENTS

The authors wish to thank Dr. J. RENDEL, Uppsala, for his stimulating interest and advice.

REFERENCES

STORMONT, C., 1952. The F–V and Z systems of bovine blood groups, *Genetics*, **37**, 39–48.
STORMONT, C., 1962. Current status of blood groups in cattle, *Ann. N. Y. Acad. Sci.*, **97**, 251–268.
RENDEL, J., 1958. Studies of cattle blood groups. I. Production of cattle iso-immune sera and the inheritance of 4 antigenic factors, *Acta Agric, Scand.*, **8**, 40–61.

DISCUSSION

A. FIORENTINI: In our laboratory (IT/Mi) we use 4 reagents in the system F/V and in the last comparison test we have found that one animal was not reacting to F_1 and V_1 but only to V_2 and F_2.

XITH EUROPEAN CONFERENCE ON ANIMAL BLOOD GROUPS AND BIOCHEMICAL POLYMORPHISM
Warsaw, July 2nd–6th, 1968

Subtypes in the R′/S′ Cattle Blood Group System

L. ERHARD, W. SCHLEGER, J. SCHRÖFFEL and J. STUKOVSKY

Institut für Blutgruppen und Resistenforschung der Tierzuchtforschung e.V., München, G.F.R.
Lehrkanzel und Institut für Tierzucht und Züchtungshygiene, Tierärztliche Hochschule, Wien, Austria
Plemenářsky vývojový ustav, Hradiško, Czechoslovakia
Központi Mesterséges Termékenyítesi Föállomás, Vercsoport Laboratóriuma, Budapest, Hungary

SUMMARY. The results of the 1967 reference test showed two types of R′ antibodies. We propose as an analogy to other subtypes relationships two antigenic factors R'_1 and R'_2.

INTRODUCTION

In the 1967 reference test performed in Munich 4 anti R′ sera (R′(A), R′(Cz), R′($D_{Mü}$), R′($D_{Schö}$)) were compared. These 4 reagents gave good agreement for all specific reactions.

The serum R′(A) gave some extra reactions. The extra reactions with blood samples $D_{Gö}5$, $P_{Kr}2$ and $YU_{Lj}14$ were found to be nonspecific in the absorption test. In contrast to this the absorptions with blood samples Cz 1 and 12 showed these extra reactions to be specific. The results of absorptions with samples $H_{Hg}7$ and $D_{Mü}20$ were dubious.

The laboratories in Austria, Liběchov and Budapest were asked therefore to reproduce the absorptions performed in Munich with the blood samples Cz 1 and 2, $H_{Hg}7$ and $D_{Mü}20$.

Positive results with these absorptions in Austria, Czechoslovakia and Hungary would indicate that a subtype reagent R'_2 could exist.

METHODS AND MATERIALS

Absorption tests were performed in Vienna and Budapest with anti R′(A) only. In both laboratories the samples Cz 12, $D_{Mü}20$ and $H_{Hg}7$ reacted negatively in the absorption test.

The blood sample Cz 1 showed a weak positive reaction which was nonspecific.

The absorption tests performed in Liběchov were very extensive and informative.

RESULTS

The following results were obtained in Liběchov: (1) Blood samples Cz 1 (343) and 12 (449) and also 5 other blood samples (218, 230, 239, 240, 478) reacted positively with anti R′(A) but did not react with anti R′(Cz).

(2) The strength of reaction of R′(Cz) positive cells with the two reagents was different. The anti R′(A) positive cells reacted more weakly than R′(Cz) cells, the difference being 1–2 grades of haemolysis.

(3) R′(Cz) cells absorbed all reactivity of anti R′(A) serum. In contrast to this it was very difficult to absorb anti R′(A) with R′(A) positive but R′(Cz) negative cells. In one case this serum gave weakly positive reactions with some R′(Cz) positive erythrocytes, after four successive absorptions of this type.

Some of the absorptions performed in Liběchov were repeated at the Munich laboratory.

The absorption tests in Munich were performed with 6 R′(A) positive blood samples from Liběchov including the cells Cz 1 and 12 and two R′(A) positive cells of Munich included the cell $D_{Mü}$ 20 of the 1967 reference test.

The repeated tests carried out at Munich gave very good agreement.

The R′(Cz) positive cells again absorbed all reactivity of anti R′(A). But 6 out of 7 R′(A) positive cells absorbed only that part of the R′(A) specificity which did not react with R′(Cz) positive cells.

In one case, cell Cz (230/525), it was not possible to absorb any R′(A) specificity, because the absorption was incomplete.

In Munich all absorptions were performed at one time.

CONCLUSIONS

The absorptions of Liběchov and Munich clearly indicate two R′ antigenic patterns.

Anti R′ of Czechoslovakia contained only R'_1 antibodies and anti R′ of Austria R'_1 and R'_2 antibodies.

The results of the 1968 reference test also showed two R′ reaction patterns.

The test reagents from Cz, $D_{Mü}$ and SU have a similar pattern of reaction. In contrast to this the reagents from Austria give strong extra reactions with the blood samples A 4, B 8, Cz 1, 2, 3, and 12, F 4, $D_{Mü}$ 8 and 15, Su 5. The blood samples Cz 1, 2, 3 and 5 and $D_{Mü}$ 8 and 15 are from animals which clearly absorb all R'_2 specificity.

The results of the 1968 comparison test also indicate two R′ reaction patterns.

On the one hand reagents from Cz, $D_{Gö}$, DK, H_{Hg}, SU and perhaps $D_{Mü}$ and S (the last two reagents both missed one reaction with blood sample 38) gave the same patterns of reaction.

On the other hand the reagents of A, NL, YU_{LJ}, and perhaps I_{MI} and J react additionally with sample 22.

We think that this sample carries only the R'_2 specificity.

On basis of all these results, but especially on the basis of the 1967 reference test, repeatable absorptions, and the preliminary results of the reference and comparison tests of 1968 we are sure that two types of R' antibodies exist. We propose by analogy to other subtype relationships that the two antigenic factors concerned should be called R'_1 and R'_2.

DISCUSSION

J. BOUW: You mentioned there were two sources of antibodies in the R' reagents from Austria.

L. ERHARD: The absorption tests in Libčchov and Munich demonstrated clearly two patterns of antibodies in the Austrian R' reagent. We have observed the same situation in any of the I_2, T_2 and P_2 reagents. These contained also two patterns of antibodies, namely I_1+I_2, T_1+T_2, and P_1+P_2.

J. BOUW: Are you sure it was not solely anti R_2?

L. ERHARD: On the basis of the absorption test I am sure that the Austrian reagent anti R' contains both R'_1 and R'_2 antibodies.

XITH EUROPEAN CONFERENCE ON ANIMAL BLOOD GROUPS AND BIOCHEMICAL POLYMORPHISM
Warsaw, July 2nd–6th, 1968

Blood Group Gene Frequencies in Two Spanish Cattle Breeds

J. L. SOTILLO, A. RICO, R. SARAZA and P. HERNANDEZ

Sección de Genética del Patronato de Biología Animal, Madrid, Spain

SUMMARY. The erythrocyte antigenic structure of two important cattle breeds, Spanish Friesian and Rubia Gallega, was typed for the first time in Spain. The blood antigenic factors were studied by means of blood typing reagents from foreign laboratories.

Frequency percentages of these antigenic factors are presented and show the lack of some factors in the Rubia Gallega cattle breed. Gene frequencies at the FV locus are also presented. χ^2 test indicates the possibility of a quick genetic improvement of the animals studied. Presently, we are producing test sera in our laboratory by immunization of some animals used in this work. Blood groups will be used in Spain for parentage control at the stations, where A. I. techniques are applied.

INTRODUCTION

Research on cattle blood groups in Spain began in very recent times (1967). Early work concentrated on the typing of some Spanish breeds with a full gamut of antigens from foreign laboratories. At the same time, gene frequencies at FV locus were studied, and immunizations were started to produce antisera within our own breeds.

This work is a initial study on two very interesting Spanish cattle breeds: Spanish Friesian, breed for dairy production, and Rubia Gallega breed for beef production.

The Spanish Friesian is very common across the country but the principial herds are in Northern and Central Spain. This breed originates from Dutch, North American and Canadian cattle, imported to Spain several years ago. It has a Herd Book and is carefully selected in particular farms and State Breeding Stations by means of A. I. performance control, etc. However there are also unselected herds which have enabled us to examine a wider variety of individual animals and to include extra antigens in the immunizations for preparation of antisera.

The Rubia Gallega cattle breed is a subspecies located in the North-West of Spain with an important Selection Centre in Fuentefiz (Orense) which is controlled by the Bureau of Animal Husbandry. Similarly, as with the other breed, there is a Herd Book, but there are also many unselected herds*.

* References are given for a more detailed study of both breeds.

TABLE 1

The occurrence of some different blood group factors in Spanish Friesian and Rubia Gallega breeds expressed in cattle number and percentage

Locus (system)	Blood group factor	Spanish Friesian $n = 279$		Rubia Gallega $n = 221$	
		f	%	*f*	%
A	A	153	54.8	208	94.1
	H	27	9.7	12	5.4
	Z′	18	6.4	—	—
B	B	180	64.5	118	53.4
	B′	9	3.2	13	5.9
	B″	—	—	78	35.3
	G	117	41.9	37	17.6
	G′	117	41.9	78	35.3
	G″	153	54.8	14	6.3
	I	81	29.0	15	6.8
	I′	99	35.5	11	5.0
	K	36	12.9	27	12.2
	K′	18	6.4	—	—
	O	135	48.4	65	29.4
	O′	63	22.6	79	35.7
	P	18	6.4	26	11.8
	P′	54	19.3	—	—
	Q	27	9.7	15	6.8
	Q′	153	54.8	—	—
	T	81	29.0	38	17.2
	Y	108	38.7	—	—
	Y′	36	6.4	—	—
	A′	117	41.9	39	17.6
	D′	63	22.6	11	5.0
	E′	234	83.9	52	23.5
	J′	27	9.7	—	—
C	C	126	45.2	208	94.1
	C′	63	22.6	39	17.6
	R	207	74.2	52	23.5
	W	144	51.6	115	52.0
	X	189	67.7	104	47.0
	E	117	41.9	—	—
	L′	27	9.7	—	—
FV	F	261	93.5	195	88.2
	V	126	45.2	117	52.9
L	L	99	35.5	24	10.8
M	M	36	12.9	—	—
S	S	108	38.7	40	18.1
	U	9	3.2	—	—
	U′	108	38.7	143	64.7
	H	225	80.6	28	12.7
	H′	36	12.9	—	
Z	Z	171	61.3	76	34.4
R′S′	R′	18	6.4	—	—
T′	T′	54	19.3	14	6.3

MATERIAL AND METHODS

We took samples 279 Spanish Friesian and 221 Rubia Gallega cows. In both cases an at random sampling was made among private farms — in which animals are not selected and A. I. is not used — to study the effect of selection (the study of gene frequencies) and to ensure full representation of the characteristic blood antigens of these breeds (frequency percentages).

All animals were tested with the following reagents: A, H, Z′, B, B′, B″, G, G′, G″, I, I′, K, K′, O, O′, P, P′, Q, Q′, T, Y, Y′, A′, D′, E′, J′, C, C′, R, W, X, E, L′, F, V, L, M, S, U, U′, H, H′, Z, R′, and T′. We have used the Stormont technique (modified). Readings were made every 30 and 60 min after previous shaking of haemolysis tubes.

RESULTS

Percentage frequency of some antigenic factors in these cattle breeds is presented in the Table 1.

Gene frequencies at the FV locus in both breeds are shown in the following Table:

	Spanish Friesian				Rubia Gallega			
	phenotypes			Total	phenotypes			Total
	F	FV	V		F	FV	V	
Observed	144.00	117	18.00	279	104	91.00	26.00	221
Calculated	148.67	110	20.33	279	102	96.20	22.80	221

CONCLUSIONS

The frequency percentages obtained in the Spanish Friesian cattle breed are similar to those found in other countries. Gene frequencies at FV locus are $p = 0.73$, $q = 0.27$.

Rubia Gallega cattle lack the following antigens: Z′, K′, P′, Q′, Y, Y′, E, L′, U, H′, and R′. Gene frequencies at the FV locus are $p = 0.68$, $q = 0.32$.

χ^2 test applied to both breeds indicates a lack of selection in these animals allowing rapid genetic improvement to be made.

Typing performed in these animals enables us to chose 40 of them carrying the full range of antigens in particular those of the B-system.

ACKOWLEDGEMENTS

We are very grateful to Dr. J. Bouw (Wageningen, The Netherlands) and Dr. F. Grosclaude (Jouy-en-Josas, France) who gave us the first samples of typing reagents and advised us in our studies on blood groups.

Blood Group Studies in Simental Cattle in Poland

Ewa SŁOTA, J. RAPACZ and Aleksandra BARINOW
Immunogenetics Laboratory, Zootechnical Institute, Balice n/Cracow, Poland

SUMMARY. The aim of this work was the study of blood group antigens of Simental cattle in Poland. The tests include 1200 breeding cattle from the counties of Rzeszów Province, where the mixture of native and imported Simentals from Austria, Switzerland, G. F. R. and Rumania occurred.

As a result of genotype analysis it was possible to establish 75 alleles with frequencies from 16.53% to 0.1%.

These alleles were compared with phenogroups of Swiss and German Simentals described in the literature.

The number of alleles showed, that these cattle are not genetically uniformed. Bulls used in recent years have had a marked influence on genetic structure of this breed.

The phenogroups which were found in Simental cattle in Poland suggest, that alleles from other breeds like Black and White and Polish Red are present in this breed.

INTRODUCTION

Evidence that blood groups in cattle exhibit breed differences, was shown by FERGUSON (1941). The large number of antigens transmitted in complexes, especially in the B system and their simple mode of inheritance, allows the study of genetic differences, structure and origin of breeds (ERHARD and SCHMID, 1965; GROSCLAUDE, 1965).

Blood groups can be very useful in study of relationship between breeds and their participation in development of new breeds. The process of development of phenogroups in the B system, started much earlier, before divergence into breeds took place. The aim of this work is to characterize Simentals in Poland on the basis of blood groups, using for this purpose gene frequencies in the B system.

MATERIAL AND METHODS

Simental cattle which occur in Poland, are a mixture of the native breed with imported Simental from Austria, Switzerland, G. F. R. and Rumania (TRAUTMAN, 1961). Material used for this study originated from State Farms (32%), Central

Breeding Farms (19.5%), Zootechnical Institute (15%), Cooperative Farms (5%), and private farms (27.5%), in Rzeszów province from the counties of Sanok, Lesko, Brzozów and Ustrzyki Dolne. In the counties Sanok and Ustrzyki Dolne Simental compose 30% of all cattle and 20% in the counties of Brzozów and Lesko. Material for this work comes from 1236 individual animals.

Blood samples of all individuals were tested with the following reagents: A_1, A_2, B, G_1, G_2, K, I_1, I_2, O_1, O_x, P, Q, T_1, Y_1, Y_2, Y′, A′, B′, D′, G′, E'_1, E'_2, I′, J′, K′, O′, P′, G''_1, G''_2 C_2, R, W, X_1, X_2, L′, F, V, J_1, J_2, L, M, S_1, S_2, U_1, U_2, U′, U″, Z, Ba4, Ba5, which were obtained from Immunogenetics Laboratory at Balice and had been standarized in the International Comparison Test (E.S.A.B.R.)

RESULTS

The genotypes of tested cattle were established on the basis of their relation and gene frequencies were estimated. Incomplete pedigree and lack of test data for some parents did not allow the establishment of genotypes for 12% of tested cattle. Most individuals were the offspring of 20 bulls.

The following list gives names of bulls and numbers of offspring sired by each of them.

No.	Name of bull	No. of offspring	No.	Name of bull	No. of offspring
1.	Multino 136K	97	11.	Ummel 13K	26
2.	Max 137K	94	12.	Sum 10G	16
3.	Major 135K	76	13.	Milo 23G	16
4.	Prunk 82K	63	14.	Łysy 39K	14
5.	Samba 57K	63	15.	Brutus 5K	13
6.	Hamster 48K	60	16.	Granit 15Wst	12
7.	Peter 58K	55	17.	Dollar 15K	11
8.	Hamster 81K	47	18.	Ursik 50K	11
9.	Plato 22G	33	19.	Urs 8K	10
10.	Ami 17K	32	20.	Delfin 56K	9

As shown — above 10 bulls sired about 75% of the offspring and the remainder did not have any significant influence on the genotypes of the studied material.

Table 1 shows frequency of alleles, which were found in Simental cattle in Poland. Alleles are given in percentage since number of individuals of unknown genotypes is limited to 12%.

Established genotypes were compared with the phenogroups of Simental cattle from Switzerland (MÜLLER, 1960) and G.F.R. The lists of phenogroups for the Swiss and German cattle were provided by the courtesy of Prof. WEBER, Berne and Dr. SCHMID, Munich, respectively.

There were some difficulties in comparison, especially with cattle from Switzerland, since they were tested by smaller numbers of reagents, for example: cattle were not tested for blood factors: O_x, I_2, O′, G′, B′, and the frequent phenogroup O_xO' in Polish cattle could not be compared with the Swiss.

Table 3 presents frequency of a few selected phenogroups in Black and White and Simental cattle.

DISCUSSION

The genetic characterization of Simental cattle using blood groups, was the subject of this study.

As shown in Table 1, 75 different alleles were found and most frequent are the following phenogroups: O_1I', O_1, G_1O_xA', O_xO' and A′B′.

Simental cattle are bred in Austria, Switzerland and G.F.R., and as imported cattle they could have influence on this breed in Poland. For this reason it seems suitable to compare the alleles, which were found in Polish cattle with the available list of alleles for Swiss and German cattle. As shown in Table 2, many alleles are common to all these cattle. It is interesting to note, that some alleles which are frequent in Poland: $O_xY_1A'G'G_1''$, $ET_1A'B'G_2''$, $I_1QA'I'$, $G_2T_1Y_1Y'A'B'D'G'$ are not present in Swiss and German cattle. On the contrary there were 130 alleles in German and 15 in Swiss cattle (MÜLLER, 1960) which are absent in Simental cattle in Poland. This is the best example of the use of blood groups, to show differences not only between breeds but also within a breed.

In Table 3, where frequency of alleles are compared, it is interesting to see, that some alleles, common in Simental are very rare in Black and White cattle. For example the allele $BO_xY_1A'G'G_1''P'$ which occurs in Simental in 0.24% occurs in Black and White in 4.69%. Two alleles $G_2Y_2E_1'$ and I_2 which have a frequency exceeding 30% in Black and White are present in 0.35% of Simental cattle. On the contrary the allele G_1O_xA' which occurs in Simental in 5.2% is present in Black and White in 0.01% only and allele A′B′ with a frequency of 2.98% in Simental is absent in Black and White cattle.

On the basis of origin we compared the following groups of Simental bred in Poland; imported from Switzerland, Austria, G.F.R., Rumania and native Polish. No significant differences occur between these groups. Small numbers of imported cattle from Switzerland (35 individuals), G.F.R. (24) and Rumania (19) does not allow of such a comparison. Only the Simental imported from Austria are in large numbers. The comparison shows that some alleles: $Y_1Y'G'G_1''$, $BG_2KO_xY_2A'O'$, $BO_xY_1A'G'G_1''P'$ and $BO_xY_1Y'A'G'G_1''P'$ found only in the native group do not occur in the imported Simentals, but they are frequent in Black and White and Polish Red cattle. This suggests that these two breeds have had some influence on native Simentals.

TABLE 1

Frequencies of alleles in the B system of Simental cattle in Poland

No.	Alleles	%	No.	Alleles	%
1	b	16.53	39	$BG_2KO_xG'O'G_1''$	0.40
2	O_1I'	5.93	40	BG_2KO_xO'	0.36
3	O_1	5.93	41	P	0.36
4	G_1O_xA'	5.20	42	BG_2O_1	0.36
5	O_xO'	5.04	43	$BG_2KQG'O'G_1''$	0.32
6	$A'B'$	2.98	44	$G_2O_xO'G_2''$	0.32
7	$O_xY_1A'G'G_2''$	2.62	45	$BO_xY_1A'G'G_1''P'$	0.24
8	$O_1T_1K'G_2''$	2.62	46	BPY_1D'	0.20
9	$G'I'G_1''$	2.62	47	PI'	0.20
10	$O_1T_1G_2''$(nb K′)	2.06	48	O_xA'	0.20
11	$I'Y_2I'$	1.98	49	$Y_2Y'O'$	0.20
12	$I_1G_1''E_1'$	1.90	50	$Y_1Y'G'G_1''$	0.20
13	BI_1Q	1.90	51	$BG_2KO_xY_2A'O'$	0.16
14	$O_xY_2A'D'E_1'$	1.77	52	$O_xA'O'E_1'$	0.16
15	I'	1.65	53	$G_2Y_2E_1'$	0.16
16	$BT_1A'B'G_2''$	1.65	54	G_1''	0.16
17	BO_1	1.61	55	BY_2D'	0.16
18	$I_1QA'I'$	1.41	56	BQ	0.16
19	$O_1T_1Y_2G_2''$	1.25	57	$I_2O_xY_1D'A'$	0.16
20	$BI'P'$	1.01	58	$BO_xA'P'$	0.16
21	BG_2KO'	1.01	59	O_1A'	0.16
22	Q	0.01	60	$QG'G_1''$	0.12
23	$G_2T_1Y_2Y'A'B'D'G'$	0.97	61	Y_2	0.12
24	$I_1G'G_1''$	0.89	62	Y_2Y'	0.12
25	$O_1T_1G'K'O_1''$	0.85		rare occurring	<0.12
26	$O_1T_1I'G_2''$	0.81	63	PY_2	
27	$PQI'E_1'$	0.81	64	G_1O_xP	
28	$O_xT_1J'O'$	0.77	65	G_1G_1''	
29	$G'G_2''$	0.73	66	QI'	
30	$D'G'I'$	0.69	67	$Y_1G'I'G_1''$	
31	$O_1T_1I'K'G_2''$	0.60	68	BE_1'	
32	I_2Y_2A'	0.58	69	Y_1Y'	
33	BO_xQT_1A'	0.56	70	$BY_1Y'A'P'$	
34	I_1	0.52	71	$I_2G'G_1''$	
35	$PG_1''E_1'$	0.52	72	$O_xT_1A'B'G_2''$	
36	$O_xQA'O'E_1'$	0.52	73	I_2	
37	$PI'G_1''E_1'$	0.48	74	G_1	
38	$BG_2KG'O'G_1''$	0.44	75	$BO_xY_1Y'A'G'G_1''P'$	

TABLE 2

Comparison of phenogroups of the B system in three groups of Simental Polish, German and Swiss cattle

No.	Alleles	P	G	Sw	No.	Alleles	P	G	Sw
1	b	+	+	+	40	O_1T_1G''	+	+	+
2	BG_2KO'	+	+	+	41	$O_1T_1G'K'G_1''$	+	—	+
3	BG_2KO_xO'	+	+	+	42	$O_1T_1I'G_2''$	+	+	+
4	$BG_2KO_xY_2A'O'$	+	+	+	43	$O_1T_1I'K'G_2''$	+	+	+
5	$BG_2KO_xG'O'G_1''$	+	—	+	44	O_1I'	+	+	+
6	$BG_2KG'O'G_1''$	+	—	+	45	O_1A'	+	+	—
7	$BG_2KQG'O'G_1''$	+	—	+	46	$O_xY_1A'G'G_1''$	+	—	—
8	BG_2O_1	+	+	—	47	$O_xY_2A'D'E_1'$	+	—	+
9	BI_1Q	+	—	+	48	$O_xT_1A'B'G_2''$	+	—	—
10	BO_1	+	+	+	49	$O_xT_1J'O'$	+	—	—
11	BO_xQT_1A'	+	—	—	50	$O_xQA'O'E_1'$	+	—	+
12	$BO_xY_1A'G'G_1''P'$	+	—	—	51	O_xA'	+	+	+
13	$BO_xY_1Y'A'G'G_1''P'$	+	—	—	52	$O_xA'O'E_1'$	+	—	—
14	$BO_xA'P'$	+	—	—	53	O_xO'	+	—	
15	$BT_1A'B'G_2''$	+	—	—	54	P	+	—	—
16	$BY_1Y'A'P'$	+	—	—	55	PY_2	+	+	—
17	BY_2D'	+	—	—	56	$PQI'E_1'G_1''$	+	+	—
18	BPY_1D'	+	—	—	57	PI'	+	+	+
19	BQ	+	+	—	58	$PI'E_1'G_1''$	+	+	+
20	$BI'P'$	+	+	—	59	$PG_1''E_1'$	+	—	—
21	BE_1'	+	+	—	60	Q	+	+	+
22	G_1	+	+	—	61	QI'	+	+	—
23	G_1O_xA'	+	—	+	62	$QG'G_1''$	+	—	+
24	G_1O_xP	+	—	—	63	Y_1Y'	+	+	—
25	$G_2O_xO'G_2''$	+	—	—	64	$Y_1Y'G'G_1''$	+	—	—
26	$G_2Y_2E_1'$	+	—	—	65	$Y_1G'I'G_1''$	+	—	+
26	$G_1T_1Y_2Y'A'B'D'G'$	+	—	—	66	Y_2	+	+	—
28	I_1	+	+	+	67	Y_2Y'	+	+	—
29	I_1Y_2I'	+	+	+	68	$Y_2Y'O'$	+	+	—
30	$I_1QA'I'$	+	—	—	69	$A'B'$	+	—	+
31	$I_1G_1''E_1'$	+	—	+	70	$D'G'I'$	+	—	+
32	$I_1G'G_1''$	+	+	+	71	$G'I'G_1''$	+	—	+
33	I_2	+			72	$G'G_2''$	+	+	
34	I_2Y_2A'	+	—	—	73	$G'G_1''$	+	+	
35	$I_2O_xY_1A'D'$	+	—	—	74	I'	+	+	+
36	$I_2G'G_1''$	+	—		75	G_1''	+		
37	O_1	+	+	+					
38	$O_1T_2Y_2G_2''$	+	+	—					
39	$O_1T_1K'G_2''$	+	+	+					

P — Poland, G — G.F.R., Sw — Switzerland.

TABLE 3

Comparison of phenogroups of the B system in Simental cattle and Black and White cattle

No.	Alleles	Frequencies in breeds %	
		Simental	Black and White
1	$BO_xY_1A'G_1'G''P'$	0.24	4.69
2	$BO_xY_2A'G'G_1''P'$	—	4.69
3	$G_2Y_2E_1'$	0.16	13.35
4	I_2	<0.2	17.93
5	O_1I'	5.93	<0.01
6	G_1O_xA'	5.2	<0.01
7	$A'B'$	2.98	—

The large number of alleles which were found in Simental cattle in Poland is due to the mixture of different groups which took part in development of these cattle.

The number of individual Simental cattle examined is sufficiently large to be representative of the whole breed in Poland.

REFERENCES

ERHARD, L. and SCHMID D. O., (1965). Blood group studies on Pinzgau cattle. Blood groups of animals, Proc. of the 9th Eur. An. Blood Gr. Conf., Prague, 42–47.

GROSCLAUDE, F., (1965). Studies on the S blood group system in French cattle breeds. Blood groups of animals, Proc. of the 9th Eur. An. Blood Gr. Conf., Prague, 79–85.

MÜLLER, E., (1960). Contribution à l'étude des groupes sanguins de la race tachetée rouge du Simental, *Zeit. für Tierz. und Zücht.*, **74**, 2, 89–105.

RAPACZ, J., DOLA, L. and JAKÓBIEC, J., (1965). Blood group studies on B groups in Polish Red Cattle. Blood groups of animals, Proc. of the 9th Eur. An. Blood Gr. Conf., Prague, 39–42.

TRAUTMAN, J., (1961). Simental cattle in Poland, *Rocz. Nauk Roln.*, **95D**, 1–196.

DISCUSSION

J. BOUW: In your report you used the word "breed" for the Simental animal. Do you not think the word "breed" stays in contrast with the data presented by you?

J. RAPACZ: We do not think that the Simental cattle in Poland is a pure breed. It is a mixture as we said in our paper.

L. ERHARD: The reason of the differences of frequency of the B-alleles is that all high local cattle races are cross-breed between original Simental bulls from Switzerland and come from many primitive local races.

XITH EUROPEAN CONFERENCE ON ANIMAL BLOOD GROUPS AND BIOCHEMICAL POLYMORPHISM
Warsaw, July 2nd–6th, 1968

Blood Group Studies on the B-alleles of Red and White Lowland Cattle

Lidia DOLA, K. PAWŁOWSKI and Teresa KIJOWSKA-PETRYK

Department of Cattle Breeding, College of Agriculture, Cracow, Poland

SUMMARY. Eight hundred and ten blood samples of Lowland Red and White cattle from the south Wrocław and Opole regions were investigated. This represents a 0.3% proportion of the total population of the breed.

Blood group phenotypes were identified by the use of 44 reagents (24 reagents of B-system). In the tested cattle we were to establish 90 different B-alleles, of which 15 had a frequency above 1%, and 5 had a frequency above 5%: b (18.03%), G_2Y_2E' (17.43%), O' (7.55%), O_1 (6.78%) and G'' (6.51%).

In 667 of the samples we were able to establish both B-alleles, in 118 samples only one B-allele and in 25 samples the B-alleles could not be identified.

The frequency of the antigens in the A, C, FV, J, SU and Z systems was established and is recorded as a percentage.

INTRODUCTION

The history of Lowland Red and White cattle in Poland is neither long nor full, since in the interwar period flocks of this breeds were not numerous. It was only in the region south of Bielsko and Cieszyn that these cattle were bred by small breeders.

With the return of the Lower Silesia and the Opole districts to Poland a considerable number of Lowland Red and White cattle were taken over, a part of Eastern-Friesian origin being of dairy type, those from Westphalia and Schleswig Holstein of mixed milk or meat type, and the so-called Kłodzkie cattle, a cross of Silesia Red cows to Simental and Red and White bulls, were a dual-purpose breed.

As the result of the development of artificial insemination and the import of bulls from the German Federal Republic and The Netherlands, and due to favourable climatic conditions and adequate nutrition the breed has been stabilized into complex animals of sturdy build with a strong skeleton, and considerable musculature. The average weight of the cows is 600–650 kg with milk yield in recorded animals during the years 1957–1963 of an average of 3000 kg milk with 3.53% of fat.

Recent years have witnessed a marked increase of interest in this breed chiefly on account of carcase characteristics such as, a comparatively early maturity due to a rapid growth rate when supplied simultaneously with optimal amounts of fodder and pasture.

The number of the Lowland Red and White cattle amounts to 300,000 and constitutes 5% of the total cattle population. There are reasons to suppose that both the number and the area of distribution of this breed will increase considerably in the future.

METHODS AND MATERIALS

Blood group research on Lowland Red and White cattle began in 1963, as service work (parentage test) for County Animal Appraisal Stations, and up to 1967—527 animals have been tested. In 1967 — 283 additional cattle from farm No. 10 in Piława Dolna were typed directly by our Department. The results of all this testing were analysed and form the substance of this report.

A total of 810 cattle were tested, 756 animals being from Wrocław province, and only 54 animals from Opole province. The greater part (550 animals) were owned by individual breeders.

The ratio of 810 animals to the total number of the Lowland Red and White cattle is small, as low as 0.3%. However, it seems to be representative, because the animals under investigation came from all districts of the Wrocław province, where the cattle are bred. Moreover, the test animals represent the pedigree breeding stock and as such exert considerable influence on general breeding.

Blood groups were identified by the generally accepted method of the hemolytic test, using 42 test sera: $A\,Z'$; $B\,G_1\,G_2\,I_1\,I_2\,I_2^0\,K\,O_1\,O_x\,P\,Q\,T_1\,Y_2\,A'\,B'\,D'\,E'\,G''\,I'\,J'\,K'\,O'\,Y'$; $C_2R_1\,R_2\,W\,X_1\,X_2\,L'$; $F\,V$; J; L; M; $S_1\,S_2\,U_1\,U'$; Z and also test sera Kr3 and Kr5, which do not have international designations however unpublished data suggest that these two test sera belong to the B-system.

In majority of cases the B-genotype was established on the basis of tests of the sampled animals and their parents.

The genetic equilibrium was established for the FV-system and on the basis of this finding for the tested population.

Frequency of the B-alleles was calculated by the NEIMANN–SØRENSEN allocative method (1958). Phenotype frequencies were calculated for the systems A, C and SU, and the gene frequencies for the systems FV, J, L, M and Z.

RESULTS AND DISCUSSION

The results of this study are summarized in six Tables. Table 1 shows division of the tested population into B-genotypes.

The results show a great number of the animals with known B-genotype. Among

TABLE 1

Classification of B-alleles

Category	Number of animals	%
Both B-alleles determinable	667	82.34
Only one B-allele determinable	118	14.57
Unclassifiable phenotypes	25	3.09
Total	810	100.00
Heterozygotes	646	79.75
dom./dom.	481	59.38
dom./rec.	165	20.37
Homozygotes	21	2.59
dom.	3	0.37
rec.	18	2.22

them 4/5 were heterozygotes, and only 1/4 proved to be heterozygotes of the dominant/recessive type. A characteristic feature is the finding of a small number of homozygotes especially homozygotes of the dominant type, constituting 1/7 only of the total number of homozygotes.

Genetic equilibrium was established on the basis of the distribution of blood group factors in the FV-system.

TABLE 2

Gene frequency in FV-system

Genotypes	F^F/F^F	F^F/F^V	F^V/F^V
Observed No.	663	142	5
Expected No.	665	138	7

$\chi^2 = 0.5774$, $P = 0.05$, 1 d.f.
Gene frequency: FF = 0.9062; FV = 0.0938.

The observed and expected values are similar and the chi-square test shows no statistically significant differences. The disproportion of the F and V factors is very large and is 9:1.

Confirmation of 90 B-alleles in the tested population and a low degree of homozygosity (0.085) as calculated by summing the squares of the B-allele frequencies, points to great serological variety, probably chiefly as a result of mixed origin as described in the introduction. The findings show similarities to results of investigations on the genealogical structure of some herds of Lowland Red and White cattle. The inbreeding coefficient for this breed is only *ca.* 3.5% (PAWLIK, 1964; FĄFROWICZ, 1966). The analysis of the genealogical structure of all the bulls used by A. I. stations indicates the inbreeding coefficient to be as low as 0.7% between related groups (WĘŻYK *et al.*, 1966).

TABLE 3

B-allele frequencies in Lowland Red and White cattle

Allele	Gene frequency	Allele	Gene frequency	Allele	Gene frequency
B	0.0020	$G_2I_2O_1$	0.0025	O_xA'	0.0006
BO_1	0.0277	$G_2O_1Y_2$	0.0031	O_xO'	0.0013
BO_x	0.0006	G_2Y_2E'	0.1743	$O_xY_2E'D'Kr2O'$	0.0094
BQ_1	0.0051	$G_2Y_2D'E'O'$	0.0019	$O_xE'D'G'Kr20'$	0.0006
BO_1D'	0.0026	$G_2Y_2E'I'$	0.0013	P	0.0014
BO_1Y_2D'	0.0343	$G_2E'(Kr2)$	0.0090	PY_2	0.0006
$BO_1Y_2D'G'$	0.0006	$G_2O_1T_1Y_2E'$	0.0006	PI'	0.0087
$BO_1Q_1E'O'$	0.0013	G_2O'	0.0013	Q_1	0.0021
BG_2O_1	0.0043	$G_2A'E'Kr2K'Kr3$	0.0006	Q_1E'	0.0064
$BG_2Y_2O_xO'$	0.0128	I_1	0.0027	Q_1Y_2D'	0.0006
BG_2K	0.0013	$I_1E'Kr2(A)$	0.0011	Kr2	0.0651
$BG_2KO_xE'O'$	0.0044	$I_1G'Kr2$	0.0013	Kr20'	0.0020
$BG_2KO_1Y_2O'$	0.0006	$I_1Q_1E'K'$	0.0032	T_1B'	0.0006
$BG_2KY_2G'Kr2Kr3$	0.0019	$I_1J'K'O'$	0.0025	Y_2	0.0077
$BG_2KY_2G'Kr20'$	0.0044	I_2	0.0076	Y_2D'	0.0007
$BG_2KA'G'Kr20'K'Y'$	0.0013	$I_2E'Kr2$	0.0006	$Y_2G'I'$	0.0013
$BG_2KO_xY_2A'E'G'Kr21'$	0.0006	$I_2Y_2J'K'O'$	0.0025	$Y_2G'Kr2$	0.0117
$BG_2KO_xA'O'Y'$	0.0006	$I_2O_1T_1B'K'$	0.0006	$Y_2G'Kr2Y'$	0.0261
$BO_xQ_1A'Kr3$	0.0013	$I_2J'K'O'$	0.0019	$Y_2G'E'Kr2Y'$	0.0050
$BO_xG'Kr2Kr3$	0.0006	O_1	0.0678	$Y_2D'E'O'$	0.0198
$BO_xY_2G'Kr2Kr3$	0.0101	$O_1T_1B'E'Kr2$	0.0013	$Y_2D'E'G'Kr20'$	0.0220
BI_1	0.0006	O_1A'	0.0057	Y_2O'	0.0013
BI_1Q_1	0.0139	$O_1E'(Kr2)$	0.0013	Y_2Y'	0.0019
$BO_xI_1T_1Kr3$	0.0006	$O_1E'G'Kr2$	0.0013	O'	0.0755
BY_2	0.0007	O_1D'	0.0026	E'Kr2	0.0331
$BY_2G'I'Kr2Kr3$	0.0013	$O_1D'Y'$	0.0006	I'	0.0342
$BY_2G'Kr20'$	0.0025	O_1O'	0.0025	b	0.1803
G_1	0.0033	O_1Q_1	0.0296	$Y_2D'E'G'$	0.0082
G_1O_1	0.0157	O_x	0.0032	PQ_1E'	0.0006
G_2O_1	0.0034	O_xKr2	0.0026	G'Kr2	0.0034

TABLE 4

Gene frequencies in the J, L, M and Z-systems

	Blood group systems							
	J		L		M		Z	
Gene	J	j	L	l	M	m	Z	z
Number of animals	199	611	161	649	55	745	440	370
Frequency	0.1315	0.8685	0.1049	0.8951	0.041	0.959	0.3242	0.6758

TABLE 5

The frequency of various A, C, and SU-factors in Lowland Red and White cattle

	Blood factor															
	A	Z′	a	C_2	R_1	R_2	W	X_1	X_2	L′	c	S_1	S_2	U_1	U′	su
Number of animals	376	4	434	435	19	453	489	59	423	97	94	126	567	5	54	215
%	46.42	0.05	53.36	53.7	2.35	55.93	60.37	7.28	52.22	11.97	11.6	15.55	70.0	0.62	6.67	26.54

Out of the 90 B-alleles only 17 have a frequency greater than 1%. The following B-alleles are found in the greatest frequency: b–G_2Y_2E'–O′–O_1–G″–BO_1Y_2D'–I′–E′G″–O_1Q–BO_1–$Y_2G'G''Y'$–$Y_2D'E'O'$–G_1O_1–BI_1Q–$BG_2Y_2O_xO'$–$Y_2G'G''$–BO_xY_2G' G″Kr3.

Gene frequencies in the simpler systems indicate that the factors L and M are rare in the breed.

Phenotype frequency of factors of the remaining blood group systems A, C, SU indicates, that the most frequent are factor S_2 of the SU system, in contrast to factor U_1, and factors W, X_2 and R_2 of C-system are also frequent.

It is interesting that factor Z′ of A-system, characteristic of the Jersey breed was found in four of the tested animals.

TABLE 6

Gene frequencies of some common B-alleles in two populations of Red and White cattle

Allele	Gene frequency	
	M.R.Y. BOUW (1960)	Lowland Red and White
P	32.29	18.03
G_2Y_2E'	0.79	17.43
BOY_2D'	9.45	3.43
PI′	0.37	0.87
GE′	0.70	0.90
BO_1	1.25	2.77
Y_2	5.11	0.77
BIO_xQ_1	1.03	1.39
O_xO	0.97	0.13
B	0.47	0.20
F	86.10	90.62
V	13.90	9.38

The comparison of the results of our own studies with those obtained by BOUW (1960) in a population of 538 bulls of the Red and White cattle breed in The Netherlands (M.R.Y.) shows great difference between the two populations in number and frequency of alleles. In our cattle 90 B-alleles were established whereas in the Dutch population there were only 27. Certain differences were also found in the frequency of the factors of the FV-system. For a comparison of some common B-alleles, their frequency and the frequency of F and V antigens see Table 6.

The comparison of these data confirm on one hand the difference and the greater variability of our Lowland Red and White cattle to the Dutch breed and on the other hand indicate a considerable influence of imported bulls from The Netherlands on the structure of the Polish breed.

CONCLUSION

The tested population of Lowland Red and White cattle reveals great serological variation expressed by a large number of B-alleles and a low degree of homozygosity, probably because of diversity of origin of this breed in Poland.

REFERENCES

BOUW, J., 1960. The genetical composition of the Dutch cattle breeds as determiner by the frequencies of blood groups, *Zeit. für Tierzüchtung u. Züchtungsbiol.*, **B 7–4**, 248–266.

FĄFROWICZ, B., 1966. Struktura genealogiczna stada zarodowego nczb w POHZ Szczytna Śląska (Genealogical structure of the Lowland Red and White breed in the State Farm of pedigree breeding at Szczytna Śląska). Praca magisterska, Kraków.

PAWLIK, ST., 1964. Struktura genealogiczna stada rasy nizinnej czerwono-białej w POHZ Głogówek (Genealogical structure of Lowland Red and White breed in the State Farm of pedigree breeding at Głogówek). Praca magisterska, Kraków.

RABEK, A., 1957. Hodowla bydła nizinnego czerwono-białego w Polsce (The breeding of Lowland Red and White cattle in Poland), *Przegląd Hodowlany*, **9**, 79–83.

WĘŻYK, S. *et al.*, 1966. Analiza genetyczna grup krewniaczych buhajów rasy nczb eksploatowanych w PZUZ (The genetical analysis of related Red and White bull groups in A.I. stations). Wyniki Oceny Wartości Użytkowej Zwierząt, 3, PWRiL, 8–10.

XITH EUROPEAN CONFERENCE ON ANIMAL BLOOD GROUPS AND BIOCHEMICAL POLYMORPHISM
Warsaw, July 2nd–6th, 1968

The Frequency of B-alleles in the Polish Red Cattle from South Region of Poland

Lidia DOLA, K. PAWŁOWSKI, M. ORMIAN and Halina KASZYCKA
Department of Cattle Breeding, College of Agriculture, Cracow, Poland

SUMMARY. The number and frequency of the B-alleles in Polish Red cattle were determined by testing the breeding flocks of cattle from the southern regions of Poland. 1958 blood samples were tested by using 24 B-system test sera during the period 1963–1967.

We succeeded in establishing 141 B-alleles. In the FV-system genetic equilibrium was not found. Because of this the B-alleles frequency is given as a percentage of this total number of B-alleles.

The following alleles exhibit the highest frequencies: I_2 (9.11%), b (8.65%), Y_2Y' (3.61%), $O_xB'E'O'$ (7.89%), $G_2E'G''O'$ (7.81%), $I_1E'G'G''$ (5.98%), Y_2 (3.49%), I′ (3.33%).

INTRODUCTION

The origin of Polish Red cattle, is to be sought in the shorthorned autochtonic cattle which at one time occupied the region of middle Europe. In 19th Century-Poland the breed was maintained by peasantry and serfs since the rich landowners used to maintain rather fashionable imported breeds and their crossings. The original domestic cattle were small sized cattle of multiple purpose, the chief characteristics were perfect health, high feeding capacity and optimal feed utilization. Although the good qualities of these cattle were appreciated by KUROWSKI and OCZAPOWSKI the credit for initiating breeding and investigations is given to HOLDEFLEISS, WILKENS and ADAMETZ. The first Polish Association of Red Cattle Breeders came into existence in 1894 in Cracow. Breeding in the interwar period tended to keep the breed pure and was aimed at producing uniformity as well as at improvement in the conformation of different varieties, increase in milk yield and in percentage of fat. Even so the starting material possessed some admixture of foreign breeds and in particular of Red Duns, Anglers and Friesians. Prior to World War II the average output of the breed amounted to 2500 kg of milk with 3.8% fat the best individual yearly yield being 7000 kg of milk with fat reaching 6.5%. In addition, Polish Red cattle displayed high grazing capacity, high slaughter yield, and meat quality (KONOPIŃSKI, 1949).

Following World War II the area occupied by the Red Polish cattle was diminished and restricted to the regions of North- and South-East with the poorer climatic and soil conditions (the Białystok Province and those parts of the Olsztyn and Warsaw Provinces adjacent to it South-East part of the Lublin Province (a valley variety), and the Cracow and Rzeszów Provinces (a mountain sub-variety). New conceptions in the breeding policy and a very popular tendency to improve the breed by crossing it with imported breeds, e.g. Red Duns and in some regions with Jersey and Simental and greater use of artificial insemination brought about rapid diminition of the pure breed.

Actual pure breed output can be hardly spoken of since it must be considered in conjunction with the Crosses Red Polish × Red Duns and Red Polish × Jersey.

Blood group investigations of many authors, among others those NEIMANN-SÖRENSEN (1956), RENDEL (1958), demonstrate that the antigen composition of red corpuscles indicates the existence of the breed similarities or dissimilarities.

Blood group investigations in Red Polish cattle were initiated by SPRYSZAK (1960) and were also carried out by GASPARSKI *et al.* (1960) and RAPACZ *et al.* (1965). This study is a continuation of the previous ones.

METHODS AND MATERIALS

Blood group samples from Polish Red cattle were used as the experimental material. They were sent to our Laboratory for investigation of parentage. Studies were carried out during the years 1963–1967. A total of 1958 blood samples made up of samples from 1629 animals from the Cracow Province, 222 animals from the Rzeszów Province, 69 animals from the Lublin Province and 38 animals from the Kielce Province were tested. 90% of the samples came from individual farms. Among the experimental material there were samples of 1074 progeny from 775 cows and 109 bulls. Up to 80% of progeny was young bulls. All the sires except 2 were utilized by A. I. stations. The number of progeny from one bull varied from 1 to 65 animals, and only 9 bulls had more than 21 tested progeny.

All the experimental material was examined by the hemolytic test using following test sera: A Z′; B G_1 G_2 K I_1 I_2 I_2^0 O_1 O_x P_1 Q T_1 Y_2 A′ B′ D′ E′ G′ G″ I′ J′ K′ O′ Y′; C_2 R_1 R_2 W X_1 X_2 L′; F V; J; L; M; S_1 S_2 U_1 U′; Z; and Kr3 and Kr5.

Due to the lack of genetic equilibrium in the tested population for the factors in the FV-system, the percentage frequency only of the B-alleles was calculated. For the remaining blood group systems A, C, J, L, SU and Z the phenotype frequency only was estimated.

RESULTS AND DISCUSSION

In the tested cattle 141 different B-alleles were established. The position is summarized in Table 1, the percentage occurrence of different B-alleles being estimated. The most frequent alleles (above 3%) were the following: I_2 (9.11%), b (8.65%),

$O_xB'E'O'$ (7.89%), $G_2E'G''O'$ (7.81%), $I_1E'G'G''$ (5.98%), Y_2Y' (3.61%), Y_2 (3.49%) and I′ (3.33%).

The alleles O_1, O_xA', $G_2O_1Y_2Y'$, $BG_2KO_xA'O'$, T_1B', BO_1, G_2O_1, $BO_xY_2A'E'G'G''Kr3$ and $I_2D'E'G''$ were those with a frequency of 1–3%.

The other 124 alleles had a frequency below 1%. In the 6 B-alleles the antigen Kr3 was established and in 2, the antigen Kr5. Both these reagents were produced in our Laboratory but so far they do not have any international designation. The antigen Kr3 appeared in following alleles: $BO_xY_2A'E'G'G''Kr3$ (1.07%), $BO_xY_2A'E'K'Kr3$ (0.10%), $BO_xY_2A'E'G'G''Y'Kr3$ (0.10%), QKr3 (0.12%), $T_1B'Kr3$ (0.15%) and BKr3 (3.5%). The factor Kr5 appeared more seldom and was established in 2

TABLE 1

Classification of B-alleles

Category	Number of animals	%
Both B-alleles determinable	1541	78.70
Only one B-allele determinable	350	17.88
Unclassifiable phenotypes	67	3.42
Total	1958	100.00
Heterozygotes	1479	75.54
dom./dom.	1219	62.26
dom./rec.	260	13.28
Homozygotes	62	3.17
dom.	22	1.13
rec.	40	2.04

alleles only — $O_1Y_2K'Y'Kr5$ (0.15%) and $O_xY_2I'K'Y'Kr5$ (0.12%). In all these cases the sires of progeny with the factor Kr5 had a confirmed B-genotype, and the dams had at least 3 tested progeny. Table 2 shows a division of the tested samples according to the B-alleles possessed. It was possible to determine both B-alleles nearly 4/5 animals in nearly 4/5 of the animals tested. As many as 3/4 of the animals appeared to be heterozygotes. Only 3.17% were homozygotes, and the recessive homozygotes were twice as frequent.

As many as 141 different B-alleles were established in the tested population. This is many more than in the studies of SPRYSZAK (1960), GASPARSKI *et al.* (1960), and RAPACZ *et al.* (1965). The data are summarized in Table 3.

Considering that SPRYSZAK (1960) studied the population of only 94 animals and estimated 42 B-alleles using 21 reagents, GASPARSKI *et al.* (1960) studied 225 animals including 116 progeny from 10 bulls and estimated 52 B-alleles using 16 reagents, and RAPACZ established 132 B-alleles in a large population study it can be concluded, that Polish Red cattle are high variable serologically. The results of this report are not readily comparable with those of the other authors, because different test sera were used and in the previous studies smaller numbers of animals

TABLE 2

Gene frequency at B-system in Polish Red Cattle

Allele	Frequency	Allele	Frequency	Allele	Frequency
b	8.651	BO_1D'	0.102	O_xY_2A'	0.051
B	0.356	GO_1Y_2	0.153	O_xA'	2.417
BG	0.051	$GO_1E'O'$	0.051	$O_xA'K'$	0.127
$BG_1I_1O_1A'$	0.458	GO_xA'	0.153	$O_xA'K'E'O'$	0.153
$BGKO_xY_2O'Y'$	0.076	GI_1	0.865	$O_xB'E'O'$	7.888
BGKQA′E′G′G″	0.178	$GI_1E'O'G''$	0.051	$O_xB'K'$	0.051
$BGKOY_2A'O'$	0.127	GY_2E'	0.178	$O_xE'J'K'$	0.280
$BGKOY_2E'G'G''O'$	0.102	GE′O′	0.076	O_1I'	0.102
$BGKQY_2E'G'G''$	0.127	GE′I′O′	0.025	$O_xJ'K'O'$	0.433
BGKQE′G′G″O′	0.051	GE′O′G″	7.812	P	0.967
$BGKO_xE'G'G''$	0.127	GO′	0.153	PQ	0.102
BGKE′G′G″O′	0.051	GI′	0.076	PQE′I′	0.102
$BG_1I_1O_xY_2A'D'E'K'Y'$	0.433	GO_1Y_2Y'	1.781	PY_2	0.305
$BGKO_xA'O'$	1.679	I_1	0.178	PI′	0.382
$BGKO_xA'E'G'G''$	0.153	I_1O_1	0.025	PE′I′G″	0.051
BI_1	0.407	I_1O_1Q	0.051	Q	0.407
BI_1O_1	0.051	$I_1O_xQA'B'$	0.026	QKr3	0.127
BI_1Q	0.840	I_1Q	0.102	QE′	0.204
$BI_1QJ'K'O'$	0.229	$I_1QE'K'$	0.127	QI′	0.204
BI_1E'	0.051	$I_1Y_2A'G'G''K'O'$	0.025	T_1Y_2B'	0.051
BI_2	0.076	$I_1E'G'G''$	5.980	T_1B'	1.628
BI_2Q	0.051	$I_1E'I'G''$	0.026	$T_1B'O'$	0.076
BO_1	1.578	I_1G''	0.102	$T_1B'I'$	0.051
BO_1Q	0.662	I_2	9.110	$T_1B'I'O'$	0.076
$BO_1T_1A'D'E'K'$	0.127	I_2QI'	0.051	$T_1B'Kr3$	0.153
$BO_xQA'B'E'G'G''$	0.026	$I_2D'E'G''$	1.043	Y_2	3.486
BO_1Y_2	0.025	$I_2E'G''$	0.025	$Y_2D'I'$	0.102
BO_1Y_2D'	0.153	O_1	2.723	$Y_2D'E'O'$	0.051
$BO_1Y_2A'E'G'Y'G''$	0.153	O_1Q	0.127	$Y_2D'E'G'G''$	0.305
$BO_xY_2A'E'G'G''Kr3$	1.069	$O_1QE'O'$	0.127	$Y_2E'G'G''$	0.305
$BO_xY_2A'E'G'G''Y'Kr3$	0.102	$O_1T_1E'K'G''$	0.814	$Y_2G'Y'G''$	0.560
$BO_xY_2A'E'K'Kr3$	0.102	O_1T_1G''	0.025	Y_2I'	0.051
$BO_xA'O'G''$	0.076	O_1Y_2	0.382	$Y_2O'Y'$	0.382
$BO_xA'E'J'K'O'$	0.051	O_1Y_2Y'	0.076	Y_2Y'	3.613
BO_1E'	0.051	$O_1Y_2A'E'K'Y'$	0.102	A′E′	0.102
$BO_1E'K'$	0.051	$O_1Y_2E'J'K'$	0.025	E′	0.331
$BQT_1B'E'$	0.051	$O_1Y_2D'G'I'$	0.051	E′G′G″	0.051
BQE′G″	0.076	$O_1Y_2K'Y'Kr5$	0.153	E′G′G″O′	0.025
BQI′	0.026	$O_1A'J'K'$	0.076	E′G′G″Y′	0.127
BQG′	0.178	$O_1A'O'$	0.153	E′G′G″I′	0.178
$BY_2E'G''O'$	0.076	$O_1D'G'G''$	0.051	E′O′	0.534
$BPY_2G'Y'$	0.025	O_xQA'	0.127	E′G″O′	0.636
BG″	0.026	$O_xQD'E'G''$	0.102	E′G″	0.789
BKr3	0.356	$O_xQE'O'G''$	0.051	G″	0.229
G	0.153	$O_xQJ'K'O'$	0.076	G′G″	0.356
GO_1	1.094	$O_xT_1A'I'O'$	0.153	I′	3.333
GO_1Q	0.585	$O_xY_2I'K'Y'Kr5$	0.127	O′	0.611

TABLE 3
Number of B-alleles found in investigated population of Polish Red cattle

Study	No. of animals	B-alleles	Reagents
SPRYSZAK (1960)	94	42	21
GASPARSKI *et al.* (1960)	225	52	16
RAPACZ *et al.* (1965)	1289	132	24
Our investigation	1958	141	24

were used. The most readily comparable are the data of RAPACZ *et al.* (1965), but he does not give B-allele frequencies, and the 52.33% of the tested material had a phenotype only assigned. It can be stated however that in the population reported upon in this study lacked 31 alleles which were to be found in individual animals of the RAPACZ population and also two alleles found more frequently by RAPACZ. The cattle under investigation differ form the population examined by RAPACZ by 43 alleles. Some differences are also apparent in the frequency of the more common B-alleles detected in both populations.

The allele I′, which in the studies of RAPACZ occupies frequency position 13, in this population was found to be the most frequent allele. The reverse situation was

TABLE 4
The most frequent B-alleles in two populations of Polish Red cattle

B-alleles	RAPACZ *et al.* (1965)		Our investigation	
	No. of alleles	%	No. of alleles	%
b	199	7.75	340	8.68
Y_2	110	4.28	137	3.50
$G_2E'G''O'$	102	3.97	307	7.84
$I_1E'G'G''$	100	3.89	235	6.00
$O_xB'E'O'$	99	3.85	310	7.91
Y_2Y'	56	2.18	142	3.63
O_1	30	1.17	107	2.73
I′	26	1.01	131	3.34
G	24	0.93	43	1.10
BO_1	18	0.70	62	1.58
B	12	0.47	14	0.36
GO_1Y_2Y'	12	0.47	70	1.79
I_2	10	0.39	358	9.14

found with the allele Y_2 which is not in the position of 2nd most frequent but in that of 7th in this study. The differences in frequency of other alleles are of less significance. Due to the hypothetical character of this comparison it is possible that the changes in the frequency of the most frequent alleles are the outcome of an increase of the number of progeny from sires with the alleles I_2 and $O_xB'E'O'$.

TABLE 5

Gene frequency at FV-system in Polish Red cattle

Genotypes	F^F/F^F	F^F/F^V	F^V/F^V
Observed No.	1307	626	25
Expected No.	1341	559	58

$\chi^2 = 19.638$, $P = 0.01$, 1 d.f.
Gene frequency: FF = 0.8274; FV = 0.1726.

To estimate genetic equilibrium in the studied cattle, the expected order of genotypes was calculated and compared with the observed order, of factors of the FV-system as summarized in Table 5. The great difference between the observed and expected order of alleles (with $\chi^2 = 19.64$) indicates a lack of gene equilibrium in the studied population. The comparison of these results with those of SPRYSZAK (1960) indicates, that the frequency of the F factor in our study is 17% higher.

TABLE 6

Frequency of tested factors in A, C, J, L, SU and Z-systems in Polish Red cattle (%)

	Antigens										
	A	Z'	a	C_2	R_1	R_2	W	X_1	X_2	L'	c
Number of animals	1128	8	829	1504	233	838	1424	172	775	231	19
%	57.61	0.05	42.34	76.81	11.89	42.79	72.72	8.78	39.58	11.80	0.97

	Antigens										
	S_1	S_2	U_1	U'	su	J	j	L	l	Z	z
Number of animals	160	1412	251	267	260	506	1452	1064	894	1623	335
%	8.17	72.11	12.82	28.96	13.28	25.84	74.16	54.24	45.66	82.89	17.11

In addition, the phenotype frequency of the factors of the system A, C, J, SU, L and Z is presented in Table 6. The factor Z' of the A-system was established in the 8 animals. This allele is characteristic of the Jersey breed. 7 animals with this factor came from the Cracow Province and it is possible that this factor results from the crossing of bulls of the Jersey breed with our Polish Red cattle.

In the system C the factors C_2 (79.8%) and W (72.7%) are frequent. The rare factor R_1 of this system appeared with twofold higher frequency than in the study of SPRYSZAK (1960). The factor S_2 (72.1%) of the SU-system and factor Z of the Z-system both appeared to be frequent.

The frequent appearence of the antigens U_1 (12.8%) and U′ (29.0%) of the SU-system seems to be a characteristic of the Polish Red cattle.

CONCLUSIONS

Finding of 141 different B-alleles in Polish Red cattle points to great serological differentiation in this breed. Only the complex origin of the breed can be responsible for this diversity. The characteristic blood group factors for the Polish Red cattle seem to be those of the most frequent alleles of the B-system, and also the factors U_1 and U′ of the SU-system and the very frequent factor Z of the Z-system.

REFERENCES

GASPARSKI, J. *et al.*, 1960. Blood groups in cattle and their heredity-frequency of blood group genes in Polish Red cattle from the Cracow Region: B-alleles, *R.N.R.*, **76-b-3**, 565–568.

KONOPIŃSKI, T., 1949. Hodowla bydła, Ed. II, I.N.W. "Polska", Poznań, Vol. I, 286–321.

NEIMANN-SØRENSEN A., 1956. Blood groups and breed structure as exemplified by three Danish breeds, *Acta. Agr. Scand.* **2**, 216.

RAPACZ, J. *et al.*, 1965. Blood group studies on B-groups in Polish Red cattle, Proc. of the 9th European Animal Blood Group Conference, Prague, 39–42.

RENDEL J., 1958. Studies of cattle blood groups. I. Production of cattle iso-immune sera and the inheritance of antigenic factors, *Acta Agr. Scand.*, **8**, 40.

SPRYSZAK, A., 1960. Badania grup krwi u bydła rasy czerwonej polskiej (Serological examinations of Polish Red cattle breed), *R.N.R.* **76-B-1**, 1–23.

SZCZEKIN-KROTOW, Wł., 1965. Polskie bydło czerwone w woj. białostockim (Polish Red cattle breed in Białystok Province), PWRiL., Warszawa, 1–8.

DISCUSSION

J. BOUW: You found a frequency for the B-group I_2 of 9%; are you sure this group I_2 was present in this cattle before the import of Dutch F.H. cattle?

L. DOLA: It could be supposed that the high frequency of group I_2 is the outcome of the in crease of the number of progeny from the sires with the allele I_2, but the influence of Dutch F.H cattle can't be excluded too.

XITH EUROPEAN CONFERENCE ON ANIMAL BLOOD GROUPS AND BIOCHEMICAL POLYMORPHISM
Warsaw, July 2nd–6th, 1968

Bovine Red Cell Survival Studies in Haemoglobin and Transferrin Variants

L. P. NEETHLING, D. R. OSTERHOFF and I. S. WARD-COX

Section of Radiation Biology, Veterinary Research Institute and Department of Zootechnology, Faculty of Veterinary Science, University of Pretoria, Onderstepoort, South Africa

SUMMARY. Using Chromium-51 labelling techniques the apparent half-life time of autologous bovine erythrocytes was determined in 50 Afrikaner cows possessing all possible haemoglobin and transferrin types.

The results obtained indicate that no association exists between the haemoglobin genotype and the erythrocyte survival time. However, a definite relationship between the transferrin genotypes and the red cell survival time could be established.

INTRODUCTION

In the search for physiological factors which could be related to inherited blood components and in view of previous findings (NEETHLING and OSTERHOFF, 1966) it was decided to study the life span of bovine erythrocytes. Since a relationship exists between red cell turnover rate and basal heat production in various species (RODNAN *et al.*, 1957) one could expect to find differences between animals within the same species on grounds of their varying genetic make-up. A through study of the literature revealed a possibility of the existence of such a relationship (FITZSIMMONS *et al.*, 1967; MCSHERRY *et al.*, 1966; SCHNAPPAUF *et al.*, 1966).

METHODS AND MATERIALS

From a large experimental herd of Afrikaner cattle 50 cows of various ages were selected according to their transferrin and haemoglobin types. No differentiation was made between the alleles $Tf^{D}1$ and $Tf^{D}2$ nor between $Tf^{A}1$ and $Tf^{A}2$. These cows were separated from the herd and kept in a group under identical conditions.

In order to label the erythrocytes *in vitro* the following procedure was adopted: 30 ml of blood were collected from each animal using heparin as anticoagulant. The blood samples were centrifuged and the cells washed once with 40 ml ACD solution (3.0 g trisodium citrate, 0.015 g dihydrogen sodium phosphate containing

2 moles water of crystallization, 0.20 g glucose made up to 100 ml with dist. water). The excess ACD was removed and one ml of an isotonic solution of sodium chromate was added (specific activity greater than 20 mC/mg Cr: Philips Duphar, Petten, The Netherlands). This corresponded to an activity of 100 μC of the isotope Cr-51 per blood sample. The blood was shaken carefully and incubated for 30 min at 37°C.

Thereafter the samples were washed twice with saline to remove all excess activity and reconstituted with saline to the original haematocrit value. The autologous labelled blood was reinjected into the jugular vein of the animals. After 30 min a 5 ml sample of blood was collected in heparin from the opposite vein for counting purposes. Thereafter the animals were bled at regular intervals for a period of at least 45 days in order to establish the life span of the erythrocytes.

The counting was performed with a Philips Automatic Welltype Scintillation Detector (Type PW 4003) equipped with a $1\frac{3}{4} \times 2$ in. NaI/Tl crystal.

When the activity per ml erythrocytes was plotted on semilogarithmic graph paper against time a typical two-component curve resulted. From this curve the second component was taken as the apparent half-life of the erythrocytes. A straight line was fitted to all experimental points by the method of least squares from which the half-life was calculated.

RESULTS

The calculated half-life of the erythrocytes of Afrikaner cows varied between 11.9 days and 16.3 days. There seems to be a relationship between the half-life of erythrocytes and the transferrin type of the animal. The results pertaining to this relationship are set out in Table 1.

TABLE 1

The relationship between transferrin types and half-life of erythrocytes

Transferrin genotypes	No. of animals	Apparent half-life of erythrocytes, days	Standard deviation, days
Tf^D/Tf^D	8	15.5	0.9
Tf^A/Tf^A	7	14.0	1.1
Tf^A/Tf^E	9	13.8	1.1
Tf^A/Tf^D	9	13.5	1.2
Tf^D/Tf^E	9	13.4	1.1
Tf^E/Tf^E	8	12.7	1.2

The different values obtained for the half-life of erythrocytes for the various transferrin genotypes were subjected to the Bonferroni-*t*-statistical test (MILLER, 1966). The values of $t^{(a/2k)}$ (44) with $a = 0.05$ and $k = 15$ are given in Table 2.

TABLE 2

t-values of difference between half-lives of erythrocytes in various transferrin types

Transferrin types	Transferrin types					
	DD	AA	AE	AD	DE	EE
DD	—	2.54	2.94	4.46	3.63	4.61
AA		—	0.33	0.83	1.00	2.20
AE			—	0.54	0.71	1.90
AD				—	0.18	1.38
DE					—	1.21
EE						—

The critical value indicating statistical significance is $t^{(0.05/15)}(40)$ equal to 3.12.

The experimental animals were furthermore selected in such a way that they formed three groups according to the haemoglobin types being most prevalent in Afrikaner cattle: Hb AA, Hb AB and Hb AC. However, no interdependence between half-life of erythrocytes and haemoglobin types could be established.

DISCUSSIONS AND CONCLUSIONS

The results obtained indicate the existence of a relationship between transferrin types and the apparent half-life of erythrocytes in cattle. To confirm that such a relationship is indeed valid the data were subjected to a rigorous statistical analysis by using the Bonferroni statistics. The differences obtained between the half-lives of red blood cells of phenotypes DD and phenotypes AD, DE and E are statistically significant beyond doubt. The difference between phenotypes DD and AE borders on the significant level. However, considering the fact that only 40 degrees of freedom were used instead of 44 degrees of freedom in calculating the critical *t* value, this difference would most probably become significant. All other differences did not reach the level of significance. It should be noted that all types containing the transferrin allele Tf^E differ significantly from the transferrin phenotype DD. This indicates a unique function of the Tf^E allele compared to the other alleles.

The difference in the half-life of erythrocytes in different animals from the same breed has been noticed before by various authors (McSHERRY *et al.*, 1966; SCHNAPPAUF *et al.*, 1966) but no explanations for these differences were advanced. We believe that these discrepancies could solely or partially be ascribed to the different transferrin types of these animals.

The apparent half-life of erythrocytes in Table 1 is arranged in decreasing order which corresponds to our previously established order regarding the relative iron binding ability (RIBA) of the various genotypes (NEETHLING and OSTERHOFF, 1966). The only difference in the order of RIBA-values compared with the order for the half-lives is an exchange of genotype Tf^A/Tf^A with genotype Tf^D/Tf^D. Since there is no significant difference in the half-life of erythrocytes between transferrin genotypes Tf^D/Tf^D and Tf^A/Tf^A as indicated in Table 2, this exchange of order is not serious and can be neglected.

We are, however, more concerned with the special features attached to the transferrin allele Tf^E, also noted by previous authors (ASHTON, 1959; ASHTON, and LAMPKIN, 1965; CARR *et al.*, 1965; OSTERHOFF and NEETHLING, 1968). In the data presented here, the animals possessing the Tf^E alleles have the shortest red cell half-life. This in turn implies that these animals have in fact a higher basal metabolic rate (RODNAN *et al.*, 1957) than animals possessing the other transferrin alleles. This is in complete agreement with our previous findings that animals possessing the Tf^E allele have by far the poorest food conversion rate (NEETHLING and OSTERHOFF, 1966). A high metabolic rate in general means amongst other things, less storage of reserve nutrients. A low RIBA-value might therefore be expected. This again is in agreement with our findings that the transferrin allele Tf^E actually had the lowest relative iron binding ability. It must be stressed that this low RIBA-value is of secondary importance in relation to the high metabolic rate of the Tf^E animals.

The remarks appearing in the literature regarding possible advantageous properties of animals possessing the Tf^E allele concerning climatic and ecological tolerance, can now be viewed in a different light. Stresses of various kinds may be overcome more easily by animals possessing higher metabolic rates than those with lower metabolic rates. This is inherent in hormonal regulation of metabolism by the thyroid and adrenal glands (WRIGHT, 1952). It is interesting to note a considerable deviation of transferrin frequencies from normal values in a group of animals exhibiting hypothyroidism. In fifteen available animals, described in detail elsewhere (SCHULZ, 1962) we observed a decrease of 37% in the Tf^E frequency (0.267–0.167) and an increase of 61% in the Tf^D frequency (0.332–0.533).

The results of ASHTON (1959) and OSTERHOFF and NEETHLING (1968) seem to be valid since the Tf^E allele satisfies the requirements as outlined above. There appears to be no doubt that the so-called better adaptability of the Tf^E animals is in fact manifested in their greater ability to resist stress than animals possessing other transferrin alleles.

ACKNOWLEDGEMENTS

The Chief, Veterinary Research Institute, Onderstepoort, is thanked for permission to publish this paper.

REFERENCES

ASHTON, G. C., 1959. β-Globulin alleles in some Zebu cattle, *Nature*, **184**, 1135–1136.

ASHTON, G. C. and LAMPKIN, G. H., 1965. Serum albumin and transferrin polymorphism in East-African cattle, *Nature*, **205**, 209–210.

CARR, W. R., CONDY, T. B. and BURROWS, P. M., 1965. Transferrin polymorphism of indigenous cattle in Rhodesia and Zambia. (Typed copy obtained by authors).

FITZSIMMONS, W. M., SANSOM, B. F., HALL, J. G., STEWART, J. S. S., SELLWOOD, S. A. and HARNESS, E., 1967. Blood transfusion and red cell survival in dizygotic cattle twins, *Br. Vet. J.*, **123**, 397–402.

MCSHERRY, B. J., VAN DREUMEL, A.A. and ROBINSON, G. A., 1966. Apparent half-time of bovine red cells labelled *in vitro* with chromate-Cr-51 and auto-transfused into 10 normal calves, *Can. Vet. Jour.*, **7**, 176–179.

MILLER, R. J., JR., 1966. Simultaneous statistical inference, McGraw-Hill, New York.

NEETHLING, L. P. and OSTERHOFF, D. R., 1966. Radio-isotope studies on transferrins in cattle. Polymorphismes biochimiques des animaux, Paris, 261–266.

OSTERHOFF, D. R. and NEETHLING, L. P., 1968. Recent studies on cattle transferrins, *J. S. Afr. Med. Ass.* (In press).

RODNAN, G. P., EBAUGH, F. E. and SPIVEY-FOX, M. R., 1957. The life span of the red blood cell, the red blood cell volume in the chicken, pigeon and duck as estimated by the use of $Na_2(Cr\text{-}51)O_4$, *Blood*, **12**, 355–366.

SCHNAPPAUF, H. P., JOEL, D. D. and CRONKITE, E. P., 1966. Bestimmung des Gesamterythrozytenvolumens und der Scheinbaren Lebensdauer der Erythrozyten mit Chrom-51 beim Rind, *Zbl. vet. Med.*, Reihe A, **13**, 231–238.

SCHULZ, K. C. A., 1962. Proceedings of the 2nd Congress of the South-African Genetic Society, 90.

WRIGHT, S., 1952. Applied physiology, Oxford University Press.

XITH EUROPEAN CONFERENCE ON ANIMAL BLOOD GROUPS AND BIOCHEMICAL POLYMORPHISM
Warsaw, July 2nd–6th, 1968

A New Bovine Hemoglobin Type in Poland

H. BALBIERZ and Maria NIKOŁAJCZUK

Laboratory of Immunopathology, Department of Obstetrics and Gynecology, College of Agriculture Wrocław, Poland

Hitherto published results of studies on bovine hemoglobin in Poland report the incidence in adult animals of two hemoglobin types, i.e. slowly migrating HbA and HbB with faster migration rate in the electric field (EWY and WÓJCIK, 1961; WIŚLIŃSKI, 1961).

These two hemoglobin types of adult cattle are to be found in a number of races from various parts of the world (BANGHAM, 1957; CABANNES and SERAIN, 1955; HUISMAN, 1966; NIKOŁAJCZUK, 1961; NIKOŁAJCZUK *et al.*, 1962; SCHMID, 1962). Three other hemoglobin variants were reported to have been discovered recently. VELLA (1958); CROCKETT *et al.* (1963) detected a new hemoglobin type with migration rate intermediate to that of HbA and HbB, referred to as hemoglobin C. In 1965, EFREMOV and BRAEND found the African Mutur cattle to carry a new hemoglobin type wtih migration rate slower than that of HbA; it was termed HbD. In the same year, NAIK and SANGHVI (1965) identified in Indian cattle the hemoglobin referred to as Hb-Khillari.

OUR OWN INVESTIGATIONS

Material

While sampling blood from calves slaughtered in abbattoirs we found an individual whose hemoglobin revealed in electrophoretic separation an additional band situated on the cathode side of the band typical of HbA. Repeated separations on starch gel confirmed the fact that we have accidently discovered a new hemoglobin type in Poland.

Methods

Starch gel electrophoresis. Starch was hydrolysed in our laboratory. Electrolyte (A): Tris 20.2; boric acid 1.5; EDTA — Na_2 2.0/1 l.; buffer (B): Tris 9.196; citric acid 1.05/1 l.; gel buffer: 1/5 of electrolyte volume (A)+4/5 of buffer volume (B);

separation: 5.5 V/cm for 14 hr. Micro Immunoelectrophoretic Analysis (IEA): Agar — "Davis standard Agar", veronal buffer pH 8.2; 100 V for 1 hr and 15 min. Hemolysates with concentration of 5 g% were diluted with salt solution in the ratio of 1:4 and subjected to electrophoretic separation; after completion of the procedure three rabbit antisera to bovine hemoglobin types HbA, HbB and HbF were introduced. The slides were stained with benzidine and ponceau red.

Double diffusion. Sections taken from individual Hb zones separated by starch gel electrophoresis were transported to agar gel and the antiserum to bovine fetal hemoglobin type was put into central reservoir.

RESULTS

The electrophoretic separation of the hemolysate containing the new type of hemoglobin and of the control hemolysates revealed the pattern shown in Figs. 1 and 2.

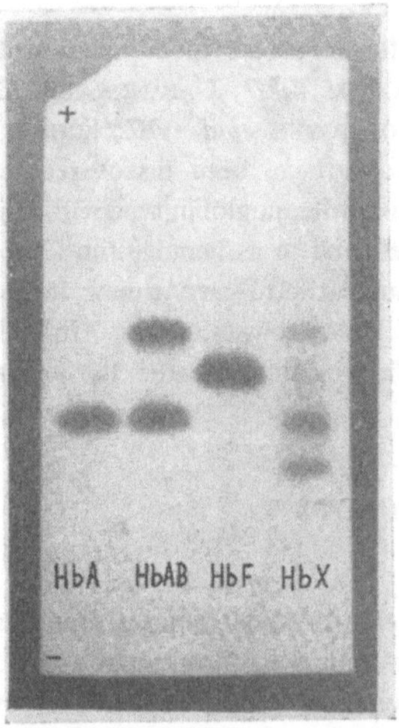

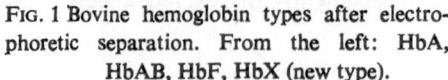

FIG. 1 Bovine hemoglobin types after electrophoretic separation. From the left: HbA, HbAB, HbF, HbX (new type).

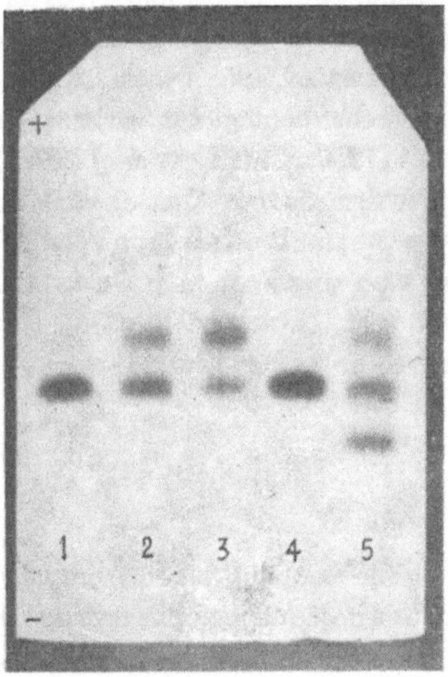

FIG. 2. Electrophoregram of blood hemolysate in calves. From the left: 6-week, 3-week, 1-week, 4-month, 19-day old (HbX).

The positions on the extreme right of both starch plates, show the separation of the hemolysate containing the new hemoglobin type. Nearest to the cathode is the Hb band hitherto unknown in Poland. The next remaining bands oriented towards the anode accordingly correspond to HbA and HbF* (Fig. 1); the frontal band next to anode, the one with the fastest migration rate shows a localization similar to that of control hemolysates obtained from the blood of calves aged about ten days or more (Fig. 2).

The IEA results for the new hemoglobin type are shown in Fig. 3. Benzidine staining, securing the differentiation of precipitation lines, revealed the details visible in Fig. 4.

The data obtained by double diffusion with reaction against HbF and the individual starch gel zones of Hb are shown in Fig. 5.

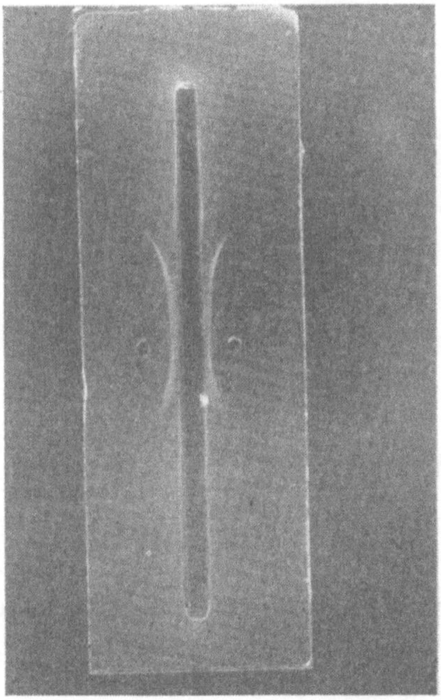

FIG. 3. IEA of the new hemoglobin with antiserum against HbF; rouge ponceau staining.

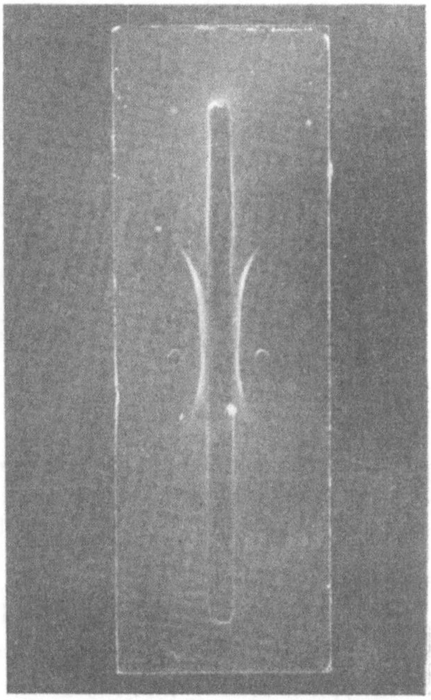

FIG. 4. The same: benzidine staining.

DISCUSSION

The band shown on the starch gel, as a result of electrophoretic separation, and having the lowest electrophoretic mobility may belong either to a new physiological hemoglobin type hitherto unknown in Poland or represent a pathological hemoglobin. As the calf having this hemoglobin was in perfect health, we are inclined

* When slaughtered the calf was 18-days old, weighing 47 kg; it was black-coloured and in excellent health. The failure to find its owner left the origin and parents of the calf undiscovered.

to accept the former hypothesis suggesting that the calf in question contained a physiologic hemoglobin type unfamiliar in this country.

The concurrent presence in the calf's hemolysate of the band characteristic of HbA proves that the calf was a heterozygote HbA/HbX. The slowest band detected referred to as X, does not seem to correspond to the band D of BRAEND and EFREMOV. The distance of separation from band A is greater than would be expected for band D*. The hemoglobin band with the highest migration rate is also worthy of comment**.

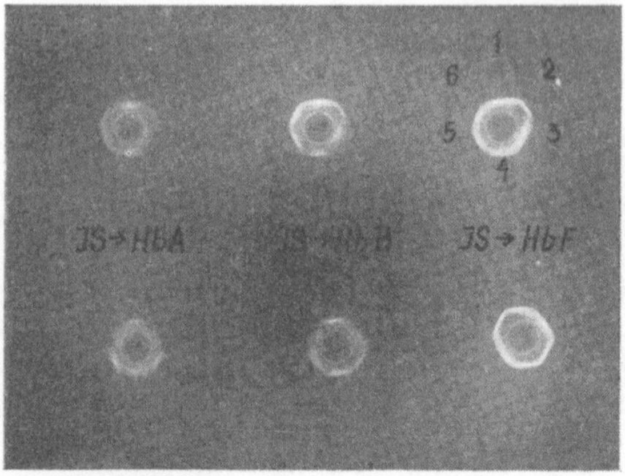

FIG. 5. Triple diffusion; peripheral grooves contain gel slices of various bovine hemoglobin types; central grooves: antisera directed against HbA (on the left), HbB (in the middle), HbF (on the right).

Our observations suggest that this band occurs in calves of 10 days or more; it is generally lacking in new-born ones; it can also be found — in addition to the band typical of HbF and HbA — in 3-week old calves.

The results obtained with the IE analysis using different antisera provide an additional evidence for the idea (BALBIERZ and NIKOŁAJCZUK, 1966) that antigenic differences between hemoglobin types are either nonexistent or so negligible as to be undetectable by the methods used. The IE preparations stained with ponceau red exhibit 3 arcs, one of them being conspicuous and the two others less so. Benzidine staining revealed that only two of the three possessed peroxidase properties. The double diffusion test showed antigenic consistence between the particular zones of the hemolysate.

We propose the term HbE for this new variant of hemoglobin.

---

* Detailed studies concerned with this hemoglobin fraction will be dealt within separate publication.

** We thank Dr. EFREMOV and Dr. BRAEND for their kind suggestions.

REFERENCES

BALBIERZ, H. and NIKOŁAJCZUK, M. A., 1966. Propriétés antigéniques de l'hémoglobine bovine. Polymorphismes biochimiques des animaux, Institut National de la Recherche Agronomique, Paris.

BANGHAM, A. D., 1957. *Nature*, **179**, 467.

CABANNES, R. and SERAIN, C., 1955. *C. r. Scienc. Soc. Biol.*, **149**, 7.

CROCKETT, J. R., KOGER, M. and CHAPMAN, H. L., 1963. *J. Anim. Sci.*, **22**, 173.

EFREMOV, G. and BRAEND, M., 1965. *Biochim. J.*, **97**, 267.

EWY, Z. and WÓJCIK, K., 1961. *Acta Phys. Polonica*, **12**, 441.

HUISMAN, T. H. J., 1966. Polymorphismes biochimiques des animaux, Institut National de la Recherche Agronomique, Paris. Xe Conf.

NAIK, S. N. and SANGHVI, L. D., 1965. Proc. 9th Europ. Conf. Anim. Blood Group Res., Prague.

NIKOŁAJCZUK, M. A., 1961. Observation sur les hemoglobines des bovides, Comptes rendus du 8e Congrès de la Société Européenne l'Hématologie, Wien.

NIKOŁAJCZUK, M. A., COQUELET, EYQUEM, A. and DE TRAVERSE, P. M., 1962. *Annales de L'Institut Pasteur*, **103**, 9, 421–432.

SCHMID, D. O., 1962. *Zentralbl. f. Vet. Med.*, **2**, 705.

WIŚLIŃSKI, M., 1961. *Acta Phys. Polonica*, **11**, 642.

XITH EUROPEAN CONFERENCE ON ANIMAL BLOOD GROUPS AND BIOCHEMICAL POLYMORPHISM
Warsaw, July 2nd–6th, 1968

Immunoelectrophoretical Study on Seminal Plasma and Spermatozoa from Bulls

L. VESELSKÝ

Laboratory of Physiology and Genetics, Czechoslovak Academy of Sciences, Liběchov, Czechoslovakia

SUMMARY. By means of immunoelectrophoresis we have observed the formation of antibodies and immunoelectrophoretic differences between spermatozoa and seminal plasma of bulls and the corresponding antibodies. We have found that bull seminal plasma forms with rabbit antiserum against seminal plasma a precipitation arc which passes through the β_1 region and reaches the albumin region, further three arcs passing through the β_1 region and reaching the α_1 region and then one arc which is placed in the β_1 region and reaches region α_2 and yet another one which may be found in the β_2 region. Ejaculated bull spermatozoa form with rabbit antiserum against ejaculated spermatozoa a continuous arc from the antigen reservoir up to the albumin region. This continuous arc, however, splits in the α_1 and albumin regions. Another precipitation arc is formed in region β_1. Ampullar spermatozoa of bulls form with rabbit antiserum against ampullar spermatozoa similar reactions. Spermatozoa from the tail of the epididymis of bulls form with rabbit antiserum against spermatozoa from the tail of the epididymis a precipitation arc which is continuous up to the α_1 region only. Another arc is formed in region β_1. Cross reactions between individual antigens were also investigated and their reciprocal immunological relationship was detected.

INTRODUCTION

Immunological properties of spermatozoa are changed to a considerable extent by contact with secretions of the accessory genital glands. LANDSTEINER (1899) made the observation that spermatozoa can evoke the formation of antibodies. These problems were later studied by a number of authors, e.g. MUDD and MUDD (1929), HENLE (1938). Antigenicity of seminal plasma was investigated by ROSS (1946), WEIL *et al.* (1956) and WEIL and FINKLER (1958). EDWARDS (1960), WEIL (1960), MATOUŠEK (1964) found that some seminal plasma antigens can be detected in the testis or the epididymis. WEIL and RODENBURG (1960) showed that spermatozoa acquire some antigens after contact with the fluids of accessory genital glands and that they have these antigens in common with the seminal plasma. Bull spermatozoa acquire blood group antigen J in this way (MATOUŠEK, 1961; SCHMID *et al.* 1964).

PERNOT (1956) and PERNOT and SZUMOWSKI (1958) dealt with immunoelectrophoretic study of spermatozoa and genital gland fluids in males and found that seminal plasma of bulls precipitated with heterologous antibodies against seminal plasma in the albumin and β-globulin regions. MATOUŠEK (1962) showed that bull seminal plasma formed five precipitation arcs with antibodies obtained by immunizing rabbits with this fluid. He also obtained antibodies against the fluid of the tail of the epididymis and the head of the epididymis from bulls. MARUTA and MOYER (1967) found by means of immunoelectrophoresis that testicular and epididymal tissue of guinea pigs contain eight antigens, two of which are tissue specific, four of them are specific for epididymal spermatozoa and epididymal fluid and two components are the same as those of the seminal plasma.

MATERIAL AND METHOD

Microimmunoelectrophoretic method, described by SCHEIDEGGER (1955) with tris buffer pH 8.9 according to ARONSSON and GRØNWALL (1957) was used for investigations. Antibodies were obtained by immunizing rabbits without adjuvant as described by BRUMMERSTEDT-HANSEN (1961). Ejaculated spermatozoa were collected by means of an artificial vagina. Epididymal and ampullar spermatozoa were obtained directly from these organs after the animal had been killed. Spermatozoa were washed 3–6 times in normal saline before use. The separation of protein components of spermatozoa and seminal plasma of bulls was compared with the separation of protein components in the blood serum of bulls. Antisera yielding the maximum number of precipitation arcs were used.

RESULTS

Bull seminal plasma with antibodies against seminal plasma forms one arc which passes through the β_1 region and reaches the albumin region, a further three arcs which pass through the β_1 region and touch the α_1 region and one arc situated in the β_1 region and reaching region α_2 and finally one arc in β_2 region. The cross reaction of seminal plasma with antiserum against ejaculated spermatozoa is manifested by a precipitation arc passing through the β_1 region and reaching continuously the albumin region. The cross reaction of seminal plasma with antiserum against ampullar spermatozoa of bulls is manifested by one arc which stretches from the antigen reservoir, continuously through the albumin region. The cross reaction of seminal plasma with antiserum against spermatozoa from the epididymis from bulls is manifested by one arc which stretches, from the antigen reservoir, continuously through the α_2 region.

Ejaculated spermatozoa of bulls with antiserum against ejaculated spermatozoa form a continuous precipitation arc from the antigen reservoir up to the albumin region. This continuous arc splits in the α_1 and albumin regions. Another precipi-

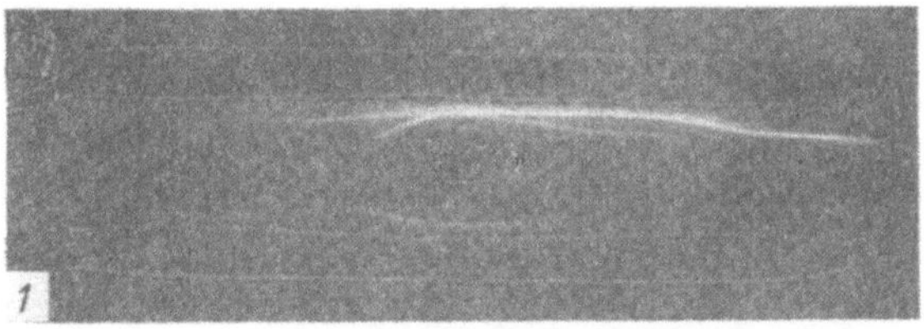

FIG. 1. Above: precipitations of bull seminal plasma with antiserum against seminal plasma from bulls. Below: cross reaction of bull seminal plasma with antiserum against ejaculated bull spermatozoa.

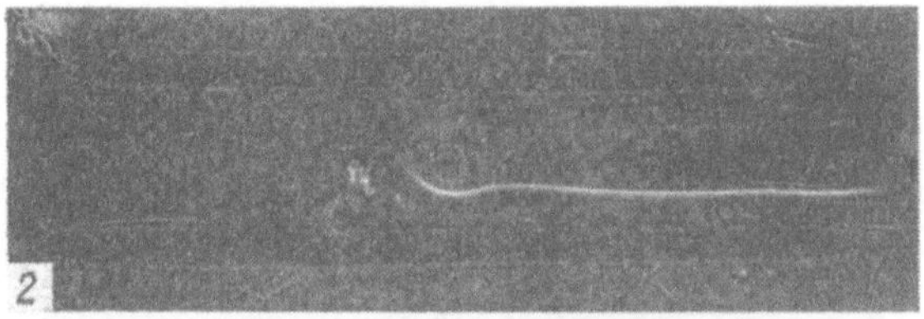

FIG. 2. Above: precipitations of ejaculated bull spermatozoa with antiserum against ejaculated spermatozoa. Below: cross reactions of ejaculated spermatozoa with antiserum against seminal plasma from bulls.

tation arc may be found in β_1 region. Similar precipitation arcs may be observed in ejaculated spermatozoa of bulls with antisera against seminal plasma of bulls. The cross reaction of ejaculated bull spermatozoa with antiserum against ampullar spermatozoa forms an arc which proceeds from the antigen reservoir to the albumin region. The cross reaction between ejaculated spermatozoa of bulls and the antiserum against spermatozoa from the tail of the epididymis of bulls forms a precipitation arc which proceeds from the antigen reservoir through the α_2 region and reaches α_1 region.

Ampullar bull spermatozoa with antiserum against ampullar bull spermatozoa form a precipitation arc which proceeds from the antigen reservoir through the α_2 and α_1 regions and reaches the albumin region, and another arc may be found in the β_1 region. Ampullar spermatozoa cross-react with antiserum against bull seminal plasma in one arc which passes through the α_2 and α_1 regions and reaches the albu-

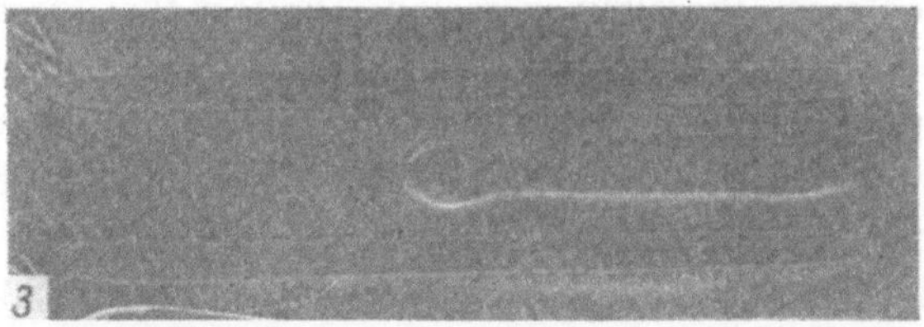

FIG. 3. Above: precipitations of ampullar bull spermatozoa with antiserum against ampullar spermatozoa. Below: cross reactions of ampullar bull spermatozoa with antiserum against bull seminal plasma.

min region. They also cross-react with antiserum against ejaculated spermatozoa of bulls, the cross reaction being manifested by two arcs, situated in the β_1 and α_2 regions. An analogous picture is produced by the cross reaction between ampullar spermatozoa and antibodies against spermatozoa from the tail of the epididymis of bulls.

Spermatozoa from the tail of the epididymis form with antiserum against spermatozoa from the tail of the epididymis a precipitation arc which passes from the antigen reservoir through the α_2 region into α_1 region, and another arc which is situated in the β_1 region. Antiserum against seminal plasma forms the same cross reaction with spermatozoa from the tail of the epididymis.

FIG. 4. Above: precipitations of spermatozoa from the tail of epididymis of bulls with antiserum against spermatozoa from the tail of the epididymis. Below: cross reactions of spermatozoa from the tail of epididymis with antiserum against ejaculated bull spermatozoa.

Cross reaction with antiserum against ejaculated spermatozoa manifests itself by an arc which reaches the α_1 region from the antigen reservoir. Cross reaction with antiserum against ampullar spermatozoa shows an arc in the α_1 region.

DISCUSSION

By means of immunoelectrophoresis we have found that spermatozoa from the tail of the epididymis are able to form antibodies. Here precipitation is manifested by an arc which proceeds from the antigen reservoir to the α_1 region. Further antigens attach themselves to spermatozoa in the ampulles of vas deferens.

Ampullar and ejaculated spermatozoa with corresponding antibodies form precipitations which reach the albumin region. The fact that the precipitation arc is continuous along its whole length, splitting only in the α_1 and albumin regions, may prove that some antigens on bull spermatozoa have different electrophoretic mobility but are antigenically very much alike. GRABAR (1958) referred to a similar close relationship in γ-globulin and albumin in the blood serum, where these precipitation lines are composed of several components. We can also supplement the findings of PERNOT and SZUMOWSKI (1958) that bull seminal plasma forms with antiserum against seminal plasma precipitations not only in the albumin and β-globulin regions but also in the α-globulin region.

It is possible to assume that the strongest antigens, attached to spermatozoa from the seminal plasma, are components belonging to the α_1-and β_1-globulin and albumin regions with respect to their electrophoretic mobility.

CONCLUSION

By means of immunoelectrophoresis we have observed the formation of antibodies which can be produced by spermatozoa from the tail of the epididymis, ampullar and ejaculated spermatozoa of bulls. Spermatozoa gain, by contact with genital gland fluids, some antigens which correspond to antigens of the seminal plasma.

REFERENCES

ARONSSON, T. and GRØNWALL, A., 1957. Cited by MICHALEC *et al.*, 1959. Elektroforeza na papíře a jiných nosičich, 1–135, NČSAV, Praha.

BRÜMMERSTEDT-HANSEN, E., 1961. Immunoelectrophoretic investigations on the blood serum of adult pigs, *Acta Vet. Scand.*, **2**, 254–262.

EDWARDS, R. G., 1960. Complement-fixing activity of normal rabbit serum with rabbit spermatozoa and seminal plasma, *J. Repr. Fert.*, **1**, 268–282.

GRABAR, P., 1958. The use of immunochemical methods in studies on proteins, *Advances in Protein Chemistry*, **12**, 1–33, Academic Press Inc., New York.

HENLE, W., 1938. The specificity of some mammalian spermatozoa, *J. Imm.*, **34**, 325–336.

LANDSTEINER, K., 1899. Cited by WEIL, A. J., 1961. Antigens of the adnexal glands of the male genital tract, *Fert. Ster.*, **12**, 538–543.

MARUTA, H. and MOYER, D. L., 1967. Immunologic studies of the antigens of guinea pig semen, *Fert. Ster.*, **18**, 649–658.

MATOUŠEK, J., 1961. Antigenic factor J in some body fluids of cattle, *Folia Biol.* (Praha), **7**, 252–257.

MATOUŠEK, J., 1962. The study of the antigenic characteristics of sperm from bulls, The 8th Animal Blood Group Conference in Europe, Ljubljana, Yugoslavia.

MATOUŠEK, J., 1964. Antigenic characteristics of spermatozoa from bulls, rams and boars. III. Absorptions analyses, precipitins and fructolysis in relation to the antigenicity of bull spermatozoa, *J. Repr. Fert.*, **8**, 13–21.

MUDD S. and MUDD, E. B. H., 1929. The specificity of mammalian spermatozoa with special reference to electrophoresis as a means of serological differentiations, *J. Imm.*, **17**, 39–51.

PERNOT, E., 1956. Recherches sur les constituants antigéniques des spermatozoides de cobayes, *Bull. Soc. Chim. Biol.*, **38**, 1041–1054.

PERNOT, E. and SZUMOWSKI, P., 1958. Étude electrophorétique et immunoélectrophorétique des proteins du plasma séminal de taureau, *Bull. Soc. Chim. Biol.*, **40**, 1423–1434.

ROSS, V., 1946. Precipitin reactions of human seminal plasma, *J. Imm.*, **52**, 87–96.

SCHEIDEGGER, J., 1955. Une micro-méthode de l'immunoélectrophorèse, *Int. Arch. Allergy Appl. Immun.*, **7**, 103–109.

SCHMID, D. O., CONNEALY, P. N. and STONE, W. H., 1964. Blood group antigens on bull spermatozoa, Meeting of the Midwestern Section A.S.A.S., November 27–28, Chicago, Illinois.

WEIL, A. J., KOTSEVALOV, O. and WILSON, L., 1956. Antigens of human seminal plasma, *Proc. Soc. Exp. Biol. Med.*, **92**, 606–610.

WEIL, A. J. and FINKLER, A. E., 1958. Antigens of rabbit semen, *Proc. Soc. Exp. Biol. Med.*, **98**, 794–797.

WEIL, A. J., 1960. Immunological differentiation of epididymal and seminal spermatozoa of the rabbit, *Science*, **131**, 1040–1041.

WEIL, A. J. and RODENBURG, J. M., 1960. Immunological differentiations of hunam testicular (spermatocele) and seminal spermatozoa, *Proc. Soc. Exp. Biol. Med.*, **105**, 43–45.

XITH EUROPEAN CONFERENCE ON ANIMAL BLOOD GROUPS AND BIOCHEMICAL POLYMORPHISM
Warsaw, July 2nd–6th, 1968

Studies on Ox Red Cell Membranes: Detection of Red Cell Antigens in Ox Red Cell Protein

R. L. SPOONER and A. H. MADDY

Cattle Blood Typing Service, A.R.C. Animal Breeding Research Organization, Edinburgh, Great Britain
Department of Zoology, University of Edinburgh, Great Britain

SUMMARY. Ox red cell membranes have been extracted with butanol. The presence of F, V, J, L, R′, S′, and T′ has been detected in the protein fraction but all activity for other blood group antigens was lost. Protein has been used for isoimmunization in cattle and produced high titre antibodies specific for the injected antigen.

INTRODUCTION

The biochemistry of blood group antigens in man has been the subject of intensive investigation over the last 20 years (WATKINS, 1967).

Little work has been performed however, on the chemistry of ox red cell antigens, other than the work on the soluble J substance which is similar to human A and probably a glycoprotein (BEDNEKOFF *et al.*, 1958; HAYASHI *et al.*, 1958). The fractionation of ox red cell membranes by cold butanol extraction described in a previous paper (MADDY, 1966) provided a starting material for initial investigations on the chemistry of ox red cell antigens. Using inhibition tests it has been possible to identify certain antigens in the protein fraction. Some of these antigens have been found to be highly immunogenic when injected into cattle.

MATERIALS AND METHODS

Inhibition tests

Titrations of inhibitors were prepared in 0.05 to 1 ml quantities and then distributed to the wells of plastic microtitre plates (Cook Engineering). To these dilutions of inhibitor were added 1 drop quantities of the appropriate blood grouping reagents at a concentration of 2 haemolytic doses, and after thorough mixing the plates were allowed to stand at room temperature for 4 hr, then 1 drop of a mixture containing ox red cells possessing the antigens being tested for and complement diluted

1:4 were added to each well. Plates were covered with sellotape and incubated overnight at 28°C being shaken continuously on a mechanical shaker. The degree of inhibition of the blood typing reagent was read the following morning.

Complement. Rabbit complement obtained in bulk from a commercial slaughterer and stored in liquid nitrogen was diluted 1:4 before use.

Diluent. Inhibition tests were performed in c.f.t. diluent (oxoid).

Immunizations

Injections of protein were made in double emulsions. The double emulsion was prepared by the emulsification of the protein (approx. 1 mg/ml) in a Freund's complete adjuvant and followed by the re-emulsification of this emulsion with an equal volume of 1% Tween 80 (HERBERT, 1965). Approximately 2 ml of protein was injected I/M.

Butanol extraction of protein. Samples were collected in A.C.D. from the abattoir and A.B.R.O. experimental herd and extracted as described previously (MADDY, 1966).

Blood typing reagents. All of the reagents used in this experiment were bovine isoantisera prepared either by the injection of washed bovine red cells or bovine whole blood. Most of the reagents produced in this way required absorption to make them monospecific.

RESULTS

Twenty-one blood samples have been tested and the results are shown in the Table. This shows the number of animals with the relevant red cell antigens and the number with these same antigens in the protein. In the early experiments the only reagents used were those reacting with antigens detectable on the intact red cells. As the experiment proceeded it was found that the antigens F, V, J, L, R′, S′ and T′ were present in the protein, and thereafter all cells were tested for each of these reagents. Testing with titrations of the reagents for the A, B, C, Z, and S systems was discontinued after 12 cells had been found uniformly negative, but proteins are now being screened undiluted against these reagents. From the results for F, V, J, L, R′, S′, and T′ it can be seen that the antigens are only found in the protein if they are also in the red cell membrane, and moreover they are always found in the protein if they are on the red cells, except for T′ where is some suggestion that this antigen may be detected in the protein when not detectable on the red cell.

A typical protocol of an inhibition test is shown in the Figure, where it can be seen in the inhibition for R′, T′, F and V that the antigen is only found when it is present in the red cell, and can be detected to a dilution of 1:64 with R′, T′ and V and 1:16 with F. There is virtually no inhibition of any of the other reagents.

Number of animals with and without the specified red cell antigens and the occurrence of these antigens in the protein extracted from the red cell membranes

Red cell antigen	Status	Number of animals	
		red cell	protein
F	+	19	19
	—	2	2
V	+	12	12
	—	9	9
	+	9	9
	—	5	5
L	+	8	8
	—	9	9
R′	+	3	3
	—	7	7
S′	+	Not tested	11
	—	Not tested	1
T′	+	6	6+4
	—	7	3

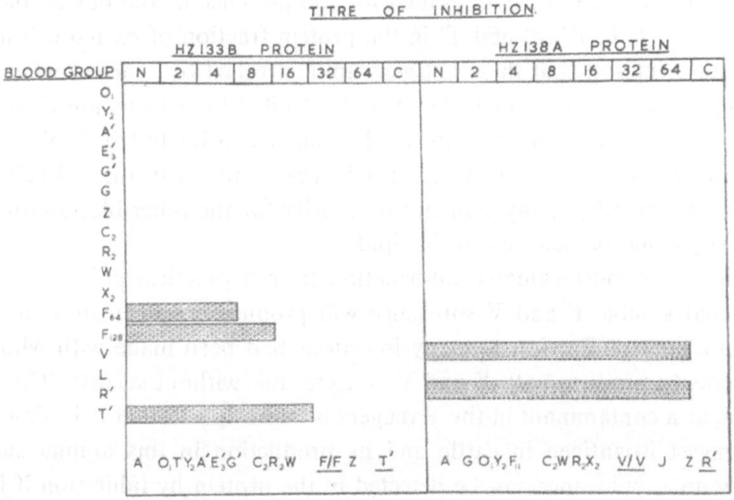

A typical protocol of an inhibition test with the inhibition of F, V, R′ and T′ in the cells having these antigens. There is no inhibition of the other reagents.

Lipid

Extraction of red cells with butanol yields soluble protein, lipid and a minor interfacial component. These lipid and interfacial components were tested in 2 cases in a search for the antigens not present in the protein. They were uniformly negative. In one experiment lipid and protein were mixed. This merely served to reduce the inhibitory titre of the protein by an amount proportional to the volume of lipid added.

Immunizations

Attempts have been made to produce antisera by injecting these soluble protein extracts into cattle in Freund's complete adjuvant. Protein from an FF animal injected into a VV animal has produced a pure anti F reagent which did not need any absorption. It gave complete haemolysis of F positive cattle cells to a dilution of 1:256.

Injection of VV protein into an FF animal in the early stages of immunization produced a pure V reagent, but after further immunization anti A was produced also. It would thus appear that some A substance is left in the protein. It is, however, at too low a concentration to be detected by the inhibition test, but because it is an efficient antigen it is able to stimulate anti A production.

DISCUSSION

In this paper evidence is presented for the presence of the bovine blood groups antigens F, V, J, L, R′, S′ and T′ in the protein fraction of ox red cell membranes. It is of some interest that these antigens are controlled by genes at what at the moment appear fairly simple loci. Antigens controlled by a locus are either all in the protein or none are in the protein. In the animals so far tested F, V, J, R′, S′ and T′ have always been present in the protein when present in the red cell, but on no occasion has there been any evidence of activity for the other blood group antigens. No activity could be detected in the lipid.

Some of the most valuable information from a practical point of view was the finding that soluble F and V substance will produce excellent antibodies when injected in adjuvant. Previously many injections had been made with whole red cells in attempts to produce both F and V antisera, but without success. The production of anti A as a contaminant in the V reagent is interesting in that A is almost certainly the strongest isoantigen in cattle and its production in this animal suggests that although no A substance can be detected in the protein by inhibition if it had been there at a very low concentration, it might not have been detected directly but could have been sufficient to stimulate antibody production. The complete loss of the antigens of the B, C, and S loci suggests that in these very much more complex loci the variability may depend on spatial configurations which collapse when the membrane is extracted by this relatively mild procedure.

REFERENCES

BEDNEKOFF, A. G., STONE, W. M., IRWIN, M. R. and LINK, K. P., 1958. Chemical studies on the J substance of cattle serum, Proc. Xth Int. Congr. Genet., **2**.

HAYASHI, J. A., STONE, W. M., LINK, K. P. and IRWIN, M. R., 1958. The J substance of cattle. V. Immunochemical studies of the J substance from bovine gastric mucosa, *J. Immunol.*, **81**, 82–90.

HERBERT, W. J., 1965. Multiple emulsions; a new form of mineral-oil antigen adjuvant, *Lancet Oct.*, **16**, 771.

MADDY, A. H., 1966. The properties of the protein of the plasma membrane of ox erythrocytes, *Biochim. Biophys. Acta*, **117**, 193–200.

WATKINS, W. M., 1967. "Blood-group substances" from "Specificity of cell surfaces" (Eds. Davis and Warren), Prentice-Hall, Inc., 257–279.

DISCUSSION

J. RENDEL: Have you ever tried using your blood grouping antisera in immunoelectrophoresis against the protein extracts?

R. SPOONER: No, we have not yet done this but intend to in the near future.

XITH EUROPEAN CONFERENCE ON ANIMAL BLOOD GROUPS AND BIOCHEMICAL POLYMORPHISM
Warsaw, July 2nd–6th, 1968

Studies on IgG Allotypes in Cattle

R. L. SPOONER
Cattle Blood Typing Service, A.R.C. Animal Breeding Research Organization, Edinburgh, Great Britain

SUMMARY. Three bovine isoantisera to IgC are described. They detect three different specificities. Evidence for the identification of the antigen is presented but so far no evidence for the inheritance of these characteristics has been obtained.

INTRODUCTION

The first demonstration of precipitating isoantibodies to rabbit serum proteins was described by SCHÜTZE in 1902. The rediscovery of isoprecipitins, this time in agar by OUDIN (1956) was however the beginning of a rapid development of the field of serum protein allotypes (KELUS and GELL, 1967). Until recently there has been little success in producing isoantisera to bovine serum proteins, but the existence of such isoantibodies has now been shown. The first report of γ globulin allotypes was by MILLOT (1966) and a recent report has described precipitating antibodies to a bovine macroglobulin which may be IgM (RAPACZ *et al.*, 1968).

This report describes some characteristics of isoantisera reacting with bovine IgG.

MATERIALS AND METHODS

Immunizations

Cattle 6 monthly I/M injections of approximately 200 mg of a Na_2SO_4 ppt of whole serum in Freund's complete adjuvant.

Rabbits I/M injections of 1 ml bovine serum in adjuvant at monthly intervals for 4 months and bled 2 weeks after the last injection.

Immunoelectrophoresis

The micro immunoelectrophoretic technique of SCHIEDEGGER (1955) was used with minor modifications. Both gel and tank buffer were used at pH 8.6, $i = 0.05$ in 1.25% agar.

D.E.A.E. cellulose chromatography

Stepwise elution from DE 11 (Whatman) using buffers pH 6.3, 0.0175 M; pH 5.7, 0.04 M; pH 5.3, 0.1 M; and pH 4.8, 0.3 M. Samples were concentrated to the original volume by pervaporation and dialysed against isotonic saline before use.

G.200 Sephadex chromatography

Serum was eluted from G.200 Sephadex using a buffer of 0.1 M tris, 0.5 M NaCl pH 7.5, from a column of 60×2.2 cm using upward flow.

IgM globulin preparation

Serum was dialysed against 2 changes of water for 4 hr. The euglobulin precipitate was removed by centrifugation and then redissolved in saline. The redissolved euglobulin was then passed through G.200 Sephadex yielding an electrophoretically pure IgM globulin in the first elution peak.

RESULTS

Initially 6 cattle were immunized with protein from 2 other cattle. 4 were injected with a Na_2SO_4 fraction of serum, and 2 with ferritin antiferritin precipitates thoroughly washed to remove absorbed serum proteins. All 6 cattle produced precipitating antibodies against both immunizing sera but only 2 were investigated further, and these had been injected with the Na_2SO_4 fraction of serum from U.74 A. Serum from one of the reacting animals was then used as antigen for injecting into 3 other cattle. One of these animals has so far produced an isoprecipitin. Sera from 2 animals from the first series and the one from the second series were examined further by immunoelectrophoresis. All of the sera produced precipitation lines in the IgG region.

When the immunizing serum was fractionated on G.200 Sephadex the sera reacted only with fractions obtained from the second peak of protein eluted from the column.

On D.E.A.E. cellulose the reacting protein was only in the first elution peak, a peak which contained only IgG globulin (Fig. 1).

As a further check IgM globulin was prepared and this is also shown in Fig. 1 to be electrophoretically pure. When the bovine sera were tested against these fractions they were found to react only with the IgG as is also shown in Fig. 1.

It was found that the dimensions of the gel were critical for the production of precipitates at optimal proportions, and that there was a marked variation between the different sera. An example of an Ouchterlony plate is shown in Fig. 2. Clearly unless the optimal proportions are worked out for each antiserum the presence of antibodies can be overlooked.

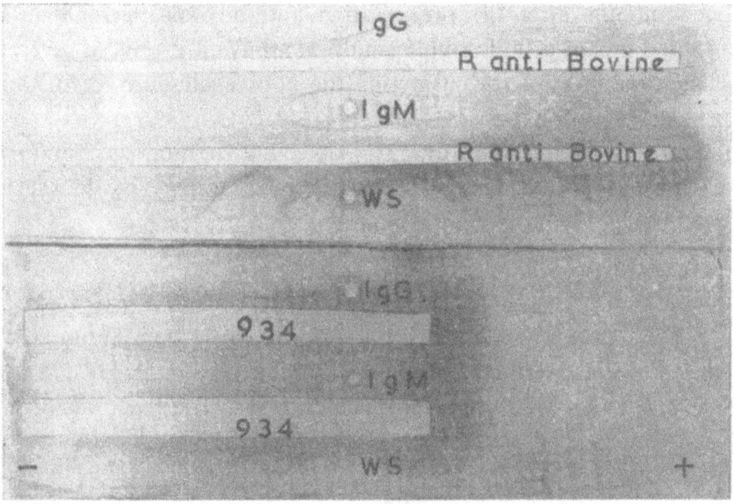

FIG. 1. Immunoelectrophoresis of U.74 A and its IgG and IgM fractions developed with a rabbit antiserum and with a bovine isoantiserum. The development of albumin is poor in this slide as the rabbit antiserum has been diluted 1:4 and the precipitates have not formed well in antigen excess.

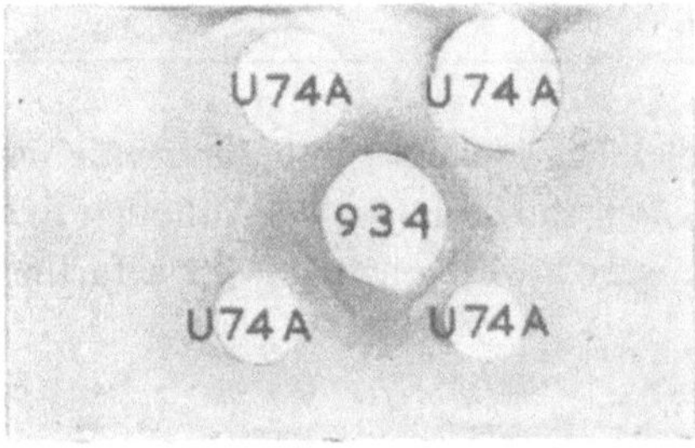

FIG. 2. To show how the size of antibody and antigen wells are critical and the precipitate will not form in antigen excess.

Progress of the immune response

As will be seen from the methods section the course of immunization was very prolonged but other characteristics of the response are interesting. The production of precipitating antibodies was transitory. In two of the animals it built up to a maximum over a period of one month and then gradually declined over the subsequent montlr. The third animal being investigated is now producing antibody, again during the 5th month.

When the precipitating antibody had declined and was no longer detectable there was nevertheless evidence that the animals were highly immune. Intramuscular injection of 10 ml of the Na_2SO_4 fraction of serum produced a severe localized Arthus type reaction and a generalized malaise, the animal not eating for a week.

Recently, 5 months after the precipitating antibody disappeared an intravenous injection of 0.2 ml antigen in double emulsion adjuvant produced severe generalized anaphylactic type reactions. Considering the poor availability of the antigen in an emulsion such a reaction was not expected.

So far no means have been found of detecting the non-precipitating antibodies which are most certainly present in these animals. It may be that such antibodies become increasingly tissue bound as immunization proceeds.

Frequency of the different antigens

Two of the sera have been tested on some 30 cattle sera. Serum 934 reacts with 90% of these — whilst serum 937 only reacts with the donor. DZ.178 A reacts with the donor (937) and with 2 other sera including 934. The reactions of these animals

Cross reactions of the three anti-IgG sera with each other and their immunizing sera

Antiserum	Immunizing serum	Antigen			
		937	934	DZ.178A	U.74 A
937	U.74 A	—	—	—	+
934	U.74 A	+	—	+	+
DZ. 178 A	337	+	+	—	—

are shown in the Table. The identity of the antigenic determinant in U.74 A and 937 detected by 934 is well shown in Fig. 3. Although the proportions are not optimal for 937 it can be seen to be reacting with a further determinant in U.74 A.

Inhibition tests

The first 2 antisera were incubated with positive and negative sera prior to running in immunoelectrophoresis. Positive serum removed all reactivity. Negative serum did not affect the reaction.

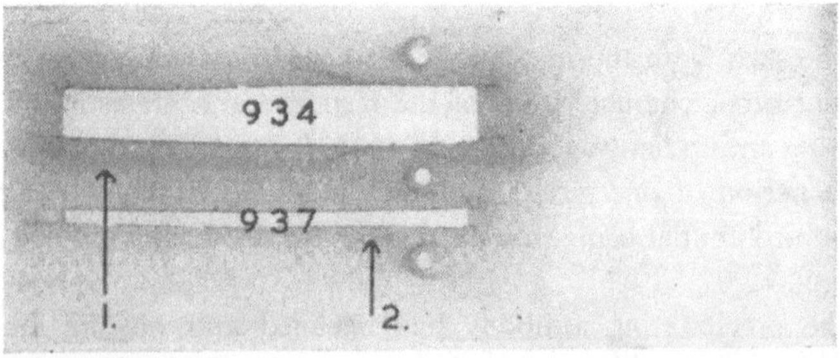

FIG. 3. Antigen in wells U.74 A. 934 and 937 in troughs. The reaction of 934 with 937 gives a line of identity with its reaction with U.74 A. There is a faint precipitate of 937 with U.74 A but the conditions are not optimal for this reaction.

DISCUSSION

In this paper the production of isoantibodies reacting with bovine IgG has been described. Up to now we do not have any evidence for the inheritance of these characteristics, merely the information that 1 is common in the population and the other 2 rare.

No evidence has been presented, other than structural for the antigen being IgG. It would be more direct evidence to be able to detect IgG fixed on cell surfaces. Attempts have been made to do this using bovine anti sheep cell antibody on sheep cells, but although positive antiglobulin reactions were obtained with the sera described above they were also obtained with some other sera from animals injected with whole blood, and these animals showed no precipitating antibody. Further experiments are being performed to investigate these reactions which appear very similar in nature to those reported by MILLOT (1966).

Clearly the investigation of IgG and IgM allotypes in the domestic animals is of the greatest importance. Amongst other things it is essential information if one wants to trace the source of antibody in the foetus or very young animal. With a series of genetic markers on γ globulin one would have the means of ascertaining if this globulin was foetal or maternal in origin. Such information will be of particular value to those studying the pathogenesis of diseases of the newborn.

ACKNOWLEDGEMENT

I would like to thank Dr. Andrew KELUS for his most helpful advice.

REFERENCES

MILLOT, P., 1966. Hetero and isoantiglobulin agglutination reactions in cattle (Les réactions d'agglutination hétéro et iso-antiglobuliniques des bovins), Proc. Xth Eur. Congr. Anim. Blood Grps. (Paris), 471–475.

OUDIN, J., 1956. Specific precipitation reaction between sera from animals of the same species (Réaction de précipitation specifique entre des sérums d'animaux de même espèce), *Compt. Rend. Acad. Sci.*, **242**, 2489–2490.

RAPACZ, J., KORDA, N. and STONE, W. H., 1968. Serum antigens of cattle I. Immunogenetics of A macroglobulin allotype, *Genet.*, **58**, 387–398.

SCHÜTZE, A., 1902. Über weitere Anwendungen der Präzipitine (Further applications of precipitins), *Dtsch. med. Wschr.*, **28**, 804–806.

KELUS, A. S. and GELL, P. G. H., 1967. Immunoglobulin allotypes of experimental animals, *Progr. Allergy*, **11**, 141.

XITH EUROPEAN CONFERENCE ON ANIMAL BLOOD GROUPS AND BIOCHEMICAL POLYMORPHISM
Warsaw, July 2nd–6th, 1968

Serum Amylase Isozymes in French Charolais Cattle

J. GASPARSKI

Department of Veterinary Bacteriology, University of Guelph, Guelph, Ontario, Canada

SUMMARY. The presented report constitutes part of the research on the blood traits of French Charolais cattle. Altogether 532 imported animals were tested. The serum samples of 30 North American Charolais were tested for comparison. Gene frequencies for Am^A, Am^B, and Am^C were estimated from the phenotypes, assuming that the serum amylase isozymes were controlled by multiple codominant alleles at a single autosomal locus. The gene frequencies of amylase isozymes in French Charolais were significantly different from Holstein–Friesian and Aberdeen Angus. Serum amylase types may be added to blood groups for parentage tests.

INTRODUCTION

In the midst of the investigations on the blood proteins of man and many different animal species, serum amylases merit an important genetical position.

ASHTON (1965) showed that amylase polymorphism in cattle was controlled by three autosomal codominant alleles Am^A, Am^B, and Am^C. He suggested that the allele Am^A is a feature of Brahman and Africander cattle and may be absent in the common British breeds of cattle. HESSELHOLT *et al.* (1966) showed that two codominant alleles Am^1 and Am^2 were present in Danish breeds of dairy cattle. GASPARSKI and STEVENS (1968) showed that amylase polymorphism in common breeds of dairy and beef cattle in Canada was controlled by three autosomal codominant alleles.

The study which is reported here was undertaken to determine the number of amylase alleles present in French Charolais cattle imported to Canada.

MATERIALS AND METHODS

Plasma samples of 532 (427 females and 105 males) French Charolais cattle imported in 1966, 1967, and 1968 were tested. For comparison, plasma samples of 30 North American Charolais bulls were also tested. North American Charolais have been a closed herd for at least 20 years.

Electrophoretograms were produced in a discontinuous buffer system as described by GASPARSKI and STEVENS (1968). Gel buffer — tris and citric acid, pH 6.85; vessel buffer — boric acid and sodium hydroxide, pH 8.08; starch concentration — 13%; voltage gradient — 10 V/cm 2 hr and 20 V/cm 1 hr: staining — sodium acetate, glacial acetic acid and *p*-phenylenediamine dihydrochloride, pH 5.7; distaining — methanol, distilled water and glacial acetic acid (5:5:1); washing — distilled water.

RESULTS

In French Charolais cattle six different types of serum amylases were observed, similar to other breeds of cattle in Canada as described by GASPARSKI and STEVENS (1968). These phenotypes were designated A, B, C, AB, AC, and BC. The observed phenotype frequencies are listed in Table 1, for French Charolais imported in 1966, 1967, and 1968 (females and males), North American Charolais (bulls) and two other breeds of domestic cattle, namely Holstein-Friesian (bulls) and Aberdeen

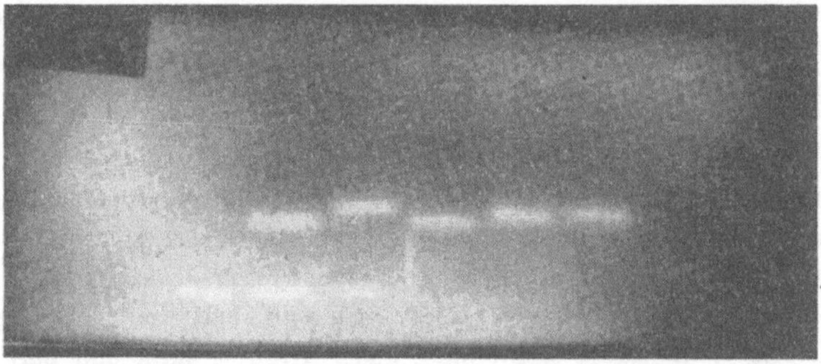

C BC AC B AB A

Serum amylase isoenzyme phenotypes in cattle.

Angus (bulls). Gene frequencies for Am^A, Am^B, and Am^C were estimated from the phenotypes, assuming that the serum amylase isozymes were controlled by multiple codominant alleles at a single autosomal locus.

In Table 2 amylase phenotype frequencies are compared between French Charolais and North American Charolais, French Charolais and Holstein-Friesian, as well as between French Charolais and Aberdeen Angus.

DISCUSSION AND CONCLUSIONS

The observed frequencies of the six amylase phenotypes agreed with the expected numbers based upon estimated gene frequencies in French Charolais (cows and bulls), North American Charolais (bulls), Holstein-Friesian (bulls) and Aberdeen Angus

TABLE 1

Serum amylase phenotypes and estimated gene frequencies in French Charolais and in three breeds of Canadian cattle

Breeds	No. of animals	Females or males	By three allele hypothesis											
			phenotypes							χ^2	P*	estimated gene frequencies		
			obs. exp.	AA	AB	AC	BB	BC	CC			Am^A	Am^B	Am^C
French Charolais	427	F	obs.	1	29	26	119	184	68	0.91	>0.8	0.07	0.53	0.40
			exp.	1.90	30.00	23.08	119.09	182.74	70.11					
	105	M	obs.	0	5	3	37	42	18	1.15	>0.5	0.04	0.58	0.38
			exp.	0.15	4.60	3.08	34.86	46.68	15.63					
North American Charolais	30	M	obs.	1	5	1	9	11	3	1.39	>0.5	0.13	0.57	0.30
			exp.	0.54	4.53	2.40	9.63	10.20	2.70					
Holstein-Friesian	206	M	obs.	11	19	38	12	58	68	2.65	>0.2	0.19	0.25	0.56
			exp.	7.58	19.37	44.50	12.38	56.86	65.31					
Aberdeen Angus	53	M	obs.	0	2	1	4	22	24	2.03	>0.5	0.03	0.30	0.67
			exp.	0.04	0.91	2.01	4.83	21.43	23.78					

* 3 degrees of freedom.

TABLE 2

Comparison tests between French Charolais and North American Charolais, French Charolais and Holstein-Friesian as well as French Charolais and Aberdeen Angus

Comparison between	No. of animals*	Phenotypes						χ^2	P
		AA	AB	AC	BB	BC	CC		
French Charolais and	105	0 (0.78)	5 (7.78)	3 (3.11)	37 (35.78)	42 (41.22)	18 (16.33)	7.20	>0.2**
North American Charolais	30	1 (0.22)	5 (2.22)	1 (0.89)	9 (10.22)	11 (11.78)	3 (4.67)		
French Charolais and	105	0 (3.71)	5 (8.10)	3 (13.84)	37 (16.54)	42 (33.76)	18 (29.05)	67.79	<0.001**
Holstein-Friesian	206	11 (7.29)	19 (15.90)	38 (27.16)	12 (32.46)	58 (66.24)	68 (56.05)		
French Charolais and	105	—	5 (4.65)	3 (2.66)	37 (27.25)	42 (42.53)	18 (27.91)	21.12	<0.001***
Aberdeen Angus	53	—	2 (2.35)	1 (1.34)	4 (13.75)	22 (21.47)	24 (14.09)		

* All bulls.
** 5 degrees of freedom.
*** 4 degrees of freedom.

(bulls) breeds. These observations indicate that the sample populations are in Hardy–Weinberg equilibrium at the amylase locus.

Within French Charolais cattle phenotype frequencies did not differ between females and males. Also French Charolais males and North American Charolais males demonstrated nearly the same phenotype frequencies.

Amylase isozyme frequencies of French Charolais differed significantly from Holstein-Friesian and Aberdeen Angus cattle.

Since serum amylase is polymorphic in all breeds studies, it is a useful system to include in parentage tests.

ACKNOWLEDGEMENTS

I wish to thank most sincerely and to express my deep gratitude to Prof. Dr. R. W. C. Stevens for his valuable advice and help.

This study was supported in part by the National Research Council Grant No. A-2287, the Canadian Charolais Association, the Ontario Association of Animal Breeders, and the Ontario Department of Agriculture and Food.

REFERENCES

Ashton, G. C., 1965. Serum amylase (thread protein) polymorphism in cattle, *Genetics*, **51**, 431–437.

Hesselholt, M., Larsen, B. and Nielsen, P. B., 1966. Studies on serum amylase systems in swine, horses and cattle, Royal Vet. and Ag. College Yearbook 1966, Copenhagen, pp. 78–90.

Gasparski, J. and Stevens, R. W. C., 1968. Bovine serum amylase isozymes in several breeds of domestic cattle, *Canadian Journal of Genetics and Cytology*, X, 1, 148–151.

XITH EUROPEAN CONFERENCE ON ANIMAL BLOOD GROUPS AND BIOCHEMICAL POLYMORPHISM
Warsaw, July 2nd–6th, 1968

Serum Ceruloplasmin Polymorphism in Cattle

J. SCHRÖFFEL, A. KŮBEK and V. GLASNÁK
Immunogenetic Laboratory, Animal Breeding Institute, Hradiško, Czechoslovakia
Laboratory of Animal Physiology and Genetics, Liběchov, Czechoslovakia

Ceruloplasmin enzyme has been found in blood sera of most animals investigated (SEAL, 1964), it shows considerable variability in its oxidase activity according to the individual animal species.

Genetical variability of ceruloplasmin has been observed in man (cf. MORELL and SCHEINBERG 1960; SCHREFFLER *et al.*, 1967). IMLAH (1964) found genetically determined polymorphism in pigs. Two types, demonstrable on starch gel, each being represented by 3 fractions, are the products of two codominant alleles Cp^1 and Cp^2 (GRAETZER *et al.*, 1965).

We have observed genetically determined heterogeneity of ceruloplasmin in cattle also (SCHRÖFFEL, 1966). In contrast to the ceruloplasmin of pigs, in cattle each type is represented by one fraction only. The intensity of colouring of the two slower fractions, designated B and C is weaker than that of fraction A and their mobility is subject to a certain variability. This phenomenon made EBERTUS (1967) assume that the mobility of the slower fractions is non-specific. He therefore proposed the hypothesis of a two-allele codominant system.

MATERIAL AND METHOD

The investigated samples of blood serum and plasma were taken from those animals tested in the "Paternity Test Programme". Blood samples were collected into citrate anticoagulant solution with the addition of Rivanol (Acrynolinum lactatum 0.04 g/1000 ml solution). Other modifications of the solution were also used, e.g. an addition of NaCN, formaldehyde or streptomycin.

Chosen samples of blood serum and plasma were divided into several groups and exposed to different conditions of storage: thawing once, thawing several times, storing at +5°C, storing at laboratory temperature at a period of up to 11 days, respectively. Electrophoretic separation was carried out in starch gel according to KRISTJANSSON (1963). We used our own hydrolysed potato starch in a concentration of 13.5–15 g/100 ml of buffer.

Ceruloplasmin was detected by means of a solution of *p*-phenylendiamine (PPD) in acetate buffer, pH 5.7, $\mu = 0.1$ (0.09 g/100 ml of buffer). The bottom part of the gel was incubated 45 min at 37°C. The PPD solution was then removed and acetate buffer substituted and after further 45 min of incubation the individual samples were evaluated.

RESULTS AND DISCUSSION

The mobility and intensity of fraction A is relative stable, whereas the mobility of the more diffuse fractions B and C is variable. When thawing samples several times, or storing them at laboratory temperature without freezing, fractions B and C migrate more quickly, but they are less marked in colour and intensity. Similar changes also occur when samples are stored at −18°C for a long period. In blood serum samples the denaturation process proceeds with the more rapidity than in plasma samples.

The anticoagulant solution with the addition of Rivanol proved to be the best. The presence of cyanide inhibited ceruloplasmin activity to such an extent that in a number of samples it was impossible to determine the Cp phenotype. This finding is in accordance with that of SPEYER and CURZON's (1968).

To eliminate all mentioned effects of storage, we have used for genetic analysis, either fresh plasma samples or samples after short storage at −18°C. Phenotypes found are shown in the Figure.

Electrophoretic analysis of more than 1200 animals and data from 416 matings prove our previous hypothesis of 3 types of ceruloplasmin (SCHRÖFFEL, 1966), the synthesis of which is determined by 3 codominant alleles Cp^A, Cp^B and Cp^C (Table 1).

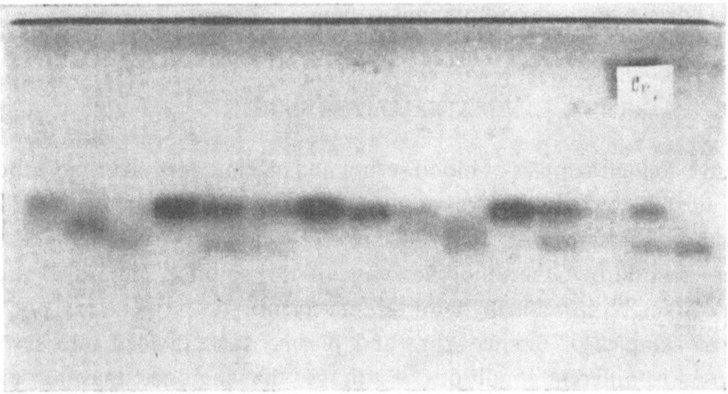

Starch gel electrophoretic pattern of cattle ceruloplasmin. Phenotypes (from left): AA denat., BB, CC denat., AA, AC, AC, AA, AA, AB, BC, AA, AC, AB denat., AC, and CC. Stained with *p*-phenylendiamin. Samples designated "denat." were stored at −18°C for a long period.

TABLE 1

Ceruloplasmin phenotypes of offspring from 416 matings

Mating	No.	Phenotypes of offsprings					
		AA	AB	BB	BC	CC	AC
AA×AA	123	123					
AA×AB	60	38	22				
AA×BB	5		5				
AA×BC	8		3				5
AA×CC	25						25
AA×AC	100	55					45
AB×AB	2	1	1				
AB×BB	0						
AB×BC	2			1			1
AB×CC	5				3		2
AB×AC	27	4	6		6		11
BB×BB	1			1			
BB×BC	1			1			
BB×CC	1				1		
BB×AC	3		1		2		
BC×BC	1			1			
BC×CC	1					1	
BC×AC	5		2			1	2
CC×CC	1					1	
CC×AC	23					12	11
AC×AC	34	7				7	20
Total	416	228	40	4	12	22	122

Owing to the low frequency of allele Cp^B in both investigated breeds, we have not succeeded in obtaining a sufficient number of matings, of some classes for study. It was therefore impossible to complete statistical elaboration of the data. Phenotypes observed always corresponded to those combinations expected in accordance with the types of the parents.

A group of 392 and 206 parental animals of the Bohemian Red Spotted and the Slovakian Spotted cattle, respectively, was used to calculate the gene frequency of individual alleles of the ceruloplasmin system:

	Cp^A	Cp^B	Cp^C
Bohemian Red Spotted breed	0.636	0.068	0.296
Slovakian Spotted breed	0.685	0.029	0.286

The distribution of observed phenotypes of ceruloplasmin, transferrin and serum amylase in the investigated group of animals of both breeds agrees with values expected according to the Hardy–Weinberg law. These results suggest that the Cp locus segregates independently of the Tf and Am loci. The segregation of Tf and Cp loci in the Red Spotted Cattle is illustrated in Table 2.

TABLE 2

Distribution of Cp and Tf phenotypes of Red Spotted cattle and their mutual relationship compared with distribution expected according to the Hardy–Weinberg's law

			Tf						Total
			AA	AD	DD	DE	EE	AE	
Cp	AA	obs.	11	64	78	5	0	4	162
		exp.	11.18	58.92	77.59	7.93	0.20	3.02	158.80
	AB	obs.	1	14	13	5	0	0	33
		exp.	2.37	12.50	16.47	1.68	0.04	0.64	33.73
	BB	obs.	0	1	0	0	0	0	1
		exp.	0.13	0.67	0.88	0.09	0.00	0.03	1.79
	BC	obs.	1	5	10	2	0	0	18
		exp.	1.10	5.82	7.66	0.78	0.02	0.30	15.69
	CC	obs.	0	16	16	3	0	1	36
		exp.	2.42	12.72	16.75	1.72	0.04	0.65	34.33
	AC	obs.	15	47	72	8	0	0	142
		exp.	10.40	54.77	72.13	7.38	0.20	2.80	147.60
Total		obs.	28	147	189	23	0	5	392
		exp.	27.60	145.40	191.48	19.58	0.50	7.44	392.00

	χ^2	D.F.	$< P <$
Segregation for Tf locus	1.9548	3	0.50–0.70
Segregation for Cp locus	1.0145	3	0.70–0.80
Linkage Tf–Cp loci	26.3765	29	0.50–0.70

REFERENCES

EBERTUS, R., 1967. Untersuchungen über Coeruloplasmin-Polymorphismus beim Rind, *Fortpfl. Haust.*, **3**, 265–270.

GRAETZER, M. A., HESSELHOLT, M., MOUSTGAARD, J. and THYMANN, M., 1965. Studies on protein polymorphism in pigs, horses and cattle, Proceedings of the 9th European Animal Blood Groups Conference, Prague, 1964.

IMLAH, P., 1964. Inherited variants in serum ceruloplasmin of the pig, *Nature*, **203**, 658.

KRISTJANSSON, F. K., 1963. Genetic control of two pre-albumins in pigs, *Genetics*, **48**, 1059–1063.

MORELL, A. G. and SCHEINBERG, I. H., 1960. Heterogeneity of human ceruloplasmin, *Science*, **131**. 930.

SEAL, U. S., 1964. Vertebrate distribution of serum ceruloplasmin and sialic acid and the effects of pregnancy, *Comp. Biochem. Physiol.*, **13**, 143–159.

SCHRÖFFEL, J., 1966. Polymorphe Serummerkmale bei Landwirtschaftlichen Tieren. Presented at the International Symposium on Animal Blood Groups, Tupadly, u Mělníka, September 1966.

SHREFFLER, D. C., BREWER, G. J., GALL, J. C. and HONEYMAN, M. S., 1967. Electrophoretic variation in human serum ceruloplasmin: a new genetic polymorphism, *Biochemical Genetics*, **1**, 101–115.

SPEYER, B. E. and CURZON, G., 1968. The inhibition of ceruloplasmin by cyanide, *Biochem. J.*, **106**, 905–911.

XITH EUROPEAN CONFERENCE ON ANIMAL BLOOD GROUPS AND BIOCHEMICAL POLYMORPHISM
Warsaw, July 2nd–6th, 1968

Carbonic Anhydrase Types of Cattle Red Cells

G. SARTORE

Istituto di Fisiologia e Chimica Biologica, Facoltà di Medicina Veterinaria dell'Università di Torino, Torino, Italy

SUMMARY. Studies of gene-controlled multiple forms of enzymes are particularly important with regard to the problems of the physiological activity of the enzymes present. With this in mind, the research reported here was undertaken to study the different types of cattle red cell carbonic anhydrase.

The results show that the enzyme synthesis is probably controlled by three genes, whose quantitative effects are additive in a simple way.

INTRODUCTION

It has recently been shown that the red cell carbonic anhydrase of different animal species, including man, is not a single enzyme but occurs as genetically controlled variants (SHAW *et al.*, 1962; BARNICOT *et al.*, 1964; TUCKER *et al.*, 1966). With reference to cattle, it has been found that inherited differences in red cell carbonic anhydrase can be demonstrated by starch gel electrophoresis and histochemical staining techniques. When this method is used, the enzyme exhibits two molecular forms which give rise to three phenotypes, indicated by F, FS and S in order of decreasing electrophoretic mobility (SARTORE, 1965). These patterns have been explained by assuming that the molecular forms CA–F and CA–S are controlled by two codominant autosomal genes, coded CA^F and CA^S (SARTORE, 1966; SARTORE *et al.*, 1968).

Although gene frequencies show breed differences, CA^F is always relatively rare. Furthermore, the appearance of the stained gels suggests that the total carbonic anhydrase activity might be different from phenotype to phenotype. In order to examine this point in more detail, the enzyme activity has been assayed quantitatively in blood derived from a series of animals of different breeds, whose CA phenotypes were determined by the starch gel method. Appropriate tests have also been performed to see whether these isoenzymes are differently inhibited by some specific inhibitors.

MATERIAL AND METHODS

Heparinized blood samples were collected from:

(1) 114 Double-muscled Piedmont cattle,

(2) 31 Piedmont cattle,

(3) 16 Aosta Red Pied cattle,

(4) 89 Italian Friesian cattle.

The Friesian samples were obtained from privately owned herds. The samples from the indigenous cattle of Piedmont were collected from the local slaughterhouse. All animals were apparently healthy and older than 4 months in age.

In order to identify the phenotypes, the washed red cells were subjected to alcohol–chloroform treatment to denature the hemoglobin according to the method suggested by ROUGHTON and BOOTH (1946). The extracts were subjected to starch gel electrophoresis and stained with α-naphthyl acetate, using fast blue BB salt as dye coupler and with nigrosin. Optimum resolution of the zones of enzymatic activity was obtained by using an electrolyte buffer (pH 8.7): 0.10 M NaOH and 0.30 M boric acid, and a gel buffer (pH 6.8): 0.0050 M citric acid and 0.0144 M tris(hydroxymethyl)-aminomethane.

Enzymatic activity has been assayed on total blood, diluted to 1/12.800, according to the method of WILBUR and ANDERSON (1948), and by doubling the quantities of the reagents as suggested by MANUNTA (1952).

As specific inhibitors of the activity, acetazolamide 10^{-7} M, sodium azide 10^{-5} M and sulfanilamide 10^{-7} M were used.

RESULTS

Identification of CA phenotypes

By denaturing the hemoglobin, very clear electrophoretic patterns have been obtained with both staining techniques.

The three CA phenotypes determined by the alleles CA^F and CA^S were present in all the breeds examined. In the double-muscled Piedmont animals, in addition to the known phenotypes, two new enzyme patterns were found. Each of these patterns was characterized by the presence of a lightly stained zone associated with either CA–F or CA–S and migrating slower than CA–S (Fig. 1).

On the basis of inhibition by acetoazolamide 10^{-4} M, this zone was classified as carbonic anhydrase and labelled $CA\text{–}S_{Piedmont}$. According to the suggestions of ASHTON *et al.* (1967) the new phenotypes have been provisionally named $FS_{Piedmont}$ and $SS_{Piedmont}$.

Enzymatic activity

The activity has been expressed as ratio between the times of the uncatalysed and catalysed reactions. This ratio was corrected:

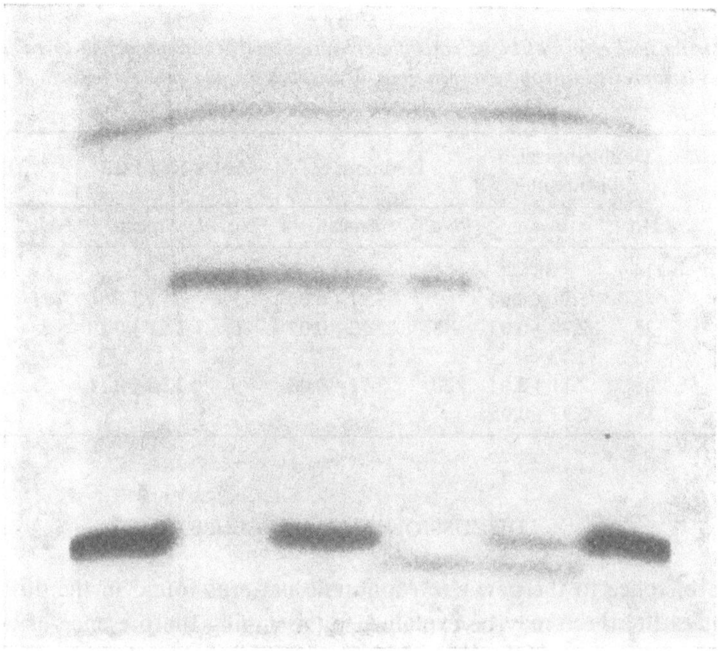

FIG. 1. Photograph of cattle red cell CA types from a gel stained with nigrosin. From left, types: S, F, FS, $FS_{Piedmont}$, $SS_{Piedmont}$, and S.

TABLE 1

Means, standard errors and standard deviations of blood carbonic anhydrase activity from animals of known phenotypes. The activity is expressed as ratio between times of uncatalysed and catalysed reactions and is corrected for an hematocrit value of 100

Type	No. of animals	Mean and standard error	Standard deviation
F	23	5.25±0.07	0.34
FS	60	6.20±0.08	0.63
$FS_{Piedmont}$	2	4.53	
S	152	7.58±0.07	0.85
$SS_{Piedmont}$	13	5.05±0.12	0.45

(a) for a mean hematocrit value of 100, irrespective of the breed; the results are shown in Table 1 and in Fig. 2;

(b) for the mean hematocrit value of each breed; the results are shown in Table 2.

Effect of inhibitors

The inhibition has been expressed as percent increase of time over the catalysed reactions as suggested by WILBUR and ANDERSON (1948). The results are indicated in Table 3.

TABLE 2

Means and standard errors of blood carbonic anhydrase activity according to breed and phenotype. The activity is expressed as ratio between times of uncatalysed and catalysed reactions and is referred to the mean hematocrit values of each breed

Breed type	Double-muscled Piedmont		Piedmont		Aosta Red Pied		Friesian	
	No.	mean	No.	mean	No.	mean	No.	mean
Hematocrit	114	38	31	34	16	35	89	32
F	8	1.91±0.04	3	1.91±0.06	1	1.78	11	1.72±0.03
FS	23	2.29±0.04	8	2.12±0.07	6	2.21±0.10	23	2.06±0.04
$FS_{Piedmont}$	2	1.73						
S	68	2.81±0.04	20	2.71±0.06	9	3.05±0.11	55	2.42±0.03
$SS_{Piedmont}$	13	1.92±0.05						

DISCUSSION AND CONCLUSION

With reference to the new electrophoretic patterns found in the double-muscled Piedmont cattle, these may be explained by assuming the presence of a third allele at the CA locus, responsible for zone CA–$S_{Piedmont}$. The homozygote phenotype $S_{Piedmont}$ (not yet found) will presumably show the single zone CA–$S_{Piedmont}$ and theory suggests that it will be rather uncommon (perhaps 1 in 200 of the population).

Concerning the activity, it will be seen from Fig. 2 that there is considerable variation in the level of activity between animals of any given type. However this does not obscure certain striking differences between the types. Thus the average level of activity in type S is greater than in type F and the average level in type FS

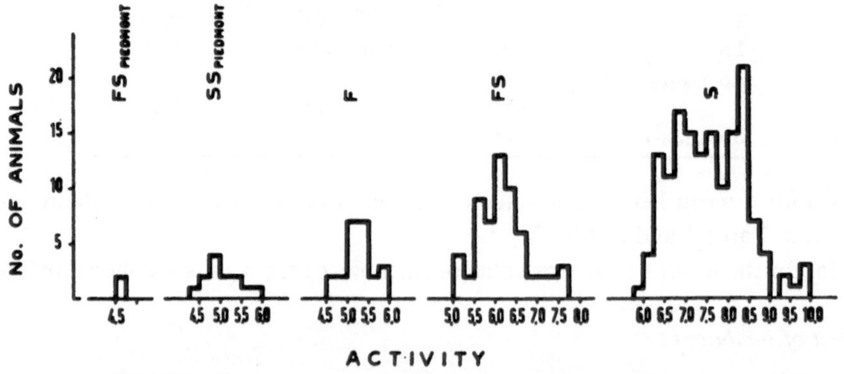

FIG. 2. Distribution of blood CA activity from animals of different phenotypes. Activities are expressed as ratios between times of uncatalysed and catalysed reactions, for blood diluted to 1/12.800 and referred to a hematocrit value of 100.

is intermediate between the two. Disregarding the mean of type $FS_{Piedmont}$ on account of the limited number of animals, a significant difference exists between the means of types F, FS and S. Also the mean of type $SS_{Piedmont}$ is significantly different from that of types S and SF, but not from that of type F.

Accepting the genetical hypothesis put forward to account for the different phenotypes, and assuming that all the carbonic anhydrase activity observed, is determined by the three genes CA^F, CA^S and $CA^S_{Piedmont}$, one might conclude from the mean values shown in Table 1, that the quantitative effects of these genes are additive in a simple way. Using the activity as reported here as an arbitrary unit, the average activity attributable to each of the genes will be approximately: CA^S 3.79 units; CA^F 2.62 units; $CA^S_{Piedmont}$ 1.26 units. It may be noted that this ratio is very close to 3:2:1 and it is tempting to think that this may have some special significance with reference to the molecular masses of these isoenzymes. Similar results were obtained by AUGUSTINSSON and OLSSON (1961) for pig plasma arylesterase and by SPENCER *et al.* (1964) for human red cell acid phosphatase. It will be interesting to see if this relation is confirmed by further research.

TABLE 3

Effect of specific inhibitors on CA types, expressed as percent increase of time over the catalysed reactions, for blood diluted to 1/12.800

Type	No.	Acetoazolamide 10^{-7}M	Sodium azide 10^{-5}M	Sulfanilamide 10^{-7}M
F	4	34±4.1	36±4.2	24±3.4
FS	4	52±3.9	55±3.3	64±3.8
S	4	65±4.0	62±2.1	66±3.2
$SS_{Piedmont}$	4	50±2.5	60±2.1	58±3.3

Concerning the activity, the last point to be considered refers to the consistent differences of activity within phenotypes between breeds, as shown in Table 2. These differences appear to be largely influenced by differences in the mean hematocrit values between breeds, but it seems also likely that the mean red cell carbonic anhydrase activities of the different phenotypes are influenced by the different genetic background of each breed.

With reference to the effects of specific inhibitors, Table 3 shows that the substances used have similar influence on the various CA types and that consistent differences in inhibition are found in types F and S. These results appear rather intriguing and they deserve further investigation.

ACKNOWLEDGMENT

The author wishes to thank Dr. L. SAVIOLO and Mr. F. CERRI for kindly supplying the Italian Friesian blood samples.

REFERENCES

Ashton, G. C., Gilmour, D. G., Kiddy, C. A. and Kristjansson, F. K., 1967. Proposal on nomenclature of protein polymorphism in farm livestock, *Genetics*, **56**, 353–362.

Augustinsson, K. B. and Olsson, B., 1961. Genetic control of arylesterase in the pig, *Hereditas*, **47**, 1–22.

Barnicot, N. A., Jolly, C., Huehns, E. R. and Moor-Jankowski, J., 1964. A carbonic anhydrase variant in the baboon, *Nature*, **202**, 198–199.

Manunta, G., 1952. Anidrasi carbonica e secrezione gastrica sotto l'influenza del desossicorticosterone, *Giornale di Biochimica*, **1**, 233–236.

Roughton, F. J. W. and Booth, V. H., 1946. The manometric determination of the activity of carbonic anhydrase under varied conditions, *Biochem. J.*, **40**, 309–319.

Sartore, G., 1965. Studi sulla struttura del locus Es-II in bovini, *Atti Soc. Ital. Scienze Veter.*, **XXI**, 228–231.

Sartore, G., 1966. Ricerche su un nuovo polimorfismo genetico riguardante una esterasi degli eritrociti bovini, *Atti Ass. Genet. It.*, **XI**, 217–222.

Sartore, G., Stormont, C., Morris, B. G. and Grunder, A. A., 1968. Electrophoretic forms of esterases in red cell of cattle and bison, *Genetics*. (In press).

Shaw, C. R., Syner, F. N. and Tashian, R. E., 1962. New genetically determined molecular form of erythrocyte esterase in man, *Science*, **138**, 31–32.

Spencer, N., Hopkinson, D. A. and Harris, H., 1964. Quantitative differences and gene dosage in the human red cell acid phosphatase polymorphism, *Nature*, **201**, 299–300.

Tucker, E. M., Suzuki, Y. and Stormont, C., 1966. Esterases in the blood of sheep, Proc. Xth Europ. Anim. Blood Group Conf., Paris, 313.

Wilbur, K. M. and Anderson, N. G., 1948. Electrometric and colorimetric determination of carbonic anhydrase, *J. Biol. Chem.*, **176**, 147–154.

XITH EUROPEAN CONFERENCE ON ANIMAL BLOOD GROUPS AND BIOCHEMICAL POLYMORPHISM
Warsaw, July 2nd–6th, 1968

Glucose-6-Phosphate Dehydrogenase Study in Indian Zebu Cattle

S. N. NAIK and A. J. BAXI

Cancer Research Institute, Tata Memorial Centre, Parel, Bombay, India
Blood Group Reference Centre, Haffkine Institute, Parel, Bombay, India

SUMMARY. Glucose-6-phosphate dehydrogenase was studied in 1160 bullocks and 126 cows of 6 different Indian breeds. The cattle showed both normal and deficient animals. The deficiency varied from 29.5% in Malvi to 50.0% in Kankrej. In other breeds the deficiency showed was Dangi 39.55%; Gir 31.0%; Khillari 35.72%; and Rathi 31.75%. The inheritance study from 123 matings constituting 10 sire families showed the possibilities of incomplete penetrance with variable expression in females. The probable role of G-6-PD deficiency in the protection against tick infestation and the possible role of other alternate enzymes in the glucose metabolism is suggested.

INTRODUCTION

During the last decade the study of the erythrocytic glucose-6-phosphate dehydrogenase (G-6-PD) deficiency in man has revealed polymorphism in both normal and defective variants. The study of this enzyme attracted the international attention because some of the deficient subjects exhibited haemolytic crisis when oxidant drugs including anti-malarial compounds were administered to them. This disorder is known to be a sex-linked character and the deficient subjects are resistant to malarial infections like the sickle-cell trait (MOTULSKY and CAMPBELL-KRAUT, 1961).

G-6-PD has been demonstrated in the erythrocytes of horse, mule, donkey, dog, pig and rabbit by BAXI *et al.* (1963) NAIK *et al.* (1963) reported the enzyme in the Jersey cattle. KHANOLKAR *et al.* (1963) reported for the first time total absence of G-6-PD in sheep and goat. An unusual mammalian variant of this enzyme linked with NAD was identified in the Indian Spotted Deer while the Barking Deer had the usual NADP linked enzyme (NAIK *et al.*, 1964). Thus, the enzyme was reported to be present in all animals of some species like horse, rabbit and buffalo (unpublished observations); other species like Jersey cattle, dog and pig had both normal and deficient animals; while the third group, which included sheep and goat, had virtually total absence of erythrocytic enzyme.

With the large number of recognized cattle breeds in India, it was considered worthwhile to study this enzyme in them to know its frequency and mode of inheritance. This report deals with the study of G-6-PD in Indian cattle.

MATERIALS AND METHODS

Blood samples (2 ml) were collected in sterile EDTA tubes with aseptic precaution from bullocks belonging to Dangi (220), Gir (200), Kankrej (230), Khillari (210), and Malvi (200) breeds. There were also 126 cows of Rathi breeds. Thus in all 1186 animals from 6 different breeds were examined. The samples were tested on the same day or the next day in which case they were kept at 4 to 6°C. Sires, dams and calves from 123 matings constituting 10 sire families were screened for G-6-PD to discover its mode of inheritance.

The samples were subjected to the dye decolourization test of MOTULSKY and CAMPBELL-KRAUT (1961) designed for the erythrocytic G-6-PD study in man. The decolourization time was noted every 5 to 10 min after 45 min of setting up the tests, and the time of decolourization was recorded. The usual time of decolourization varied from 50 to 120 min. Those which did not decolourize after more than 120 min were taken as deficient.

RESULTS

Table 1 presents the results in detail regarding the number of animals examined, their sexes and percentage deficiency in different breeds. The deficiency in bullocks varied from 29.5% in Malvi to 50.0% in Kankrej, while Dangi had 39.55%, Gir 31.00% and Khillari 35.72%. Rathi cows had 31.75% deficiency. Repeated examinations of the same animal showed consistant results. The enzyme was found fully developed to the adult level in new-born calves.

TABLE 1

G-6-PD deficiency in different Indian cattle breeds

Cattle	Breeds	No. examined	Animals		Deficiency, %
			normal	deficient	
Dangi	Bullocks	220	131	89	39.55
Gir	Bullocks	200	138	62	31.00
Kankrej	Bullocks	230	115	115	50.00
Khillari	Bullocks	210	135	75	35.72
Malvi	Bullocks	200	141	59	29.50
Rathi	Cows	126	86	40	31.75

G-6-PD enzyme is known to be a sex-linked character with incomplete penetration in man (CHILDS *et al.*, 1958). This enzyme was shown to be determined by a gene on X-chromosome in Equidae (MATHAI *et al.*, 1966). However, the nature of its inheritance in cattle is not yet known. Therefore, a preliminary study of the inheritance of this enzyme in 10 sire families constituting 123 matings was undertaken. The data are presented in Table 2. The matings were classified into 4 groups

TABLE 2

The inheritance of the G-6-PD from the sire family study

Type of matings		No. of matings	Calves			
			normal		deficient	
sire	dams		male	female	male	female
Normal	×Normal	38	20	17	1	0
Normal	×Deficient	20	6	7	3	4
Deficient	×Normal	47	11	13	14	9
Deficient	×Deficient	18	3	2	10	3

depending on the status of the enzyme in sires and dams in order to evaluate its sex-linked nature. The enzyme status in calves was also recorded in groups arranged according to sex. The 38 matings between normal sires and dams gave rise to 20 male and 17 female normal calves and one male G-6-PD deficient calf; whereas from 20 matings between normal sires and deficient dams there were 6 male and 7 female normal calves and also 3 male and 4 female deficient calves. In the third group, there were 47 matings between deficient sires and normal dams giving rise to 11 male and 13 female normal calves and also 14 male and 9 female deficient ones. In the last group, there were 18 matings between deficient sires and dams, which resulted in 3 male and 2 female normal calves and also 10 male and 3 female deficient calves. Some of the observed discrepant results from these matings could only be explained on the basis of incomplete penetrance in female because the possible illegitimacy was excluded by control blood group and haemoglobin studies of the families.

DISCUSSION

This preliminary study has revealed G-6-PD polymorphism in cattle in contrast to horse and rabbit (BAXI *et al.*, 1963). The frequency of the enzyme deficiency was in general high with interbreed variations. Interbreed differences were evaluated by the χ^2 test of significance. Highly significant differences between Kankrej and other breeds were shown as follows: Malvi ($\chi^2 = 21.13$); Gir ($\chi^2 = 15.94$); Rathi ($\chi^2 = 11.04$); Khillari ($\chi^2 = 9.12$); and significant differences as follows: Dangi ($\chi^2 = 4.10$); with one degree of freedom. The Dangi breed was significantly different to Malvi ($\chi^2 = 5.56$) and Gir ($\chi^2 = 4.06$). Other interbreed differences were not significant. The 31.75% deficiency found in Rathi cows was quite a high frequency for a sex-linked character when compared to that among males of the other breeds except Kankrej. The Kankrej and Rathi breeds from the arid region and the Dangi from the mountainous Western Ghats, showed high frequency of deficiency and this might have something to do with their adaptation to environment. The G-6-PD deficiency might also afford protection against the Trypanosomes as suggested by

KHANOLKAR *et al.* (1963) or it might have something to do with the resistance to tick infestation. Study of more breeds from these regions and planned biological experiments could elucidate the correct picture.

The only Western cattle breed studied for this enzyme was Jersey (NAIK *et al.*, 1964), which had 40.64% deficiency but no sex-linkage was reported. The study of this enzyme in other Western breeds and also in African Zebu cattle could throw more light on the process of natural selection in cattle.

If G-6-PD deficiency were to be a sex-linked character in cattle as in Equidae (MATHAI *et al.*, 1966), matings between normal sires and deficient dams should produce only normal female and deficient male calves. But on the contrary there were normal males and deficient females as well (see Table 2). Similarly from the matings of deficient sires and dams, both male and female calves with normal enzyme activity were produced. Such offspring are only possible on the basis of incomplete penetrance in heterozygous females with variable expression of G-6-PD as in man (CHILDS *et al.*, 1958) or because of illegitimacy. The dye decolourization technique probably could not effectively detect the heterozygous females. This might also be one of the reasons why the present data showed almost the same frequency of deficiency in bullocks and cows. Further work to assess the suitability of this test in cattle along with other biochemical, cytochemical and quantitative methods recommended for G-6-PD study is in progress.

Since G-6-PD is the key enzyme in the hexosemonophosphate oxidative pathway in the red cell metabolism it would be worthwhile to investigate which alternative enzymes take over its functions in deficient cattle.

ACKNOWLEDGEMENT

We are grateful to Dr. P. M. WATSARAJ, Kandivli Cattle Breeding Farm; Dr. S. A. MALANDKAR, I/C. Bandra Slaughter House and Dr. CHAVAN, Government Milk Colony, Aarey, for their kind permission to collect the blood samples from the cattle.

REFERENCES

BAXI, A. J., NAIK, S. N. and BHATIA, H. M., 1963. Erythrocytic glucose-6-phosphate dehydrogenase in various animal species, *Curr. Sci.*, **32**, 405–406.

CHILDS, B., ZHINKHAM, W., BROWNE, E. A., KIMBRO, E. L. and TORBERT, J. V., 1958. A genetic study of a defect in glutathione metabolism of the erythrocyte, *Bull. John Hopkins Hosp.*, **102**, 21–37.

KHANOLKAR, V. R., NAIK, S. N., BAXI, A. J. and BHATIA, H. M., 1963. Studies on haemoglobin variants and glucose-6-phosphate dehydrogenase in Indian sheep and goat, *Experimentia*, **19** 472.

MATHAI, C. K., OHNO, S. and BEUTLER, E., 1966. Sex linkage of the G-6-PD gene in Equidae, *Nature*, **210**, 115–116.

MOTULSKY, A. G. and CAMPBELL-KRAUT, J. M., 1961. Proceedings of the Conference on Genetic Polymorphisms and Geographic Variations in Disease, N. Y., pp. 159–180.

NAIK, S. N., BHATIA, H. M., BAXI, A. J. and NAIK, P. V., 1963. Blood groups, haemoglobin variants and glucose-6-phosphate dehydrogenase study in the imported "Jersey" cattle, *Indian Vet. J.*, **40**, 680–685.

NAIK, S. N., BHATIA, H. M., BAXI, A. J. and NAIK, P. V., 1964. Hematological study of Indian Spotted Deer (Axis Deer), *J. Exp. Zool.*, **155**, 231–236.

XITH EUROPEAN CONFERENCE ON ANIMAL BLOOD GROUPS AND BIOCHEMICAL POLYMORPHISM
Warsaw, July 2nd–6th, 1968

Investigations on Erythrocyte Acid Phosphatase in Cattle

F. DOGRUL

Blood Group Laboratory of the Department of Animal Breeding and Genetics, Vienna, Austria

In 1949 ABUL-FADL and coworkers found by means of pH measurements individual differences in erythrocyte acid phosphatase (SEP) in human blood. ANGELETTI and coworkers and also GEORGATSOS (1965) supported these results by chromatographic studies.

In 1963 HOPKINSON and coworkers applied the horizontal starch gel electrophoresis and were able to show on the basis of their results that SEP variation represents an hereditary polymorphism. They found various phenotypes which could be explained by the action of three codominant alleles. In the Vienna laboratory we initiated investigations on SEP in cattle.

METHODS

The method in use in our laboratory was taken from the work done so far on human serum. We use horizontal starch gel electrophoresis with a phosphate citrate buffer system. The buffer solution for the electrode compartments is 527.5 g sodium citrate dihydrate + 43 g citric acid monohydrate in 5 l of distilled water and has a pH value of 5.8. The vessel has separated chambers (Fig. 1) which are connected by a bridge of filter paper. The cathode chamber is filled with diluted citrate buffer.

For the preparation of gels the phosphate buffer which is used consists of 2.238 g of anhydrous KH_2PO_4+3.940 g Na_2HPO_4 per 500 ml distilled water. The pH value of this buffer is 6.98.

10–12% of starch is suspended in 10 times diluted phosphate buffer, heated and poured into the gel trough ($23\times9\times0.6$ cm dimensions).

For SEP investigations cattle erythrocytes are washed three times in distilled water (1:1.5 volume), haemolysed, deep frozen three times for 15 min each time at $-15°$ centigrade, and thawed. Afterwards Schleier–Schüll filter paper No. 2727 (5×5 form) is drenched with the haemolysed solution and put into the slits, which are prepared with a spatula, at 5 cm from the cathode end of the plate. The filter paper pieces remain in the gel throughout electrophoresis. The apparatus is put

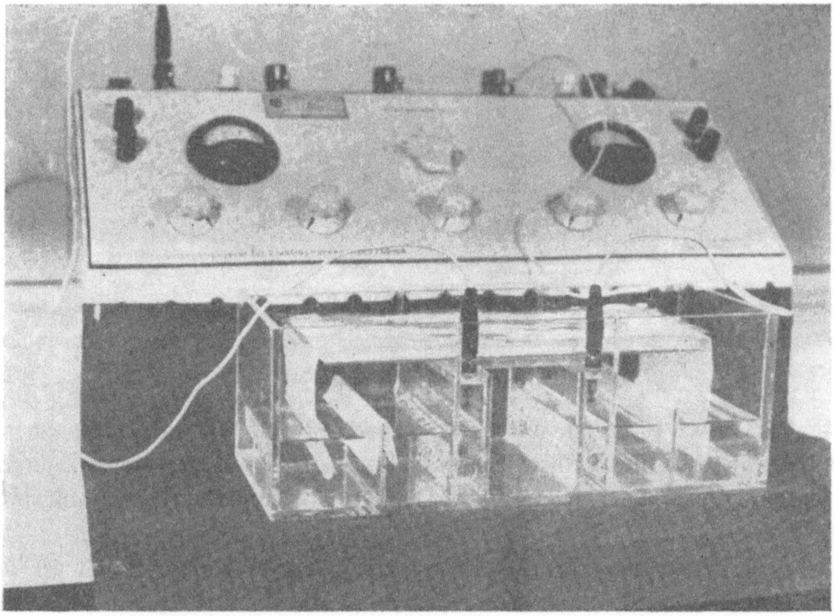

FIG. 1. Trough used for electrophoresis in SEP investigations.

under initial voltage of 240 V and 28–30 mA at +5°C. SEP migrates to the anode while the haemoglobins migrate to the cathode. After 14–16 hr time the gel is put into a 150 ml substrate in which 200 mg phenolphthalein-phosphate-di-sodium salt is dissolved. The substrate solution is composed of one part electrode buffer and 7 parts distilled water. Incubation is carried out at 37° centigrade for 3 to 4 hr. After that the substrate is poured off and a few millilitres of concentrated ammonium solution put on the plate. The enzyme activity releases phenolphthalein and the zones can be recognized by the rose-violet staining in alkaline medium.

RESULTS AND DISCUSSIONS

The blood samples investigated were from 104 cattle, 77 Braunvieh* (with two pairs of twins), 22 Fleckvieh, 1 Pinzgauer, 1 Pinzgauer × Braunvieh and 1 twin pair Fleckvieh × Pinzgauer. With the exception of the Fleckvieh all animals were from the experimental farm of the Veterinary School. Parent–offspring investigations were carried out on animals from the Veterinary School.

Blood samples from 44 cattle were studied twice, and those from six cattle three times. The blood samples which contained a preservative were washed on the following day, and haemolysed. To investigate whether older bloods or haemolysed

* These cattle breeds will have English names but I do not recognize them. Braunvieh may be Brown Swiss and Fleckvieh, a spotted breed.

bloods show a decrease in enzyme activity or colour intensity we investigated bloods haemolysed on the same day as well as on the following two days. The results showed two types of acid phosphatase which differ in their speed of migration, a slow-moving type with an intensive rose-violet colour which shall be denoted as "type 1" and a fast-migrating type with a weaker intensity of colour, which occurs together with

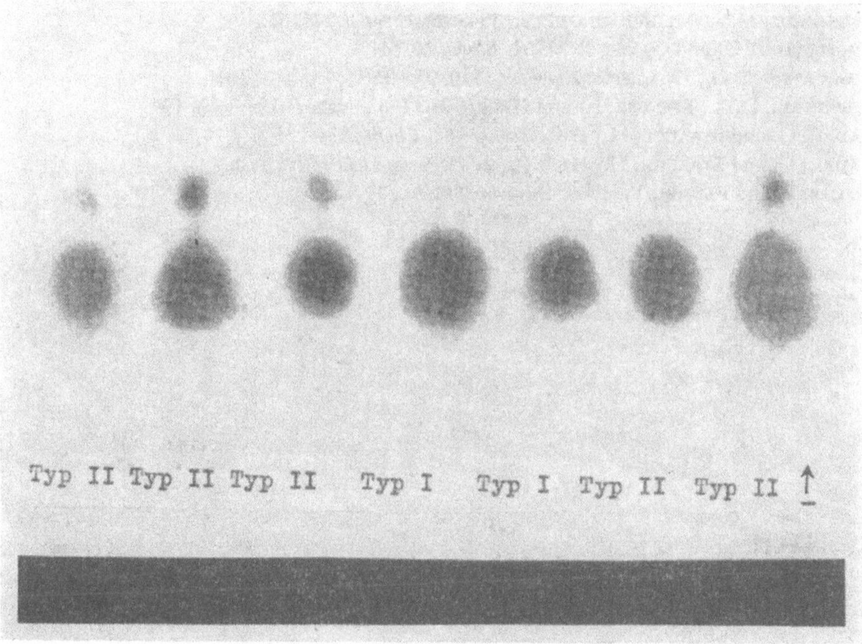

FIG. 2. SEP phenotypes found in cattle.

type 1 and which shall be denoted as "type 2" (Fig. 2). Bloods which were more than one week old and which were stored in the refrigerator still showed the faster migrating fraction in addition to the other fraction. Sometimes fast-moving fractions which, however, could not be seen in repeated determinations, occurred in fresh blood samples. In the haemolysed samples which were stored in the refrigerator the faster migrating zones with weaker colour intensity could be observed after the second day of storage. No differences between breeds, with regard to the occurrence of types could be observed. The results showed a more frequent occurrence of type 2 (79 animals) as compared to type 1 (24 animals). The genetics of this polymorphism will be studied in further investigations.

CONCLUSION

By use of horizontal starch gel electrophoresis and a phosphate/citrate buffer system, the occurrence of an acid phosphatase variant in cattle erythrocytes was investigated. In samples from 103 cattle of various breeds two types of SEP were found. Type 2 appears to be much more frequent than type 1.

REFERENCES

ABUL-FADL, M. A. M. and KING, E. J., 1949. *Biochem. J.*, **45**, 51.
ANGELETTI, P. U. and GAYLE, R., 1962. *Blood*, **20**, 51.
GEORGATSOS, J. G., 1965. *Arch. Biochem.*, **110**, 354–356.
HOPKINSON, D. A., SPENCER, N. and HARRIS, H., 1963. *Nature* (London), **199**.
RADAM, G. and STRAUCH, H., 1966. *Zschr. klin. Chem.*, 234–235.
RADAM, G. and STRAUCH, H., 1967. *Dtsch. Z. ges. gerichtl. Med.*, **60**.
SPEISER, P. and PAUSCH, V., 1968. *Wr. klin. Wschr.*, **1**, 5–15.

XITH EUROPEAN CONFERENCE ON ANIMAL BLOOD GROUPS AND BIOCHEMICAL POLYMORPHISM
Warsaw, July 2nd–6th, 1968

On the Occurrence of a New Transferrin Allele in Two Hungarian Cattle Breeds

G. KOVÁCS

Bloodgrouping Laboratory, University of Veterinary Science, Budapest, Hungary

INTRODUCTION

Some years ago KRISTJANSSON (1962) and later ASHTON (1965) and JAMIESON (1965) succeded in subdividing the bovine transferrin D into D_1 and D_2 types by using various starch-gel electrophoretic techniques.

These new results seemed to be very promising, as in most of the European and also in the American cattle breeds the frequency of the allele Tf^D proved to be relatively high in comparison to other alleles. Due to this fact the variability of the Tf phenotypes was very limited in the above breeds, which consequently decreased the usefulness of transferrin types in parentage control.

After the detection of a faster and a somewhat slower D type — called D_1 and D_2 respectively — some of the research workers (ASHTON, 1965; KRISTJANSSON and HICKMANN, 1965; MAKARECHIAN and HOWELL, 1966) tried to develop such new electrophoretic techniques as would provide better separation in the transferrin region so that more perfect characterization of the Tf locus in various breeds of cattle could be achieved.

In order to compare the frequencies of transferrin genes and to subdivide the TfD type in the Hungarian Grey (Steppe) and Hungarian Spotted cattle the investigation of large populations were undertaken.

MATERIAL AND METHOD

Plasma samples were collected from Hungarian Spotted herds and from a Hungarian Grey cattle herd, the only one to be found in the country. The number of samples was 2650 and 363 respectively.

The plasma samples were analysed by using the electrophoretic method described by ASHTON (1965). This technique was used unaltered except that the gel-tray was not divided into independent units, the gel was 0.4 cm deep and it was prepared from a starch hydrolysed in our own laboratory. The Figure shows the results obtained by the above technique which are compared with those produced with another method described by MAKARECHIAN and HOWELL (1966), we realized that ASHTON's technique proved to be the better for the detection of the new Tf type which had been found in the sample. Whilst that of MAKARECHIAN and HOWELL made possible a better distinction between the D_1 and D_2 types.

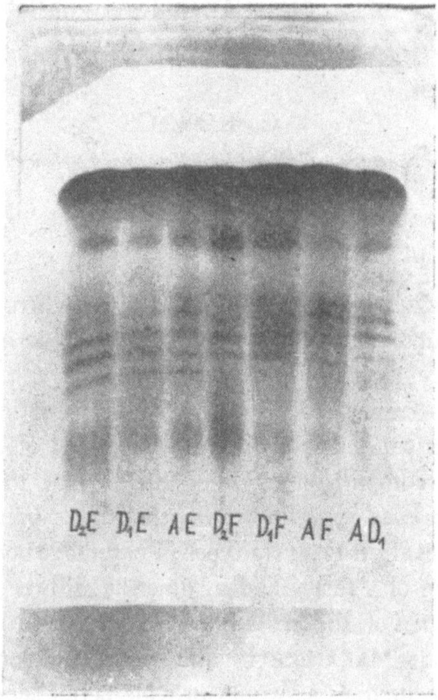

RESULTS AND DISCUSSION

In the course of these investigations we found, very rarely, a new transferrin type different from those described so far in the improved European cattle breeds.

This new Tf type shows a somewhat slower electrophoretic mobility than D_2 and it can be clearly identified only if separation is so perfect, that D_1 and D_2 can also be reliably distinguished.

Having studied the literary data it seems likely that this new Tf type is identical with the TfF type found by ASHTON (1959) in some zebu cattle, and also by MAKAVEYEV (1966) in the Bulgarian Rhodope cattle of brachicerous type and in the Bulgarian Grey Iskar cattle although it has not as yet been compared. The new Tf type was first observed in the Hungarian Grey (Steppe) cattle which can be considered as the original primitive cattle breed of Hungary brought from Asia by the Hungarians when they settled in the Carpathian Basin (the contemporary Hungary) some one thousand years ago. Thus this finding is also a further proof for the Eastern origin of this cattle breed.

Later on we also found this new Tf type in the Hungarian Spotted breed which has a considerable Simental background. It should be mentioned however that the occurrence of this new Tf type in the Hungarian Spotted breed can be explained by the fact that Hungarian Grey cattle were used in the development of the Hungarian Spotted breed some eight decades ago when Hungarian Grey cows were mated to Simental bulls, through several generations.

The frequencies of the Tf alleles observed in the two Hungarian cattle breeds are shown in the Table.

Breed	Allele	Frequencies
	A	0.388
	D_1	0.146
Hungarian Grey (Steppe) cattle	D_2	0.391
	E	0.051
	F	0.023
	A	0.197
	D_1	0.167
Hungarian Spotted cattle	D_2	0.610
	E	0.025
	F	0.001

REFERENCES

ASHTON, G. C., 1959. Beta-globulin alleles in some Zebu cattle, *Nature*, **184**, 1135–1136.

ASHTON, G. C., 1965. Serum transferrin D alleles in Australian cattle, *Aust. J. Biol. Sci.*, **18**, 665–670.

ASHTON, G. C. and LAMPKIN, G. H., 1965. Serum albumin and transferrin polymorphism in East African cattle, *Nature*, **205**, 209.

JAMIESON, A., 1965. The genetics of transferrins in cattle, *Heredity*, 3, 419–441.

KRISTJANSSON, F. K., 1962. Recent research in serum protein polymorphisms of livestock, Proc. 8th Annual Blood Group Conference, Ljubljana, Yugoslavia (Mimeo).

KRISTJANSSON, F. K. and HICKMAN, C. G., 1965. Subdivision of the allele Tf^D for transferrins in Holstein and Ayrshire cattle, *Genetics*, **52**, 627–630.

MAKARECHIAN, M. and HOWELL, W. E., 1966. Improved technique for separation and identification of bovine beta globulins by starch-gel electrophoresis, *Canad. J. Biochem.*, **44**, 1089–1091.

MAKAVEYEV, Ts., 1966. Polymorphism of the serum transferrins and erythrocytic antigens in the Rhodope Shorthorn cattle (*Bos taurus brachyceros*), *Zsivotnovödni Nauki*, **2**, 229.

DISCUSSION

G. C. ASHTON: Do you not think the frequency of Tf D_2 type is very high in your cattle? Perhaps the highest that has ever been reported.

G. KOVÁCS: We have noticed it but for the time being we cannot give any explanation on this.

J. MOUSTGAARD: How many animals have you examined and how many times have you found the new Tf type.

G. KOVÁCS: 2650 Hungarian Spotted and 363 Hungarian Grey cattle were examined out of which 6 Hungarian Spotted and 17 Hungarian Grey individuals possessed the new transferrin allele

J. MOUSTGAARD: Isn't it surprising that you have never found the homozygous form?

G. KOVÁCS: At first sight it is really surprising that we have not found the new transferrin type in homozygous form but if we consider its low frequency than this problem seems to be more understandable. However it cannot be excluded that some other circumstances may also be responsible for this.

XITH EUROPEAN CONFERENCE ON ANIMAL BLOOD GROUPS AND BIOCHEMICAL POLYMORPHISM
Warsaw, July 2nd–6th, 1968

Polymorphism of Albumins, Post-albumins, Transferrins and β-lactoglobulins in Estonian Black-spotted and Yakutian Cattle of Siberia

D. K. BELAYEV and Irina FOMICHEVA

Institute of Cytology and Genetics, Siberian Department of the U.S.S.R. Academy of Sciences, Novosibirsk, U.S.S.R.

INTRODUCTION

Investigation of the genetic polymorphism of blood serum and milk proteins in cattle was started in a number of laboratories of the U.S.S.R. in 1965. The large diversity of stud and indigenous breeds reared under the wide range of climatic and ecological conditions provided the material for the study of biochemical polymorphism and its probable correlation with the adaptive and productive traits.

MATERIAL AND METHODS

We have studied the biochemical polymorphism of blood serum and milk proteins in the common black-spotted breed and also in indigenous cattle of Yakutia, living under the extremely severe climatic conditions of the Zapoliarye (KHARITONOVA, FOMICHEVA, in press). Protein polymorphism was identified by starch gel electrophoresis according to modifications of SMITHIES' technique (1959) as described by ASHTON (1965a) and GAHNE (1963). Reference blood sera were obtained from the Czechoslovakian Academy of Sciences (Laboratory of Dr. MATOUŠEK) and reference β-lactoglobulin samples were provided by Dr. MICHALAK of the Polish Academy of Sciences.

RESULTS

In the cattle breeds tested the following alleles were found: 4 alleles of Tf-locus (Tf^A, Tf^{D1}, Tf^{D2} and Tf^E); 2 alleles of Pa-locus (Pa^F and Pa^S); 1 allele of albumin (Al^A) and 2 alleles of β-lactoglobulin (Lg^A and Lg^B). The gene frequencies are shown in Table 1.

TABLE 1

Gene frequencies in Estonian black-spotted and Yakutian cattle

System	Alleles	Breed: Estonian black-spotted ($n = 156$)	Breed: Yakutian ($n = 152$)
Transferrin	A	0.560	0.069
	D1	0.167	0.457
	D2	0.247	0.418
	E	0.026	0.056
Albumin	A	1.000	1.000
	B	0.000	0.000
Post-albumin	F	0.270	0.745
	S	0.730	0.255
β-lactoglobulin	A	0.486	0.071
	B	0.514	0.929

Alleles Tf^A and Tf^{D2} were most frequent in the Estonian black-spotted breed whereas among the Yakutian cattle the Tf^A allele occurs very rarely and there was not a single homozygote Tf^A/Tf^A. The Yakutian cattle had the highest frequency for Tf^{D1} and Tf^{D2}, and the Tf^E frequency did not exceed that of typical for European cattle. Both Lg^A and Lg^B have the same frequency in the Estonian breed and Lg^B is most frequent in the Yakutian one.

Correlations between Tf-type and productive traits were analyzed; differences in the distribution of butterfat percentage with respect to Tf-types were established.

Heterozygotes Tf^A/Tf^{D1} had significantly higher butterfat percentage than homozygotes Tf^{D1}/Tf^{D1} ($P = 0.95$) and in Tf^A/Tf^{D2} the percentage exceeded that of Tf^{D2}/Tf^{D2} ($P = 0.97$). For the Estonian cows of the genotype Tf^{D2}/Tf^{D2} the butterfat percentage was lower than in Tf^A/Tf^A ($P = 0.91$). The difference among other genotypes within two breeds were not significant for this trait.

Table 2 schematically represents diverse Tf-types and their corresponding values of butterfat percentage. According to this Table all the homozygotes are situated along the diagonal and the lowest values of trait in each column will be distributed along the same diagonal. This Table represents the mean (X) for each line, and the total mean ($\bar{x}$) for all the line-means. Deviation of $\bar{x}$ in every line represents the influence exerted by the Tf-allele on the percentage value. If the effect of the alleles is additive then for the genotype Tf^A/Tf^A, for instance, the expected milk butterfat percentage should be equal to $\bar{x}$+doubled deviation for Tf^A, i.e. $3.73+0.08+0.08$ etc. (see the bottom of Table 2). It was found that butterfat content in homozygotes was lower than expected from the additive effect of Tf-alleles. The same effect was observed in respect to milk yield in Estonian homozygotes Tf^{D1}/Tf^{D1} and Tf^{D2}/Tf^{D2}.

It is of interest that the same tendency was seen in the Yakutian cattle.

TABLE 2

The distribution of butterfat percentages with respect to Tf-type (Estonian black-spotted cattle)

Tf	A	D1	D2	The mean for the line X	Deviation for $\bar{x}$
A	3.75	3.85	3.84	3.81	+0.08
D1	3.85	3.63	3.66	3.71	−0.02
D2	3.84	3.66	3.55	3.68	−0.05
				The general mean $\bar{x} = 3.73$	

Tf	Butterfat percentages		
	expected	observed	
AA	3.73+(+0.08+0.08) = 3.89	3.75	lower
AD1	3.73+(+0.08−0.02) = 3.79	3.85	higher
AD2	3.73+(+0.08−0.05) = 3.76	3.84	higher
D1D1	3.73+(−0.02−0.02) = 3.69	3.63	lower
D1D2	3.73+(−0.02−0.05) = 3.66	3.66	equal
D2D2	3.73+(−0.05−0.05) = 3.63	3.55	lower

The duration of the lactation period in Estonian homozygotes Tf^A/Tf^A and heterozygotes Tf^{D1}/Tf^{D1} was lower, whereas in homozygotes Tf^{D1}/Tf^{D1} and Tf^{D2}/Tf^{D2} it was higher than expected.

Studies on a greater number of herds of various breeds will, possibly, confirm ASHTON's suggestion on the selective superiority of heterozygotes (1965b).

No correlation was found between Tf^E and butterfat percentage in the studied population of Yakutian cattle: butterfat percentage amounted to 6–7%, and the frequency of Tf^E did not exceed the usual level.

The high frequency of Tf^{D1} and Tf^{D2} is of note in Yakutian cattle. According to JAMIESON's summary of Tf-types (1966) the frequency of these alleles is high in animals living in mountain regions of Europe with severe climatic conditions and poor grazing. It is possible that adaptability and Tf^{D1} and Tf^{D2} alleles are correlated in cattle of the northern hemisphere.

REFERENCES

ASHTON, G. C., 1965a. Serum transferrin D alleles in Australian cattle, *Austr. J. Biol. Sci.*, **18**, 665–670.

ASHTON, G. C., 1965b. Cattle serum transferrins: a balanced polymorphism? *Genetics*, **52**, 983–997.

GAHNE, B., 1963. Inherited variations in the post-albumins of cattle serum, *Hereditas*, **50**, 126–135.

JAMIESON, A., 1966. The distribution of transferrin genes in cattle, *Heredity*, **21**, 191–218.

SMITHIES, O., 1959. An improved procedure for starch gel electrophoresis, *Biochemical J.*, **71**, 585–587.

XITH EUROPEAN CONFERENCE ON ANIMAL BLOOD GROUPS AND BIOCHEMICAL POLYMORPHISM
Warsaw, July 2nd–6th, 1968

Albumins, Transferrins, Serum Amylase and Blood Groups in Bulgarian Water Buffalo

Ts. MAKAVEYEV

Agricultural Research Institute "Obraztsov Chiflik", Rousse, Bulgaria

SUMMARY. Serum samples of 150 Bulgarian native Water buffalo cows, Indian buffalo (Murah) and crosses between them were studied by starch gel electrophoresis.

It was established that Water buffaloes have characteristic genetic polymorphism of serum albumins, transferrins and amylases.

Buffalo albumins show three phenotypes-AlbAA, AlbBB and AlbAB. The system is controlled by two alleles-Alb^A and Alb^B.

Buffalo transferrins show three phenotypes-TfBB, TfCC and TfBC. The system is controlled by two alleles-Tf^B and Tf^C.

Buffalo amylases also show three phenotypes-AmAA, AmBB and AmAB. The system is controlled by two alleles-Am^A and Am^B.

The presence of certain antigens on the erythrocytes of Water buffalo have been established by means of cattle reagents.

INTRODUCTION

During the past years genetic polymorphism of a number biochemical serum systems has been established in cattle.

The buffalo (*Bos bubalus*) is a primitive animal not yet investigated in detail for genetical control of biochemical serum systems.

LOYPETRA (1962) is the only author describing polymorphism of transferrins and the presence of erythrocyte antigens in Buffalo — the Thailand Water buffalo.

The presence of erythrocyte antigens in Indian Water buffalo was demonstrated by means of cattle reagents by CHET RAM *et al.* (1964). At present there is no accepted nomenclature and no definite studies of genetically determined albumins, transferrins and serum amylases of buffaloes.

METHODS AND MATERIALS

The horizontal starch gel electrophoresis technique was used. Commercial potato starch was hydrolysed by the method of SMITHIES (1955). Gel trays 24 × 12.5 × 0.5 cm were used for the albumins and 22 × 10.5 × 0.5 cm trays for the transferrins and amylases. Insertion lines were 6 cm from the cathode end of the trays for the albumins, and 2 cm for the transferrins and amylases. Insertions were made on Whatman No. 3 filter paper. The bridges of Whatman No. 3 filter paper overlapped the gels for the albumins by 3 cm, and for the transferrins and amylases by 1 cm at each end. The buffer solutions used for the albumins were based upon the buffer systems described by ASHTON and LAMPKIN (1965) with a small modification. Instead of mixing 450 ml. buffer solution and 50 ml of the electrolyte we mixed these solutions 1:1. The gel buffer pH = 5.9–6.

For the transferrins and amylases we used the method and buffer systems for cattle transferrins and amylases described by KRISTJANSSON (1962) and ASHTON (1965). A hundred and three Bulgarian Water buffaloes, 25 Indian buffaloes (Murah) and 22 crosses between them were used in our investigation.

RESULTS

Three albumin phenotypes-AlbAA (fastest), AlbBB (slowest) and AlbAB were found in Bulgarian Water buffaloes. The first two are homozygote with one band, and the third heterozygote with two bands. In Indian buffalo and in the crosses we found only AlbBB and AlbAB phenotypes.

We established that buffalo differ from cattle by the number of transferrin phenotypes as well as by the electrophoretical mobility of the transferrin bands. We observed in Bulgarian Water buffalo three transferrin phenotypes only TfBB (fastest), TfCC (slowest) and TfBC. The first two phenotypes are homozygote with three bands, and the third is heterozygote with four bands. We observed only TfCC and TfBC phenotypes in a small number of Indian Water buffaloes and in the crosses between Bulgarian and Indian buffaloes.

Three serum amylase phenotypes were found in Bulgarian Water buffalo-AmAA (fastest), AmBB (slowest) and AmAB. The first two phenotypes are homozygote with one band, and the third is heterozygote with two bands. They migrate more slowly than the cattle amylases.

We have established agreement between observed and expected phenotypes of all investigated serum systems in the three buffalo breeds with the exception of albumin phenotypes in the crosses (Table 1).

We have found a common tendency of gene frequencies of albumin, transferrin and amylase alleles in the Bulgarian and Indian buffaloes, which is probably due to the common origin of these buffalo breeds.

Blood samples of 150 Water buffaloes from Bulgaria, India and their crosses were examined for the presence of erythrocyte antigens by means of 35 cattle reagents. Reagents A_1, B, G, G_2, Q, Y_2, I, O, P, T_1, T_2, B′, D′, E'_3, F′, G′, J′, I′, K′ O′, Y′, C_1, C_2, X_1, X_2, W, R_1, F, V, J, L, S_1, U, U′ and Z were used. Bulgarian Water buffaloes reacted with 28 cattle reagents. The percentage of samples reacting in high frequency with the reagents was: A_1 (92.31%), Y′ (86.54%), W (80.47%), E'_3 (73.58%), P (64.42%),

TABLE 1

The observed and expected albumin, transferrin and amylase phenotypes among the three investigated groups of buffaloes

Investigated breeds	Serum systems in buffaloes								
	albumins			transferrins			amylases		
	AA	BB	AB	BB	CC	BC	AA	BB	AB
Bulgarian buffalo									
Observed	20	28	53	3	74	26	50	5	46
Expected	21.4	29.1	50.2	2.5	73.5	27	52.8	7.7	40.5
Indian buffalo (Murah)									
Observed	—	19	6	—	22	3	19	—	5
Expected	0.1	19.4	5.3	0.1	22.1	2.8	21.6	0.2	4.5
Indian × Bulgarian									
Observed	—	7	15	—	21	1	11	—	11
Expected	2.5	9.5	9.9	0.0	23.2	0.9	12.3	1.3	8.2
Bulgarian buffalo	$\chi^2 = 1.4974$			$\chi^2 = 0.1404$			$\chi^2 = 1.8421$		
Indian buffalo (Murah)	$\chi^2 = 0.3250$			$\chi^2 = 0.0207$			$\chi^2 = 0.3913$		
Indian × Bulgarian	$\chi^2 = 5.8800$			$\chi^2 = 1.1380$			$\chi^2 = 2.6861$		

	Gene frequencies					
	Alb^A	Alb^B	Tf^B	Tf^C	Am^A	Am^B
Bulgarian buffalo	0.4604	0.5396	0.1553	0.8447	0.7228	0.2772
Indian buffalo (Murah)	0.1200	0.8800	0.0600	0.9400	0.8958	0.1042
Indian × Bulgarian	0.3409	0.6591	0.0227	0.9773	0.7500	0.2500

T_1 (48.08%) and Y_2 (47.12%). The same group of buffaloes did not react with the reagents Q, O, B′, F, S_1, U and Z. The Indian buffaloes reacted with 17 cattle reagents. The percentage of samples reacting in high frequency with the reagents was O_1 (88%), Y′ (36%), T_1 (36%), E_3' (36%) P (28%) and A_1 (20%). This group of buffaloes did not react with the reagents Q, I, B′, F′ J′, K′, O′, C_1, C_2, X_1, X_2, R_1, F, V, J, L and Z. The crosses (Indian × Bulgarian buffalo) reacted with 19 cattle reagents. They reacted with the reagents O_1 (100%), T_1 (95.45%), E_3' (77.20%), A_1 (72.73%), Y′(72.73%), W (63.66%) and P (50%) in high frequency. They did not react with the reagents B, I, O, B′, F′, J′, K′, C_2, X_1, X_2, R_1, F, V, L, S_1, U and Z.

DISCUSSION AND CONCLUSIONS

Bulgarian and Indian (Murah) buffaloes possess characteristic specific genetic polymorphism of serum albumins, transferrins and amylases.

The three transferrin phenotypes established in Thailand Water buffaloes by

LOYPETRA (1962) are probably not identical to these of the breeds investigated by us because the transferrin types observed by LOYPETRA moved faster.

The presence of erythrocyte antigens in investigated buffaloes is in agreement with the results of LOYPETRA (1962) and CHET RAM (1964).

On basis of our results we wish to propose our nomenclature for the genetically determined albumin, transferrin and amylase phenotypes in water buffaloes.

REFERENCES

ASHTON, G. C. and LAMPKIN, G. H., 1965. Serum albumin and transferrin polymorphism in East African cattle, *Nature*, **205**, 209–210.

ASHTON, G. C., 1965. Serum amylase (thread protein) polymorphism in cattle, *Genetics*, **51**, 431.

CHET RAM *et al.*, 1964. Studies on Indian bovine blood groups, I. Buffalo blood antigenic factors detected through cattle blood groups reagents, *The Indian Veterinary and Animal Husbundry Journal*, **34**, 84–88.

KRISTJANSSON, F. K., 1962. Recent research in serum protein polymorphism of livestock, Eighth Animal Blood Groups Conference in Europe, Ljubljana.

LOYPETRA, P., 1962. Undersogelser over blodtyper samt haemoglobinog Serumtyper hos thailandske vandbofler, Aarsberetning, Institute for Sterilitetsforskning, 221–226.

SMITHIES O., 1955. Zone electrophoresis in starch gels: group variations in the serum proteins of normal human adults, *Biochemical Journal*, **61**, 629.

XITH EUROPEAN CONFERENCE ON ANIMAL BLOOD GROUPS AND BIOCHEMICAL POLYMORPHISM
Warsaw, July 2nd–6th, 1968

Study on the Polymorphism of Haemoglobins, Transferrins, Albumins and Serum Amylases in Bulgarian Cattle Breeds

Ts. MAKAVEYEV

Agricultural Research Institute "Obraztsov Chiflik", Rousse, Bulgaria

SUMMARY. Blood and serum samples of over 2500 cattle from five Bulgarian cattle breeds were studied by starch gel electrophoresis. Genetic polymorphism of haemoglobins, serum transferrins, amylases and albumins was established in all investigated breeds.

Gene frequencies of the alleles of above mentioned blood and serum systems in Bulgarian cattle breeds and in cattle breeds from some European and Asian countries having common origin was observed to be similar.

The presence of the allele Tf^F in a small number of animals belonging to Bulgarian indigenous Grey Iskar and Rhodope shorthorn breeds was found. The presence of this allele in our local primitive breeds and in the Zebu like and in the Turkish Anatolian black cattle supports the existing view of the Asiatic origin of the primitive breeds in Bulgaria.

INTRODUCTION

During the recent years a number of polymorphic biochemical systems with simple inheritance have been detected in cattle. Special attention has been paid to the haemoglobins and transferrins. It has been found that the frequencies of the alleles determining some serum systems in cattle vary from breed to breed. As a result of that the gene frequencies of the alleles controlling serum systems in cattle have been used in studies of the relationship among cattle breeds and of breed structure. In connection with this many investigators have studied the distribution of haemoglobins, transferrins, serum amylase and recently albumins in cattle breeds from Europe, Africa and Asia. Excellent reviews on protein polymorphism of cattle breeds are available in the papers of OGDEN (1961) and JAMIESON (1966).

Cattle breeds from Bulgaria have not previously been the object of studies of haemoglobins, transferrins, amylases and other serum systems.

METHODS AND MATERIALS

The method of starch gel electrophoresis was used. At the begining of our investigations we used vertical starch gel electrophoresis for the detection of serum transferrins as described by ASHTON (1957). Later we used horizontal gel electrophoresis. Tris-boric acid-EDTA buffer system was used for the detection haemoglobins and tris-citric acid-boric acid buffer system for the transferrins and amylases as described by KRISTJANSSON (1962). For the detection of transferrins and serum amylases we used one and the same gel. After the completion of the electrophoresis the gel was bisected horizontally. The upper portion gel was stained to detect the transferrins. The lower part was immersed into buffer solution (12.4 g sodium acetate+0.56 g acetic acid per litre distilled water) with pH = 5.9. After incubation at 37°C overnight and washing with the washing solution use for transferrins, the amylase bands appeared as semitransparent zones in the gel.

The albumins were detected by the method of ASHTON *et al.* (1965) with a small modification: instead of 450 ml buffer solution and 50 ml electrolyte for the preparing starch gel, we mixed these solutions 1:1. The gel buffer had pH = 5.9–6.

RESULTS

The gene frequencies of the haemoglobin, transferrin, amylase and albumin alleles are given in the Table.

Gene frequencies of the alleles controlling haemoglobin, transferrin, amylase and albumin phenotypes in Bulgarian cattle breeds

Alleles	Cattle breeds					
	Grey Iskar	Rhodope Shorthorn	Bulgarian Simental	Bulgarian Red	Bulgarian Brown	Sofia Brown
			Haemoglobins			
No.	275	80	421	1128	576	173
Hb^A	0.984	0.987	0.837	0.963	0.869	0.887
Hb^B	0.016	0.013	0.163	0.037	0.131	0.113
			Transferrins			
No.	265	124	318	1585	572	173
Tf^A	0.383	0.226	0.262	0.489	0.293	0.298
Tf^D	0.457	0.714	0.706	0.414	0.658	0.676
Tf^E	0.147	0.040	0.032	0.097	0.049	0.026
Tf^F	0.013	0.020	0.000	0.000	0.000	0.000
			Amylases			
No.	120	82	313	944	418	173
Am^B	0.658	0.878	0.907	0.843	0.826	0.867
Am^C	0.342	0.122	0.093	0.157	0.174	0.133
			Albumins			
No.	—	87	—	—	119	170
Alb^A	—	0.983	—	—	0.861	0.959
Alb^B	—	0.017	—	—	0.139	0.041

It was found that the haemoglobin phenotype HbBB is absent in the Bulgarian indigenous breeds — Grey Iskar and Rhodope cattle. The three haemoglobin phenotypes — HbAA, HbBB and HbAB were found in the other breeds. The allele Hb^A has higher frequency in all Bulgarian cattle breeds. Similar gene frequencies of haemoglobin and transferrin alleles in our indigenous breeds — Grey Iskar and Rhodope and in Yugoslav Grey steppe and Bush cattle was found. The same similarity was found between Bulgarian Simental and Simental from Switzerland, Bulgarian Brown cattle and Brown cattle from Switzerland and German Federal Republic. Bulgarian Red cattle also have higher gene frequency of Hb^A as in Red Danish cattle, but in contrast to the Red Danish breed have small numbers of the phenotypes HbBB and HbAB. Probably this is because Bulgarian native breeds and Simental participated in the formation of Bulgarian Red cattle.

The observed similarity of frequencies of the alleles controlling the above mentioned serum systems in Bulgarian cattle breeds and in some cattle breeds from the Balkan and European countries probably is due to common origin of these cattle breeds.

In a small number of the investigated animals of the Grey Iskar and Rhodope cattle breeds we found the transferrin phenotypes TfAF, TfDF and TfFF which indicates the presence of the allele Tf^F.

The amylase phenotypes AmBB, AmCC and AmBC were observed in all investigated breeds. The allele Am^B has higher frequency in all breeds. Since there is very little evidence concerning the distribution of amylase phenotypes in European cattle breeds (HESSELHOLT *et al.*, 1966), we compared only gene frequencies of the amylase alleles between Bulgarian Red and Danish Red cattle breeds. We observed very close tendency of gene frequencies between these cattle breeds, which results from Danish Red cattle participation in the formation of the Bulgarian Red cattle.

The albumin phenotypes have been studied only in Rhodope and Bulgarian Brown cattle. Only AlbAA(FF) and AlbAB(FS) phenotypes were found. The allele Alb^A has higher frequency.

The observed gene frequencies of the alleles of haemoglobins, transferrins and amylases are similar to the frequencies observed in cattle breeds having common origin with Bulgarian cattle breeds.

The presence of transferrin allele Tf^F found in Bulgarian native cattle breeds seems to confirm the Asiatic origin of these breeds. The allele Tf^F has been found only in Zebu and Turkish Anatolian Black and Red cattle breeds (JAMIESON, 1966).

REFERENCES

ASHTON, G. C., 1957. Serum protein differences in cattle by starch gel electrophoresis, *Nature*, **180**, 917.

ASHTON, G. C., 1965. Serum amylase (thread protein) polymorphism in cattle, *Genetics*, **51**, 431.

ASHTON, G. C. *et. al.*, 1965. Serum albumin and transferrin polymorphism in East African cattle, *Nature*, **205**, 209–210.

HESSELHOLT, M. *et al.*, 1966. Studies on serum amylase systems in swine, horses and cattle, *Royal Veter. and Agricultural College Yearbook*, 78–90.

JAMIESON, A., 1966. The distribution of transferrin genes in cattle, *Heredity*, **21**, 191–218.

KRISTJANSSON, F. K., 1962. Recent research in serum protein polymorphism of livestock, Eighth Animal Blood Groups Conference in Europe, Ljubljana.

OGDEN, A. L., 1961. Biochemical polymorphism in farm animals, *Animal Breeding Abstract*, **29**, 127–138.

DISCUSSION

G. C. ASHTON: It will be interesting to compare the Tf^F types discovered by Dr. Makaveyev and Dr. Kovács with the Tf^N discovered by Gahne and Braend in Iceland cattle, as well as with Tf^F from Indian cattle.

TS. MAKAVEYEV: It will be interesting for me to compare the Tf^F type discovered in our Bulgarian Iskar cattle with the Tf^N discovered by Dr. Gahne and Dr. Braend in Iceland cattle. But I would like to inform you that I have sent a serum sample to Dr. Bouw and Dr. Oosterlee from The Netherlands and I have received answer that the Tf^F found in Bulgarian Iskar cattle is identical with Tf^F discovered in Indian cattle.

XITH EUROPEAN CONFERENCE ON ANIMAL BLOOD GROUPS AND BIOCHEMICAL POLYMORPHISM
Warsaw, July 2nd–6th, 1968

Polymorphism of Transferrins in Cattle of the Polish Red Breed and in Zlotnicka Pigs

Anna MADEYSKA, Elżbieta SKŁADANOWSKA and M. ŻURKOWSKI

Institute of Experimental Animal Breeding, Polish Academy of Sciences, Warsaw, Poland

SUMMARY. Investigations have been carried out on 936 individuals of the Polish Red cattle and on 219 pigs of Zlotnicka breed. In cattle, together with the recognized transferrin types controlled by three alleles, with frequencies of: $q_A = 0.4208$, $q_D = 0.5123$, and $q_E = 0,0668$, a band migrating more slowly than the band E was observed and is presumably the product of the allele Tf^F. A considerable variability of transferrins in the Zlotnicka pigs examined was observed. Further the existence of 4 alleles Tf^A, Tf^B, Tf^C and Tf^D has been confined. These alleles determine the following phenotypes: AA, BB, AB, AC, BC and BD. Frequency of these alleles was as follows: $q_A = 0.2808$, $q_B = 0.6507$, $q_C = 0.0411$, $q_D = 0.0274$.

INTRODUCTION

Investigations on polymorphism of transferrins in cattle, started by ASHTON (1957), have been followed up by numerous scientists. This has resulted not only in the discovery of new types of transferrins (ASHTON, 1959; OSTERHOFF *et al.*, 1964; GAHNE, 1961; JAMIESON, 1965), but also in showing some differences amongst breeds of cattle throughout the world.

ASHTON (1960) and KRISTJANSSON (1960) found transferrin types Tf^A and Tf^B in the blood serum of pigs. KING in 1962 found an additional fraction, described by him as type Tf^C. KRISTJANSSON in 1963 described type Tf^D. SCHRÖFFEL, who has investigated pig populations in Czechoslovakia, in 1966 described both type Tf^C and type Tf^D, the last in the form of heterozygotes Tf^A/Tf^D and Tf^B/Tf^D only. The frequency of the allele Tf^D was very low.

MATERIAL AND METHODS

In the present investigations material was obtained from cows of various peasant farms as well as from the State Farms in Białystok and Warsaw Provinces. In all, 936 individuals of the Polish Red cattle were examined. The pig material originated from pigs of Zlotnicka breed of the State Breeding Farm Parcz in the Olsztyn Province and from the Experimental Farm in Popielno. Two hundred and nineteen pigs were examined.

Distribution of transferrins has been carried out with application of the generally accepted method of horizontal electrophoresis (SMITHIES, 1955; POULIK, 1957; KRISTJANSSON, 1959). Borate buffer at pH 8.6 was used in the electrode compartments (18.55 g H_3BO_3+4.00 g NaOH per 1000 ml distilled H_2O). The buffer used in preparation of the gel was tris buffer prepared according to KRISTJANSSON (1959).

The gel was made from commercial potato flour previously hydrolysed. A very good separation was obtained with a gel thickness of 0.4 cm. Glass plates were of dimensions 14×18×0.6 cm. To separate the beta-globulin fraction in the blood serum of pigs a voltage drop of 19.5 V/cm was applied and the duration of electrophoresis was 3 hr.

To separate transferrins in cattle, the voltage drop was 25 V/cm and the time of electrophoresis 2 hr. To stain amido black was used with icy acetic acid and H_2O in proportion 5:5:1. To decolour, ethanol with distilled H_2O and icy acetic acid in the same proportion 5:5:1 was used.

RESULTS

Table 1 presents the distribution of observed and expected phenotypes which were noted in the Polish Red cattle.

TABLE 1

Distribution of observed and expected phenotypes in the Polish Red cattle

	Tf types						
%	AA	DD	EE	AD	AE	DE	Number
Observed	172	244	6	401	43	70	936
Expected	168.5	245.6	4.2	403.6	52,6	64.0	935.8

TABLE 2

% frequency of phenotypes in cattle

	AA	DD	EE	AD	AE	DE
%	18.38	26.07	0.64	42.84	4.59	7.48

The differences between observed and expected figures were not statistically significant ($\chi^2 = 3.05$, $n = 5$, $p < 0.05$).

Tables 2 and 3 show the frequency of % phenotypes and alleles in cattle.

In cattle, the Tf^A and Tf^D alleles were in high frequency. Eighty-six percent of observed phenotypes were AA, DD and AD.

TABLE 3
% frequency of genotypes in cattle

	A	D	E
%	0.4209	0.5123	0.0668

TABLE 4
Distribution of observed and expected phenotypes in pigs of Zlotnicka breed

Zlotnicka	Tf types						
	AA	BB	AB	AC	BC	BD	Number
Observed	11	80	98	3	15	12	219
Expected	17.2	92.7	80.0	5.0	11.7	7.8	214.4

Pigs

The distribution of observed and expected transferrin phenotypes in the population of Zlotnicka pigs examined is shown in Table 4.

The populations of pigs of Zlotnicka breed examined are in genetical equilibrium.

Tables 5 and 6 present the frequency of phenotypes and frequency of alleles of transferrins in Zlotnicka pigs.

TABLE 5

	%					
	AA	BB	AB	AC	BC	BD
Zlotnicka	5.02	36.53	44.75	1.37	6.85	5.48

TABLE 6
% frequency of genotypes in pigs of Zlotnicka breed

	A	B	C	D
Zlotnicka	0.2808	0.6507	0.0411	0.0274

In the above investigations on the polymorphism of transferrins in pigs of Zlotnicka breed the considerable variability should be noted. The existence of very rarely observed alleles Tf^C and Tf^D in the breed was shown.

Of the possible 10 phenotypes controlled by 4 alleles, in pigs of Zlotnicka breed 6 phenotypes have been found, but as yet, the phenotypes CC, DD, AD, and CD have not been discovered. This is mainly a result of the low frequency of Tf^C and Tf^D alleles.

REFERENCES

ASHTON, G. C., 1957. Serum protein differences in cattle by starch gel electrophoresis, *Nature*, **180**, 917.

ASHTON, G. C., 1959. Beta globulin alleles in some zebu cattle, *Nature*, **184**, 1135.

ASHTON, G. C., 1960. Beta globulin polymorphism and economic factors in dairy cattle, *J. Arg. Sci.*, **54**, 321.

GAHNE, B., 1961. Studies of transferrins in serum and milk of Swedish cattle, *Animal Production*, **3**, 135.

JAMIESON, A., 1966. The distribution of transferrin genes in cattle, *Heredity*, **21**, 2, 191.

KING, J. W. B., 1962, cited by Kristjansson, F. K., Contrib. VIIIth. Conf. Animal Blood Groups Ljubljana, Yugoslavia.

KRISTJANSSON, F. K., 1960. Genetic control of two blood protein in swine, *Canad. J. Genet. Cytol.* **2**, 295.

KRISTJANSSON, F. K., 1963. Genetic control of two prealbumins in pigs, *Genetics*, **48**, 1059.

OSTERHOFF, D. R. and HEERDEN van, J. R., 1964. Transferrin types in South African cattle breeds, p. 301, IXth E.A.B.G.C., Prague, Czechoslovakia.

POULIK, D. M., 1957. Starch gel electrophoresis in a discontinuous system of buffers, *Nature*, **180**, 1477.

SMITHIES, O., 1955. Zone electrophoresis in starch gel; group variations in the serum proteins of normal human adults, *Biochemical Journal*, **61**, **629**.

SCHRÖFFEL, J., 1966. New genetic variants of transferrins and haptoglobins, *Nature*, **210**, 1274.

XITH EUROPEAN CONFERENCE ON ANIMAL BLOOD GROUPS AND BIOCHEMICAL POLYMORPHISM
Warsaw, July 2nd–6th, 1968

Milk Protein Polymorphism of Cattle Breeds in Czechoslovakia

J. MÁCHA and Žofie MÜLLEROVÁ
Department of Genetics, University of Agriculture, Brno, Czechoslovakia

SUMMARY. Using electrophoresis on starch gel with urea and mercaptoethanol added, four polymorphic systems of milk proteins in Bohemian spotted, Slovakian spotted, Pinzgau, Ayrshire cattle breeds were studied. The authors describe the distribution of phenotypes, allele frequency, the verification of the existence of linkage between the casein loci, the determination of genotypes in bulls, and the possible use of protein polymorphism in the determination of dizygotic and monozygotic twins.

INTRODUCTION

From the experiments performed by a number of authors (e.g. ASCHAFFENBURG and DREWRY, 1955; MOUSTGAARD *et al.*, 1960; ASCHAFFENBURG, 1963; KIDDY and JOHNSTON, 1964; THOMPSON *et al.*, 1964; ČUPERLOVIČ *et al.*, 1964; COMBERG *et al.*, 1964, THYMANN and LARSEN, 1965; WOYCHIK, 1965; GROSCLAUDE *et al.*, 1964, 1965, 1966), it is possible to form a picture of the distribution of phenotypes and the frequency of the corresponding alleles in the four previously known polymorphic milk proteins of several European cattle breeds. GROSCLAUDE *et al.* (1964, 1965, 1966), LARSEN and THYMANN (1966) report results demonstrating the existence of linkage between three loci controlling casein synthesis. In the present paper we describe the results of our studies of milk protein polymorphism in cattle breeds raised in Czechoslovakia.

METHODS AND MATERIALS

Phenotyping of proteins was carried out by means of electrophoresis of skimmilk on starch gel to which urea and mercaptoethanol had been added according to the method described by THYMANN and LARSEN (1965). Our modification of this method consisted in using the conventional equipment, but instead of high-voltage, low-voltage electrophoresis was employed (300 V). The pH of the buffer (continuous system) was raised to 9.2. The duration of electrophoresis was 18–21 hr. This modified method permits separation all four polymorphic proteins simultaneously, with the difference that in β-lactoglobulin only two variants, i.e. A and B, can be distinguished.

The study of protein polymorphism focused on the distribution of phenotypes and alleles in the different cattle breeds, the determination of genotypes in bulls, gene linkage, and on the use of polymorphism in the determination of dizygotic and monozygotic twins.

RESULTS

In Table 1 are presented the distribution of the phenotypes in α-, β-, and ϰ-caseins and the frequencies of the corresponding alleles, and also the results of the statistical treatment of the differences between the observed and the assumed frequencies of phenotypes.

Our observations were made on 1138 cows of the Bohemian spotted breed, 296 cows of the Slovakian spotted breed, 265 cows of the Pinzgau breed, and 178 cows of the Ayrshire breed. In α-casein, allele A was missing in all breeds. The first three breeds differred from the Ayrshire breed in having a higher frequency of allele α_{s1}-Cn^{C}. In β-casein, the otherwise rare alleles β-Cn^{B} and β-Cn^{C} are present in the first three breeds, although not with very high frequency. In ϰ-casein, the frequency of both alleles is almost balanced in both indigenous breeds; the same is approximately true for β-lactoglobulin. In the Bohemian spotted breed a statistically significant deviation from the genetic equilibrium was shown by a preponderance of homozygotic over heterozygotic individuals. A deviation in the same direction was also found in the Pinzgau breed in β-lactoglobulin, and in the opposite direction (i.e. an excess of heterozygotes) in ϰ-casein.

In the Bohemian spotted breed we deduced the genotype of 27 bulls on the basis of phenotyping of the milk from 251 mother–daughter pairs. Out of this number, in α-casein 15 bulls, in β-casein 7 bulls, i.e. 5 of genotype β-Cn^{A}/β-Cn^{B} and 2 of genotype β-Cn^{A}/β-Cn^{C}, in ϰ-casein 21 bulls and in β-lactoglobulin 16 bulls were heterozygotic. In at least two casein systems of the descendants of 14 heterozygotic bulls inheritance of combinations of casein types in the progeny was observed in the mother–daughter pairs to verify the hypothesis of the existence of linkages between the casein loci. From Table 2 it can be seen that the progeny of 1 bull often demonstrate all four possible types of gametes and very often 3 of 4 possible types of gametes in all three known combinations of casein systems. Similar results were obtained when examining the segregation in the individual casein systems and in β-lactoglobulin, which means that all systems developed in all combinations of the four types of gamete. The inference from these results of the existence of linkages partly between the casein loci, partly between the casein and β-lactoglobulin loci is not possible. The assumption of linkages between the loci was however statistically verified by using the test of association between the phenotypes of the three caseins and β-lactoglobulin in 1138 cows of the Bohemian spotted breed. Association

TABLE 1

Distribution of phenotypes

Breed		α_{s_1}-casein			β-casein						κ-casein			β-lactoglobulin		
		B	C	BC	A	B	C	AB	AC	BC	A	B	AB	A	B	AB
Bohem. spotted	obs.	821	29	288	998			94	43	3	349	264	525	275	298	565
$n = 1138$	ass.	818.3	26.3	293.4	999.5	2.07	0.47	90.9	43.1	1.96	328.6	243.6	565.4	273.1	296.1	568.8
		$\chi^2 = 0.385$			$\chi^2 = 0.657$						$\chi^2 = 5.87$; $P<0.05$			$\chi^2 = 0.050$		
Slovak. spotted	obs.	250	0	46	225	2	4	28	35	2	89	70	1.37	56	91	149
$n = 296$	ass.	251.6	1.8	42.6	222.5	0.95	1.7	29.25	39.0	2.6	83.8	64.8	147.4	57.6	92.5	145.9
		$\chi^2 = 2.081$			$\chi^2 = 4.901$						$\chi^2 = 1.474$			$\chi^2 = 0.134$		
Pinzgau	obs.	196	6	63	194	2	1	37	30	1	112	19	134	33	133	100
$n = 265$	ass.	195.1	5.3	64.6	195.5	1.7	1.0	36.0	28.3	2.5	120.9	27.9	116.2	25.3	126.5	113.2
		$\chi^2 = 0.136$			$\chi^2 = 1.102$						$\chi^2 = 6.23$; $P<0.05$			$\chi^2 = 4.21$; $P<0.05$		
Ayrshire	obs.	174		4	177			1			138	3	37	7	108	63
$n = 178$	ass.	174.1	0.02	3.88	176.9	0.002		1.068			137.5	2.6	37.9	8.3	109.4	60.3
		$\chi^2 = 0.024$			$\chi^2 = 0.007$						$\chi^2 = 0.085$			$\chi^2 = 0.342$		

Frequency of alleles

Breed	α_{s_1}-Cn^B	α_{s_1}-Cn^C	β-Cn^A	β-Cn^B	β-Cn^C	κ-Cn^A	κ-Cn^B	Lg^A	Lg^B
Bohem. spotted	0.848	0.152	0.937	0.043	0.020	0.537	0.462	0.490	0.510
Slovak. spotted	0.922	0.078	0.867	0.057	0.076	0.532	0.468	0.441	0.559
Pinzgau	0.858	0.142	0.859	0.079	0.062	0.675	0.325	0.309	0.691
Ayrshire	0.989	0.011	0.997	0.003	—	0.879	0.121	0.216	0.784

TABLE 2

Inheritance of casein type combinations in descendants of bulls heterozygotic in a minimum of two casein systems

Bull		Phenotype of father casein			Distribution in progeny casein					
		α_{s_1}	β	$\varkappa$	α_{s_1}	β	α_{s_1}	$\varkappa$	β	$\varkappa$
MK	34	BC	AA	AB			10 BB			
							2 CA			
MK	37	BC	AA	AB			2 CA	2 BA		
								1 CB		
Prut	75	BC	AA	AB			1 BB	2 BA		
							2 CA			
Robus	12	BC	AA	AB			2 BB	1 BA		
							2 CA	2 CB		
Robus	13	BC	AA	AB			3 BB			
							3 CA			
Rych	14	BC	AA	AB			1 BB	1 BA		
								3 CB		
Lord	20	BC	AA	AB			3 BB	1 BA		
							1 CA			
Rd	159	BC	AA	AB			3 BB			
							2 CA			
Hodr	39	BC	AA	AB			1 BB	3 BA		
								1 CB		
Rd	216	BB	AB	AB					3 AA	
									4 BB	
Rd	166	BC	AB	AB	3 BB	3 BA	1 BB	6 BA	3 AA	3 BA
					1 CA			1 CB	1 BB	1 AB
Prut	113	BC	AB	AB	4 BB	3 BA	1 BB	6 BA	4 AA	3 BA
					2 CA			1 CB	1 BB	1 AB
Rd	162	BC	AB	AB		6 BA	1 BB	3 BA	3 AA	1 AB
					1 CA				1 BB	
Venas	31	BC	AC	AB	2 BA	5 BC	3 BB	2 BA	1 AA	
						2 CA			2 CB	
14					7 BB	12 BA	30 BB	27 BA	14 AA	6 BA
					4 CA		13 CA	9 CB	7 BB	3 AB
					2 BA	5 BC			1 AA	
						2 CA			2 CB	

was displayed by the phenotypes of α- and β-casein, α- and ϰ-casein, β- and ϰ-casein, and also by the phenotypes of α-casein and β-lactoglobulin. These results may point to the possibility of linkage between the mentioned loci.

The possible use of milk protein polymorphism as an aid to breeding practice was also examined by typing of the milk from 15 pairs of cow-twins, in which the blood groups (in 13 systems) were ascertained at the same time in order to decide whether the twins were monozygotic or dizygotic. On the basis of the results of protein typing we determined that 9 pairs of twins were dizygotic and in 6 pairs we noted the possibility that they were monozygotic. When comparing these results with the blood typing, we obtained identical results in 9 cases, in 3 cases it was demonstrated by blood typing that the twins concerned were dizygotic. In the 3 other cases, however, the assumption of monozygosity from blood typing evidence had to be changed to dizygosity as a result of protein typing evidence.

DISCUSSION

When comparing the distribution of phenotypes and the frequencies of alleles established by us in the breeds under study with the results of similar observations made by the previously cited authors, the following points emerge: in both spotted breeds from Czechoslovakia and also in the Pinzgau breed relatively considerable variability of α_{s1}-Cn and β-Cn loci is found. A relatively elevated frequency of the allele α_{s1}-Cn^C in these breeds resemble the situation found in the Jersey, Guernsey, Montbéliarde, Normande, and Buša breeds. A relatively high frequency of allele β-Cn^C especially in the Slovak spotted and the Pinzgau breeds is a major difference from most European cattle breeds, in which this allele is missing. In the κ-Cn locus and in the Lg locus considerable variability occurs in all previously tested breeds; in κ-casein, the Ayrshire, Guernsey, and Friesian breeds form a group displaying an elevated frequency of the allele κ-Cn^A, while the other cattle breeds, including those studied by us, exhibit a medium-high frequency of this allele (near 0.5). In β-lactoglobulin the frequency of allele Lg^A in previously tested breeds ranges between 0.10 and 0.61, while in both of our spotted breeds it is near 0.5.

We do not consider the discrepancy in the results of the two approaches we used for verifying the assumption of linkage between the loci controlling milk protein synthesis to be fundamental. What we deem decisive are the results of the establishment of segregation of protein types in the progeny of heterozygotic bulls. These results which do not permit the inference of the existence of linkages between the casein loci are at variance with those reported by GROSCLAUDE *et al.* (1964, 1965) and LARSEN and THYMANN (1966) who both found in analogous tests the occurrence of only two types of gametes, i.e. during segregation in α_{s1}- and in β-casein types α-$_{s1}Cn^B$–β-Cn^B or α_{s1}–Cn^C–β-Cn^A, in α_{s1}- and κ-casein types α_{s1}-Cn^B–κ-Cn^B or α_{s1}-Cn^C–κ-Cn^A, in β- and κ-casein types β-Cn^A–κ-Cn^A or β-Cn^B–κ-Cn^B, from which they conclude that the loci α_{s1}-Cn, β-Cn and κ-Cn are in close linkage. In order to elucidate this discrepancy it will be necessary to extend the studies to a larger number of descendants.

The result of the establishment of association between the phenotypes of caseins agree with the results of the association test reported by LARSEN and THYMANN (1966). Yet, GLASNÁK (1967) reports that in 337 cows of the Bohemian spotted breed no association between the phenotypes α_{s1}- and β-casein, α_{s1}- and κ-casein, α_{s1}-casein and β-lactoglobulin, nor κ-casein and β-lactoglobulin was found; it was found only between the phenotypes β- and κ-casein and between κ-casein and β-lactoglobulin. By experimental testing of smaller blood samples from Bohemian spotted cattle we obtained in certain cases also contradictory results. For this reason we do not consider the results of the association test to be reliable in the inference of the existence of linkages between the loci, especially in those cases where at least one of the loci exhibits a weak segregation as is the case with α_{s1}-Cn and β-Cn.

REFERENCES

ASCHAFFENBURG, R., 1963. Inherited casein variants in cow's milk, *J. Dairy Res.*, **30**, 251.

ASCHAFFENBURG, R. and DREWRY, J. 1955., Occurrence of different β-lactoglobulins in cow's milk, *Nature*, **176**, 218.

COMBERG, G., MEYER, H. and GRÖNING, M., 1964. Die Beziehungen der Beta-Lactoglobulintypen beim Rind zu Erstkalbealter, Milchmenge and Fettgehalt, *Züchtungskunde*, **36**, 248.

ČUPERLOVIČ, M., KOVÁCS, G. and THYMANN, M., 1964. Kazein polymorfi hos jugoslavisk og ungarsk kvaeg, Aarsberetning, Institut for Sterilitetsforskning.

GLASNÁK, V., 1967. Polymorfizmus bielkovin mlieka hovadzieho dobytka a ošípaných, Dissertation Paper, Liběchov.

GROSCLAUDE, F., GARNIER, J., RIBADEAU-DUMAS, B. and JENNET, R., 1964. Étroite dépendance des loci controlant la polymorphisme des caséines α_s et β, *C. R. hebd. Séanc. Acad. Sci.* (Paris), **259**, 1569–1571.

GROSCLAUDE, F., PUJOLLE J., GARNIER, J. and RIBADEAU-DUMAS, B., 1965. Déterminisme génétique des caséines du lait de vache; étroite liaison du locus κ-Cn avec loci x_s-Cn et β-Cn, *C. R. hebd. Séanc. Acad. Sci.* (Paris), **261**, 5229–5232.

GROSCLAUDE, F., PUJOLLE, J., RIBADEAU-DUMAS, B. and GARNIER, J., 1966. Analyse génétique du groupe de loci de structure synthétisant les caséines bovines, Xth European Conference on Animal Blood Groups and Biochemical Polymorphism, Paris, 415–420.

KIDDY, C. A. and JOHNSTON, J. O., 1964. Genetic polymorphism in caseins of cow's milk. I. Genetic control of alpha-casein variation, *J. Dairy Sci.*, **47**, 147.

LARSEN, B. and THYMANN, M., 1966. Studies on milk protein polymorphism in Danish cattle and the interaction of the controlling genes, *Acta Vet. Scand.*, **7**, 189–205.

MOUSTGAARD, J., MÖLLER, I. and KARSKOW-SÖRENSEN, P., 1960. Beta-lactoglobulintypen hos kvaeg, Aarsberetning — cit. Glasnák 1967.

THOMPSON, M. P., KIDDY, C. A., JOHNSTON, J. O. and WEINBERG, R. M., 1964. Genetic polymorphism in caseins of cow's milk. II. Confirmation of the genetic control of beta-casein variation. *J. Dairy Sci.*, **37**, 378.

THYMANN, M. and LARSEN, B., 1965. Maelkeproteinpolymorfi hos dansk kvaeg, Aarsberetning, Institut for Sterilitetsforskning.

WOYCHIK, J. N., 1965. Phenotyping of kappa-caseins, *J. Dairy Sci.*, **38**, 495.

CHAPTER 2

BLOOD GROUPS AND BIOCHEMICAL POLYMORPHISM IN PIGS

XITH EUROPEAN CONFERENCE ON ANIMAL BLOOD GROUPS AND BIOCHEMICAL POLYMORPHISM
Warsaw, July 2nd–6th, 1968

Complement Activity of Blood Serum in Pigs

Nina O. SUKHOVA

Laboratory of Immune Genetics, Siberian Research Institute of Animal Breeding, Novosibirsk, U.S.S.R.

SUMMARY. Investigations performed on 1050 pigs of the principal breeds kept in Siberia have shown that the complement activity of the blood of each pig breed is characterized by certain indices depending on age and conditions of storage of serum functional state.

INTRODUCTION

The bactericidal action of one of the blood serum components was first noted by BUCHNER (1889). The isolated component was called alexine and subsequently renamed complement. Interest aroused by the complement has remained unabated since BUCHNER's discovery. It reached a peak after publications of papers by PILLEMER and his co-authors on the properdine system and its influence on the non-specific resistance to disease of the organism (PILLEMER *et al.*, 1956). Complement is known to be a part of this system. The role of complement in the organism is not limited to immunological functions. It takes part in regulation of erythropoiesis and in a number of other physiological functions (ADAMOV and NIKOLAYEV, 1968).

Following the observations of HATT on the genetic origin of complement activity of blood serum we have conducted observations on the changes in complement activity in blood serum of pigs kept in Siberia in order to determine the possibility of using such information as an aid to selective breeding. A study was made of the level of serum complement activity dependent on the animals' age and their breed.

MATERIAL AND METHODS

The complement was studied in 1050 animals of the Large White, Sibirskaya Severnaya, Kemerovo, Lacombe, Landrace and Black and White breeds as well as in their cross-breeds. The pigs of Black and White breed and those of the Black and White: Large White cross-breed were obtained by fertilizing the sows of the Black and White breeding group with a quantitatively equivalent mixture of semen of Black and White and Large White hogs. The origin of pigs was determined according to the antigenic properties of the blood by using a panel of polyvalent serums differing in specificity. The panel containing 40–50 serums used guaranteed a 98 per cent determination of paternity.

The principal material was obtained from experimental animals of the pig-breeding section of Siberian Research Institute of Animal Husbandry. The serum was examined not later than 24 hr after sample bleeding from the marginal vein of the ear. Complement activity was revealed by haemolytic titration of the serum (PREDTECHENSKY, 1964). The test volume was limited to 1.25 ml. Erythrocyte density in the haemolytic system after a tentative dilution to 1 per cent with 1:10 physiological solution corresponded to mark 280 on the erythrocyte scale of a 065 model photoelectric erythrohaemometer.

The findings were mathematically processed and the values M, Mo, Me, mM, V% calculated.

The degree of difference between two juxtaposed distribution series was calculated by means of the formula

$$t = \frac{M_1 - M_2}{\sqrt{m_1^2 M_1 + m_2^2 M_2}},$$

for a value of t greater than or equal to 3 differences were considered to be real.

RESULTS OF EXPERIMENTS

Observations showed that complement activity of blood serum in adult pigs is contained within the limits of 0.030–0.150 (Table).

A small number of animals showed even lower titres. The hogs of Lacombe and Landrace breed possessed lower complement titre compared with the hogs of Large White and Sibirskaya Severnaya breeds. This difference was also noted in

Complement level in blood serum of pigs in some breeds kept in Siberia

	Indices of complement activity in the blood serum of pedigree animals					
	Large White	Sibirskaya Severnaya	Kemerovo	Black and White	Lacombe	Landrace
	Hogs					
M	0.070	0.071	0.075	0.078	0.083	0.096
Me	0.070	0.075	0.075	0.080	0.090	0.090
Mo	0.050	0.075	0.075	0.090	0.090	0.090
mM	0.006	0.006	0.006	0.009	0.007	0.006
V%	41	27	29	26	22	21
	Sows					
M	0.071	0.070	0.072	0.072	0.080	0.080
Me	0.070	0.060	0.070	0.070	0.070	0.070
Mo	0.070	0.050	0.050	0.050	0.070	0.070
mM	0.002	0.012	0.004	0.004	0.004	0.010
V%	20	40	27	27	20	28

younger animals. The difference was increased when rearing conditions were not optimal (in the first ten day period after weaning, when kept in the cold, etc.). Complement activity of pig blood serum was not constant during the process of growth. In new-born pigs complement was absent. In one-day-old pigs the complement titres were 0.150–0.130, and reached 0.030–0.070 by the age of 12–14 days. The complement indices during lactation were influenced not so much by the breed as by the milk supply available. At the age of 4–6 months the complement activity of the pig blood serum reaches, and sometimes surpasses the level distinctive of adult animals. Under optimum conditions of rearing pigs of Large White × Landrace, Large White × Lacombe crosses did not show as high a titre as was shown by the pigs of Large White breed.

In sows with young especially in the second half of gestation the complement titre increased considerably ($t > 4$). At first sight this seems to contradict published findings reporting decrease in complement in pregnant guinea pig sand humans (BOULANGER *et al.*, 1954; REZNIKOVA, 1967). However these findings suggest to us a possible mechanism of complement activity during gestation. Humans and guinea pigs with haemochorial placentation allowing unhindered penetration of globulin fractions through the placenta show a decrease in complement level. But in pigs characterized by placentation of a diffuse type the placenta barrier blocks the passage not only of antibodies but also of complement. As a result of hormonal activation of complement during gestation its level increases but in humans and guinea pigs this is compensated for by a process of absorption of complement components.

Taking into consideration the globulin character of blood serum complement and the present concept that the processes of general protein synthesis and those of immune globulin synthesis are interdependent we can assume that in the near future the complement activity of blood serum may be used as a marker to help solve problems connected with the understanding of growth and developement in the animal organism.

ACKNOWLEDGEMENT

I am very grateful to post-graduates V. BEKENEVA, U. BOBKOVA, N. SEMYONOVA, F. TARAKANOWA for providing serum from experimental animals for investigation.

REFERENCES

ADAMOV, A. K. and NIKOLAYEV, N. I., 1968. On the role of normal and autoimmune antibodies in regulation of micromolecular composition of animal cells (in Russian), *JMEI*, No. 4, 75–80.

BOULANGER P., RICE, CH. E. and PLUMMER, P. I. G., 1954. Parallel studies of complement and blood coagulation, XII Effect of Pregnancy and sex Hormones, *The Cornell Veterinarian*, No. 2, XLIV, 2, 191–198.

BUCHNER, H., 1889. Immunität und Immunisiring, *Münch. Wochenchr.*, No. 2–3.

PILLEMER, L., BLUM, D., LEPOW., I., WURZ, L. and TODD, E., 1956. The properdine system and immunity, *J. Exp. Med.*, **103**, 1.

PREDTECHENSKY, V. E., 1964. Handbook on clinical laboratory studies (in Russian), *Medicine*, p. 959.

REZNIKOVA, A. C., 1967. Complement and its importance in immunological reactions (in Russian), *Medicine*, p. 270.

XITH EUROPEAN CONFERENCE ON ANIMAL BLOOD GROUPS AND BIOCHEMICAL POLYMORPHISM
Warsaw, July 2nd–6th, 1968

Dextran Test in the Study of Blood Groups in Pigs

J. HOJNÝ and J. HRADECKÝ

Laboratory of Physiology and Genetics of Animals, Czechoslovak Academy of Sciences, Liběchov, Czechoslovakia

SUMMARY. In our paper we describe a survey of the serological character of antibodies, detecting blood factors on pig erythrocytes and emphasize that ample use can be made of the dextran test.

We compare titres of several antibodies of the same specificity by means of the direct and indirect agglutination test, the dextran test and the haemolytic test against each of the 44 blood factors. Reagents are divided into several groups according to the nature of individual test sera: anti A, Bb, Da, all reagents of the E system (Ea, Eb, Ed, Ee, Ef, Eg, Eh) and Fa, Ka, Kd, Ke, La, Li, and exceptionally also Ga, Gb, Kb and Lg are complete agglutinins. Another group of antibodies, giving excellent reactions in the dextran test, comprises the majority of reagents of the K system and also most, antibodies detecting blood factors of the L system and the O system. The dextran test also considerably increases the titre of many complete antibodies.

The numerous group of typical incomplete antibodies reacting in the main only in the indirect antiglobulin test: anti Ca, Ha, Ia, Ib, Ja, Ma, Mc, M(d), Na, Nb, but also Ga, Gb, Hb and Hc.

Least represented are reagents of haemolytic character examples of which we obtained only as anti (Ac, Ap), Fa, Hb, Ka, Kd and Kb.

The concluding note mentions new blood factors Fb and Md.

INTRODUCTION

The determination of a large number of antigens on pig erythrocytes is based on the production of monospecific isoimmune antibodies and their utilization in the optimal serological test. The direct agglutination test, the haemolytic and indirect antiglobulin tests have been considered the most reliable for blood group testing in pigs. Other methods have also been tried, mainly the dextran test (ANDRESEN, 1963; and others). The dextran test has not been more widely used because it caused justified apprehension of unspecific reactions. A decrease of dextran concentration and a simplified procedure (HOJNÝ and HÁLA, 1965) yielded very good results, which made us use this method more intensively. It even helped to detect some new blood factors (HOJNÝ *et al.*, 1966).

In our present work we make use of materials gathered, a large number of antisera of different specificity and also suplicate sera to give a survey of the serological character of blood group reagents and to emphasize the importance of the dextran test; its advantage and the possibility of its wider use.

MATERIALS AND METHODS

For the comparison of antibody titres in 4 different serological tests we have used monospecific antisera prepared against 44 blood factors of pigs in the Laboratory of Liběchov. Antisera were prepared by immunizing sows of the Large White, Cornwall, Landrace and Black and White-Přeštice breeds and their crosses. In parallel tests we have also compared against most blood factors 2–5 specific antibodies obtained from different animals. Exceptionally we have also used some reagents obtained by exchange with foreign laboratories (G.F.R., G.D.R., U.S.A., U.S.S.R.).

Direct agglutination test, haemolytic and indirect Coombs tests were carried out according to ANDRESEN (1957), the modified dextran test was performed as described by HOJNÝ and HÁLA (1965) 1 drop of 6% dextran solution "Spofa" was added to 2 drops of reagent and 1 drop of 2.5 erythrocyte suspension. Results were read after 2.5-hr incubation at 37°C.

Titres were determined by means of macrotitration. The resulting titre corresponded to the maximal serum dilution at which a clear positive reaction could be identified. Evaluation was carried out on grounds of 4 tests, grade 2 being considered as a positive reaction.

RESULTS

The serological character and titre of antibodies used to determine blood groups in pigs are shown in Tables 1a and 1b. Owing to the large number of investigations performed, the Tables do not comprise all available results. We have therefore chosen characteristic couples of reagents of identical specificity, corresponding in their value to other compared reagents which were not included in the Tables.

According to their serological character, reagents may be divided into the following groups: complete agglutinins, haemolysins, incomplete antibodies reacting only in the indirect Coombs test, and antibodies reliably agglutinating positive erythrocytes in the dextran test.

(a) Anti A, Bb, Da, all reagents of the E system and anti Fa can be counted as complete agglutinins. The same character may often be observed also in anti Ka, Kd, Ke, La and Li, rarely in Ga, Gb, Kb and Lg.

(b) Haemolytic reaction was found in a small number of reagents only: anti A (Ac, Ap), Hb, Ka, Kd, Kb and Fa.

(c) Typical incomplete antibodies reacting only in the indirect antiglobulin test

TABLE 1a

Serological character and titres and antibodies used for investigations of pig blood groups

Reagents: designation	Reagents: No.	Antibody titres: agglutination test	Antibody titres: haemolytic test	Antibody titres: Coombs test	Antibody titres: dextran test
A	134/23	2	4	2	8
(c)	D 2	8	8	32	128
	D 2	N	0	4	16
(p)	(G.F.R.)	0	8	0	0
Bb	74/168	8	0	2–4	16
	Ř 57	2	0	N	8
Ca	39/67	0	0	16	0
	V 49	0	0	16	0
Da	L 044	8	0	8	64
	L 047	0	0	4	32
Ea	Fular	32	0	16	64
	V 19	32	0	16	32
Eb	189/3	16	0	8	32
	50	4	0	4	16
Ed	81/6	64	0	16	64
	49/12	64	0	32	128
Ee	909	16	0	16	32
	151	16	0	8	32
Ef	L 034	2	0	2	4
	1258	4	0	8	16
Eg	189/3	8	0	8	64
	2811	8	0	4	32
Eh	56/176	N	0	N	2

Reagents: designation	Reagents: No.	Antibody titres: agglutination test	Antibody titres: haemolytic test	Antibody titres: Coombs test	Antibody titres: dextran test
Fa	83/9	64	8	16	128
	L 1	2	0	4	4
Fb	49/25	0	0	2	2
	(U.S.S.R.)	32	0	128	128
Ga	Ř 56	4	0	2	8
	V 1	0	0	32	16
Gb	Ř 31	2	0	8	2
	64/12	0	0	64	4
Ha	Ř 46	0	0	32	0
	S 10	0	0	8	0
Hb	Ř 3	0	32	16	8
	49/25/8	0	0	32	8
Hc	72/166	0	0	16	0
	(U.S.A.)	0	N	2	2
Ia	42/71	0	0	8	0
	1669	0	0	4	0
Ib	Ř 4	0	0	32	0
	6/37	0	0	16	0
Ja	79/47	0	0	4	0
	50/146	0	0	2	0
Ka	3/34	16	8	16	32
	S 10	0	128	16	0
Kb	Ř 23	4	16	32	64
	Ř 6	0	32	32	2

TABLE 1b

Reagents		Antibody titres				Reagents		Antibody titres			
designation	No.	agglutination test	haemolytic test	Coombs test	dextran test	designation	No.	agglutination test	haemolytic test	Coombs test	dextran test
Kc	72/166	0	0	4	4	Lk	L 06	0	0	64	128
	(G.D.R.)	0	0	4	8		(G.F.R.)	0	0	4	4
Kd	V 51	32	0	16	64	Ll	L 048	N	0	0	8
	V 39	4	0	8	16		1258	0	0	8	8
Ke	66/10	16	0	8	32	Ma	17/19	0	0	16	0
							30/64	0	0	16	0
La	64/12	16	0	4	64	Mc	T 2	0	0	4	0
	1781	0	0	8	N		(G.D.R.)	0	0	8	0
Lb	L 026	0	0	8	32	M(d)	36/48	0	0	8	0
	4376	0	0	8	32		3/34	0	0	4	0
Lc	Ř 46	N	0	32	64	Na	Ř 20	0	0	16	0
	Z 6	0	0	2	4		15/1	0	0	8	0
Ld	86/162	N	0	2	16	Nb	1506	0	0	8	0
	Ř 6	N	0	4	4		V 25	0	0	8	0
Lf	19/56	0	0	N	2	Oa	Ř 10	0	0	8	32
	(G.F.R.)	2	0	4	8						
Lg	Ř 48	N	0	32	128	Ob	85/124	N	0	N	2
	L 1	0	0	32	32						
Lh	Ř 6	N	0	4	4						
	L 06	N	0	16	32						
Li	L 02	32	0	16	64						
	(G.F.R.)	0	0	32	32						
Lj	V 22	0	0	2	4						
	1432	0	0	8	0						

are: anti Ca, Ha, Ia, Ib, Ja, Ma, Mc, Md, Na and Nb. Also anti Ga, Gb and Hc are usually of this type.

(d) Our main interest was devoted by the fourth group, showing often very high titres in the dextran test. The majority of reagents of the K system and most of the L (Lb, Lc, Ld, Lf, Lg, Lh, Lj, Lk and Ll) and O systems reagents belong here. The dextran test also increases the titres of many antibodies, classified in the first group (e.g. A, Bb, Da, Ef, Eg and others) and facilitates their use at higher dilutions.

Worth mentioning is antibody anti-Ap (normal cattle serum) which haemolyses, at a negligible titre difference, erythrocytes of Ac and Ap phenotypes.

Among test sera we mention for the first time antibody Fb which according to previously obtained results, detects blood factors antithetical to Fa. Our serum is weak; it can be used only in the dextran test. An antibody of the same specificity and of high titre has previously been prepared at the Kharkov Laboratory in the U.S.S.R.

There is a new antibody, designated Md, observed among antibodies reacting solely in the Coombs test. The genetic classification of blood factor Md has yet to be finally proved. We may however assume that factor Md is genetically determined by alleles Mc^{cd} and M^{d}.

DISCUSSION

The division of antibodies into 4 groups according to serological character need not be regarded as generally valid or unchangeable. We know that in some laboratories abroad antibodies Ha or Ma have, in contrast to ours, a lytic character, and we could add a number of other examples. It was our aim, however, to show that there is a difference in the character of antibodies which are used to determine blood groups in pigs and to emphasize the importance of choosing the right serological test. We wish to stress especially those results demonstrating the advantage of the dextran test which is still being neglected. We are not against preferential use of direct agglutination and haemolysis. The wide spectrum of determined blood group factors has shown, however, that the simple dextran test can often replace the elaborate Coombs test. The dextran test may be also used successfully in the case of weak complete antibodies when a small amount only is available, since higher dilutions may be used. Non-specific reactions need not be feared if sera are cross tested in the dextran test be forehand. Seemingly unspecific reactions are caused by the presence of antibodies which have to be removed by subsequent absorption. The reliability of the dextran test finally depends on the dextran solution itself. After testing several foreign trade marks, the dextran of Czechoslovak production "Spofa" proved the most satisfactory, used as a 6% solution. Its molecular weight is in the range of 75,000.

CONCLUSION

In this report we give the serologic characteristic of 44 different blood group reagents of pigs by means of parallel comparison of titres of a greater number of reagents with the same specificity in 4 different tests. On grounds of results obtained we recommend that the dextran test is applied more frequently.

REFERENCES

ANDRESEN, E., 1957. Blodtypeforskningen hos svinet, *Raport Den Kgl. Veterinaer- og Landbohøjs koles afd. f. Fysiologi, Endokrinologi og Blodtypeforskning*, **1**, 21.

ANDRESEN, E., 1963. A study of blood groups of the pig, Munksgaard, Copenhagen, pp. 30–32.

HOJNÝ, J. and HÁLA, K., 1965. Blood group system O in pigs. Blood groups of animals, Edited by J. Matoušek, Publishing House of the Czechoslovak Academy of Sciences, Prague, pp. 163–168.

HOJNÝ, J., GAVALIER, M., HRADECKÝ, J. and LINHART, J., 1966. New blood factors in pigs. Polymorphismes biochimiques des animaux. Xe Congrès européen sur les groupes sanguins et le polymorphisme biochimique des animaux, Paris. pp. 151–158.

DISCUSSION

J. MATOUŠEK: The most important thing in dextran test is the character of dextran. Several types of dextran were compared but the best results were acquired with Spofa dextran. The same results were confirmed by Wageningen Laboratory.

XITH EUROPEAN CONFERENCE ON ANIMAL BLOOD GROUPS AND BIOCHEMICAL POLYMORPHISM
Warsaw, July 2nd–6th, 1968

Studies on Blood Groups in Wild Boar*

I. WIATROSZAK

College of Agriculture in Poznań, Department of Animal Husbandry, Laboratory of Research in Animal Blood Groups, Poznań, Poland

SUMMARY. Studies were carried out on 488 wild boars from Wielkopolska Province. The blood cell antigens of these boars were determined using 36 pig test sera belonging to 12 blood group systems. Very distinct differences in blood cell antigen structure were found between the wild boar and the domestic pig, especially in blood group systems A, F and G. The blood samples from 82.17 per cent of the tested wild boars gave positive reaction with test serum anti-A, but none of the tested individuals reacted positively with test sera anti-Fa or anti-Gb.

Comparison of the results obtained with those of BUSCHMANN (1965) shows great similarity in blood antigenic structure of wild boars in Poland and in G.F.R. Statistically significant differences were found only in the number of animals reacting positively to test sera anti-A, anti-Ef and anti-La.

INTRODUCTION

Previous serological investigations (MADEYSKA, 1962; BUSCHMANN, 1965) had shown that pig test sera could be used for determination of blood cell antigens in wild boars. The aim of this study was to characterize the blood cell antigens of the wild boar of Wielkopolska Province, by using a large number of pig test sera.

MATERIAL AND METHODS

Blood samples from 488 wild boars were taken from animals of both sexes shot by the members of the Polish Hunters' Association.

They were tested with 36 pig test sera belonging to 12 blood group systems, applying the methods conventionally used for determining blood cell antigens in pig. We calculated the frequencies of animals reacting positively to individual test sera, and frequencies of phenotypes in individual blood group systems.

The results obtained were compared with those of BUSCHMANN (1965) for wild

* This research has been financed in part by a grant made by the United States Department of Agriculture under P. L. 480.

boar, and with those of our investigations on the White Zlotnicka pig — one of pig breeds reared in Wielkopolska.

RESULTS AND DISCUSSION

The investigations revealed highly significant differences in blood group factors between wild boar and domestic pig. Table 1 shows the numbers of wild boars whose blood cells reacted positively to individual test sera.

TABLE 1

Blood groups in wild boars, N = 488

Reagent	Frequency, %	Reagent	Frequency, %
A	82.2±1.7	Ib	30.9±2.0
O	27.9±2.0	Ja	56.8±2.0
Ba	100	Ka	10.0±1.4
Bb	0.6±0.3	Kb	100
Ca	2.5±0.7	Kd	0.4±0.3
Ea	7.8±1.2	La	26.4±2.0
Eb	40.0±2.2	Lb	94.1±1.0
Ed	100	Lc	93.4±1.1
Ee	96.1±2.8	Ld	30.9±2.0
Ef	32.8±2.1	Lf	6.8±1.1
Eg	95.7±0.9	Lg	94.1±1.0
Eh	95.7±0.9	Lh	24.2±1.9
Fa	0	Li	94.1±1.1
Ga	100	Me	1.4±0.5
Gb	0	Po_{15}(Nb)	79.3±1.8
Ha	4.9±1.0	Po_4(Ob)	97.1±0.8
Hb	0	Po_1	0
Ia	94.5±0.1	Po_2	0

Table 2 presents a comparison of our results with BUSCHMANN's data concerning wild boar in G.F.R. There is great similarity in blood cell antigen structure of wild boars in both countries. Significant differences were found only in the frequencies of animals positively reacting to test sera anti-A, anti-Ef, and anti-La. This findings suggest that wild boars from various regions of Europe may differ in the antigenic structure of their blood cells.

Table 3 presents the frequencies of phenotypes in individual blood group systems in wild boars.

Data in the Tables 1–3 point to differences between the wild boar and the domestic pig in blood group systems G, F, A, and E.

Table 4 compares the frequencies of phenotypes in these systems in the wild boar and in the White Zlotnicka pig.

TABLE 2

Comparison of results of testing wild boars in Poland and in G.F.R.

Reagent	Positive animals in Poland, %	Positive animals in G.F.R., %*	Differences −	+	Significance
A	82.2±1.7	97.5±1.0		15.3	33.56***
Ba	100 ±	99.2±0.6	0.8		1.62
Ea	7.8±1.2	9.2±1.9		1.4	0.45
Eb	40.0±2.2	32.8±3.0	7.2		3.52
Ee	96.1±2.8	97.5±1.0		1.4	0.87
Ef	32.8±2.1	23.4±3.3	9.4		5.24**
Fa	0	0		0	
Ga	100	100		0	
Gb	0	0		0	
Ha	4.9±1.0	0		0	
Kb	100	100		0	
La	26.4±2.0	37.4±3.1		11.0	9.15***

* Calculations of frequencies and standard deviations are based on BUSCHMANN's published data.
** Significant at 5% level.
*** Significant at 1% level.

TABLE 3

Phenotypes in blood group systems in wild boars

Blood group system	Phenotype	Frequency, %
A	A(A+0+)	10.0±1.4
	A(A+0−)	72.1±2.0
	A(A−0+)	14.8±1.6
	A(A−0−)	3.0±0.8
B	B(a+b−)	99.4±0.3
	B(a+b+)	0.6±0.4
	B(a−b+)	0
C	C(a+)	2.5±0.7
	C(a−)	97.5±0.7
E	E(a+b−d−e+f−g+h−)	0
	E(a+b+d+e+f−g+h−)	0.4±0.3
	E(a+b−d+e+f+g+h+)	2.5±0.7
	E(a+b−d+e+f−g+h+)	4.9±1.0
	E(a−b+d+e−f−g+h−)	3.9±0.9
	E(a−b+d+e+f+g+h+)	7.4±1.1
	E(a−b+d+e+f−g+h+)	28.3±2.0
	E(a−b−d+e+f+g−h+)	4.3±0.9
	E(a−b−d+e+f+g+h+)	18.6±1.8
	E(a−b−d+e+f−g+g+)	29.7±2.0

TABLE 3 (*cont.*)

Blood group system	Phenotype	Frequency, %
F	F(a+)	0
	F(a−)	100
G	G(a+b−)	100
	G(a+b+)	0
	G(a−b+)	0
H	H(a+b−)	4.9±0.9
	H(a−b+)	0
	H(a−b−)	95.0±0.9
I	I(a+b−)	69.0±2.0
	I(a+b+)	25.4±1.9
	I(a−b+)	5.5±1.0
J	J(a+)	56.8±2.2
	J(a−)	43.2±2.2
K	K(a+b−d−)	0
	K(a+b+d−)	9.6±1.3
	K(a+b+d+)	0.4±0.3
	K(a−b+d−)	90.0±1.4
	K(a−b−d−)	0
L	L(a+b−c−d+f−g−h+i−)	5.9±0.6
	L(a+b+c+d+f−g+h+i+)	18.2±1.7
	L(a+b+c+d−f−g+h−i+)	1.6±0.6
	L(a+b+c−d+f+g+h−i+)	0.6±0.4
	L(a−b+c+d−f−g+h−i+)	67.4±2.1
	L(a−b+c+d+f+g+h−i+)	6.1±1.0
M	M(c+)	1.4±5.4
Po_{15}(Nb)	N(b+)	79.3±1.8
	N(b−)	20.7±1.8
Po_4(Ob)	O(b+)	83.0±1.7
	O(b−)	17.0±1.7
Po_1	Po_1(+)	0
Po_2	Po_2(+)	0

The most interesting observations in our study are as follows: A high percentage of wild boars reacted positively with test serum anti-A. Reaction with anti-A is rare in our domestic pig. None of the tested wild boars reacted positively with test sera anti-Fa and anti-Ga.

These findings confirm the results of BUSCHMANN's investigations. In the G system all wild boars so far tested have been a/a homozygotes.

TABLE 4

Comparison of frequency of phenotypes in systems A, E, F, G in the wild boar and in the White Zlotnicka pig

Blood group system	Phenotypes	Wild boar N = 488	White Zlotnicka N = 484	Differences −	Differences +	χ^2
A	A(A+0+)	10.0±1.4	0	10.0		51.2**
	A(A+0−)	72.1±2.0	5.6±1.0	66.5		173.0**
	A(A−0+)	14.8±1.6	31.8±2.1		17.0	39.7**
	A(A−0−)	3.0±0.8	62.6±2.2		59.6	391.2**
E	E(a+b−d−e+f−g+h−)	0	4.1±0.9		4.1	20.6**
	E(a+b+d+e+f−g+h−)	0.4±0.3	18.4±1.8		18.0	92.6**
	E(a+b−d+e+f+g+h+)	2.5±0.7	3.3±0.8		0.8	0.62
	E(a+b−d+e+f−g+h+)	4.9±1.0	0	4.9		29.4**
	E(a−b+d+e−f−g+h−)	3.9±0.9	48.8±2.3		9.8	252.8**
	E(a−b+d+e+f+g+h+)	7.4±1.1	22.9±1.9		15.5	45.8**
	E(a−b+d+e+f−g+h+)	28.3±2.0	0.8±0.4	27.5		146.8**
	E(a−b−d+e+f+g−h+)	4.3±0.9	1.6±0.6	2.7		5.9*
	E(a−b−d+e+f+g+h+)	86.6±1.8	0	86.6		99.6**
	E(a−b−d+e+f−g+h+)	29.7±2.0	0	29.7		169.0**
F	F(a+)	0	8.1±1.2		8.1	41.0**
	F(a−)	100	91.9±1.2	8.1		41.0**
G	G(a+b−)	100	29.1±2.1	70.9		534.4**
	G(a+b+)	0	50.4±2.3		50.4	328.5**
	G(a−b+)	0	20.4±1.8		20.4	111.1**

* Significant at 5% level.
** Significant at 1% level.

Wild boars and the native domestic pigs are descended from common ancestral stocks. The wild boar lives in severe natural conditions, and is subjected to environmental selection pressure. Interference by man is limited mainly to hunting, not a selection criterion applied in modern breeding.

It seems possible that the results obtained and those of further immunogenetic studies on wild boars, may reveal relationship between blood group factors and other physiological features in the pig.

CONCLUSIONS

(1) Comparison of the results obtained with those of BUSCHMANN (1965) reveals great similarity of blood cell antigen structure in wild boars in Poland and in German Federal Republic.

(2) Very distinct differences were found between the wild boar and the domestic pig, especially in the blood group systems A, E, F, G.

(3) All wild boars tested have carried only the G^a antigen in the G blood group system.

REFERENCES

BUSCHMANN, H., 1965. Blood groups serological studies. Blood groups of animals, Proceedings of the 9th European Animal Blood Group Conference in Prague, pp. 131–133.

MADEYSKA, A., 1962. Wstępne badania nad grupami krwi dzików (Initial studies on blood groups in wild boar), *Postępy nauk rolniczych*, **4**, p. 55–57.

DISCUSSION

P. IMLAH: I wish to call attention to an abstract by JENSEN *et al.* published in *J. Anim. Sci.* **26**, 885, 1967. These authors studied blood group loci and serum protein polymorphism in swine to determine their relationship with various quantitative traits. There was a significant excess of significant F values using least square analyses. However, the variation explained by the loci was less than one percent of the within litter variation.

A. JAMIESON: Referring to the report on linkage studies presented by Dr. Andresen at the Paris Conference in 1966 I would like to ask you if cytogenetic studies have been undertaken in connection with the blood group studies of the wild pigs.

I. WIATROSZAK: Cytogenetic studies have not been performed.

E. ANDRESEN: At the Paris Conference I mentioned that linkage studies using wild pigs might be particularly useful for mapping certain chromosomes in pigs. However, the proposed short cut is only possible if the chromosomal evolution suggested by BUSTAD (*Scientific American*, **214**, 94, 1966) is true. If so, one should expect that certain loci which are located on two non-homologous chromosomes in the domesticated pigs are genetically linked in the wild pig. On the other hand, I have not seen further support in the literature for the possibility that the European wild pig should have only 36 chromosomes.

L. FESÜS: Have you tested the serum samples of these wild pigs for serum polymorphism too?

I. WIATROSZAK: In the reported study on wild boar we tested only blood groups. We intend to study polymorphism in serum proteins in our further work.

J. HOJNÝ: We also tested blood groups in wild boar, but in 30 animals only. Our results were similar to those presented by Dr. Wiatroszak however we did not find any animal possessing the phenotype A(A+O+) in A blood group system.

I. WIATROSZAK: In our work we paid special attention to the presence of the phenotype A(A+O+) in a few wild boars because we did not find this phenotype in our investigations in domestic pigs, carried out on over 6000 animals. Maybe blood groups are inherited in different way in wild boar than in domestic pig. To explain this problem we plan immunogenetic investigations on hybrids obtained by crossing wild boar with domestic pig. In Poland we have conditions enabling this study.

XITH EUROPEAN CONFERENCE ON ANIMAL BLOOD GROUPS AND BIOCHEMICAL POLYMORPHISM
Warsaw, July 2nd–6th, 1968

Gene Frequency of Blood Groups from Different European Landrace Pigs

F. MAJOR, H. DINKLAGE and R. GRUHN
Institute of Animal Husbandry and Animal Genetics, University of Göttingen, G.F.R.

SUMMARY. By means of blood group factors, the authors studied the extent to which Danish, Swedish, Czechoslovakian, Russian, Dutch, Hungarian and German Landrace pig populations differed in their genetic make-up. The historical development of the various populations is expressed both in the gene frequency and the similarity index. There is a very close relationship between the Danish, Swedish, Czechoslovakian and Russian Landrace populations and there is also a very close but different relationship between the German, Dutch and Hungarian Landrace populations.

INTRODUCTION

During the last decades there has been throughout Europe an increasing distribution of the Landrace pig, bred about 1900 in Germany and Denmark. By using frequencies of blood group genes calculated within the A, B, D, E, F, G, H, I, K, L, M and N systems, an attempt was made to determine whether there are differences or similarities in the genetic make-up of the Danish, Swedish, Dutch, Czechoslovakian, Russian, Hungarian and German Landrace pig populations.

For this purpose, 199 Danish, 108 Swedish, 120 Dutch, 80 Hungarian (crossbreds between the Dutch Landrace pig and the Hungarian Large White meat type pig) and 205 German Improved Landrace pigs were tested by means of the monofactorial sera available at the Göttingen Blood Group Research Laboratory (BRUCKS and DINKLAGE, 1967). For the Russian (TIKHONOV *et al.*, 1966) and Czechoslovakian Landrace populations (GAVALIER *et al.*, 1966) data from the literature were used. For the determination of the genetic make-up of the populations the following parameters were used: gene frequency, genetic equilibrium, similarity index and degree of homo- and heterozygosis.

GENE FREQUENCY AND SIMILARITY INDEX

From the calculation of gene frequency it followed that there are differences between populations in the frequency of the various alleles. These differences are most distinct in the A, E, G, I, L and M systems. At the same time it may be observed that the Landrace populations studied can be divided into two groups. The Danish, Swedish, Czechoslovakian and Russian Landrace pigs form one group, and the German and Dutch Landrace pigs form another. This situation emerges very clearly from the gene frequencies determined in the E, G and H blood group systems (see Table 1).

TABLE 1

Allele frequency in the E, G and H blood group systems in six European Landrace pig populations

Alleles	Populations					
	German	Dutch	Danish	Swedish	Czechoslovakian	Russian
E^{aeg}	0.298	0.287	0.010	0.065	0.079	0.196
E^{bdg}	0.269	0.255	0.407	0.449	0.449	0.332
E^{bdf}	0.000	0.008	0.040	0.000	0.015	0.012
E^{edg}	0.282	0.325	0.244	0.227	0.062	0.186
E^{efd}	0.151	0.125	0.299	0.259	0.395	0.274
G^{a}	0.678	0.695	0.241	0.348	0.218	0.282
G^{b}	0.322	0.305	0.759	0.652	0.782	0.718
H^{c}	0.274	0.336	0.187	0.159	0.196	—

The simple comparison of gene frequencies indicates where there are relationships and differences between populations. The relationship between different breeds and populations can be numerically expressed by the similarity index introduced by MAIJALA and LINDSTRÖM (1966). This index which indicates the degree of similarity varies between 0 and 1. Since the E system provides the best information, all indices were calculated for this system and are shown in Table 2.

TABLE 2

Similarity indices (r) of the E system in seven European Landrace pig populations

Population	German	Dutch	Danish	Swedish	Czechoslovakian	Russian
Hungarian	0.762	**0.815**	0.799	0.761	0.553	0.718
German		**0.995**	0.792	0.832	0.743	0.925
Dutch			0.767	0.811	0.657	0.894
Danish				**0.987**	0.930	0.926
Swedish					**0.935**	**0.949**
Czechoslovakian						0.923

On the basis of the determined gene frequencies and on the basis of the similarity of the calculated indices the Landrace populations under study may be divided into the following groups:

Landrace Populations

German	Danish
Dutch	Swedish
Hungarian	Czechoslovakian
	Russian

DEGREEE OF HOMO- AND HETEROZYGOSIS

If, as far as a genetic system is concerned, a population fulfils the conditions of the Hardy–Weinberg law, there will always be a certain percentage of homo- and heterozygous animals. With few exceptions, this was the case in the study here reported. The maximal degree of hetetozygosis is attained when all the alleles of a system occur with the same relative frequency. In most of the populations, a good approach to or agreement with the highest possible degree of heterozygosis was observed with the E, G and I blood group systems. The L system, however, is an exception as the proportion of homozygotes is considerably higher than would be theoretically possible according to the number of alleles. Whether this is solely due to the random genetic drift or to a selection advantage of the L^{bcgi} carriers is not yet sure.

In order to get more insight into the degree of homo- and heterozygosis of the populations, the four closed blood group systems E, G, I and L were used to determine the degree of homo- or heterozygosity. The results of this investigation showed that the Danish Landrace population is most strongly homozygous and the Hungarian Landrace population is most strongly heterozygous.

The present investigations show that the historical development of the various Landrace populations is reflected by the blood group frequencies. Consequently the value of continuing blood group research for evolutionary studies becomes evident.

REFERENCES

BRUCKS, R. and DINKLAGE, H., 1967. Die Bedeutung der verschiedenen Blutgruppensysteme für den Elternschaftsnachweis beim Deutschen veredelten Landschwein und beim Deutschen Edelschwein (The value of various blood group systems for the control of parentage in the German Improved Landrace pig and the German Large White pig), *Der Tierzüchter*, **19**, 23, 797–798.

GAVALIER, M., HOJNÝ, J., HRADECKÝ, J., LINHART, J. and SCHRÖFFEL, J., 1966. Blood groups and serum proteins in pigs, Xth Eur. Conf. Anim. Blood Groups and Bioch. Polym., Paris, 159–164.

MAIJALA, K. and LINDSTRÖM, G., 1966. Frequencies of blood group genes and factors in the Finnish cattle breeds with special regard to breed comparison, *Ann. Agr. Fenniae*, **5**, 76–93.

MAJOŘ, F., 1968. Untersuchung über die verwandtschaftlichen Beziehungen zwischen verschiedenen europäischen Landrasse-Populationen mit Hilfe von Blutgruppenfaktoren (Investigations of the relationship between different European Landrace pig populations by the use of blood group factors), Thesis, Göttingen.

TIKHONOV, V. N., BURLAK, Z. K., GORELOV, I. G. and CEIDO, M. A., 1966. Immunogeneticki analiz mezhduporodnoi i vnutriporodnoi populatsionnoi differentsiatsii u svinei. (Immunogenetical analysis of between- and within-breed differentiation of populations in pigs), *Genetika*, **12**, 106–119.

XITH EUROPEAN CONFERENCE ON ANIMAL BLOOD GROUPS AND BIOCHEMICAL POLYMORPHISM
Warsaw, July 2nd–6th, 1968

Blood Groups and Serum Proteins in the Hungarian Mangalica Pig

L. FÉSÜS

Bloodgrouping Laboratory, University of Veterinary Medicine, Budapest, Hungary

SUMMARY. 430 Hungarian Mangalica pigs originating from three different herds were tested for blood antigens (Bb, Da, Ea, Eb, Ed, Ee, Ja, Ka, La, Mc, Hpl) and 360 of those for serum protein systems (Tf, Hp, Cp, Am). Starch-gel electrophoresis was carried out according to KRISTJANSSON (1963).

The following gene frequencies were observed B^b: 0.023, D^a: 0.065, E^{ae}: 0.035, E^{de}: 0.616, E^{bd}: 0.384, J^a: 0.606, K^a: 0.000, L^a: 0.012, M^c: 0.041 and Hp^1: 0.046. Tf^B: 1.000, Hp^1: 0.850, Hp^2: 0.15, Cp^2: 1.000, Am^2: 1.000.

We could not compare the results of our blood group studies with those of the former authors because different reagents were used, but we compared the results of our electrophoretic studies with those of HISTRIC *et al.* (1967).

INTRODUCTION

The Hungarian Mangalica pig was developed in the course of the last century and was bred in very poor environments. After the introduction of modern breeding systems this breed of pig gradually disappeared and today it can be found in very limited numbers only. In Yugoslavia this breed is known as White Mangalica (JOVANOVIČ and STOJANOVIČ, 1966).

The blood groups and serum proteins of this breed are described in this study.

MATERIAL AND METHODS

Blood samples were collected from 430 Hungarian Mangalica pigs of three different herds and tested for the presence of blood group antigens on their red cells. Only 360 of these samples were subjected to starch-gel electrophoresis (KRISTJANSSON, 1963).

The following methods for the estimation of blood group gene frequencies were applied in the present work:

(1) The simple gene counting method was used for estimating the frequencies of B^b, D^a, J^a, L^a, M^c and Hp^1.

(2) The gene counting allocation method was applied in connection with the E system.

RESULTS

Table 1 shows the blood group gene frequencies found.

Table 2 shows the observed frequencies of the transferrin, hemopexin, caeruloplasmin and amylase alleles.

TABLE 1

Alleles	Frequencies	Alleles	Frequencies
B^b	0.023	J^a	0.606
D^a	0.065	K^a	0.000
E^{ae}	0.035	L^a	0.012
E^{de}	0.616	M^c	0.041
E^{bd}	0.384	Hp^1	0.046

TABLE 2

Number of animals	Gene frequencies				
	Tf^B	Hp^1	Hp^2	Cp^2	Am^2
360	1.000	0.850	0.150	1.000	1.000

DISCUSSION

Blood group determinations were carried out with regard to 16 red cell antigens (Ac, Bb, Da, Ea, Eb, Ed, Ee, Gb, Ha, Ja, Ka, Kb, La, Mc, Oa, and Hpl) but considering the results of the Third Pig Blood Grouping Comparison Test (1968) the reactions with 11 reagents should be taken into account (anti-Bb, anti-Da, anti-Ea, anti-Eb, anti-Ed, anti-Ee, anti-Ja, anti-Ka, anti-La, anti-Mc, and anti-Hpl).

The allele E^{de} occurred in the highest frequency and we could not observe any reactions with our anti-Ka reagent.

JOVANOVIČ and STOJANOVIČ (1964 and 1966) carried out the same investigations on the blood groups of the two types of Yugoslavian Mangalica pigs. We could not compare the results of the present study with those of the former ones because different reagents were used by the two laboratories.

Determining the transferrins, hemopexins, caeruloplasmins and amylases very little variation was found. No polymorphism was found for transferrin, caeruloplasmin and amylase loci, and only two hemopexin alleles (Hp^1 and Hp^2) were present in this material.

HISTRIC *et al.* (1967) found nearly the same results in Yugoslavian Mangalica pigs, however they also observed the presence of the allele Am^3 in their breed, but at a very low frequency ($q^{Am^3} = 0.01$).

REFERENCES

HISTRIC, V., HESSELHOLT, M. and STOSIC, D., 1967. Serum protein polymorphism in some Yugoslavian pig breeds, *Acta Vet. Scand.*, **8**, 43–49.

JOVANOVIČ, V. and STOJANOVIČ, Z., 1964. Study of pig blood groups in Vojvodina, Blood Groups of Animals, Prague, 1965, p. 149.

JOVANOVIČ, V. and STOJANOVIČ, Z., 1966. Examination of blood types in two autochtonous types of the pig race "Mangalica", Polymorphismes Biochimiques des Animaux, Paris, 1966, pp.147–149.

KRISTJANSSON, F. K., 1963. Genetic control of two pre-albumins in pigs, *Genetics*, **48**, 1059–1063.

XITH EUROPEAN CONFERENCE ON ANIMAL BLOOD GROUPS AND BIOCHEMICAL POLYMORPHISM
Warsaw, July 2nd–6th, 1968

An Immunogenetic Comparison of Blood Antigens and their Value in Selection of Breeding Stock

Nina O. SUKHOVA

Laboratory of Immune Genetics, Siberian Research Institute of Animal Breeding, Novosibirsk, U.S.S.R.

SUMMARY. A comparative study of the antigenic properties of pig blood has been made using polyvalent antisera. The study was carried out as an aid to selection of breeding stock.

INTRODUCTION

Immunogenetical research has assumed an importance in animal husbandry. Immunological methods are used to identify animals, to assign paternity, to determine the zygosity of twins and to assist in rapid diagnosis of haemolytic disease of the new-born animal. Immunogenetics has also been applied to theoretical problems bearing an animal breeding. Correlations have been attempted between genotype and economic value in the search for the best genotype. The mechanism of the maintenance of genic equilibrium in populations and other problems have been examined.

To attempt such investigations, it is necessary to carry out immunogenetic research on a variety of species of domestic livestock to develop new techniques and to attempt to improve old ones.

MATERIAL AND METHODS

Animals of the Large White, Sibirskaya Severnaya, Landrace and Lacombe breeds were studied. Polyvalent antisera of different specificities were used to test the experimental animals. Such antisera were prepared in the following way. Aliquots of serum corresponding in number to the number of experimental animals were prepared. Each aliquot was subjected to absorption by the erythrocytes of a single experimental animal until all reactivity with these erythrocytes was exhausted.

Special care was taken to prepare the initial polyvalent serum. Such antisera represented a mixture of 25 to 30 "crude" sera obtained from interbreed immunizations. Before being mixed to make a polyvalent serum the individual sera were tested with erythrocytes of known antigenic constitution*. This examination revealed the numbers of antibodies contained in a serum. The reaction of erythrocytes to complete and incomplete antibodies was determined by cross-reaction of the antisera. The methods of testing used were those of ANDRESEN (1963).

Because individual antibodies in the initial polyvalent serum would not be present equally and because mixture of individual sera to give a polyvalent antiserum would result in dilution of antibody and thus in reduction of titre special care was given to the evaluation of the reaction. After a visual evaluation the negative reactions were microscopically examined with a 100 and 200-fold magnification. The reaction intensity was evaluated by means of a five mark grading system which took into consideration the degree of serum dilution. After extraction of specific antibodies by the blood of the experimental animals the sera were used as a control.

Experimental studies with these polyvalent antisera in immunogenetics and breeding were carried out by post-graduate workers I. S. LISINA and G. K. GORELOVA on the farms of the Novosibirsk region.

RESULTS

(1) It was established that each serum aliquot obtained from the same polyvalent antiserum differed in specificity.

(2) The use of a series of aliquots from a polyvalent antisera against the erythrocytes of a population (herd) enabled the degree of antigenic resemblance between the blood of individual animals to be established.

(3) The analysis of antisera obtained from the blood of experimental animals shows that an increase in the phenotypic variety of the blood used for absorption leads to a reduction (in the subsequent cross-breed reaction) of the serum prepared.

DISCUSSION AND CONCLUSION

Studies which use polyvalent antisera are not antagonistic to the "classical" methods used for investigations of the polymorphism of the blood and tissue elements of domestic livestock. They may in fact have particular relevance to immunogenetic studies of certain problems. Absorbed sera may be used in cattle-breeding to secure a rapid identification of the antigenic properties of the blood on an animal and when necessary allow individual identification so avoiding accidental confusion. In the absence of animals tested for blood groups isoimmune antisera may be obtained from suitable donors and recipients selected by absorption studies.

It is likely that the future importance of polyvalent antisera will be in studies

* Testing of the blood groups of these animals was performed by the Liběchov Laboratory for Physiology and Genetics of Animals, Czechoslovak Academy of Sciences.

concerned with breeding for productivity. Analysis of the findings on the connection of blood group genotypes and the economic usefulness of animals demonstrates the advantage of individuals heterozygotic at a number of loci over those animals homozygous at these loci. The most convincing data were obtained to studies of chickens (BRILES, 1954; BRILES and KRUEGER, 1955; GILMOUR, 1958). The known facts confirm the conception of heterosis as an effect caused by the combination of the qualitatively different genetic information of the parents. SCHULTZ and BRILES (1953) indicate a possible summation of the favourable effect of heterozygosis at a number of antigenic loci. Such effects would be especially important in the type of study visualized.

Observations carried out on 508 pigs by Lisina showed high vitality and energetic growth in offspring of matings in which the boars and sows presented clear-cut antigenic differences.

The most effective immunogenetic investigations will undoubtedly be those using simultaneous polyvalent and monovalent antisera. We believe they are likely to reveal the biological effects of individual loci and their combinations.

ACKNOWLEDGEMENT

I would like to express my deep gratitude to Dr. MATOUŠEK and M. Sc. J. HOJNÝ for the blood group testing of pedigree animals in their laboratory which allowed us to conduct immuno-genetic investigations. I am equally thankful to M.Sc. I. LINHART for his help in elaborating the methods of our study.

REFERENCES

ANDRESEN, E., 1963. A study of blood groups of the pig, Munksgaard, Copenhagen.

BRILES, W. E., 1954. Evidence for overdominance of the B blood group alleles in the chicken, *Genetics*, **39**, 361.

BRILES, W. E., and KRUEGER, W. F., 1955. The effect of parental B blood group genotypes on hatchability and livability in Leghorn inbred lines, *Poultry Sci.*, **34**, 5, 1182.

GILMOUR, D. G., 1958. Maintenance of segregation of blood group genes during inbreeding in chickens, *Heredity*, **12**, 1, 141.

SCHULTZ, F. T. and BRILES, W. E., 1953. The adaptive value of blood group genes in chickens, *Genetics*, **38**, 1, 34.

XITH EUROPEAN CONFERENCE ON ANIMAL BLOOD GROUPS AND BIOCHEMICAL POLYMORPHISM
Warsaw, July 2nd–6th, 1968

Immunogenetic Analysis of Intrabreed Structure of Genealogically Related Breeds

V. N. TIKHONOV, Z. K. BURLAK, I. G. GORELOV, and M. A. CHEIDO
Institute of Cytology and Genetics of the U.S.S.R. Academy of Sciences and Siberian Animal Breeding Institute, Novosibirsk, U.S.S.R.

SUMMARY. Intrabreed differentiation according to gene and genotype frequencies of blood group systems occurring in linebreeding and in conventional breeding methods of improvement of breeds was studied. An immunogenetic analysis was undertaken involving 20 antigens of 9 genetic blood group systems of intrabreed structures of 14 lines of major populations of the Ukrainian Steppe White breed and the Ukrainian Spotted breed derived from it (766 boars and sows). Considerable differentiation between these two related breeds and also intrabreed differentiation between lines has been established. Prolonged and severe line-breeding causes increase of the intrabreed differentiation. The temporary correlations between blood groups and productivity traits are formed and maintained in this way. It is suggested that for increased efficiency of the breed improvement in line-breeding immunogenetic analysis should be applied.

INTRODUCTION

Line-breeding is the most progressive method of breed improvement. Lines are classified as genealogical, stud (zootechnical) and inbred. In animal selection line-breeding is not so efficient as expected because of the absence of objective criteria for genetic similarity between the line's foundation stock and its line-progeny. Besides, there are no objective criteria of line homogeneity. Immunogenetic investigation of inbred lines of Landrace pigs and genealogical lines of Berkshire and Large White pigs (TIKHONOV *et al.*, 1966) has shown that analysis of blood groups is an efficient method of studying the genetics of intrabreed structures and the processes of inbreeding. The present report gives some results of an immunogenetic analysis of breeds which have been improved for many years by the line-breeding method.

MATERIAL AND METHOD

The animals used in this investigation were boars and sows of the Ukrainian Steppe White breed (USW) and the Ukrainian Steppe Spotted breed (USS), derived from it, from the main stud farm of these breed in Askanija-Nova (Ukrainian S.S.R.).

Seven hundred sixty-six animals were studied immunogenetically by means of the monospecific sera of our laboratory specimens of which have been tested in all the three international Pig Blood Grouping Comparison Tests (1964, 1966 and 1968) and by several new sera. Determination of 20 blood group antigens of 9 genetic systems was carried out by the direct agglutination and indirect Coombs tests. Gene and genotypic frequencies were investigated in 8 stud lines of USW (342 pigs) and 6 stud lines of USS (424 pigs) separately. The same characteristics were also studied in all these animals whilst classifying them into genealogical lines (which ignores the degree of genetic similarity with the foundation stock of the line resulting from inbreeding and selection for body conformation).

RESULTS

The Ukrainian Steppe White breed (GREBEN, 1962) created by crossing indigenous Ukrainian pigs with the Large White breed introduced from England differs considerably from the latter by blood type (TIKHONOV, 1967). This breed has now retained blood group characteristics inherited from the indigenous pigs over a considerable period of time. The antigen Fa detected in the USW breed is as a rule, absent in Large Whites, and the alleles of the E system were found to be less variable in USW than in Large Whites. Pigs with red and black spots though rarely found in the improved USW breed are so coloured because of the influence of the indigenous breed. The USS breed was created from a foundation stock of these spotted animals with subsequent crossings to the Berkshire and Mangalica breeds (GREBEN, L. and GREBEN, E., 1961). Statistically significant immunogenetic differences were established in almost all the genetical blood group systems of the related USW and USS breeds. The differences were the greatest in systems B, E, F, G.

Immunogenetic analysis produces evidence that the gene pool of USS is poorer and more homogeneous than that of USW. Only 3 alleles and 6 genotypes were found in the E system of the USS population studied, and all USS lines possess these genotypes, while the USW lines have 4 alleles and 7 genotypes, and greater diversity of combinations within lines. All the USS and USW lines have characteristic antigens.

Immunogenetic comparison of the animals performed during classifying into stud and genealogical lines shows, that the blood group analysis is able to reveal the mechanisms through which foreign genetic peculiarities are introduced into a line, i.e. the ways in which the line becomes contaminated with extraneous features.

Blood group differentiation between stud lines is more marked than between genealogical ones under high selection pressure. Some stud lines differ significantly from the genealogical lines of the same name by gene frequencies and higher homozygosity. For instance, the stud line Askania differs significantly from the corresponding genealogical line by the gene frequency of the allele L^a and frequency of homozygous genotypes Eedg/edg (4.25 and 33.3%, respectively, $P < 0.001$).

Comparison of main sires of different generations within one line reveals what tendencies arise as a result of changes in the gene pool of the line in the course of crossing with other lines.

DISCUSSION AND CONCLUSIONS

Line-breeding results in the interbreed differentiation into genealogical structures-lines differing by gene and genotype frequency of blood group systems. This genetic differentiation occurs automatically as a result of selection and inbreeding provided that isolation of lines is sufficient.

Line-breeding leads to forming and maintaining of temporary correlations between blood groups and productivity traits within lines. This process may be accelerated if the line-breeding is accompanied by the immunogenetic analysis.

The blood groups as genetical markers may help to check the "purity" of the lines and thus assist their consolidation. Immunogenetic analysis of the sires used for maintenance of a line enables forecasts of the possible changes in the immunogenetic characteristics of the immediately subsequent generations of the line.

It is concluded that blood groups may serve as a reliable objective criterion for the determination of the genetic similarity between the animals of a line and its foundation stock and for establishing the homogeneity of lines. Thus, line-breeding under immunogenetic control should considerably accelerate breed improvement.

ACKNOWLEDGEMENT

We are grateful to Prof. Dr. L. Greben and our colleagues at the Animal Breeding Institute "Askanija-Nova" for kindly providing us with pig blood samples and zootechnical records.

REFERENCES

Greben, L., 1962. Ukrainian Steppe White pig breed, Kiev, 252 (in Russian).

Greben, L. and Greben, E., 1961. Ukrainian Steppe Spotted breed of pigs, Kiev, 90 (in Russian).

Tikhonov, V. N., 1967. Ispolzovanie grupp krovi pri selektsii zhivotnikh (The blood groups in animals and their application to animal breeding), Moscow, 393.

DISCUSSION

J. Rapacz: Did you carry out electrophoretic analysis of the injected material which was ammonium sulphate fraction.

V. N. Tikhonov: We did not carry out electrophoretic analysis of the injected material.

J. Rapacz: What kind of agar did you use?

V. N. Tikhonov: The bacto-agar "Difco" was used.

J. Rapacz: When did you read your results?

V. N. Tikhonov: The results were read after 2, 3 and 5 days and after the preparing of stained samples too.

XITH EUROPEAN CONFERENCE ON ANIMAL BLOOD GROUPS AND BIOCHEMICAL POLYMORPHISM
Warsaw, July 2nd–6th, 1968

Blood Group System A in the Reproduction of Pigs

J. MATOUŠEK

Laboratory of Physiology and Genetics of Animals, Czechoslovak Academy of Sciences, Liběchov, Czechoslovakia

SUMMARY. Neither blood group factor A nor its corresponding normal antibody anti-A were detected in the fluids of seminal vesicles, tail of the epididymis or in testicle fluid of boars. The ovarian follicle fluid and the cervical mucus were chosen from the genital tract fluids of sows for the study of the mentioned antigenic factor and antibody. Antigenic factor A was present in a significantly lower amount in the ovarian follicle fluid than in the serum. In the cervical mucus the relation was the opposite. Anti-A antibody was present in ovarian follicle fluid and serum in the same titre, but was not detected in the cervical mucus. Anti-A, present in ovarian follicle fluid of sows, did not increase *in vitro**agglutination of spermatozoa of A-positive boars nor did it shorten the period of their motility. Fertility of A-negative sows mated to A-positive boars was not decreased. The number of offspring after two boars, one of which was A-positive and the other A-negative, and both of which were mated during one oestrus, was not changed in favour of the progeny of A-negative fathers.

INTRODUCTION

Some papers have been published lately dealing with possible causes of sterility of women due to incompatible mating in the AB0 blood group system (TYLER *et al.*, 1967; SCHWIMMER *et al.*, 1967). The mechanism of this effect is (however) not known. GERSHOWITZ *et al.* (1958) detected blood group antibodies anti-A and anti-B in the vaginal secretion of women. It is therefore possible that by the binding of these antibodies onto the corresponding antigen present on the spermatozoa of men of the secretor type (EDWARDS *et al.*, 1964; BOETTCHER, 1965), the male genital cells may be in some way disturbed. Even foetuses in their initial stages of development may be attacked by antibodies contained in the fluids of the female genital organs.

There are of course papers (WHITELAW *et al.*, 1962) which dispute the importance of incompatibility in the AB0 system with respect to fertility or sterility of married couples. The present report is devoted to study of the analogous situation in the blood group system A of pigs.

METHODS

Genital tract fluids from boars (seminal vesicle fluid, fluid from the tail of the epididymis and the testes) and ovarian follicle fluid from sows were obtained in a similar way to that described for bulls (MATOUŠEK, 1968) and cows (MATOUŠEK, 1965). Cervical mucus was drawn from the vagina by means of a rubber tube and hypodermic syringe.

The presence of blood factor A on erythrocytes was studied by means of antiserum anti-Ap (HOJNÝ and HÁLA, 1965) according to the method described therein. Serologically active substances of the antigenic factor A in the serum, the seminal vesicle fluid, the fluid of the tail of the epididymis, the testes of boars and ovarian follicle fluid of sows were investigated by means of the inhibition test (PODLIACHOUK and EYQUEM, 1956). The normal agglutination and indirect incomplete tests were used for the detection of anti-A antibodies in all investigated fluids (DUNSFORD and BOWLEY, 1955).

The haemagglutinating factor present in the seminal vesicle fluid of boars and interfering with the detection of the A factor and anti-A antibody, was removed by absorption with ejaculated spermatozoa from bulls or rams. Absorption was carried out three times consecutively with thrice washed spermatozoa at a ratio of 1:1. Preliminary separation of the haemagglutination factor from other proteins in this fluid was performed by dialysis and fractionation by means of continuous electrophoresis (SEDLÁKOVÁ, 1966). Three protein fractions without the haemagglutination factor were analysed (B — coagulate after dialysis, α^A and α^B — first two fractions obtained by continuous electrophoresis of supernatant proteins).

The effect of anti-A antibody, present in the ovarian follicle fluid of sows on ejaculated spermatozoa of A positive and A negative individuals, was studied by the agglutination and immobilization method (QUINLIVAN, 1966). Follicle fluids were previously incubated at a temperature of 56°C for 30 min.

Three hundreds and forty-seven pairs of parents from the breeding herd were used to investigate the effect of incompatibility in the A system on fertility. Besides that, 58 A negative sows were mated, during one oestrus, with one A positive and one A negative boar. The order of boars was changed at random. Parentage determination of the piglets was carried out by blood and serum group tests (GAVALIER *et al.*, 1966; MATOUŠEK and SCHRÖFFEL, 1967).

RESULTS AND DISCUSSION

Blood group factor A has not been found in any sample of fluid of the tail of the epididymis or testicle fluid from 42 boars. Nor was antibody anti-A found. Our previous investigation (MATOUŠEK *et al.*, 1966) did not detect the primary presence of the A antigen on epididymal spermatozoa.

Direct study of the antigenic factor A and the antibody anti-A in the seminal vesicle fluid of boars was hindered by the presence of the agglutination factor (BOURSNELL and COOMBS, 1966). After its absorption by ram or bull spermatozoa, the results with respect to the antigenic factor A and its corresponding antibody were negative. Neither did preliminary analyses of protein fractions of this fluid obtained by dialysis and continuous electrophoresis, and without haemagglutination activity, point to the presence of A factor or anti-A antibody. These results, however, must be considered with reserve because it is not possible to exclude that absorption and fractionation of the seminal vesicle fluids affects the A antigen molecules, and even those of the anti-A antibody.

TABLE 1

Antigenic factor A and antibody anti-A in ovarian follicle fluid and serum of sows

Body fluid	No. of animals	No. of animals with A factor	No. of inhibitory units of A substance	No. of animals with anti-A	Titre of anti-A
Follicle fluid	250	51	6.53±2.76	44	5.14±2.36
Serum of same animals	250	90	14.82±5.86	44	4.77±2.27
Cervical mucus	12	6	76.66±30.66	0	0
Serum of same animals	12	6	10.66±3.22	6	6.66±1.78

As regards genital organs of sows, the ovarian follicle fluid and cervical mucus were studied. It follows from the results (Table 1) that the follicle fluid contains both the antigenic factor A and the antibody anti-A. The titre level of antibody in the follicle fluid corresponds to the level of this antibody in the serum. The presence of factor A in follicular fluid, however, not only does not correspond to its presence in the serum but its quantity is also significantly lower in follicle fluid.

It is difficult to explain this phenomenon. We do not know the molecular size of antigen A, which could be a decisive influence in its passive transport from the

TABLE 2

Agglutination and loss of motility of boar spermatozoa in the medium of ovarian follicle fluid of sows with and without anti-A

Follicle fluid with	No. of fluids	Titre of spermatozoa agglutination		Loss of motility of spermatozoa, min	
		3 boars A positive	3 boars A negative	3 boars A positive	3 boars A negative
Anti-A	5	0–8	0–16	60–160	60–180
Without anti-A	4	0–16	0–16	80–160	60–180

blood vessels into the interior of the follicle, nor do we know anything of the selective functions of follicular cells. Some conjectures might also point to a possible degradation of the A antigen in the follicle itself. The same difficulties apply to the interpretation of the high content of the A antigen in the cervical mucus (Table 1). Here, however, it is not possible to speak of passive transfer of this factor but rather of its direct synthesis. Such considerations are also supported by the absence of A antibodies in this fluid.

An eventual effect of the antibody anti-A on the corresponding spermatozoal antigen would be negligible in any case. The only possibility seems to be that the follicular fluid might adhere to the ova during ovulation or reach the oviduct by

TABLE 3

Size of litter with respect to antigenic factor A after mating sows with one boar

Antigen A in ♂	Antigen A in ♀	No. of parent pairs	No. of piglets born	Average number of piglets per 1 litter
A^+	× A	46	538	11.52
a	× a	118	1335	11.31
A	× a	96	1080	11.25
a	× A	87	982	11.28

A^+ — animals with A antigen, a — without A antigen.

passive means and come into contact with the spermatozoa. Due to capacitation and other changes in spermatozoa in the female genital tract, it is evident that considerable changes of the surface composition of spermatozoa antigens and thus eventually also of the A antigen, if it is present, can occur. Nevertheless we have tried to investigate the direct influence of antibody anti-A of the follicular fluid on ejaculated spermatozoa from A positive and A negative boars. Negative results obtained (see Table 2) which show, however, that there is no effect on this follicular antibody on spermatozoa.

TABLE 4

Fertility of 58 A negative sows mated during one oestrus with two boars of type A and a respectively

No. of piglets after boars A	No. of piglets after boars a	Average number of piglets per 1 litter after boars A	Average number of piglets per 1 litter after boars a	Percentage of increase after A boars
317	234	5.46	4.03	35

The neutral effect of the blood group system A on the reproduction of pigs is also demonstrated by the results regarding fertility. The size of litter from A negative sows mated with A positive boars is not consistently smaller than that from other mating types (Table 3). The fact that, after mating A negative sows during one oestrus to two boars, one A positive and the other A negative, more piglets were born to A positive fathers, is proof of this (Table 4).

An analogous picture, with respect to the presence of natural antigen and antibody in genital fluids and their influence on reproduction, was found in cattle during investigations of the J-system (MATOUŠEK, 1968).

REFERENCES

BOETTCHER, B., 1965. Human AB0 blood group antigens on spermatozoa from secretors and non-secretors, *J. Reprod. Fertil.*, **9**, 267–268.

BOURSNELL, J. C. and COOMBS, R. R. A., 1966. A haemagglutinating factor in boar seminal plasma, *J. Reprod. Fertil.*, **11**, 139–144.

DUNSFORD, I. and BOWLEY, C. CH., 1955. Techniques in Blood Grouping, Oliver and Boyd, Edinburgh, p. 250.

EDWARDS, R. G., FERGUSON, L. C. and COOMBS, R. R. A., 1964. Blood group antigens on human spermatozoa, *J. Reprod. Fertil.*, **7**, 153–161.

GAVALIER, M., HOJNÝ, J., HRADECKÝ, J., LINHART, J. and SCHRÖFFEL, J., 1966. Blood groups and serum proteins in pigs. Polymorphismes biochimiques des animaux, X European Conference on Animal Blood Groups and Biochemical Polymorphisms, 159–164.

GERSHOWITZ, H. G., BEHRMAN, S. J. and NEEL, J. V., 1958. Hemagglutinins in uterine secretions, *Science*, **128**, 719–720.

HOJNÝ, J. and HÁLA, K., 1965. A contribution to the study of the blood group system A in pigs. Blood Groups of Animals, Proceedings of the 9th European Animal Blood Group Conference, 155–161.

MATOUŠEK, J., 1965. Antigenicity and polymorphism of the ovarian follicle fluids in cows. Blood Groups of Animals, Proceedings of the 9th European Animal Blood Group Conference, 359–368.

MATOUŠEK, J., DOSTÁL, J. and FULKA, J., 1966. Antigenicity and polymorphism of the seminal vesicle fluid in boars, Proceedings of the Xth European Conference on Animal Blood Groups and Biochemical Polymorphisms, 523–531.

MATOUŠEK, J. and SCHRÖFFEL, J., 1967. Blutgruppensubstanzen A, Na und polymorphe Merkmale im Serum und in der Follikularflüssigkeit der Ovarien von Sauen, *Fortpfl. Haust.*, **3**, 118–123.

MATOUŠEK, J., Blood group system J in cattle reproduction, *Fortpfl. Haust.* (In press).

PODLIACHOUK, L. and EYQUEM, A., 1956. Les antigénes érythrocytaires et plasmatiques des porcs, *Ann. Inst. Pasteur*, **91**, 751–758.

QUINLIVAN, W. L. G., 1966. Antigen-antibody reactions with human semen, *Fertil. Steril.*, **17**, 722–730.

SEDLÁKOVÁ, E., 1966. Frakcionace bílkovin semennych vaků kance a isolace aglutinujícího faktoru. Dissertation Thesis, Laboratory of Physiology and Genetics of Animals, Liběchov.

SCHWIMMER, W. B., USTAY, K. A. and BEHRMAN, S. J., 1967. An evaluation of immunologic factors of infertility, *Fertil. Steril.*, **18**, 167–180.

TYLER, A., TYLER, E. T. and DENNY, P. C., 1967. Concepts and experiments in immunoreproduction, *Fertil. Steril.*, **18**, 153–166.

WHITELAW, M. J., GRAMS, L. and ANTONE, M., 1962. Blood groups in fertility and sterility, *Obstetrics and Gynecology*, **20**, 317–319.

XITH EUROPEAN CONFERENCE ON ANIMAL BLOOD GROUPS AND BIOCHEMICAL POLYMORPHISM
Warsaw, July 2nd–6th, 1968

Use of Blood Groups in Pig for Double Mating

J. LINHART

Laboratory of Physiology and Genetics, Czechoslovak, Academy of Sciences, Liběchov, Czechoslovakia

SUMMARY. By the mating of two boars or the insemination of two boars of the Large White, Black and White-Přeštice, Landrace and Cornwall and Pietrain breeds to a sow of the same breed during one oestrus 149 litters were obtained. Only 69 were mixed litters.

The paternal origin of the litter was determined by means of blood groups. From the resultant mixed litters it was not possible to obtain from any of them two young sows and two young boars by one boar. It was therefore decided to test groups of four in which pairs of animals by each boar were represented. For the factors compared the differences between the single boars and single sows were found to be statistically insignificant.

In accordance with the literature the preference for gametes of one boar to those of the other was shown.

INTRODUCTION

The wide spectrum of blood groups which can be detected in pigs facilitates an objective determination of the paternity of progeny. The possibility to deciding the paternity of offspring after two sires even within a breed led to the design of matings to obtain half-sibs within one litter. By comparing the productivity of the sibs it should be possible to evaluate the sires compared.

MATERIAL AND METHOD

Unrelated animals of one production breed and from 12 breeding centers were used for the study of double mating (heterospermal insemination) with the intention of obtaining mixed litters.

The chosen pairs of boars were mated with the corresponding sows at intervals of 10–15 min one after the other. The order of boars at matings was changed.

Insemination of sows was carried out by means of a semen mixture with an equal number of spermatozoa from each boar. The paternity of piglets was determined by means of 32–43 blood group factors, supplemented in some cases by 5 systems of serum polymorphic markers.

RESULTS

196 litters from double matings and heterospermal insemination comprising 1757 offspring were identified. 135 litters with 1336 live born piglets originated from random combinations. From these random matings there were 13 litters (129 piglets) in which the fathers of 56 offsprings could not be determined (i.e. 4.2%). The use of serum polymorphic markers reduced the number of unidentified piglets to 1.05%.

The part played by individual blood group systems in the identification of offspring may be seen in Table 1.

TABLE 1

Number of identified offspring in each blood group system used

Blood systems	A	B	C	D	E	F	G	H	I	J	K	L	M	N	O
Number of offspring determined	109	75	—	49	789	23	283	144	203	41	90	770	106	148	104

Systems E and L were of foremost importance in the determination of paternity and also in numerous cases systems G, H, I, M, N and O. All litters were analysed with respect to distribution of offspring, number and sex in relation to each sire. From the total investigated, 69 litters were mixed, (i.e. 46.3%), the other 80 litters (i.e. 53.7%) came from one or the other boar.

Analysis of mixed litters showed that in most the offspring were largely from one sire (9:1,1:11 etc.). This in balance was still more pronounced with regards to sex. A somewhat higher percentage of mixed litters from heterospermal insemination may be attributed to a relatively small number of animals served in this way (Table 2).

DISCUSSION AND CONCLUSION

MINKEMA and BOUW (1961) referred to the possibility of comparing, by means of blood groups, halfbrothers within one litter. These problems were studied more closely by BUSCHMANN and KRAUSSLICH (1964). They failed to determine paternity in 24.4% of offspring from selected parental animals.

WIDDOWSON and NEWTON (1965) obtained mixed litters from heterospermal insemination and determined the origin in 93.8% progeny. Similar results were obtained by HOJNÝ (1965).

The detection of further blood group factors and their genetic classification facilitated experiments with breeding animals without previous selection of parental animals according to their blood groups.

TABLE 2

Distribution of progeny from double mating and heterospermal insemination

	Type of mating ♀	Type of mating ♂	Type of mating ♂	No. of litters after 1 boar	Number of mixed litters: in which by 1♂ both sexes were not represented	Number of mixed litters: in which both sexes were by each boar: ratio 1:1	Number of mixed litters: in which both sexes were by each boar: ratio 2:2	Total
Double mating	BW	BW	BW	10	5	2	1	18
	BW	BW	P	11	3	3	0	17
	LW	LW	LW	29	13	9	3	54
	L	L	L	6	2	4	0	12
	LW	LW	L	3	1	2	0	6
	LW	LW	C	7	5	1	0	13
	LW	L	L	5	1	0	0	6
	LW	L	C	4	2	0	0	6
	L	C	C	0	0	0	1	1
Total				75	32	21	5	133
%				56.38	24.06	15.79	3.77	100.00
Heterospermal insemination	LW	LW	LW	3	3	4	0	10
	LW	LW	C	2	2	1	1	6
Total				5	5	5	1	16
%				31.25	31.25	31.25	6.25	100.00

LW = Large White, BW = Black and White Přeštice, L = Landrace, C = Cornwall, P = Pietrain.

No irregularities in the inheritance of blood groups from parents to progeny, as reported by TIKHONOV *et al.* (1964), were noted. The percentage of mixed litters mentioned in literature (SUMPTION and ADAMS, 1961; SAISON and MOXLEY, 1966) from 45 to 67% is in agreement with our findings.

It follows from the results in this study the number of unidentified offspring is negligible due to the present stage of development of blood group studies. The relative distribution of progeny between the two sires is, from the practical aspect, unfavourable in the majority of cases.

REFERENCES

BUSCHMANN, H. and KRAUSSLICH, H., 1964. Untersuchungen über die Doppelbefruchtung von Sauen mit anschliessender Überprüfung der Abstannung an Hand des Blutgruppentestes, *Züchtungskunde*, **36**, 97–106.

HOJNÝ, J., 1965. Erytrocytární antigeny prasat chovaných v ČSSR, (Erythrocytic antigens of pigs kept in Czechoslovakia), Thesis, Laboratory of Physiology and Genetics of Animals, Liběchov, 103–120.

MINKEMA, D. and BOUW, J., 1961. Bloedgroepen bij varkens, *Tijdschr.* v. *Diergeneeskunde.*, **86**, 227–242.

SAISON, R. and MOXLEY, J. E., 1966. Polymorphismes biochimiques des animaux, Proc. Xth Europ. Conf. on Animal Blood Groups and Biochemical Polymorphisms, Paris, 171–174.

SUMPTION, L. J. and ADAMS, J. C., 1961. Multiple sire mating in swine. III. Factors influencing multiple paternity, *J. Heredit.*, **52**, 214.

TIKHONOV, V. N., MASJUKOV, I. M. and SOROKINA, L. N., 1964. Immunological methods of determining parentage in relation to the investigation of selective fertilization and chimaerism, *Sér. biol.-med. nauk.*, **2**, 117–125.

WIDDOWSON, R. W. and NEWTON, T. A., 1965. Blood groups of animals, Proc. 9th Europ. Anim. Blood Group Conf., Prague, 137–148.

XITH EUROPEAN CONFERENCE ON ANIMAL BLOOD GROUPS AND BIOCHEMICAL POLYMORPHISM
Warsaw, July 2nd–6th, 1968

Protein Polymorphism in Boar Seminal Plasma

J. DOSTÁL

Laboratory of Physiology and Genetics of Animals, Liběchov, Czechoslovakia

SUMMARY. By means of starch gel electrophoresis with the addition of urea and mercaptoethanol 3 regions of polymorphic proteins were found in seminal vesicle fluid of boars. These regions were designated K I, K II and K III. There is no difference between the pattern shown in seminal plasma or seminal vesicle fluid of the same boar.

In total, 271 samples of seminal vesicle fluid and seminal plasma of boars were studied. After analysing the phenotypes of 17 boars and their 73 sons, the genetic hypothesis was suggested that the proteins in region K I, with phenotypes A, B and AB are controlled by two codominant alleles of one locus. Proteins of the region K II, comprising phenotypes A, B, C, AB, BC and AC are controlled by three codominant alleles of one locus.

In the protein region designated K III there were 19 phenotypes differing in position, number and intensity of fractions. It has not been proved, however, whether the expression of these proteins is under genetic control or is a result of the physiological condition of the animal concerned.

INTRODUCTION

At the Xth European Conference on Animal Blood Groups and Biochemical Polymorphism, in Paris, 1966, preliminary results were presented, on the study of polymorphism of seminal vesicle fluid of boars (MATOUŠEK *et al.*, 1966). An improved starch gel electrophoretic technique demonstrated that seminal vesicle proteins and those of the seminal plasma show an identical electrophoretic pattern in the starch gel (DOSTÁL, 1968). Therefore seminal plasma, as the more accessible material, was chosen for the study of genetic determination of these proteins.

MATERIAL AND METHOD

Electrophoresis of seminal plasma proteins of boars was carried out in starch gel with urea and mercaptoethanol in a buffer consisting of acetic acid and formic acid according to ASCHAFFENBURG (1966). 271 samples were analysed by this method. Seventeen samples of seminal plasma came from sire boars, 73 samples from their male progeny and 181 samples of seminal plasma and seminal vesicle fluid from boars picked at random at different farms and slaughterhouses.

RESULTS AND DISCUSSION

As may be seen from Fig. 1, seminal plasma proteins show, in the starch gel, 3 protein groups with marked polymorphism. The groups were designated K I (fastest cathodic migration), K II (medium speed migration) and K III (slowest migration of polymorphic protein). Designations of those groups, their phenotypes and phenotypic frequencies are given in Fig. 2.

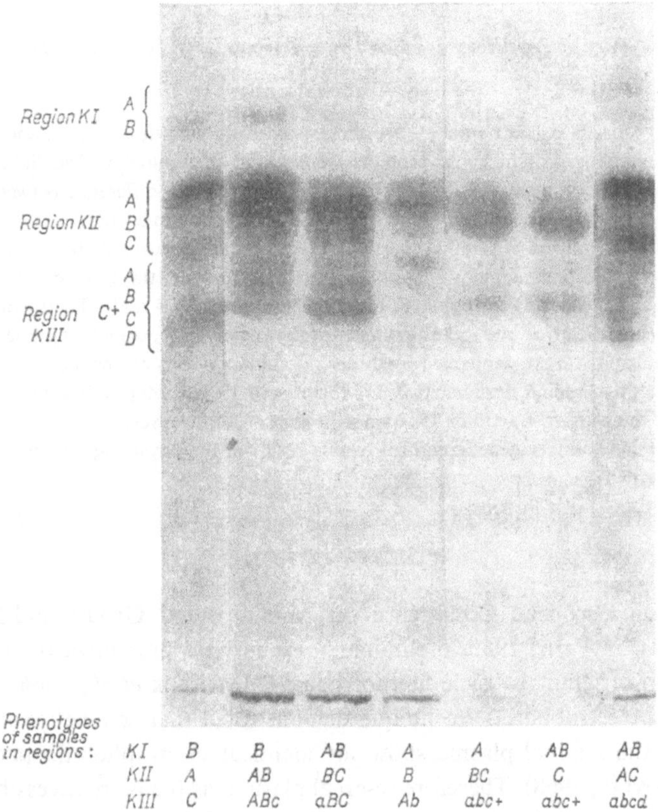

FIG. 1. Photography of starch gel electrophoresis with samples of boar seminal plasma.

Protein region K I comprises three phenotypes: A, B and AB; region K II shows 6 phenotypes: A, B, C, AB, BC and AC. Results of the genetic study of these proteins are summarized in Tables 1 and 2.

Protein region K III comprises 19 different phenotypes, differing in position, number and intensity of fractions. According to results so far obtained, it is not possible to state whether the phenotypes are the product of genetic determination or of physiological condition of the animal.

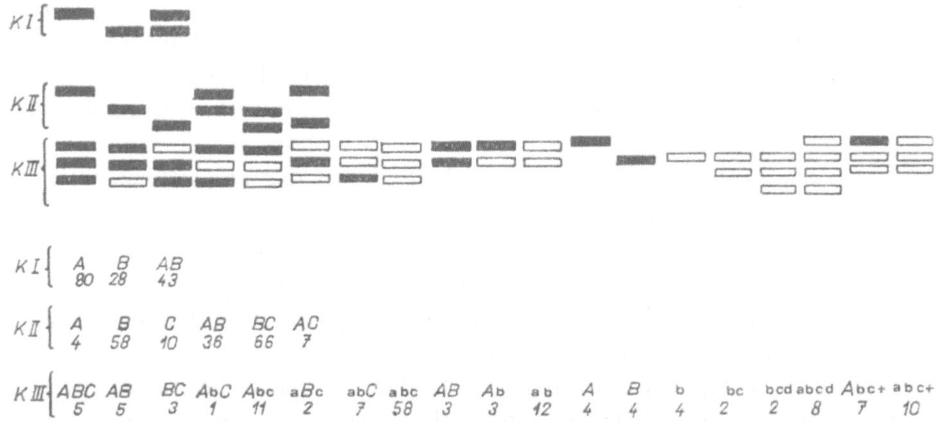

FIG. 2. Schematic illustration of phenotypes and frequencies of seminal plasma proteins of boars.

TABLE 1

Comparison of phenotypes of sire boars and sons in protein region K I

Phenotype of fathers	No. of fathers of phenotype	No. of mothers	Total of sons	No. of sons with phenotypes		
				A	B	AB
A	5	10	21	8	—	13
B	1	1	6	—	4	2
AB	11	27	46	15	6	25
Total	17	38	73	23	10	40

TABLE 2

Comparison of phenotypes of sire boars and sons in protein region K II

Phenotype of fathers	No. of fathers of phenotype	No. of mothers	Total of sons	No. of sons with phenotypes					
				A	B	C	AB	BC	AC
A	1	1	2				2		
B	5	12	26		13		3	10	
C	1	6	7			1		5	1
AB	2	2	2				1	1	
BC	6	13	29		8		7	13	1
AC	2	4	7			1	2	4	
Total	17	38	73	—	21	2	15	33	2

From the total number of 181 randomly chosen animals it was not possible to determine the protein types of region K I and K III in 30 cryptorchid boars. We assume this phenomenon to be due to the very low level of proteins in the genital organs of cryptorchid boars caused by low physiological function of the whole genital tract.

According to the results of genetic study it appears that two codominant alleles of one locus are responsible for the expression of protein phenotypes in region K I.

Individuals of phenotypes A and B are probably homozygous with genotypes K $I^{A/A}$ and K $I^{B/B}$, whereas individuals of phenotype AB would be heterozygotes with genotypes K $I^{A/B}$. There was not one animal of the 73 offspring of 17 different boars that would be excluded on this hypothesis (Table 1).

The six phenotypes in protein region K II are probably determined by three codominant alleles of one locus. Individuals of types A, B and C would then be homozygotes with genotypes K $II^{A/A}$, K $II^{B/B}$ and K $II^{C/C}$ respectively, and individuals with phenotypes AB, BC and AC would be heterozygotes with the corresponding genotypes K $II^{A/B}$, K $II^{B/C}$ and K $II^{A/C}$. Again, there was no deviation from the mentioned hypothesis (Table 2).

ACKNOWLEDGEMENT

The author wishes to thank Miss Vlasta JANATKOVÁ for her technical assistance.

REFERENCES

ASCHAFFENBURG, R., 1966. Modified procedure of starch gel electrophoresis of β-casein phenotyping, *J. Dairy Sci.*, **49**, 1284–1285.

MATOUŠEK, J., DOSTÁL, J. and FULKA, J., 1966. Antigenicity and polymorphism of the seminal vesicle fluid in boars. Polymorphismes biochimiques des animaux, Xe Congrès européen sur les groupes sanguins et le polymorphisme biochimique des animaux, Paris, 5–8 juillet 1966, 523–531.

DOSTÁL, J., 1968. A study on polymorphic proteins of seminal vesicle fluid and seminal plasma in boars, *Immunogenetics Letter*, **5**, 117–119.

DISCUSSION

G. C. ASHTON: Have you examined other biological fluids for these polymorphisms?

J. DOSTÁL: The proteins mentioned by me are typical for the seminal plasma of boars. The polymorphic protein of the seminal plasma of boars has the isoelectrical point higher than 8.6. These proteins treated on the gel with buffer with pH 8.6, migrate to the cathodic side of the gel. They are not present in the serum.

XITH EUROPEAN CONFERENCE ON ANIMAL BLOOD GROUPS AND BIOCHEMICAL POLYMORPHISM
Warsaw, July 2nd–6th, 1968

Globulin Allotyping in Pigs Using Iso-precipitins

B. G. LANG

Blood Group Research Unit, Royal (Dick) School of Veterinary Studies, Veterinary Field Station, Easter Bush, n/Roslin, Midlothian, Great Britain

SUMMARY. Iso-immune precipitins have been produced using a water–oil–water (WOW) type of emulsion adjuvant incorporating whole pig serum. The reactions are demonstrated by double diffusion (Ouchterlony) tests on microscope slides.

The characteristics of the antibodies and antigens involved have been analysed using immuno-electrophoresis (I.E.), specific staining techniques, exclusion chromatography (molecular sieving) and ion exchange fractionation techniques. Antigens have been demonstrated with α, β and γ mobilities, and to date they all fractionate in the middle (7S) peak of Sephadex G200. The γ characteristic fractionates as IgG on anion exchange chromatography and demonstrate IgG identity in other ways.

INTRODUCTION

Allotypes are different antigenic forms of a serum protein, each type being detected by iso-immune reagents, in some individuals but not all, within a species. Isotypes appear as antigenic forms common to all individuals of a species and are detected by hetero-immune reagents (DRAY *et al.*, 1962). The study of allotypes and their inheritance in combination with physico-chemical analysis has thrown considerable light on the form, structure and synthesis of immuno-globulins. Hence we know that in the human IgG molecule certain isotypic heavy (H) chain classes We (GREY *et al.*, 1964) or $\gamma 2^{b}_{1}$ (TERRY and FAHEY, 1964) or C (BAILLEUX *et al.*, 1964) and Vi or $\gamma 2^{c}$ or Z also show allotypic factors Gma or Gmf and Gmb^{1} or Gm — respectively. The remaining two isotypic H chain classes Ge ($\gamma 2^{d}$) and Ne ($\gamma 2^{a}$) show only Gm (a— f—b^{1}—) (MARTENSSON, 1966). Since we already know that Gm and Inv type genes are associated with H and L chain synthesis respectively (FUDENBERG *et al.*, 1963) it is obvious that we have a complex segregation of isotypic and allotypic groups in the sequential amino acid synthesis of IgG H chains.

Two gamma globulin iso-antigens in pigs have previously been described by RASMUSEN (1965). These were detected by a haemagglutination–inhibition system of the type used for Gm and Inv typing in man. Three phenotypes were demonstrated involving two codominant autosomal alleles Gl^{a} and Gl^{b}.

Attempts had already been made in our unit to produce iso-immune precipitins to pig immuno-globulins by injecting pigs with their own washed cells coated with incomplete (7S) blood typing antibodies. The attempt was unsuccessful. On the basis of Dr. RASMUSEN's typing a further experiment to produce precipitins was initiated.

METHODS

Immunization

Eight Gl (a— b+) pigs were divided into two groups of four. One group received a single injection of 10 ml of a double emulsion (water–oil–water) type adjuvant WOW), HERBERT (1965), incorporating whole serum from a Gl a+ pig at a final dilution of 1/200. The other group received once weekly injections of 10 ml of the same serum, diluted 1/64 with isotonic saline, for a period of 6 weeks. Blood sampling was carried out at weekly intervals for 8 weeks. Two of each group were injected intramuscularly and two subcutaneously.

Two pigs receiving only one injection of WOW produced precipitins one as early as 3 weeks after injecting. Both groups after 11 months rest were given a booster course of whole serum injections. 10 ml of serum was given followed 6 days later with 2 ml, and then a further 2 ml after 2 days. All four pigs previously injected with WOW produced precipitating antibodies. Only one pig previously receiving serum diluted 1/64 produced a weak precipitin.

Double diffusion technique

The precipitin reactions are demonstrated by double diffusion (Ouchterlony) tests on microscope slides with a 2 mm diffusion distance in 1.5% Difco (Special Agar-Noble) in 0.9% saline phosphate buffered at pH 7.2 (Fig 1).

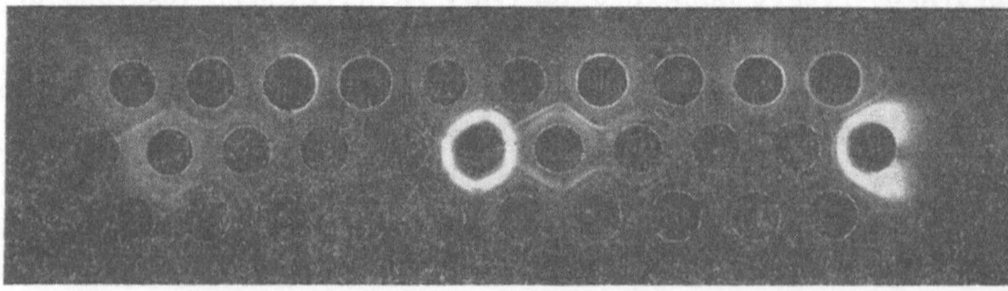

FIG. 1. Double diffusion test in agar on a microscope slide. Upper row: WSA — N, 1/2, N, 1/2, ... etc. Middle row: AA — N, 1/2, 1/4, 1/8, RC.N.AAA — N, 1/2, 1/4, 1/8 RC. N. Lower row: WSA — 1/8, 1/4, 1/8, 1/4, ... etc. WSA — whole serum antigen. AA — allotyping antibody. N — undiluted. RC — rabbit control. First well in the middle row is empty.

Immuno-electrophoresis (I.E.)

A micro technique on microscope slides was carried out on an LKB apparatus with 1% agar as above but in a Veronal/Calcium Lactate buffer pH 8.6 after HIRSHFELD (1960). The migration of albumin is checked at 20 mm using Bromophenol Blue. Antisera produced in rabbits against whole serum or recycling chromatography fractions were used to identify serum fraction isotypes (Fig. 2).

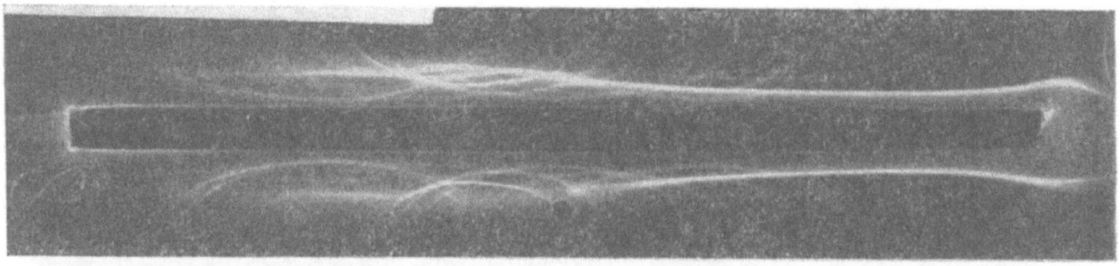

FIG. 2. Micro-immunoelectrophoresis in agar on a microscope slide. Antigen wells contain WSA: 5 lambda N (upper well), 5 lambda 1/2 (lower well). Antibody trough contains 100 lambda 1/2 rabbit isotyping control serum.

Fractionation techniques

1. *Sephadex G 200 column chromatography,* was carried out on a combined column length of 140 cm×2.5 cm according to KILLANDER and FLODIN (1962). Recycling chromatography was employed on G 200 after PORATH and BENNICH (1962). The use of 0.1 M Tris/HCl+1 M NaCl buffer was discontinued, and a 0.9% NaCl w/v aq. dist. with $\frac{M}{5}$ (Na/K) phosphate added (9:1) was substituted. Buffer exchange was carried out on Sephadex G 25 after FLODIN (1961). Concentration of fractions was performed by dialysis against 5–10% Carbowax 4000 in 0.9% saline phosphate buffered at pH 7.7, followed by pressure dialysis against saline to remove Carbowax monomers.

2. *Ion exchange chromatography.* Preparation of IgG fractions has been performed using D.E.A.E. Sephadex A50 anion exchanger either in column (SOBER *et al.*, 1956) or by batch process (BAUMSTARK *et al.*, 1964) followed by concentration/elution on C.M. Sephadex C50 cation exchanger after LEVY and SOBER (1960). The technique was modified for use with Sephadex derivatives instead of cellulose as in the original papers.

RESULTS

Five pigs produced precipitins including all four given WOW. Not all the antisera showed a single specificity (Fig. 1). However, the principal arcs formed by all five antisera appear to show a similar reaction by fusing together on one slide. In general all five antibodies react or fail to react against the same animals. There is, however, considerable quantitative variation, and the weaker antibodies sometimes fail to react.

The piglets from a —ve sow x+ve boar mating show no reaction until 2–3 weeks of age and the full reaction is not developed until 5–8 weeks of age.

The two strongest antisera were selected for further study. One (No. 22) showed a predominantly β globulin specificity on I. E. but the other (No. 9328) by manipu-

FIG. 3. As for 2, except antibody trough contains 100 lambda allotyping serum No. 9328 diluted 1/2. This slide shows the IgG allotype reaction.

lating the antigen/antibody proportions and amounts could demonstrate a predominant γ globulin characteristic with minor α and β globulin arcs. Figure 3 shows mainly the γ arc. This is thought to be an IgG allotype, confirmation being provided in a number of ways. Firstly the reaction on I. E. shows the mobility and characteristic shape and length of arc of IgG. The antigen protein fractionates in the middle (7S) peak on Sephadex G 200, and is the most cathodic peak coming off D.E.A.E. Sephadex at pH 6.5 with 0.01 M phosphate buffer or remaining attached to C.M. Sephadex under the same conditions.

The precipitates on double diffusion slides stain with Amido black, azocarmine and Schiff reaction for carbohydrate, but fail to stain with Sudan B, paraphenylenediamine, or benzidine and peroxide in the presence of aged haemoglobin. The strong Schiff reaction is to be expected as the late hyperimmune allotyping antibody shows γ mobility and fractionates as a 7S antibody on G200.

This IgG characteristic is present in colostrum from a positively reacting sow, and whereas her piglets do not show reaction prior to suckling, the 36 hr postcolostral serum samples are positive.

The β globulin reactivity is only different from the γ antigen in its mobility and optimal proportions, and in the fact that as yet it has not been detected in colostrum. However, it appears in a litter from a positive sow after suckling, fractionates in the 7S peak on G200, and appears in piglets from —ve sow x+ve boar mating at 2–3 weeks of age.

The α reactivity has a narrow range of activity, but at optimal proportions tends to fuse with the γ line which is less distinct at these proportions.

DISCUSSION

From the results obtained it would appear that WOW is remarkably efficient compared with serum injections even though the latter had three times as much protein injected repeatedly. Unlike Freunds complete adjuvant WOW does not appear to produce any physical reactions in the pig.

The fact that all five reagents even though multispecific react or fail to react with the same animals and show a reaction of identity, might indicate some relationship between the α, β and γ antigens, particularly as so far I. E. has not demonstrated any animal where the β reaction segregates from the γ reaction. It has been noted that the α reaction tends to fuse with the γ reaction. In rabbits where the b locus governs the synthesis of the L chains (FEINSTEIN *et al.*, 1963), and the a locus is associated with the Fd part of the H chains, it has been demonstrated that some allotypes are common to all immunoglobulins (FEINSTEIN, 1963; TODD, 1963) as for human Inv, while others are peculiar to one (KELUS and GELL, 1965; SELL, 1966) immunoglobulin.

In pigs on I. E. the β reactivity does not fuse with the γ arc, indeed its optimal proportions are entirely different, and since on I. E. the antibody must pass through the γ arc to reach the β arc identity of reaction does not seem probable. IgM is ruled out since the β reactivity fractionates with the γ in the middle (7S) peak on Sephadex G200, while IgM is found in the first (19S Macroglobulin) peak.

Unfortunately IgA has not been adequately characterized in pigs. If the β antigen is IgA, however, it is peculiar not to detect it in colostrum. Experiments are now in hand to fractionate various red cell typing antibodies to characterize the different classes of porcine immunoglobulin by adsorption on and elution from red cell stromata.

ACKNOWLEDGEMENTS

Although the allotypes produced do not coincide with Dr. RASMUSEN's Gla and Glb, acknowledgement is made for his typing of the pigs used, and to Dr. HERBERT for assistance in preparing WOW. The Unit was supported by a Pig Industry Development Authority Grant using laboratory and other facilities provided by Prof. A. ROBERTSON. Dr. IMLAH is thanked for supervision and critical discussion.

REFERENCES

BAILLEUX, R. E., BERNIER, G. M., TOMINAGA, K. and PUTNAM, F. W., 1964. Gamma globulin antigenic types defined by heavy chain determinants, *Science*, **145**, 168.

BAUMSTARK, J. S., LAFFIN, R. J. and BARDAWIL, W. A., 1964. A preparative method for the separation of 7S gamma globulin from human serum, *Arch. Biochem. Biophys.*, **108**, 3, 514–521.

DRAY, S., DUBISKI, S., KELUS, A., LENNOX, E. S. and OUDIN, J., 1962. A notation for allotypes, *Nature*, **195**, 785.

FEINSTEIN, A., 1963. Character and allotype and immune globulin in rabbit colostrum, *Nature*, **199**, 1197.

FEINSTEIN, A., GELL, P. G. H. and KELUS, A. S., 1963. Immunochemical analysis of rabbit gamma globulin allotypes, *Nature*, **200**, 653.

FLODIN, P., 1961. Methodological aspects of gel filtration with special reference to desalting operations, *J. Chromat.*, **5**, 103.

FUDENBERG, H. H., HEREMANS, J. F. and FRANKLIN, E. C., 1963. A hypothesis for the genetic control of synthesis of the gamma-globulin, *Ann. Inst. Past.*, **104**, 155.

GREY, H. M. and KUNKEL, H. G, 1964. H chain subgroups of myeloma proteins and normal 7S gamma-globulin, *J. Exp. Med.*, **120**, 253.

HERBERT, W. J., 1965. Multiple emulsions — A new form of mineral oil antigen adjuvant, *Lancet*, ii, 771.

HIRSCHFELD, J., 1960. Immuno-electrophoresis — Procedure and application to the study of group specific variation in sera, *Science Tools*, **7**, 18–25.

KELUS, A. S. and GELL, P. G. H., 1965. An allotypic determinant specific to rabbit macroglobulin, *Nature*, **206**, 313.

KILLANDER, J. and FLODIN, P., 1962. The fractionation of serum proteins by gel filtration, *Vox. Sang.*, **7**, 113.

LEVY, H. B. and SOBER, H. A., 1960. A simple chromatographic method for preparation of gamma globulin, *Proc. Soc. Expl. Biol. Med.*, **103**, 250.

MARTENSSON, L., 1966. Genes and immunoglobulin — Editorial, *Vox. Sang.*, **11**, 521–545.

PORATH, J. and BENNICH, H., 1962. Recycling chromatography, *Arch. Biochem. Biophys.* (Suppl.), **1**, 152–156.

RASMUSEN, B. A., 1965. Iso-antigens of gamma globulins in pigs, *Science*, **148**, 1742.

SELL, S., 1966. Immunoglobulin M allotypes of the rabbit: identification of a second specificity, *Science*, **153**, 641.

SOBER, H. A., GUTTER, F. J., WYCKOFF, M. and PETERSON, E. A., 1956. Chromatography of proteins. II. Fractionation of serum proteins on anion-exchange cellulose, *J. Amer. Chem. Soc.*, **78**.,756.

TERRY, W. D. and FAHEY, J. L., 1964. Subclasses of human gamma 2 globulin based on differences in the heavy polypeptide chains, *Science*, **146**, 400.

TODD, C. W., 1963. Allotypes in rabbit 19S protein, *Biochem. Biophys. Res. Comm.*, **11**, 170.

DISCUSSION

R. L. SPOONER: Have you investigated the globulins of your allotyping reagent as well as your antigen?

B. G. LANG: The late hyper-immune antiserum has been shown to be 7S (IgG) by a reversed immunoelectrophoresis (i.e. allotyping reagent in the antigen well and the serum antigen in the antibody trough), and by G.200 fractionation.

R. SPOONER: Are the allotypic antibodies IgG or IgM?

B. G. LANG: IgG.

XITH EUROPEAN CONFERENCE ON ANIMAL BLOOD GROUPS AND BIOCHEMICAL POLYMORPHISM
Warsaw, July 2nd–6th, 1968

Immunogenetic Study of Allotypes of Pig Serum Proteins Using Isoprecipitins

V. N. TIKHONOV, S. M. VALDMAN, and M. A. SAVINA
Institute of Cytology and Genetics, U.S.S.R. Academy of Sciences, Novosibirsk, U.S.S.R.

SUMMARY. Immunogenetic study of the polymorphism of serum proteins in pigs was carried out using anti-allotypic isoimmune sera. These sera were abtained by a direct isoimmunization with native sera in the presence of adjuvant or without it. By means of immunodiffusion methods and immunoelectrophoresis 9 serum allotypes depending on group-specific antigens with beta-globulin mobility were identified. Serum samples of 993 pigs from different breeds were tested for hereditary character and allotype allelism. The occurrence of the allotypes varies in the investigated pig breeds. The gene frequencies of allotypes in two genealogically related breeds are reported.

INTRODUCTION

Investigations on the polymorphism of serum proteins carried out in the recent decade demonstrated the presence of the group-specific allotypes in rabbits (OUDIN, 1956; DRAY and YOUNG, 1958; DUBISKI *et al.*, 1959) and in mice (KELUS and MOOR-JANKOWSKI, 1961). Among farm animals serum allotypes have been discovered only in hens by the Polish investigator D. SKALBA (1966). In all the above-mentioned investigations the isoprecipitating sera were obtained only by long term isoimmunization when antibacterial sera with homologous antigen were used as antigens.

The present paper reports the immunogenetic study of serum allotypes in pigs by means of isoprecipitins produced by direct isoimmunization with native sera.

MATERIALS AND METHODS

A group of 27 purebred adult Landrace pigs of average weight about 200 kg were used as recipients for immunization. The first group (12 pigs) was immunized by a mixture of pig serum of the same breed, but of different inbred lines. Animals of this group were immunized with Freund's adjuvant. One month after a single injection with 5 ml of serum with Freund's adjuvant 6 immunizations were made subcutaneously using the same dose of serum but without the adjuvant at weekly intervals. After a futher week the recipient sera were tested and precipitins were discovered in 3 of 12 animals. Within a few days they disappeared and were not observed again in these and other recipients in spite of repeated reimmunizations during a period of 10 months.

The second group of Landrace pig recipients (15 animals) was immunized by a mixture of sera of pigs of Large White and Estonian Baconian breeds. Immunization was made by native serum without any adjuvants with 10 ml of serum, subcutaneously in cycles of 3 injections with a day interval between the injections and week interval between the cycles. In the 5–7 days following the first cycle and then before each injection of antigen, tests for precipitin formation were carried out with the recipient sera. After carrying out immunization cycles 4 animals of 15 had antibodies: two animals had monospecific sera and each of other two had two antibodies of different specificity and titres.

Simultaneously, to obtain a heteroimmune precipitating sera which could under immunoelectrophoresis exhibit similar antigen characters to the isoimmune one, immunization of 10 rabbits with the same antigens was carried out, according to similar scheme.

Isoprecipitating sera were studied by the immuno-diffusion agar gel test of Ouchterlony. This method was used for testing the serum samples of 933 pigs. These animals belonged to the following breeds: Landrace, Siberian Severnaya and Siberian Spotted. The results were read after 2–5 days. Iso- and heteroprecipitating sera were also studied using immunoelectrophoresis. It was carried out by the micromethod of Scheidegger with veronal buffer, pH 8.6 and ionic strength 0.05. The separation time of the serum antigens was about 40 min and the potential gradient 20 V/cm.

RESULTS

The precipitating anti-allotype sera were obtained by immunization of pigs using blood sera of other animals of the same species. In the immuno-diffusion test of Ouchterlony the isoprecipitating sera identified 9 different allotypes.

The study of these allotypes by immunoelectrophoresis demonstrated that they belonged to the beta-globulin fraction. Similar allotypes could not be found by immunoelectrophoresis of heteroimmune rabbit antisera.

Genetic analysis established that the investigated allotypes were of hereditary dominant character. Twenty-four families with 243 offsprings were analysed. The allotype antigens were observed only when the respective antigen was present at least in one of the parents. Statistical analysis of purebred pigs of the above breeds showed that 5 of the allotypes found are controlled by allelic genes belonging to the same genetic system.

To determine the occurrence of the serum allotypes in different populations, blood samples of all purebred boars and sows of the main stud-farms of Siberian Severnaya breed-SSev (361 animals) and Siberian Spotted breed group-SSp (292 animals) were investigated. In these two populations allotypes 1/4333, 2/5709, 3/5481 and 4/5987 had gene frequency equal to 0.1784 and 0.4207, 0.1193 and 0.1439, 0.0930 and 0.1010, 0.0529 and 0.0366. The occurrence of the allotype 1/4333 in pig population of SSev-breed was 32.5% but in SSp-breed 66.4% i.e. the difference is significant ($0.01>P>0.001$). The investigated populations show no significant difference of occurrence of other three allotypes. The finding of greater similarity rather than difference may be explained by the fact that SSp-breed group was created in very recent time by breeding from the SSev-breed.

The established polymorphism of the group-specific serum proteins of blood may have not only theoretical but also practical importance being a genetical marker of the same type as the red blood cell antigens. The possibility of obtaining the allotypic sera in pigs by means of isoimmunization and use of the not very difficult immuno-diffusion agar gel test suggests that the serum allotypes would find wide application in biological investigations and the selection of farm animals.

The studies of the relationship between serum blood groups and productivity traits may also be of interest.

REFERENCES

DRAY, S. and YOUNG G. O., 1958. Differences in the antigenic components of sera of individual rabbits as shown by induced isoprecipitin, *J. Immunol.*, **81**, 142–149.

DUBISKI, S., DUDZIAK, Z., SKALBA, D. and DUBISKA, A., 1959. Serum groups in rabbits, *Immunology*, **2**, 84–92.

KELUS, A. S. and MOOR-JANKOWSKI, J. K., 1961. An iso-antigen (gamma-B^a) of mouse gamma-globulin present in inbred strains, *Nature*, **191**, 1405–1506.

OUDIN, J., 1956. Réaction de précipitation spécifique entre des sérums d'animaux de méme espéce, *C.R. Acad. Sci.* (Paris), **242**, 2489–2490.

SKALBA D., 1966. Antigenetic differences of hen serum proteins detected by anti-allotypic immune sera, Proc. 10th Europ. Animal Blood Group Conf., Paris, 477–480.

XITH EUROPEAN CONFERENCE ON ANIMAL BLOOD GROUPS AND BIOCHEMICAL POLYMORPHISM
Warsaw, July 2nd–6th, 1968

Pig Serum Protein Phenotyping by Electrophoresis in Thin Acrylamide Gel Slabs

P. AKROYD

Unilever Research Laboratory, Colworth House, Sharnbrook, Bedford, Great Britain

SUMMARY. An improved method of acrylamide gel electrophoresis enables 20 sera to be separated in the same gel slab in 3 hr into 20–30 discrete protein zones each. The thin acrylamide gel slab is formed inside a simple glass cell mounted vertically between electrode tanks. The resulting protein zones can be specified by their migration distance relative to the albumin front = 100. The advantages and disadvantages of acrylamide are compared with starch gels.

INTRODUCTION

Starch gel electrophoresis was the first high-resolution method of separating complex protein mixtures, and it has been widely used for serum protein phenotyping. Acrylamide gels were introduced later, but have not had the same popularity for serum phenotyping despite some advantages which are discussed below. This report describes our results using an improved method of acrylamide gel slab electrophoresis.

METHODS AND MATERIALS

The simple glass electrophoresis cell and technique have been fully described (AKROYD, 1967, 1968). A 10% acrylamide gel is prepared inside the cell by dissolving acrylamide (3.8 g) and N,N′-methylene-bis-acrylamide (0.20 g) in tris-HCl buffer (40 ml; 45.7 g tris dissolved in water is titrated to pH 8.9 with HCl and diluted to 1 l.) and adding tetramethylethylenediamine (20 μl) and ammonium persulphate (0.05 g) catalysts. The electrode tanks contain tris-glycine, pH 8.3 buffer (2.0 g tris and 9.6 g glycine in 1 l. water). Electrophoresis takes about 3 hr at 480 V; the initial current of 100 mA falls to about 50 mA at the end of the experiment. The slab is then removed from the cell and can be stained by any of the methods normally used for starch gels.

The zones in different experiments of a specified gel and buffer system can be compared by determining their relative migrations (Ra values, defined as migration distance of the zone relative to the leading edge of the albumin band = 100 in each gel).

RESULTS

The very large proteins such as β_1-lipo-protein and α_2-macroglobulins remain at the starting point since they are too large to enter the small pores of 10% acrylamide gel. γ-Globulins occupy the region Ra = 0–15; the amounts can vary considerably from pig to pig.

There are two free haptoglobin bands (Ra = 27 and 34) since these disappear and an intense, slower-migrating (Ra = 22) haptoglobin–hemoglobin complex is formed if pig or chicken hemoglobin is added to these sera before electrophoresis (Fig. 1). A benzidine-stained gel (Fig. 2) shows no zones at Ra = 27 or 34, but an

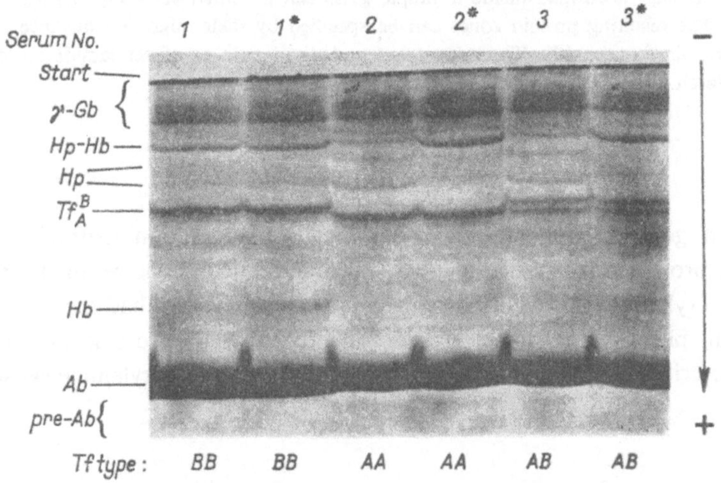

FIG. 1. Amido Black stained acrylamide gel electrophoretic patterns of blood sera of 3 pigs. *Indicates pig hemoglobin was added to the serum before electrophoresis; γ-Gb = γ-globulins; Hp–Hb = haptoglobin–hemoglobin complex; Tf = transferrin, Ab = albumin.

intense zone or doublet at Ra = 22. As with starch gels, it is seen that some sera contain much less haptoglobin than others or sometimes none. The doublet is probably due to different ratios of haptoglobin and hemoglobin in the complex. Figure 2 also shows the heme-binding globulin (hemopexin) variants which are identifiable by this technique after addition of aged hemoglobin solution to the serum before electrophoresis. Each homozygote expresses one intense band and the heterozygotes two, of moderate intensity; the Ra values are 59 (type 0), 51 (type 1), 47 (type 2) or 43 (type 3). The "fast" variants are just detectable, particularly if electrophoresis is continued for a longer time so that the heme-binding globulins migrate further

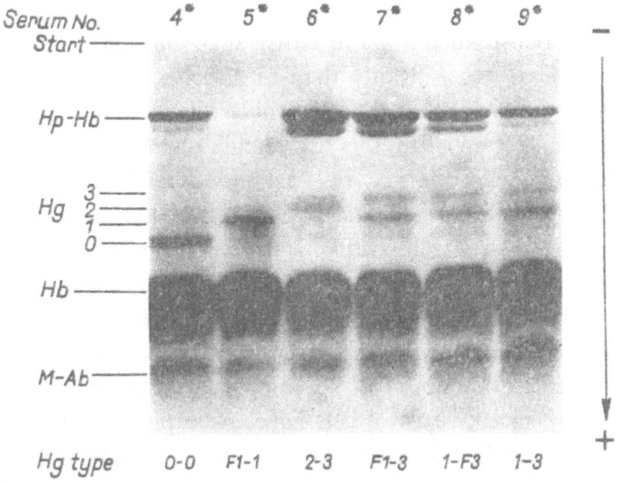

FIG. 2. Benzidine-stained acrylamide gel electrophoretic patterns of blood sera of 6 pigs. Hg = heme-binding globulin; M–Ab = methemealbumin.

down the gel slab. Any excess of free hemoglobin migrates faster as a broad band (Ra = 71–73).

Ceruloplasmin B/B expresses one zone, Ra = 32. The only A/B phenotype tested gave two zones very close together (Ra = 32 and 33). Homozygous transferrins express three zones with Amido Black stain, the fast zone staining with moderate intensity, the middle zone with high intensity and the slow zone is faint; the heterozygotes show four zones corresponding to the superimposing of the corresponding homozygote patterns. The relative migrations are: TfA/A, 48, 45 and 42; TfB/B, 45, 42 and 39; TfC/C, 44, 41 and 38. Thus transferrin phenotypes A/A, A/B and B/B are readily distinguished in this particular acrylamide system, but the C zones are only slightly slower migrating than the B zones.

There are big variations in the post-albumin zones from sera to sera, but it is not yet known if these are genetically linked. Up to 4 pre-albumin zones have been observed.

DISCUSSION AND CONCLUSIONS

The particular gel concentration and buffer system described above were chosen to give good resolution of most of the genetically controlled serum protein variant systems. For a specific separation possibly a different concentration of acrylamide of buffer would give better results (AKROYD, 1967, 1968). The gel slab technique was designed to combine the simplicity and excellent resolution of the individual acrylamide gel rod "disc" electrophoresis method of ORNSTEIN and DAVIS (1964) with the advantages of ease of intercomparison of many sera when they are examined in the same gel slab.

Acrylamide gels are completely transparent, so that the protein bands are seen, and scanned by densitometry, more easily than in starch. The molecular-sieving effect can be utilized more readily than in starch gels, since suitable gels can be prepared within the range 4–20% acrylamide. Generally we find more zones can be resolved from a serum using this acrylamide method than by starch electrophoresis. The acrylamide starting materials are more readily purified and characterized than the hydrolysed starch, and the preparation of the gel is simpler, quicker and more reproducible than starch. Unlike starch, there is no reverse migration due to endosmosis. Starch gels however have the advantages of allowing more rapid migration of the proteins, and a longer run which may show up slight differences in mobility (e.g. between Tf B and C bands). Also, starch gels can be sliced after electrophoresis so that several different stains can be compared on the same sample; acrylamide gels are not easily sliced, so extra gels have to be prepared if additional stains are to be used. Thus acrylamide electrophoresis is a valuable complement to starch, and is particularly useful where it is desired to characterize "unknown" zones of an Amido Black stained serum pattern by means of Ra values.

ACKNOWLEDGEMENT

I thank Mrs. D. H. Loane and Mr. M. Rogers for expert technical assistance.

REFERENCES

Akroyd, P., 1967. Acrylamide gel slab electrophoresis in a simple glass cell for improved resolution and comparison of serum proteins, *Anal. Biochem.*, **19**, 399–410.

Akroyd, P., 1968. Chromatographic and electrophoretic techniques, vol. II. Zone electrophoresis, 458–474, 2nd edition, Heinemann, Gt. Britain. Editor Ivor Smith.

Ornstein, L. and Davis, B. J., 1964. Disc electrophoresis, *Ann. N.Y. Acad. Sci.*, **121**, 321–349; 404–427.

DISCUSSION

G. C. Ashton: Have you tried separating Tf Dl and Tf D2 in cattle by this technique?

P. Akroyd: No, we have used only pig serum in our investigations. But one of the advantages of acrylamide is that gel concentrations of 4–20% can be used and in this way the optimum of molecular sieving properties for the particular separation is achieved.

XITH EUROPEAN CONFERENCE ON ANIMAL BLOOD GROUPS AND BIOCHEMICAL POLYMORPHISM
Warsaw, July 2nd–6th, 1968

A Study on Postalbumins in Pigs

Research note

A. KÚBEK

Laboratory of Physiology and Genetics of Animals, Czechoslovak Academy of Sciences, Liběchov, Czechoslovakia

Many genetic systems of polymorphic proteins in the serum of pigs have already been described (see SCHRÖFFEL, 1965). Proteins of the migration region in starch gel between transferrin and albumin is still subject to investigation.

During our studies of esterase we assumed that esterase region II corresponds to the migration region designated as postalbumin. We have therefore directed our attention to a more detailed study of postalbumin separation.

Our attempt was successful to a certain extent. By using buffers modified according to POULIK (1957), pH 8.46 and KRISTJANSSON (1963) we succeeded in se-

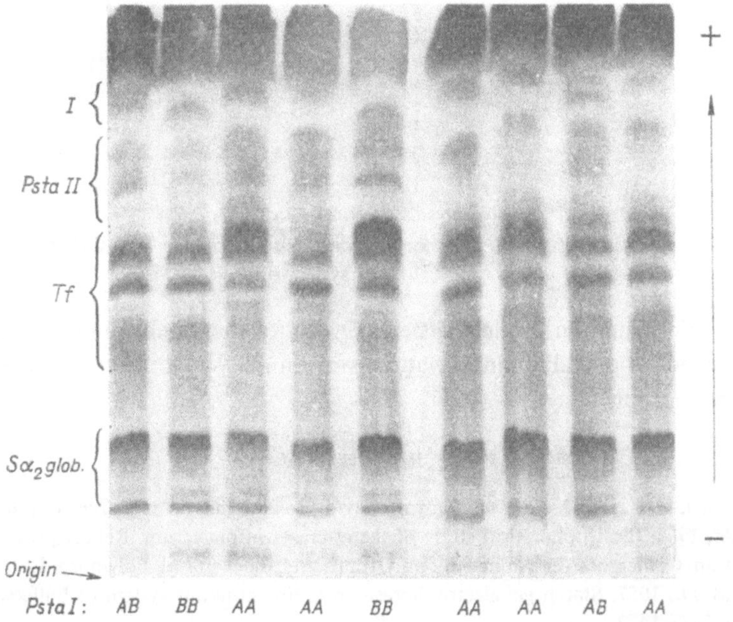

FIG. 1. Starch gel electrophoresis of pig sera showing polymorphism in postalbumin (Psta) region I, using buffer according to POULIK slightly modified.

parating postalbumins into two groups. We found however, that esterase region II corresponds neither with postalbumin polymorphism nor with the localization of fractions in the gel (KÚBEK, 1968).

In the first region situated closely behind albumin there are two fractions with different electrophoretic mobility (Fig. 1). The faster one was designated "A", the slower one "B". Family studies gave the following results:

Family study of esterase in pig serum

Mating type	No. of matings	No. of progeny	A		B		AB	
			obs.	exp.	obs.	exp.	obs.	exp.
AA×AA	1	4	4	4	—	—	—	—
AA×BB	3	12	—	—	—	—	12	12
AB×AB	5	20	4	5	7	5	9	10
AB×BB	6	24	—	—	8	12	16	12
BB×BB	1	4	—	—	4	4	—	—
Total	16	64	8	9	19	21	37	34

In the second postalbumin region (Fig. 2) which is nearer to transferrins we have also observed polymorphism of three fractions with different electrophoretic

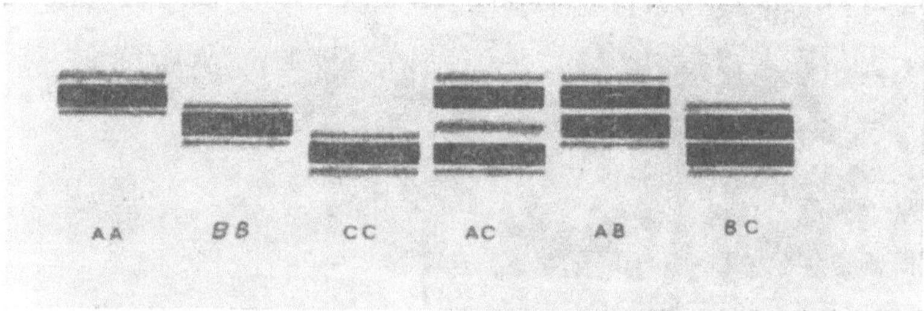

FIG. 2. Schematic picture of polymorphic postalbumin region II, using KRISTJANSSON's system.

mobility. We assume a 3 allele system. Owing to the small number of individuals so far studied for postalbumin variation we cannot yet confirm the hypothesis of genetic polymorphism.

REFERENCES

KRISTJANSSON, F. K., 1963. Genetic control of two pre-albumins in pig, *Genetics*, **48**, 1059–1063.

KÚBEK, A., 1968. Electrophoretical study of the esterases in pig serum. Report presented at XIth European Conference on Animal Blood Groups and Biochemical Polymorphism, Warsaw.

POULIK, M. D., 1957. Starch gel electrophoresis in a discontinuous system of buffers, *Nature*, **180** (4600), 1477–1479.

SCHRÖFFEL, J., 1965. Serové biochemické polymorfní znaky u hospodářskych zvířat, Thesis, Laboratory of Physiology and Genetics of Animals, Liběchov.

XITH EUROPEAN CONFERENCE ON ANIMAL BLOOD GROUPS AND BIOCHEMICAL POLYMORPHISM
Warsaw, July 2nd–6th, 1968

Enzymes in Pig Red Cell Typing

J. HARDY

Unilever Research Laboratory, Colworth House, Sharnbrook, Bedford, Great Britain

SUMMARY. Bromelin, papain, ficin and trypsin have been used with several different antibodies in pig red cell typing. Ficin and trypsin have proved unreliable since both enzymes readily produce non-specific reactions. Bromelin and papain are far more satisfactory. A one stage bromelin test parallels results obtained by the standard agglutination test with all antibodies tested in the E and L systems. Papainised red cells give quicker, clearer and stronger positives at lower dilutions of antibody than untreated cells in both haemolytic and indirect Coombs tests. In some cases a two stage papain test can be used instead of the standard antiglobulin test.

In some mixed antisera, enzyme tests can be used to differentiate between antibodies present.

Enzymes cannot be used when red cells have been collected in cyanide-containing anticoagulants and the best results are obtained when cells are tested within seven days of collection.

INTRODUCTION

Enzymes have a well established role in human red cell typing and cross-matching, where they are used either with or instead of the standard indirect Coombs test. Much of the research work has been done by KISSMEYER-NIELSEN (1960, 1963) and DYBKJAER (1964, 1965). They have found that one stage enzyme tests are best used with strong antisera where there is less likelihood of excess enzyme destroying the antibody. Two stage methods where excess enzyme is washed off before the antisera is added to the red cells, were in their experience, more reliable than one stage methods when weak antisera were being used.

MATERIAL AND METHODS

The pig red cells were collected into an anticoagulant solution of 2% sodium citrate and 0.5% sodium chloride with no added preservative.

A 0.5% solution of bromelin was prepared by dissolving concentrated bromelin powder (Takamine) in 9 parts of isotonic saline and one part of Sorenson phosphate buffer adjusted to pH 5.5 (PIROFSKY, 1959). One drop of antisera, one drop of bromelin and one drop of 5% red cells were incubated for 15 min at 37°C, centrifuged for 1 min at 1000 r.p.m. and read.

Both Low's (1955) and GOLDSMITH's (1955) papain methods were tried. The latter was found to be easier and more satisfactory. A 0.01% solution of papain (B.D.H. or Merck) in N saline was prepared and stored in small amounts −20°C until required. One volume of stock papain was diluted with 9 volumes of Sorenson's pH 7.3 buffer and two volumes of this diluted papain mixed with one volume of washed, packed red cells and incubated for 30 min at 37°C. The cells were then washed three times with saline and one volume of a 5% suspension mixed with 1 volume of antiserum. This mixture was incubated for 1 hr and then examined for agglutination. If a haemolytic antibody is used, complement should be added after 15–20 min and the cells examined for haemolysis at 15 min intervals. Alternatively the enzyme treated cells may be washed after incubation with the antisera and treated with anti-pig globulin as in the standard Coombs test.

RESULTS

Direct agglutinating antibodies

Anti Ea, Ed, Ee, Eb and Ef used in the 15 min one stage bromelin test gave results that parallelled those found by the standard agglutination test. 550 cells have been examined in this way. The only discrepancy was a pig found to be Ea positive by the enzyme technique and negative by the ordinary test. This pig was used as a donor in an immunization programme and the recipient formed anti Ea indicating the correctness of the enzyme method. The enzyme test besides being more sensitive also gives stronger and clearer results than the conventional method and antisera can be used reliably at greater dilution. An additional advantage is speed, although this is of more obvious importance in cross matching blood for human transfusion than it is in pig red cell typing. A disadvantage unless numbered centrifuge holders are available is the necessity of labelling all the tubes for centrifuging. The test is critical and if incubation is prolonged for more than 30 min no agglutination will be seen — presumably due to destruction of the antibody by the bromelin.

Other direct antisera which have been extensively tested are anti A, anti Lc+g, anti La, anti Gb and anti Na. Again very satisfactory results were obtained with anti A and anti Lc+g. No agglutination was observed with anti Gb and while agglutination was observed with anti Na the enzyme seemed to have little potentiating action and the results were no clearer than those obtained without enzyme.

Haemolytic antibodies

Bromelin treated cells were of no value when used with haemolytic antibodies and occasionally seemed to inhibit haemolysis. Washed papain sensitized red cells however gave rapid haemolysis with the appropriate antisera+complement and this applied to all the haemolytic sera tested — anti A, anti Hb, anti Ka, anti Ea, anti Ga, and an unidentified haemolytic antibody. Again these sera could be used at a lower dilution without affecting the clarity of the result.

Indirect antibodies

Enzyme tests would of course be of inestimable value if they could be used instead of the antiglobulin test.

Unfortunately we have so far found that they can replace this test only when used with indirect antibodies in the L system. Washed papainised red cells can be used with any of the L system antibodies we have so far produced, i.e. anti La, anti Ld, anti Lc+g, anti Lc, anti Lj and a mixture of anti La, h, j and d. The one stage bromelin test was found to duplicate the Coombs test when used with an indirect anti Lc+g, but it failed to react with any of the other L reagents. The two stage papain test also detected anti Ka and anti Kb and anti Fa without the addition of antiglobulin but while it produced stronger, more easily readable results with anti Ga and Gb it was still necessary to add A.P.G. to obtain agglutination. No reaction at all was obtained with the few tests we did using anti Ia and Ib. Neither reagent was very strong and possibly was destroyed by excess bromelin in the one stage test and may even have been destroyed in the papain test. Alternatively the enzyme may have completely removed the I antigens.

The differential action of antisera with enzyme treated cells can be of advantage when examining mixed antisera. Thus we found it was possible to distinguish the antibodies of the L system in a mixture of indirect anti Lc+g and indirect anti Ga. Papainised cells with Lc+g or Lg antigens were agglutinated before the addition of A.P.G. Those positive for Ga were only agglutinated after adding A.P.G.

CONCLUSIONS

Papain and bromelin may with advantage be used in pig red cell typing either to potentiate conventional tests or as a substitute for the Coombs test in the case of antibodies to the L and K systems. It is however absolutely essential that every antibody produced be tested rigorously before deciding to substitute an enzyme test for a conventional test because there is no guarantee that, for example all L reagents will react satisfactorily in this way and our experience may not be duplicated by other workers using other antisera even though these are against the same antigens.

REFERENCES

DYBKJAER, ESBEN., 1964. The use of Papain and Bromelin in one stage method in blood group serology, *Sangre*, **9**, 82–92.

DYBKJAER, ESBEN., 1965. Enzymmetoder ved Udførelse af Forligelighedsprøuer (Enzyme methods in cross-matching), *Saertryk af Ugeskrift for Laeger*, 127/38, 1188–1192.

GOLDSMITH, K., 1955. *Lancet*, **I**, 76.

KISSMEYER-NIELSEN, F., 1960. Eldon-Kort til fulstaendig forligelighedsproue, *Ugestr. loeg.*, **122**, 1575.

KISSMEYER-NIEI SEN, F., 1963. The serology of 123 anti E and 55 anti c antibodies, IXth Congress of the European Society of Haematology, Lisbon, Sangre, Barcelona, 1964, 9, 221–224.

LOW, B., 1955. *Vox Sang* (Basel), **5**, 94.

PIROFSKY, B., 1959. The use of Bromelin in establishing a Standard Cross Match, *Amer. J. of Clin. Path.*, **32**, 4, 330–356.

DISCUSSION

R. SAISON: Have you tried mixing antiserum and papain and adding the red cell suspension? This is a much simpler method than treating the red cells.

J. HARDY: Yes, we have, but this did not react in all systems; only the L and K systems.

R. SAISON: We use this method in capillary tubes which use very small amounts of serum

J. HARDY: How long does it take to read them?

R. SAISON: They are read as set up within 10 minutes.

XITH EUROPEAN CONFERENCE ON ANIMAL BLOOD GROUPS AND BIOCHEMICAL POLYMORPHISM
Warsaw, July 2nd–6th, 1968

Serum and Red Cell Enzyme Systems in Pigs*

Ruth SAISON

Department of Veterinary Bacteriology, Ontario Veterinary College, University of Guelph, Guelph, Ontario, Canada

In the past few years there have been many publications on starch gel electrophoretic studies of enzymes. These studies have been concerned with serum, erythrocyte, leucocyte, platelet and tissue enzymes. Many species and breeds are represented in these investigations. Of special interest to workers in the blood group field are those having variants, shown to be under genetic control.

Two enzymes have been investigated by me in the past year, serum alkaline phosphatase (AKp) and erythrocyte and leucocyte phosphogluconic dehydrogenase (6-PGD). A preliminary study on phosphohexose isomerase has also been done.

SERUM ALKALINE PHOSPHATASE STUDY

Many reports of polymorphism in serum AKp in several species, have appeared in the literature (ARFORS *et al.*, 1963; RENDEL and STORMONT, 1964; GAHNE, 1963; LAW and MUNRO, 1965; GARZA, 1967). RASMUSEN (1965) reported no variation in the sera of pigs of the Duroc, Landrace and Yorkshire breeds. Other workers have also failed to find polymorphism in pig sera (BAKER, 1967; WIDDOWSON, 1967).

Studies of cattle, sheep and human sera revealed a correlation between specific AKp bands and certain soluble blood group substances.

Human placental AKp has greater resistance to heat inactivation than the enzyme from other sources (NEALE *et al.*, 1965). In the presence of Mg ions the enzyme retains its activity after heating at 70°C for 30 min; other human alkaline phosphatases lose their activity after heating for 15 min at 56°C. This placental AKp band is also seen in human mother serum, but not in the infant serum. Treatment with neuraminidase affects the mobility of some AKp components (ROBINSON and PIERCE, 1964). AKp activity has also been shown to be altered by the state of nutrition and fat feeding in humans (LANGMAN *et al.*, 1966).

* This work was supported by National Research Council Grant A-4442 and the Ontario Department of Agriculture and Food.

A study was undertaken to learn whether the findings in pigs of breeds already tested hold true for pigs of other breeds and if variants similar to those of other species are present in pigs. A possible corelation between variants and the soluble blood group substances of the A–O system in pigs was investigated.

MATERIALS AND METHODS

Thirty-five Yorkshires and 11 Yorkshire-Landrace crosses from the Ontario Agricultural College (O.A.C.) herd provided one group of animals: 34 pigs of mixed breeds from the Ontario Veterinary College (O.V.C.) formed a second group. This group was comprised of Berkshire, Landrace, Tamworth and Large Black pigs and crosses of these breeds with Wessex, Yorkshire, Hampshire and Lacombe crosses. In addition 4 Berkshire gilts of the same breeding as the O.V.C. Berkshires and 4 Berkshire crosses from another source were tested. The samples from the European Pig Comparison Test were also typed. They were German Improved Landrace German Improved Landrace/German Large White crosses, Göttingen Miniature and Pietrain breeds.

Tissues were obtained from young pigs and placenta from a sow at Caesarian section. Extracts were prepared by MORTON's method (1954). Starch gel electrophoresis was carried out using a modification of the technique used for turkeys (GARZA, 1967).

RESULTS

All pigs had a slightly stained AKp band of uniform mobility. In some animals the band was difficult to detect, probably due to the low activity of the enzyme. In the sera of 3 of the animals from the O.V.C. herd, a fast band was present. It appeared to have a higher concentration of the enzyme and stained much more heavily. It was difficult to distinguish a slow band, but subsequent tests indicate that the slow component is also present. The slow type was designated S and the fast type SF. Two of the animals with the fast band were Berkshires and one a Landrace, Hampshire, Tamworth, York/Large Black cross. Of the 4 Berkshires of similar breeding and the 4 Berkshire crosses, 2 in each group had the fast band, also. Of the 40 animals from the European Comparison test, 8 were of the SF type. One of these SF animals had a fast band of slightly faster mobility than the usual fast band; one S type animal had a band of slightly slower mobility than the usual slow band.

Sera of baby pigs from 13 litters were taken at intervals from birth to 4 to 11 weeks of age. Nine litters were from S×S matings and 4 from SF×S matings. The latter 4 litters were from two matings from each of 2 sows and the same 2 boars used with each sow.

Regardless of the type of mating, the sera of all baby pigs had double AKp bands at birth, a S and F band of the same mobility as the S and F bands of mature animals. Figure 1 shows the pattern consistently present in a series of bleeds from one animal, from birth to 40 days of age. In pigs from S×S matings, the fast band had disappeared by 35 to 40 days of age.

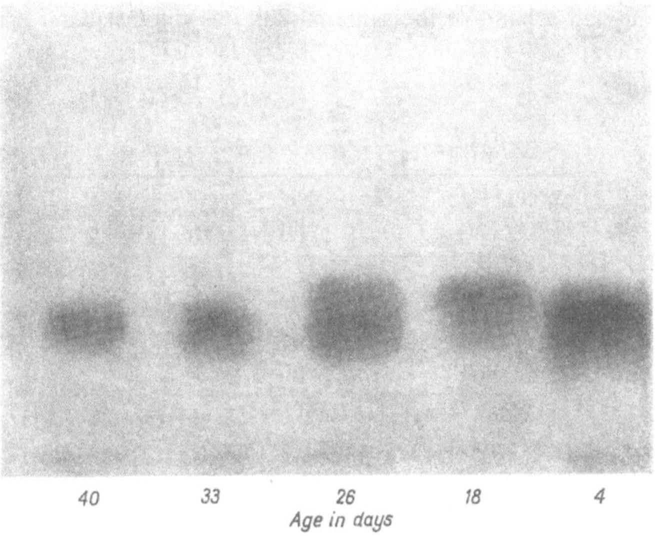

FIG. 1. Series of alkaline phosphatase zymograms from baby pig.

In litters from SF×S matings, some animals lost the fast band, but it persisted in others (Fig. 2). Table 1 shows the distribution of the phenotypes in the 4 litters The early deaths in the two litters from sow number 12 may have distorted the distribution of the phenotypes in these litters. To date we have no litters from SF×SF

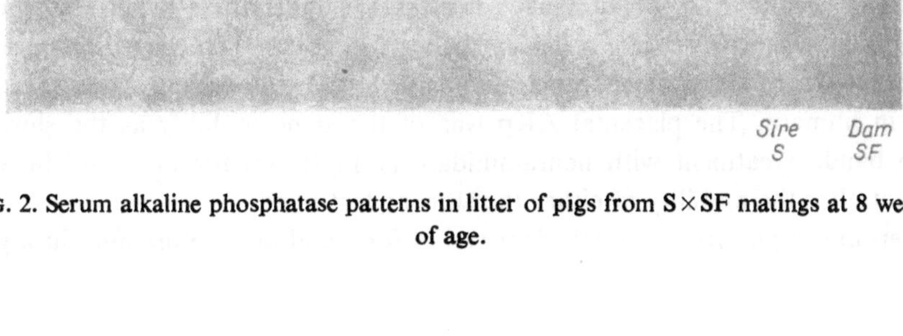

FIG. 2. Serum alkaline phosphatase patterns in litter of pigs from S×SF matings at 8 weeks of age.

matings so it is difficult to hypothesize the inheritance of the SF phenotype. In the small numbers tested, there appears to be no correlation between soluble serum substances of A and O specificity and AKp types.

From the results of electrophoretic runs using tissue extracts, the slow serum AKp band proved to be of the same mobility as that from intestine and liver. However, it was not possible to trace the origin of the fast band from these tissue extracts as none had a band of the same mobility as the fast band present in serum (Fig. 3).

TABLE 1

AKp phenotypes of pigs from SF×S matings

Mating SF♀ ×	S♂	No. in litter	Phenotypes SF	S
Berk	Berk	7	3	4
Berk	2862	7	3	4
12	2862	10	0	6*
12	Berk	10	2	6**

* Four not tested; died within first few days of birth.
** Two not tested; died within first few days of birth.

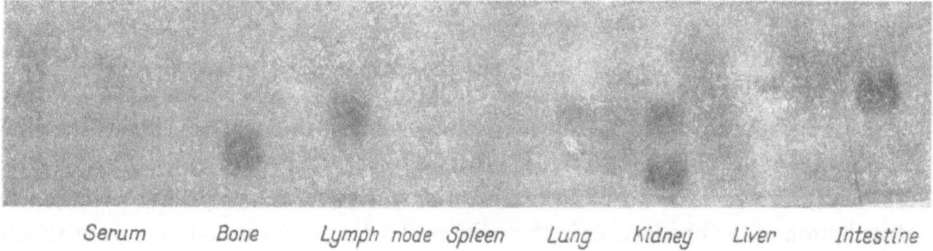

FIG. 3. Zymograms of tissue extracts from organs of young pig possessing the fast alkaline phosphatase component.

Sera of the S and SF phenotypes were heated to 56°C for 1/2 hour in a water bath, and the heated and unheated samples were tested. The activity of the fast band was slightly decreased, and all activity was removed from the slow band (Fig. 4). All tissue extracts, treated in the same manner, were completely inactivated when heated. Placental AKp of humans is heat stable under these conditions; however, pig placental AKp was not resistant to heat, nor was there any difference in the AKp pattern of AKp zymograms of pregnant and non-pregnant sows, as is the case in humans. The placental AKp was of the same mobility as the slow serum AKp band. Treatment with neuraminidase (1 mg/1 ml) for up to 47 hr at 27°C did not alter the mobility of either the slow or fast bands.

Serum samples from animals deprived of food and bled before and during depri-

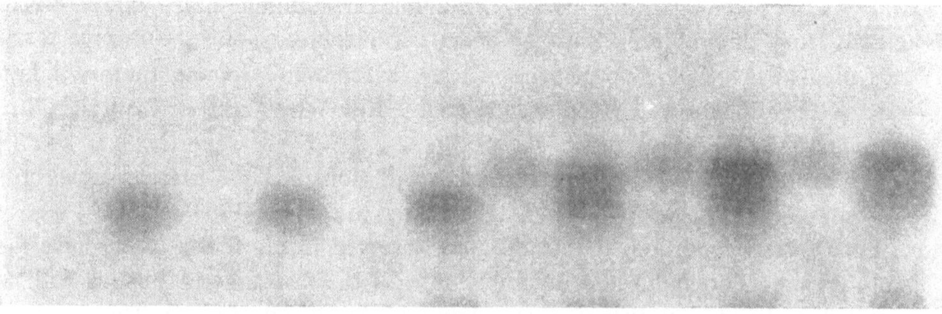

FIG. 4. Heated (56°C for 1/2 hr) and unheated sera from 3 young pigs who have lost the fast component and 3 who still have the fast component in their sera. The slow band is completely inactivated after treatment: the fast component showed some decrease in activity.

vation, and immediately after food was restored, did not show any change in electrophoretic pattern under these conditions.

Erythrocyte extracts from the O.V.C. pig panel, and leucocyte extracts from cells obtained from S and SF phenotypes, failed to show the presence of AKp under the same conditions as those used for serum and tissue enzymes. It is possible that the enzyme from white cells was not extracted by the method used.

From these results it was not possible to trace the origin of the fast AKp band in the sera of neo-natal pigs or mature pigs of the SF phenotype. It is possible that the enzyme is inactive in the tissue of origin and only becomes activated when secreted into the serum.

The limited number of litters tested have not supplied sufficient information to explain the inheritance of the fast AKp band. The fast component in mature animals appears to have the same properties as that of young pigs, and may represent a persistance of the production of this component in adults, under genetic control. The gene controlling the production of the fast component may be at a different locus to that producing the slow component; or it may be that a "switch" mechanism, similar to that controlling the production of human foetal haemoglobin, fails to act, and the animal continues to produce the neo-natal fast component. A persistence of this component in adult animals may be related to a persistence of foetal haemoglobin. This possibility will be investigated.

VARIANTS IN PIG RED AND WHITE CELL PHOSPHOGLUCONIC DEHYDROGENASE

Variants of 6-phosphogluconic dehydrogenase (6-PGD) have been described in humans, rats, cats and drosophila. Three populations of pigs were studied for the detection of this enzyme in red and white cells: the Macdonald College breeding stock of Yorkshire, Landrace, Lacombe and Large Black breeds; the Ontario Veterinary College blood group panel, including, in addition to the above breeds, animals from Hampshire, Tamworth and Berkshire breeds, and the European animals used in the 1968 pig comparison tests, which were German Improved Landrace, German Improved Landrace/German Large White crosses, Göttingen Miniature and Pietrain breeds.

Pig red and white cells were extracted with toluene, and starch gel electrophoresis was performed by the method of FILDES and PARR (1963).

Under these conditions 3 variants were detected in pig red cells: a single fast band, a single slow band and a combination of the fast and slow bands with an intermediate band equi-distant to the slow and fast bands (Fig. 5). The fast band is designated, A, the slow band, B, and the 3 band variant, AB.

FIG. 5. 6-PGD zymograms from pig red cell haemolysates.

Family studies for the inheritance of these variants show that the single fast band represents the homozygote PGD^A/PGD^A, the slow band, PGD^B/PGD^B, and the third variant, PGD^A/PGD^B. The enzyme may be made up of 2 subunits, as suggested by PARR (1966) for the human enzyme, KAZAZIAN *et al.* (1965) in drosophila and THULINE *et al.* (1967) in cats. The intermediate band found in the heterozygote probably represents a dimer of the A and B subunits.

The frequencies of the 6-PGD variants in the three populations are very similar (Table 2). Those of the 2 Canadian herds are the same and the data from both sources were pooled. Those of the genotypes in the European animals are different but occur in the same order of frequency. When the data from the 3 herds was pooled the observed and expected numbers were in almost perfect agreement. These results indicate that the genes controlling the production of the components of this enzyme are in equilibrium in the general pig population.

Table 3 shows the inheritance of the alleles for 6-PGD in 15 litters. From the tests to date, the system is composed of a pair of co-dominant, autosomal alleles which, in the heterozygous state, give rise to a hybrid substance.

TABLE 2

Frequencies of 6-PGD genotypes in 3 populations of pigs

Herd	Genotype	Frequency	No. exp.	No. obs.
Canada (Pool)	PGD^A/PGD^A	0.64	61.44	61
	PGD^A/PGD^B	0.32	30.72	31
	PGD^B/PGD^B	0.04	3.84	4
Europe	PGD^A/PGD	0.53	21.20	20
	PGD^A/PGD^B	0.40	16.00	18
	PGD^B/PGD^B	0.07	2.80	2

TABLE 3

Inheritance of 6-PGD red cell variants in pigs

Type of mating	No. of litters	No. of pigs	Exp.			Obs.		
			AA	AB	BB	AA	AB	BB
$PGD_A \times PGD_A$	8	78	78	0	0	78	0	0
$PGD_A \times PGD_{AB}$	3	33	16.5	16.5	0	14	19	0
$PGD_A \times PGD_B$	2	15	0	15.0	0	0	15	0
$PGD_{AB} \times PGD_B$	2	12	0	6.0	6	0	8	4

The white cells from 9 animals typed for the red cell enzyme, were also extracted and tested. Animals of the 3 genotypes were included. The white cell enzyme had identical patterns to that of the red cells in each animal. The enzyme appears to have a higher activity in the white cells than it has in the erythrocytes.

In addition to these two enzymes the red cell phosphohexose isomerase in pigs has been investigated. Two alleles have been recognized, each comprised of 3 bands. The fast allele has three strong bands, the slowest of which runs about 8 mm toward the anode. The slow allele also has 3 strong bands, the slowest of which runs about 3 mm toward the cathode, the middle band 3 mm toward the anode, and a third a few millimeters ahead of the middle band. The heterozygote has 5 bands, the combinations of the fast and slow types. All 3 types are present in the O.V.C. pig panel. Of the 29 animals typed for this enzyme 59% were the slow type — type B, 7% were the fast type, type A and 34% were heterozygotes, type AB.

REFERENCES

ARFORS, K. E., BECKMAN, L. and LUNDIN, L. G., 1963. *Acta genet.*, **13**, 18.
ARFORS, K. E., BECKMAN, L. and LUNDIN, L. G. *ibid.*, p. 366.
BAKER, L., 1967. Iowa State University, Ames, Iowa, U.S.A. Personal communication.
FILDES, R. A. and PARR, C. W., 1963. *Nature* (London), **200**, 890.
GAHNE, B., 1963. *Nature*, **199**, 305.
GARZA, J., 1967. M.Sc. Thesis, University of Guelph, Canada.

KAZAZIAN, H. H., YOUNG, W. J. and CHILDS, B., 1965. *Science*, **150**, 1601.
LANGMAN, M. J. S., LEUTHOLD, E., ROBSON, E. B., HARRIS, J., LUFFMAN, J. E. and HARRIS, H., 1966. *Nature* (London), **212**, 41.
LAW, G. R. J. and MUNRO, S. S., 1965. *Science*, **149**, 1518.
MORTON, P. K., 1954. *Biochem. J.*, **57**, 595.
NEALE, F. C., CLUBB, J. S., HOTCHKISS, D. and POSEN, S., 1965. *J. Clin. Path.*, **18**, 359.
PARR, C. W., 1966. *Nature* (London), **210**, 487.
RASMUSEN, B. A., 1965. *Genetics*, **51**, 767.
RENDEL, J. and STORMONT, C., 1964. *Proc. Soc. Expl. Biol. Med.*, **115**, 853.
ROBINSON, J. C. and PIERCE, J. E., 1964. *Nature* (London), **204**, 472.
THULINE, H. C., MORROW, A. C., NORBY, D. E. and MOTULSKY, A. G., 1967. *Science*, **157**, 431.
WIDDOWSON, R. W., 1967. BOCM Pig Advisory Department, Stokes Mandeville, Aylesbury, Bucks. Personal communication.

DISCUSSION

HYLDGAARD-JENSEN: Your repoıt was concerned with the stability of the alkaline phosphatase bands against the heat. Have you ever investigated the stability at low temperatures? What temperature do you use for your electrophoresis?

R. SAISON: Cold stability was not investigated. Temperature during electrophoresis is controlled by circulating water (tap-water).

The Alkaline Phosphatase System in the Pig

H. DINKLAGE

Institute of Animal Husbandry and Animal Genetics, University of Göttingen, Göttingen, G.F.R.

Starch gel electrophoresis studies of the alkaline phosphatase polymorphism were conducted in pigs using the procedure of GAHNE (1966, 1967). The following results were obtained:

(1) In all piglets up to 8 weeks of age immediately behind the insertion line there is a fraction which cannot be found in adult animals.

(2) Independent of the animal's age there were lilac coloured fractions of varying intensity below the yellow coloured albumins. An animal never showed more than one or two of these fractions.

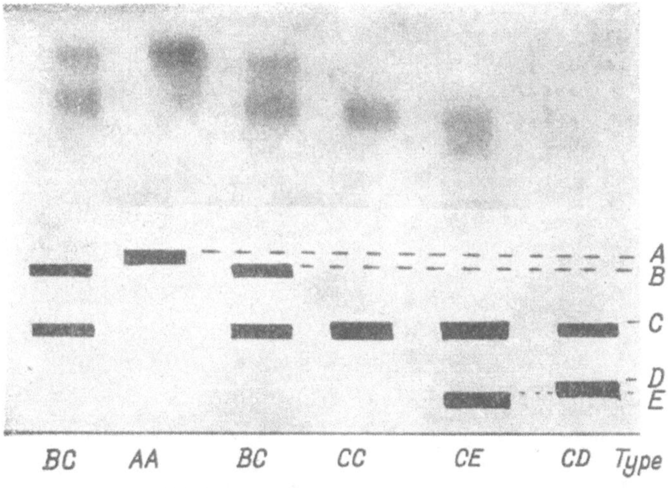

Alkaline phosphatase types in the pig.

(3) In various animals belonging to the German Improved Landrace and Göttingen Miniature pig breeds five different regions have been distinguished differing from one another by the rate of migration.

(4) Hemolytic sera are not suitable for separations as hemoglobin precipitates in the C zone and may therefore simulate a wrong type.

(5) On the basis of preliminary family studies it is suggested that alkaline phosphatase is controlled by five alleles denotated by Akp^A, Akp^B, Akp^C, Akp^D and Akp^E (see diagram).

REFERENCES

GAHNE, B., 1966. Studies on the inheritance of electrophoretic forms of transferrins, albumins, prealbumins and plasma esterases of horses, *Genetics*, **53**, 681–694.

GAHNE, B., 1967. Inherited high alkaline phosphatase activity in cattle serum, *Hereditas*, **57**, 83–99

DISCUSSION

C. STORMONT: I'd like to make a comment. This is a terrific amount of material and a very valuable one. We produced it in a very short period of time. And I think it would be of a great importance to discuss it.

XITH EUROPEAN CONFERENCE ON ANIMAL BLOOD GROUPS AND BIOCHEMICAL POLYMORPHISM
Warsaw, July 2nd–6th, 1968

Ontogenic and Familial Variation in Serum Alkaline Phosphatase of Pigs

P. IMLAH

Blood Group Research Unit, Royal (Dick) School of Veterinary Studies, University of Edinburgh, Easter Bush Field Station, Roslin, Midlothian, Great Britain

SUMMARY. Qualitative and quantitative variation in pig serum alkaline phosphatase activity is demonstrated. Polymorphic-type patterns of alkaline phosphatase activity are detectable at 1 to 3 weeks of age by starch gel electrophoresis. The patterns seen are due to a variable fast component, which disappears after 3 weeks of age, and appears to be a feature of normal development. Alkaline phosphatase in pre-colostral serum samples exhibits a different electrophoretic mobility to that shown by samples taken after 4 days of age. The change in mobility is not related to colostrum uptake. Quantitative estimations of alkaline phosphatase activity in serum appears to be highest in pre-colostral samples, then rapidly decreases over a period of 4 weeks, and subsequently decreases gradually. In studies of serum alkaline phosphatase levels taken at weekly intervals from seventeen litters for a period of 9 weeks significant differences between litters can be detected.

INTRODUCTION

Variation in serum proteins and enzymes can be caused by several factors. In recent years, biochemical polymorphism, or the study of variants which are genetically determined has attracted a lot of attention. However, it should not be forgotten that factors other than genetical ones may be involved in variation. For example, variation may be due to normal development of the animal; pathological conditions; environmental conditions or other miscellaneous effects such as deterioration due to storage and denaturation. Genetical variants of serum phosphatases have been found in cattle (GAHNE, 1963), and sheep (RENDEL and STORMONT, 1964). The following investigation was undertaken to see if genetic variants occur in serum alkaline phosphatase of pigs. The report demonstrates that although variation does occur, it is of two types, namely ontogenic and possibly familial in character.

MATERIALS AND METHODS

Seventeen complete litters were bled for serum samples at weekly intervals from birth to 9 or 10 weeks of age. The breeds involved included pure Landrace, and

Pitman–Moore miniatures, also crosses between Landrace, Large White, Saddleback and Tanworth breeds.

The electrophoresis of serum samples was carried out using the horizontal starch gel technique of SMITHIES (1955) in a discontinuous tris/citric acid and sodium hydroxide/boric acid buffer system (POULIK, 1957). For optimum demonstration of alkaline phosphatase activity the gel buffer was adjusted to pH 9 using 0.076 M tris and 0.005 M citric acid. The tank buffer was similar to the modification used by LUSH (1961). Running conditions were as described in a previous paper (IMLAH, 1964). Gels were split horizontally, and the bottom layer stained for alkaline phosphatase activity according to the technique of LAWRENCE *et al.* (1960) and GAHNE (1963), but using β-naphthyl phosphate in preference to α-naphthyl phosphate as a substrate, and incubating the gels at 37°C in a sodium borate buffer at a pH optimum of 9.8. The stain mixture was made up of 100 ml of 0.2 M sodium hydroxide/0.3 M boric acid buffer at pH 9.8; 100 mg β-naphthyl phosphate; 100 g Fast Garnet GBC salt (o-amino-azo-toluene); 0.5 g polyvinylpyrolidone; 2 g sodium chloride and 10 drops of a 10% solution of magnesium chloride. Prior to immersing the gel the buffer was heated to 37°C, then salt and PVP were added. Once the gel was immersed Fast Garnet and β-naphthyl phosphate along with the magnesium chloride were added, and the gel incubated at 37°C. Maximum staining was apparent within 1 hr. Phosphatase activity can be demonstrated by using either α- or β-naphthyl phosphate as a substrate. The enzyme releases α or β naphthyl which can be coupled to a dye of the azo-group.

The quantitative estimation of alkaline phosphatase activity in serum was based on the technique by KING and ARMSTRONG (1934). In this method phosphatase activity is determined by estimating the amount of phenol liberated when the enzyme is allowed to act on a suitable substrate substance such as phenyl phosphate. The liberated phenol is taken as a measure of the amount of enzyme present. A unit of phosphatase activity is defined as the amount of enzyme which will set free 1 mg of phenol under standard conditions of 37°C for 15 min. This is measured in units for 100 ml of serum or King Armstrong Units, which are equivalent to the milligrammes of phenol liberated from phenyl phosphate under standard conditions. All tests were carried out in duplicate and compared with a known standard solution of phenol using Folin and Ciocalteau's Phenol Reagent (BDH Ltd.), and read in a Hilger Spectrophotometer (H700/701 Uvispek).

RESULTS

This investigation was primarily to look at sera from different pigs to see if genetic variants of alkaline phosphatase exist. Material was therefore selected from complete litters where the sire and dam could be compared with their own litter all on the same gel. Figure 1*a* shows a photograph of a gel showing alkaline phosphatase activity in the sera from a complete litter including the sire, dam and seven offspring. There are two important points to observe in this photograph. First, the sire and dam show very little phosphatase activity, and second, the piglets sampled at 3 weeks of age show high activity and a polymorphic-type variation.

It is known that serum alkaline phosphatase activity is markedly increased in young children, and it has been reported by WEGGER and WESTERGAARD (1963), that alkaline phosphatase activity is high in young pigs. As the sire of the litter shown in Fig. 1*a* had been bled at various times during his life, a sample of serum had been taken from the same boar when he was 3 weeks of age. This was the same age as his litter shown in Fig. 1*a*. This sample was recovered from cold storage and run in a comparable gel to that shown in Fig. 1*a*, the only difference being, that the specimen representing the sire was his 3 weeks of age sample. A photograph of the gel is shown in Fig. 1*b*. The picture presented by the 3 weeks of age sample from the sire is very similar to that shown by the litter.

The patterns of electrophoretic variation between piglets within the same litter could be confused with polymorphic type variation at first sight. With the evidence presented in the gels shown in Fig. 1*a* and *b* however, the variation appeared to be more of an ontogenic type. To establish this hypothesis, a piglet was blood sampled at birth before taking colostrum, and then at weekly intervals for several weeks. All the serum samples obtained were run in the same gel and stained for alkaline phosphatase. A photograph of the gel obtained is shown in Fig. 2. A change in electrophoretic mobility can be observed between the pre and post-colostral samples, with the development of a fast migrating component in the 1 week of age sample. This fast component can be present at 4 days of age and gradually disappears as about 3 to 4 weeks of age. It is interesting to note, that the picture presented in this gel for the 2 weeks of age sample is very similar to that seen for the litter in Fig. 1*a* and *b*. All piglets appear to show this type of change whether they are of blood group type A or O.

To establish that the change in mobility observed occurs independently of the uptake of colostrum, a piglet was reared artificially without colostrum for 4 days, and sampled at comparable times with a litter mate reared naturally. The serum samples obtained from both piglets were run in the same gel and stained for alkaline phosphatase. A photograph of the gel is shown in Fig. 3. The change in mobility observed in Fig. 2 is also observed in Fig. 3 for both piglets: therefore this change appears to occur independently of the uptake of colostrum.

Another feature of the fast component is an association with a protein sensitive fraction. Figure 4 shows a photograph of a gel, which has been split horizontally with one half stained in nigrosine, and the other stained for alkaline phosphatase. Serial serum samples were taken from one pig pre- and post-colostrum for the ages shown, and run in the same gel. The presence of a protein sensitive component coincides with the development of the fast migrating alkaline phosphatase fraction. As the fast alkaline phosphatase fraction disappears, the protein sensitive band also disappears, but more slowly.

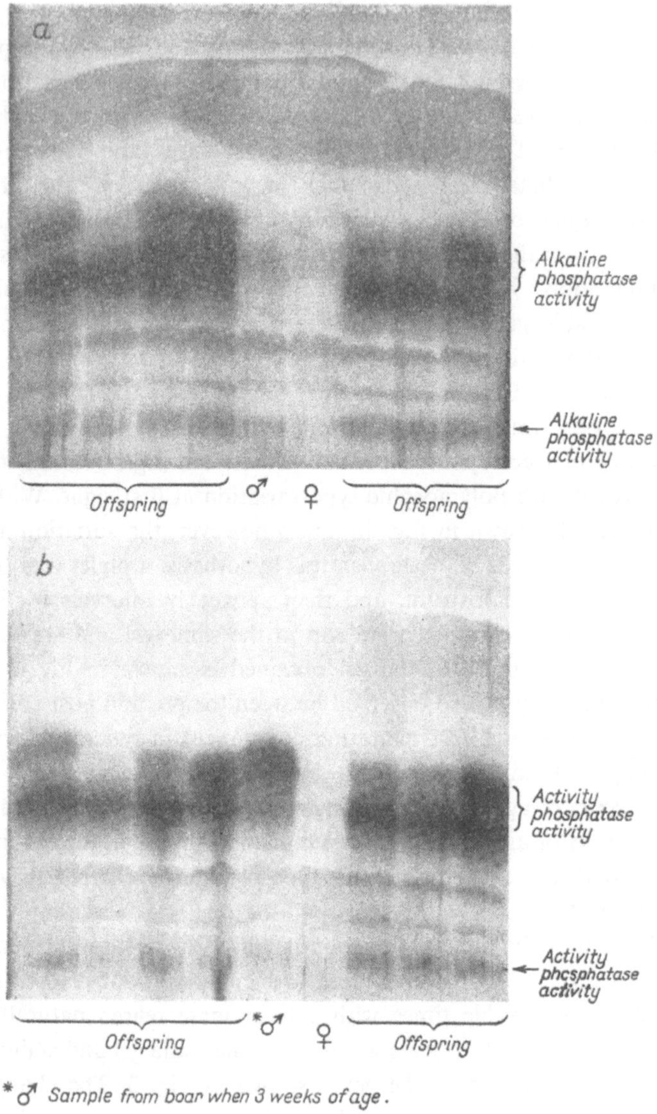

FIG. 1.

As a result of sampling individual piglets there appeared to be considerable differences in the intensity of staining reaction for alkaline phosphatase between piglets of comparable ages from different litters. Quantitative estimations of serum alkaline phosphatase activity were made. This activity was compared at weekly intervals for piglets in 17 litters from birth before colostrum through to 9 weeks of age. The number of litters and piglets sampled with their average serum alkaline phosphatase levels and range at comparable ages are shown in the Table. An anal-

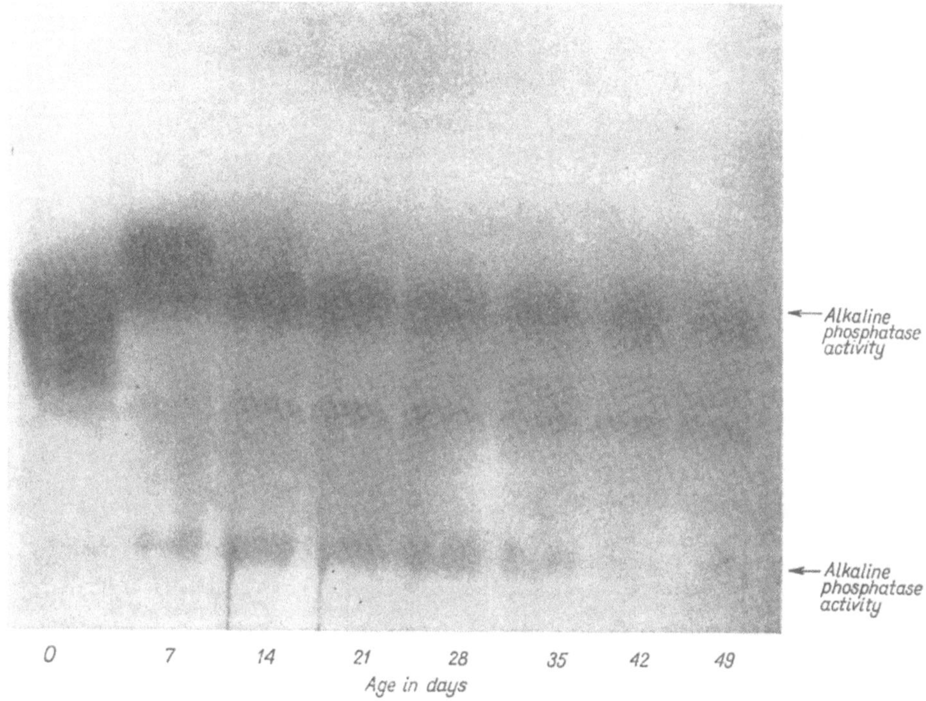

FIG. 2.

Serum alkaline phosphatase levels in pig litters at comparable ages

Age in days	0	7	14	21	35	49	63
No. of litters	10	16	13	15	9	10	9
No. of piglets	76	120	87	106	78	75	71
Mean alkaline phosphatase levels and range	179.6 (73.8– –323.9)	119.1 (47.4– –236)	86.8 (35– –183.9)	52.9 (20.9– –142.9)	30.8 (11.4– –60.7)	29.4 (9.5– 55.8)	25.2 (12.9– –51.3)
Variance ratio and degrees of freedom	12.2 $N^1/_9\ N^2/_{66}$	2.4 $N^1/_{15}\ N^2/_{104}$	8.1 $N^1/_{12}\ N^2/_{74}$	8.7 $N^1/_{14}\ N^2/_{91}$	11.8 $N^1/_8 N^2/_{69}$	3.9 $N^1/_9\ N^2/_{65}$	3.9 $N^1/_8\ N^2/_{62}$
Probability	0.001	0.01	0.001	0.001	0.001	0.001	0.001

ysis of variance was carried out on all litters at comparable ages, and the variance ratios and probability levels are shown in the same Table. The variation between litters is significantly greater than the variation within litters for the ages shown.

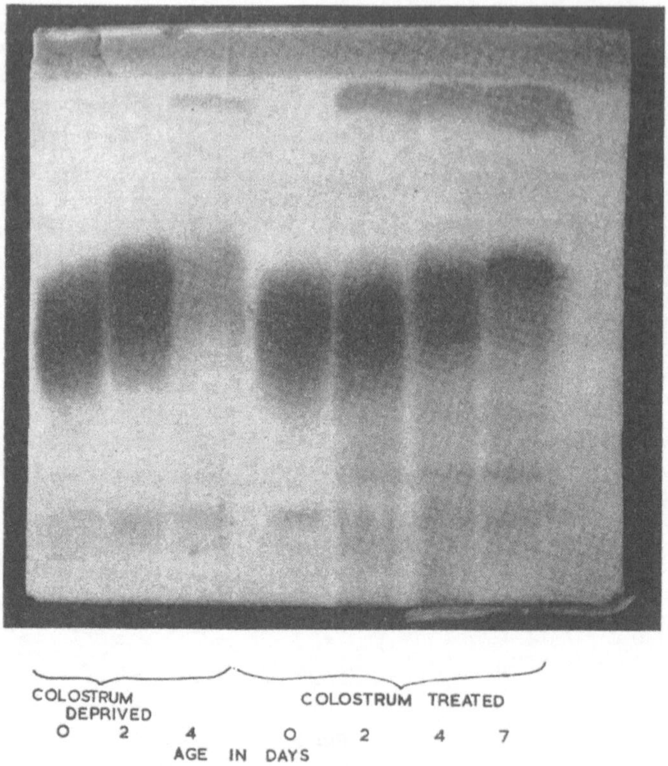

FIG. 3. Alkaline phosphatase development in a colostrum deprived and colostrum treated piglet.

DISCUSSION

Contrary to the other species no evidence for genetical variation in serum alkaline phosphatase could be found in the breeds of pig examined. The low activity of this enzyme, when run on starch gel electrophoresis may mask possible genetic variants in adult pigs, but must make any investigation of this type difficult. This is also true for serum acid phosphatases in adult pigs, which also show relatively low activity after electrophoresis. The high activity of serum alkaline phosphatase encountered in young pigs with the highest activity in the newborn animal confirms the work of WEGGER and WESTERGAARD (1963), and their values compare favourably with the figures recorded in this paper. The sudden drop in activity which occurs during the first few weeks of life might be explained by the rise in plasma volume which takes place after birth (MCCANCE and WIDDOWSON, 1959).

The significance of the difference in mobility between pre- and post-colostral samples of serum alkaline phosphatase, and the brief appearance of a fast migrating component associated with a protein sensitive fraction is not known. All the phos-

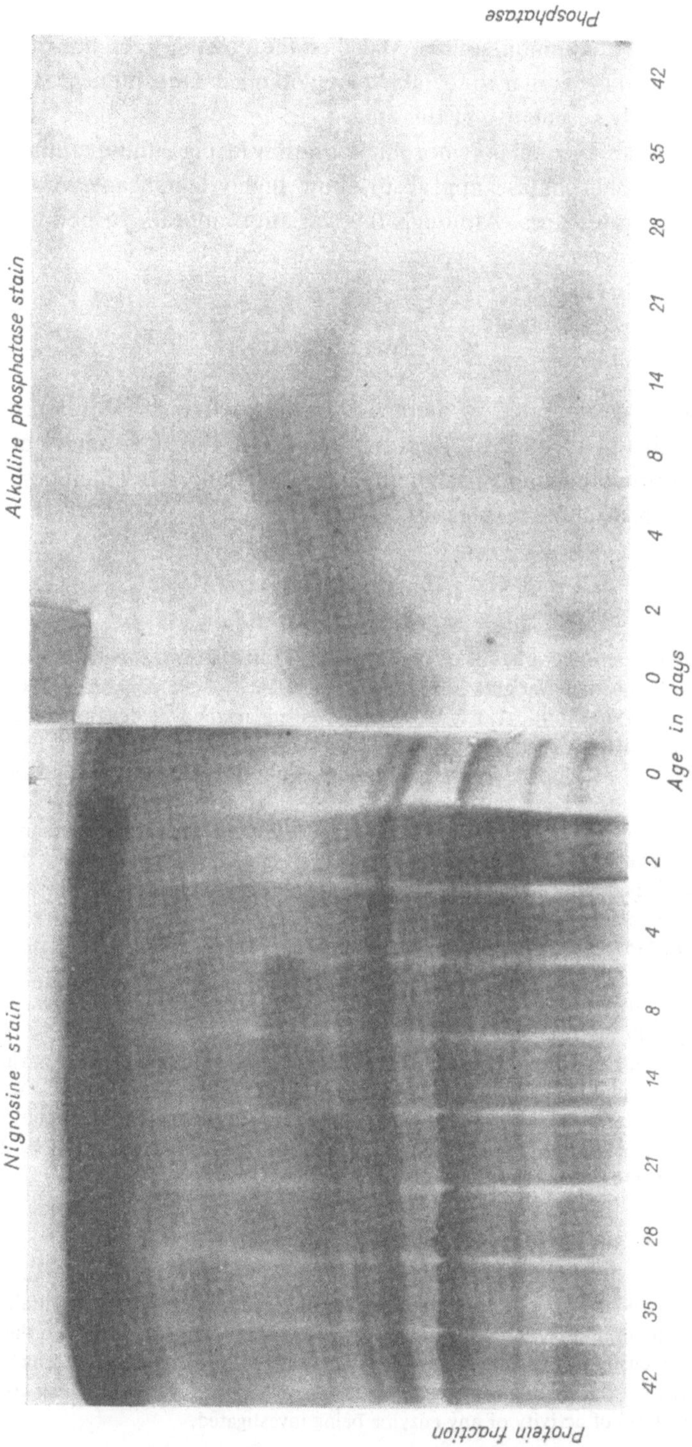
Phosphatase
Alkaline phosphatase stain
Nigrosine stain
Protein fraction
42 35 28 21 14 8 4 2 0
0 2 4 8 14 21 28 35 42
Age in days

FIG. 4.

phatase components appear to be inhibited by DL-phenylalanine which may indicate that they are of a common source. Whether the possible presence of a foetal-type alkaline phosphatase similar to foetal haemoglobin has some influence on the changes observed is purely speculative at this stage.

Despite the absence of polymorphic variation in pigs, the quantitative levels of serum alkaline phosphatase appear to show highly significant variation between litters at comparable ages. Although this variation appears to be familial in type, there is no clear cut genetical hypothesis for its control.

ACKNOWLEDGEMENTS

This work was supported by a grant from the Pig Industry Development Authority. I would like to thank Prof. A. ROBERTSON, Dean of the Faculty of Veterinary Medicine for providing animal accommodation and laboratory facilities, also Valerie A. MACPHEE for technical assistance.

REFERENCES

GAHNE, B., 1963. Genetic variation of phosphatase in cattle serum, *Nature*, **199**, 305–306.

IMLAH, P., 1964. Inherited variants in serum ceruloplasmins of the pig, *Nature*, **203**, 658–659.

KING, E. H. and ARMSTRONG, A. R., 1934. A convenient method of determining serum and bile phosphatase activity, *Canad. Med. Ass. J.*, **31**, 376–381.

LUSH, I. E., 1961. Genetic polymorphisms in the egg albumen proteins of the domestic fowl, *Nature*, **189**, 981–984.

LAWRENCE, S. H., MELNICK, P. J. and WEIMER, H. E., 1960. A species comparison of serum proteins and enzymes by starch gel electrophoresis, *Proc. Soc. Exp. Biol. Med.*, **105**, 572–575.

MCCANCE, R. A. and WIDDOWSON, E. M., 1959. The effect of colostrum on the composition and volume of the plasma of newborn piglets, *J. Physiol.*, **145**, 547–550.

POULIK, M. D., 1957. Starch gel electrophoresis in a discontinuous system of buffers, *Nature*, **180**, 1477–1479.

RENDEL, J. and STORMONT, C., 1964. Variants of ovine alkaline serum phosphatases and their association with the R-O blood groups, *Proc. Soc. Exp. Biol. N.Y.*, **115**, 853–856.

SMITHIES, O., 1955. Zone electrophoresis in starch gels: group variations in the serum proteins of normal human adults, *Biochem. J.*, **61**, 629–641.

WEGGER, I. and WESTERGAARD, U., 1963. Glutamic oxaloacetic transaminase and alkaline phosphatase in serum of pigs, Annual Report, The Royal Veterin. and Agric. Coll., Inst. F. Sterility Research, 115–126.

DISCUSSION

G. C. ASHTON: I would just like to comment that there is one possible pitfall which one ha to be aware of in an interpretation of this sort of effect and that is not putting on the same loading of enzymes. It's customary to take a certain volume of serum of fluid and of course one can get a quite different pattern by diluting serum. So it is necessary to try and establish comparisons between the same level of activity of any enzyme being investigated.

P. IMLAH: Yes, Mr. Chairman, I would agree that it is difficult to get comparable amount of enzymes and proteins in a sample, even although you are applying exactly the same volume of serum to your insert. But, in fact, the quantitative estimations which were carried out also correspond to and confirm the apparent increase in activity, which was observed in the gels. It has also been shown in children and in pigs, that serum alkaline phosphatase activity is much higher in the young. The findings reported in this paper therefore supports the work of Wegger and Westergaard. I might remind you, that there is a much lower content of protein in a pre-colostrum sample than in a post-colostrum sample, and yet the highest activity of the enzyme was observed in the pre-colostrum sample.

J. MOUSTGAARD: I just want to mention that the preliminary investigations by Wegger and Westergaard gave results very similar to yours regarding the variations between litters.

J. HYLDGAARD-JENSEN: The question concerns: Rapid moving fraction of alkaline phosphatase in pig and piglets serum. Have you looked for the tissue origin of this fraction?

P. IMLAH: No!

XITH EUROPEAN CONFERENCE ON ANIMAL BLOOD GROUPS AND BIOCHEMICAL POLYMORPHISM
Warsaw, July 2nd–6th, 1968

A Study of Lactic Dehydrogenase Isoenzymes in Pig

J. HYLDGAARD-JENSEN, S. E. JENSEN, J. MOUSTGAARD, B. PALLUDAN and M. VALENTA

Department of Physiology, Endocrinology and Blood Grouping, The Royal Veterinary and Agricultural College, Copenhagen, Denmark

SUMMARY. In the present report some results from an investigation of LDH isoenzymes in plasma, tissues and spermatozoa from pig are presented.

Three different LDH isoenzyme systems were discovered. Their characteristic behaviour is described and discussed. Polymorphism in subunit C was found. The polymorph types were found in a relatively high frequency.

Separation of LDH isoenzymes were carried out by using both agar gel and starch gel electrophoresis. In both methods certain modifications were done in order to improve sensitivity and accuracy for the estimation of LDH isoenzymes. Investigation of LDH isoenzymes in tissues, healthy and unhealthy, indicates the possibility to obtain not only quite reliable informations about changes in progression of the disease but also about the local metabolic conditions. The finding of two extra LDH isoenzyme systems is considered to further increase the specificity of such determinations in the clinical enzymology.

INTRODUCTION

In general lactic dehydrogenase (LDH) in most vertebrates exist as five isoen zymes. Each isoenzyme is considered to be a tetramer, usually composed by two-types of subunits, polypeptide monomers, A and B. The homotetramer of subunit A (LDH_5 or M-type) has the slowest migration rate and the homotetramer of subunit B (LDH_1 or H-type) the fastest migration towards the anode. The hybrid (heterotetramers) LDH_4 (A_3B or M_3H), LDH_3 (A_2B_2 or M_2H_2), LDH_2 (AB_3 or MH_3) have intermediate, distinctive migration rates.

Literature data report the findings of six to eight fractions with LDH activity in tissues and spermatozoa from different animals. The extra bands were named *x*-bands. CONKLIN *et al.* (1962) and RESSLER *et al.* (1965) suggested that LDH_x-bands are formed from A and B subunits and some other compound, furthermore ZINKHAM and BLANCO (1964) suggested that the LDH_x-band in pigeon testis is controlled by a third locus C. In pig spermatozoa evidence for a third locus C is reported (VALENTA *et al.*, 1967). Subunits form *in vivo* a new LDH isoenzyme system with subunits A and B. Furthermore a third system, which probably is not involving A, B and C subunits was found. Polymorphism of A and B subunits have been described in man (KRAUS and NEELY, 1964; DAVIDSON *et al.*, 1965) in monkeys (SYNER and GOODMAN, 1966) in fish (GOLDBERG, 1966) and in pig (HYLDGAARD-JENSEN *et al.*, 1967). Polymorphism of C subunit have been found in pigeon tests (ZINKHAM and BLANCO, 1964) and in pig spermatozoa (VALENTA *et al.*, 1967). With the exception of pig spermatozoa the frequency of polymorph types were very low.

The purpose of the presented work was to study the LDH isoenzymes in different tissues of pig and the possibility to use the changes in the isoenzyme patterns in pathological plasma and tissue extracts as a diagnostic tool. For that purpose exact methods of LDH isoenzyme determination were developed.

MATERIAL

Blood was collected by puncturing the *V. cava cran.* Tissues and spermatozoa were obtained from slaughtered pigs (HYLDGAARD-JENSEN *et al.*, 1967). The samples were homogenized in an Ultra Turrax homogenizer and centrifugated (14–16,000 r.p.m. for 1 hr). Supernatants plasma and hemolysates of washed erythrocytes were used for electrophoretic separation of LDH isoenzymes.

METHODOLOGICAL INVESTIGATIONS AND RESULTS

Agar gel electrophoresis using veronal as buffer is very often used for the estimation of LDH isoenzymes (WIEME, 1965). In a very short time an excellent separation of isoenzymes is obtained. When samples with the same activity of LDH_1 and LDH_5 were run in this technique and determined by the zymogram method (WIEME *et al.*, 1962) the intensity of LDH_5 spot was only about 50% of that of LDH_1 (Table 1). A study of this discrepancy showed that veronal is a competitive inhibitor of LDH isoenzymes (VALENTA *et al.*, 1968). The inhibition is strongest expressed for LDH_5 (Figure). Inhibition of LDH_5 may be taken off by citrate. When modifications of the original method were used (HYLDGAARD-JENSEN *et al.*, 1968) the same values for LDH_1 and LDH_5 were obtained (Table 1).

The separation of isoenzymes in polymorph systems or when several systems are present in the material, starch gel electrophoresis is usually preferred. By using a combination of a starch and a double agar gel layer technique a relatively simple and exact method for determination of LDH isoenzymes was developed (HYLDGAARD-JENSEN *et al.*, 1968). Results obtained with this method were comparable to results obtained with the agar gel technique. The amount of activity, which is put into gel, may influence the relative intensities of LDH isoenzymes (HYLDGAARD-JENSEN *et al.*, 1968). This problem disappear, when activity of each isoenzyme is kept lower than 2000 Wacker units per ml and 10 μl of this sample is put into gel. Colloids present in small amounts were shown to stimulate the formazan for-

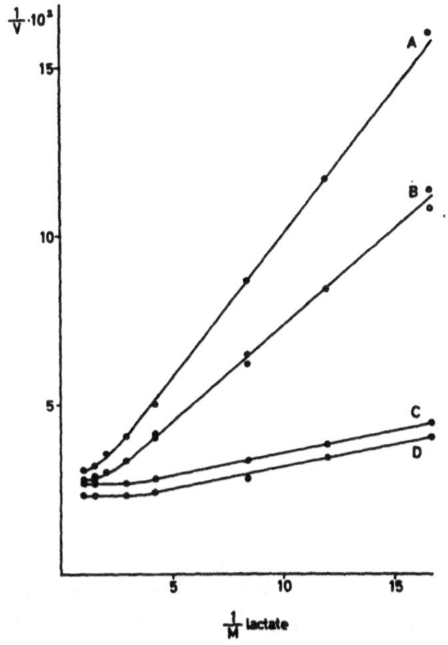

Double reciprocal plot of reaction velocity and DL-lactate concentration obtained with LDH_5 (B ○——○), muscle extract (A, B) and heart extract (C, D) in the presence of 0.06 M veronal (A, D) and without veronal (B, C).

mation. When samples with different protein contents are electrophoretically separated, single isoenzymes may be located in areas with very different protein concentrations and formazan formation is stimulated to a different degree independent of the real isoenzyme activity. When optimal concentrations of protein are added into gel the conditions for LDH isoenzyme determination may be more similar. Different aspects of the preparation of tissue extracts have also been investigated. Literature data indicate a cellular compartmentalization of LDH isoenzymes (GÜTTLER and CLAUSEN, 1967). Different tissues from pig were investigated and results indicate an overwhelming cytoplasmic localization of LDH_1 with a possible selective presence of LDH_5 in cell-nuclei.

The electrophoretic conditions may interfere with the positions of LDH isoenzymes. Using a certain technique (HYLDGAARD-JENSEN *et al.*, 1967) for starch gel electrophoresis of plasma the relative position of both LDH_2 and albumin is changed in similar way and LDH_2 is positioned close to LDH_1. In some cases this might interfere with the polymorph picture. In some old frozen samples extra fractions may appear. These findings indicate the necessity for standardized conditions.

When all above mentioned conditions are kept the true pictures of LDH isoenzymes were obtained. The isoenzyme distribution in tissues is subjected to a great variation (HYLDGAARD-JENSEN and JENSEN, 1967) and as mentioned above differences may also exist within the single cell. In pathological changed tissues or cells

TABLE 1

The percentage distribution of LDH isoenzymes in samples separated and estimated by the original agar gel electrophoresis technique (a) and by a modification of this same technique (b)

Sample		LDH_1	LDH_2	LDH_3	LDH_4	LDH_5
LDH_1+LDH_5	a	65.9	—	—	—	34.1
1:1	b	50.8	—	—	—	49.2
Heart (H)	a	92.5	7.5	—	—	—
	b	90.1	6.3	0.9	0.9	1.8
Skeletal muscle	a	3.8	2.4	2.0	1.6	90.2
(M) (M. long. dorsi)	b	0.5	0.8	0.4	2.5	95.8
H+M	a	62.3	4.3	—	0.7	32.6
1:1	b	47.9	1.8	0.4	0.9	49.4

the isoenzymes are released into the blood. The change in plasma isoenzyme pattern may indicate different disease states. Investigation of LDH isoenzymes directly in the diseased organs (biopsies) is believed to give more correct information about changes in both progression of the disease and the local metabolic conditions.

TABLE 2

The distribution of type I, II and III of a second LDH isoenzyme system in boar spermatozoa

Phenotypes	I	II	III	Total
No.	47	29	68	144
%	32.6	20.2	47.2	100

The existence of a new LDH isoenzyme system is believed to increase the applicability of such determinations within the diagnostic work. Three LDH isoenzyme systems were discovered in pig spermatozoa and more than five isoenzymes were found in many pig tissues. At present an investigation is carried out to find out if the extra bands in spermatozoa and tissue extracts are the same. A great help in such investigation is based on the facts that the occurrence of polymorph subunits controlled by C locus is relatively high (Table 2) and that the identification of isoenzymes of a third system is facilitated by not requiring phenazine methosulphate in the staining solution (VALENTA *et al.*, 1967). Homotetramers composed by C or C′ subunits are relatively thermostable and much less sensitive to mercaptoethanol inhibition than isoenzymes composed by A and B subunits. Isoenzymes of the third system have intermediate stability but are relatively stable when butanol is used for inhibition. Using α-hydroxybutyrate as substrate activity of isoenzymes composed by C subunits is practically the same, but activity of isoenzymes from the LDH_{1-5} system is strongly depressed. When α-hydroxyvalerate is used only isoenzymes from the second and third system are active.

Literature data indicate that there may exist a correlation between LDH polymorphism in man and cancer (LATNER, 1967). The study of the relationship between polymorph types of LDH and certain diseases seems therefore to be an interesting field.

In more than 700 plasma and erythrocyte samples the existence of polymorph A or B subunits was investigated. Six samples showed a double spot in LDH_1 position. Twenty-five samples exhibited small differences in mobility. The differences in mobility decreased gradually from LDH_1 to LDH_5. Changes in electrophoretic conditions involving changes in pH from 5 to 9, voltage gradient 4 to 12 V/cm and starch concentration 8 to 16% produced only small changes in mobility. Genetic analysis is presently going on in order to solve this problem.

REFERENCES

CONKLIN, J. L., DEWEY, M. M. and MAY, B., 1962. *J. Histochem. Cytochem.*, **10**, 365–366.

DAVIDSON, R. G., FILDES, R. A., GLEN-BOTT, A. M., HARRIS, H. and ROBSON, E. B., 1965. *Ann. Hum. Genet.* (Lond.), **29**, 5–13.

GOLDBERG, E., 1966. *Science*, **151**, 1091–1093.

GÜTTLER, F. and CLAUSEN, J., 1967. *Enzym. biol. clin.*, **8**, 456–470.

HYLDGAARD-JENSEN, J. and JENSEN, S. E., 1967. Aarsberetn. Inst. Sterilitetsforsk., p. 109–120

HYLDGAARD-JENSEN, J., VALENTA, M. and MOUSTGAARD, J., 1967. Aarsberetn. Inst. Sterilitetsforsk., p. 121–137.

HYLDGAARD-JENSEN, J., VALENTA, M., JENSEN, S. E. and MOUSTGAARD, J., 1968. *Anal. Biochem.* (In press).

KRAUS, A. and NEELY, C. L., 1964. *Science*, **145**, 595–597.

LATNER, A. L., 1967. Advances in Clinical Chemistry, Vol. 9, pp. 69–163, Acad. Press, N.Y.

RESSLER, N., COOK, U., OLIVERO, E. and JOSEPH, R. R., 1965. *Nature*, **206**, 828–829.

SYNER, F. N. and GOODMAN, M., 1966. *Science*, **151**, 206–208.

VALENTA, M., HYLDGAARD-JENSEN, J. and MOUSTGAARD, J., 1967. *Nature*, **216**, 506–507.

VALENTA, M., HYLDGAARD-JENSEN, J., JENSEN, S. E. and MOUSTGAARD, J., 1968. *Biochem. J.* (In press).

WIEME, R. J., 1965. Agar Gel Electrophoresis, Elsevier, Amsterdam.

WIEME, R. J., VAN SANDE, M., KARCKER, D., LOWENTHAL, A. and VAN DER HELM, J. H., 1962 *Clin, Chim. Acta*, **7**, 750.

ZINKHAM, W. H. and BLANCO, A., 1964. *Science*, **144**, 1353.

XITH EUROPEAN CONFERENCE ON ANIMAL BLOOD GROUPS AND BIOCHEMICAL POLYMORPHISM
Warsaw, July 2nd–6th, 1968

Additional Studies into Serum Amylase in Swine

M. HESSELHOLT

Department of Physiology, Endocrinology and Blood Grouping, The Royal Veterinary and Agricultural College, Copenhagen, Denmark

SUMMARY. At least 4 zones of amylase activity were distinguished after starch–agar electrophoresis of swine serum. One zone was situated in the leading edge of serum albumin and three zones in the γ-globulin area. Two-dimensional electrophoresis, starch–agar–starch gel, demonstrated that the amylase of the albumin field migrates in starch gel and exhibits gene-controlled variations, while the amylase of the γ-field apparently remain at the insertion point.

Various tissues were subjected to starch–agar and starch gel electrophoresis for the separation of amylase. The majority of tissue amylase investigated appeared in the γ-globulin field. The amylases in the follicular fluid and uterus extracts showed, however, an electrophoretic picture identical to that of serum. Finally, the tissue source of serum amylase and the state of the various amylases in serum are discussed.

INTRODUCTION

The existence of a polymorphic serum amylase in swine has been ascertained by means of starch gel electrophoresis. Genetic investigations in British pig breeds and in Danish Landrace pigs indicated that three codominant alleles, Am^1, Am^2 and Am^3, segregate in these breeds (IMLAH, 1964; GRAETZER *et al.*, 1964; HESSELHOLT *et al.*, 1966). Later a fourth variant, Am 2F, was demonstrated in one family and the existence of a fourth allele, Am^{2F}, was suggested (HESSELHOLT, 1966). The serum amylase phenotypes hitherto observed in the Danish Laboratory are demonstrated in Fig. 1.

In the course of the initial starch gel electrophoretic experiments on serum amylase in swine it became apparent that two fractions showing amylase activity could be distinguished on the gel after electrophoresis of serum and incubation of the gel. In addition to the polymorphic amylase a soft, whitish area of amylase activity around the insertion point of the serum soaked filter paper was consistently observed (GRAETZER *et al.*, 1964). Various attempts were made to induce the migration of this amylase fraction in starch gel, but without success.

In the present investigations it was decided to study porcine serum amylase by electrophoresis in starch–agar, a stabilizing medium suitable for the detection of amylase activity. Moreover, two-dimensional electrophoresis were carried out for the purpose of referring the polymorphic amylase observed by starch gel electrophoresis to the amylase zones observed by zone electrophoresis in starch–agar. Finally, attempts were made to establish the tissue origin of the polymorphic amylase by electrophoresis of extracts from various tissues.

METHODS AND MATERIALS

Starch gel electrophoresis of serum amylase was carried out as described by HESSELHOLT *et al.* (1966).

Starch–agar electrophoresis was performed in 1% agar gels to which soluble starch was added in such amounts that the starch–agar gels contained 1% soluble starch. The discontinuous buffer system of HIRSCHFELD (1960) was used. Electrophoresis was carried out on a LKB 3276 BN apparatus. Voltage gradient: 8 V/cm. Time: 1 hr. After cessation of electrophoresis the gels were incubated at 42°C for 1 hr and subsequently immersed in an iodide solution (iodide 0.5 g, potassium iodide 1 g, distilled water 1500 ml) for the detection of amylase digested starch–agar zones.

Paper electrophoresis was carried out according to the method of ARONSSON and GRØNVALL (1958).

Two-dimensional electrophoresis was performed following the principles of POULIK and SMITHIES (1958).

Tissue specimens were sampled at slaughtering. After careful cleaning under cold tap water the specimens were immediately frozen and kept at ÷20°C until use. The tissues were homogenized in physiological saline (1:1) and subsequently centrifuged at 17,000 r.p.m. for 1 hr in a refrigerated centrifuge.

RESULTS

Figure 1 illustrates the various zones of amylase activity after electrophoresis of swine serum in starch–agar compared with the mobility of albumin, α-, β- and γ-globulin. It appears from Fig. 1 that the most anodic amylase zone is situated in the leading edge of the albumin fraction and that three zones of amylase activity are observed in the γ-globulin field. In opposition to the amylase zone in the albumin field the latter three fractions exhibited pronounced quantitative variations from individual to individual. The most cathodic band seemed constantly to be more heavy than the two remaining amylase bands.

In order to refer the polymorphic serum amylase demonstrated by means of starch gel electrophoresis to the zones demonstrated by means of starch–agar electrophoresis two-dimensional electrophoresis was performed where the initial electro-

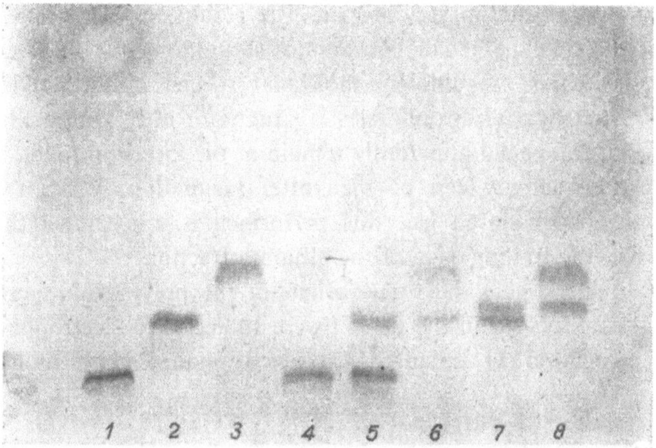

FIG. 1. Starch gel demonstrating the 8 amylase phenotypes observed in the Danish laboratory. Amylase phenotypes, 1: Am 3-3; 2: Am 2-2; 3: Am 1-1; 4: Am 3-1; 5: Am 3-3; 6: Am 2-1; 7: Am 2-2F; 8: Am 2F-1.

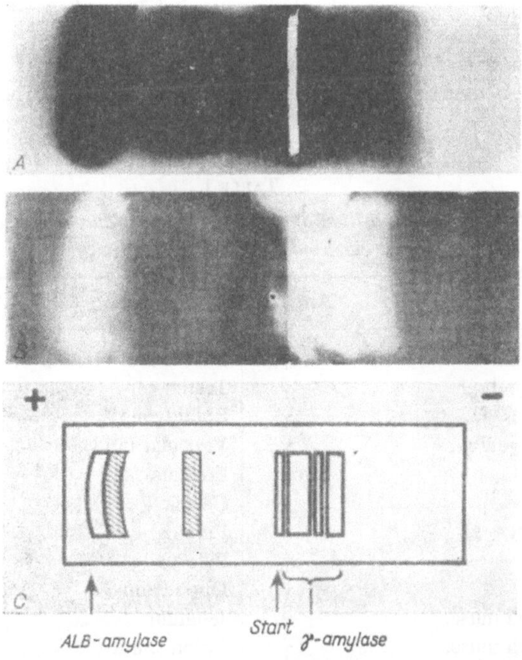

FIG. 2. Starch–agar electrophoresis of serum. *A*: Amido black-nigrosin stain. *B*: Zones of amylase activity in the gel. In the γ-field the intermediate zone is overshaded by the anodic zone. *C*: Diagram of *B*. Shaded areas indicate zones of interference (see text).

phoresis was carried out in starch–agar. The results are seen in the diagram on Fig. 3. It is seen that the genetic variants of serum amylase observed in starch gel electrophoresis in fact are different molecular species of the amylase component which after starch–agar electrophoresis is situated in the leading edge of albumin. The amylases of the γ-field apparently remain at the insertion point, where digested zones of amylase activity can be seen after incubation. When two-dimensional electrophoresis, paper–starch gel, was performed it was shown that the genetic variants were in the trailing edge of the albumin fraction.

In order to investigate the tissue origin of the polymorphic amylase, extracts from various tissues listed in the Table 1 were subjected to electrophoresis in starch-agar gels and in starch gel. In starch–agar electrophoresis very strong amylase activ-

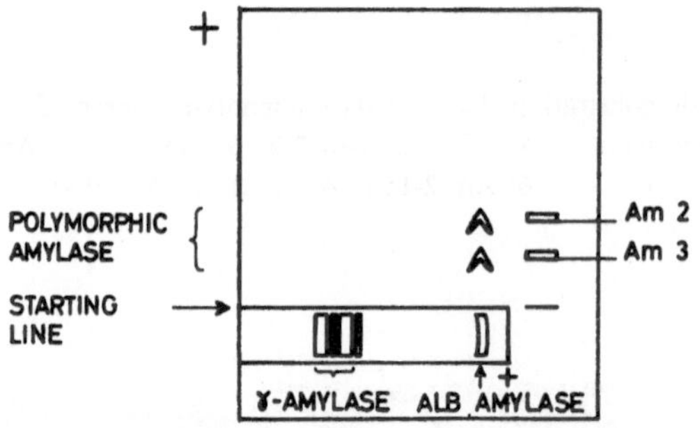

FIG. 3. Two-dimensional electrophoresis, starch–agar–starch gel, of an Am 3-2 serum. The same sample is used as reference in starch gel electrophoresis. The amylase zones only are developed.

TABLE 1

The electrophoretic position of amylases from various tissues after zone electrophoresis in starch–agar

Tissue	Amylase	Tissue	Amylase
Pancreas	γ	Uterus (not pregn.)	alb. γ
Submand. gl.	γ	Testis	γ
Parotic. gl.	γ	Epididymis	γ
Sublingual gl.	γ	Vesicular fluid	γ
Liver	γ	Prostata	γ
Kidney	γ	Cardia	γ
Supraren. gl.	γ	Fundus	γ
Lung	γ	Pylorus	γ
Heart	γ	Duodenum	γ
Striated musc.	γ	Jejunum	γ
Smooth musc.	γ	Colon spiral	γ
Lymph gland	γ	Colon sin.	γ
Erythrocytes	γ	Rectum	γ
Follicular fluid	alb. γ		

ity was observed in the γ-field after electrophoresis of pancreas, salivary glands, testis, epididymis, vesicular fluid and prostata. When the amylase activity observed in the tissues from the alimentary tract is considered the possibility cannot be excluded that the γ-activity is of blood or microbial origin. The tissue extracts from kidney, suprarenal gland, lung, heart, muscles and lymph gland were also contaminated with blood. In follicular fluid and extract from a non-pregnant uterus amylase fractions were observed, which showed an electrophoretic picture identical to that of serum. Follicular fluids from 6 follicles in one ovarium and 9 follicles from another animal showed in starch gel electrophoresis the amylase phenotypes Am 2-1 and Am 2-2, respectively. The intensity of the polymorphic albumin amylase of follicular fluid and uterus extract was identical to that of serum evaluated on the basis of broadness of bands in starch gel and starch–agar electrophoresis.

DISCUSSION AND CONCLUSION

It appears from the above studies that at least 4 amylases can be distinguished after starch–agar electrophoresis, namely one zone in the leading edge of albumin and three zones in the γ-globulin area. Two-dimensional electrophoresis, starch–agar–starch gel, demonstrated that the serum amylase of the albumin field migrates in starch gel and exhibits gene-controlled variations, while the amylases of the γ-field apparently remain at the insertion point, where digested zones of amylase activity are seen after incubation. It has been claimed that amylase activity exists in the albumin area after electrophoresis of human serum in agar or paper (Mc Geachin and Lewis, 1959; Dreiling *et al.*, 1963). It is, however, now generally agreed that the amylase of normal human serum after electrophoresis in these media resides as an unbound moiety in the γ-globulin area, and that the amylase activity observed in the albumin zone is artefactual and caused by interference phenomena, where large amounts of protein affect the starch–iodide color (Wilding, 1963; Berk and Searcy, 1965). Similar interference zones were observed in the present study (Fig. 2), but the clear transparent amylase zone in the leading edge of albumin could be distinguished from the whitish-gray interference zone of albumin. A polymorphic amylase has likewise been demonstrated in cattle (Ashton, 1965; Hesselholt and Moustgaard, 1965). From the above mentioned features the absence in humans of an amylase polymorphism similar to that in swine and cattle can be predicted, and various attempts in our laboratory to reveal amylase variations by means of the starch gel electrophoretic technique have failed.

In 1960 Ashton described a thread protein system in pigs containing two alleles. These findings were later confirmed by Schröffel and Hojný (1962). When the results from the 1st' European Comparison Test on Serum Protein Polymorphism in Swine, comprising 40 serum samples, were studied, it was found that the thread protein determinations carried out by the Czechoslovak Laboratory using Ashton's technique closely parallelled the amylase determinations of the Danish Laboratory using the technique for amylase typing described above. In ASHTON's original paper it was demonstrated by means of two-dimensional electrophoresis that the thread protein is situated in the leading edge of albumin after electrophoresis in agar and in the trailing edge of albumin after paper electrophoresis. These features are similar to the electrophoretic characterization of amylase in the present study and lends further support to the suggestion that the two genetic systems are identical.

It is appropriate to suggest that the amylase observed in the γ-globulin field after starch–agar electrophoresis correspond to the human serum amylase with γ-mobility, which is present in serum as an unbound moiety and which is subjected to quantitative variations (WILDING, 1963). Most probably these amylases are released into the blood stream from organs as pancreas, salivary glands and liver, which both in man and swine produce amylase with an electrophoretic mobility of γ-globulin. Among the tissues and secretions investigated the polymorphic albumin amylase was observed in follicular fluid and uterus extract only. Apparently the amylase concentration in these tissues are the same as that in the blood suggesting that the polymorphic amylase observed in these tissues is of blood origin. The albumin amylase of swine serum might still occur as an unbound moiety originating from an yet unknown tissue source. In this case the interesting situation emerges that this isoenzyme by starch gel electrophoresis can be resolved into various molecular species, which exhibit genetic polymorphism. Another possibility is that the albumin amylase represent an enzyme–protein complex. There seems to be a growing body of evidence pointing to the binding of some serum enzymes by other serum proteins (URIEL, 1963; LATNER, 1965). In the present case we have to consider that perhaps only one enzyme form is involved, but that it may be complexed to another protein, which exists in different genetically controlled forms.

REFERENCES

ARONSSON, T. and GRØNVALL, A., 1957. Improved separations of serum proteins in paper electrophoresis. A new electrophoresis buffer, *Scand. J. Clin. Lab. Invest.*, **9**, 338–341.

ASHTON, G. C., 1960. "Thread protein" and β-globulin polymorphism in the serum proteins of pigs, *Nature* (Lond.), **186**, 991–992.

ASHTON, G. C., 1965. Serum amylase (thread protein) polymorphism in cattle, *Genetics* (Princeton), **51**, 431–437.

BERK, J. E. and SEARCY, R. L., 1965. Isoenzymes of serum amylase in man, *Gastroenterology*, **48**, 651–653.

DREILING, D. A., JANOWITZ, H. D. and JOSEPHBERG, L. J., 1963. Serum isoamylases. An electrophoretic study of the blood amylase and the patterns observed in pancreatic disease, *Ann. Intern. Med.*, **58**, 235–244.

GRAETZER, M. A., HESSELHOLT, M., MOUSTGAARD, J. and THYMANN, M., 1964. Studies on protein polymorphism in pigs, horses and cattle. Blood Groups of Animals, Proc. 9 Europ. Anim. Blood Groups Conf, Prague, 1965, 279–293.

HESSELHOLT, M., 1966. Additional serum type alleles in pigs (In Danish, English summary), Aarsberetning. Inst. Sterilitetsforskn, Copenhagen, 75–83.

HESSELHOLT, M., LARSEN, B. and NIELSEN, P. B., 1966. Studies on serum amylase systems in swine, horses and cattle, *Aarsskr. K. Vet. Landbohøjsk*, 78–90.

HESSELHOLT, M. and MOUSTGAARD, J., 1965. Serum amylase polymorphism in cattle and its application in the parentage control (In Danish, English summary). Aarsberetn. Inst. Sterilitetsforsk, Copenhagen, 175–182.

HIRSCHFELD, J. 1960. Immunoelectrophoresis — Procedure and application to the study of group-specific variations in sera, *Sci. Tools*, **7**, 18–25.

IMLAH, P., 1964. A study of blood groups in pigs. Blood Groups of Animals, Proc. 9 Europ. Anim. Blood Groups Conf., Prague, 1965, 109–122.

LATNER, A. L., 1965. The binding of circulating enzymes by plasma proteins, West-Europ. Symp. on Clin. Chem, Vol. **5**, 121–127.

MC GEACHIN, R. L. and LEWIS, J. P., 1959. Electrophoretic behaviour of serum amylase, *J. Biol. Chem.*, **234**, 795–798.

POULIK, M. D. and SMITHIES, O., 1958. Comparison and combination of the starch-gel and filter-paper electrophoretic methods applied to human sera: Two-dimensional electrophoresis, *Biochem. J.*, **68**, 636–643.

SCHRÖFFEL, J. and HOJNÝ, J., 1962. β-globulin fractions in pig serum and their heredity (In Czech. Abstract), *Cs. fysiologie*, **11**, 277.

WILDING, P., 1963. Use of gel filtration in the study of human amylase, *Clin. Chim. Acta*, **8**, 918–924.

XITH EUROPEAN CONFERENCE ON ANIMAL BLOOD GROUPS AND BIOCHEMICAL POLYMORPHISM
Warsaw, July 2nd–6th, 1968

Electrophoretical Study of the Esterases in Pig Serum

A. KÚBEK

Laboratory of Physiology and Genetics of Animals, Czechoslovak Academy of Sciences, Liběchov Czechoslovakia

SUMMARY. Esterase in the blood serum of pigs was studied by means of starch gel electrophoresis. We found three regions with esterase activity. The second (middle) region is polymorphic, represented by three different phenotypes A, B and AB. We assume their genetic determination by two codominant alleles Es^A and Es^B.

INTRODUCTION

Genetic polymorphism of esterases has been demonstrated in different animals (SHAW, 1965). PELZER (1965) found, by means of starch gel electrophoresis, that the lysate of mouse erythrocytes also contains two types of acetylesterase, differing in electrophoretic mobility. He proved that the presence of acetylesterase was determined by two codominant alleles Ee^{1a} and Ee^{1b} from one autosomal locus. OKI *et al.* (1966) arrived at similar results when studying serum esterase in mice. During their study of esterase in rabbit erythrocytes GRUNDER *et al.* (1965) observed genetic differences and proved that esterase is determined by a pair of codominant alleles. Identical results were gained by SARTORE (1966) when he studied erythrocytary esterase in cattle. GAHNE (1966) found in horse blood plasma 6 different phenotypes of esterase and presumed their determination by 4 autosomal alleles, the first three being codominant and the fourth recessive. AUGUSTINSSON and OLSSON (1961) investigated blood plasma esterase in pigs. They found five phenotypes determined by an activity level of 0, 25, 50, 75, 100. Matings confirmed their hypothesis that arylesterase activity and its different phenotypic expressions in individuals were the product of a set of multiple alleles. This report presents the results gained by the study of qualitative differences of esterase in the blood serum of pigs.

MATERIAL AND METHOD

The serum of pigs was obtained by the usual method from the Large White, Cornwall, Black and White-Přeštice breeds from different breeding centers without exercising intentional choice. Samples were collected from pigs of different age (from 5 days to 5 years). The serum was stored at a temperature of −22°C. Prolonged storage, however, caused the decrease of esterase activity. It is best to use fresh serum for experiments. Investigations were carried out by means of starch gel electrophoresis (SMITHIES, 1955) in tris-borate buffer (KRISTJANSSON, 1963; POULIK, 1957; OKI *et al.*, 1966). Electrophoresis was performed either horizontally without cooling for 6 hr, or in a vertical apparatus with water cooling for 8–10 hr at 8–14 V/cm. We used our own hydrolysed starch at a concentration of 14.5 g/100 ml buffer. After electrophoresis, the gel was sliced into two parts. One half was stained with Amido-Black 10B for proteins and the other half for esterases with a staining solution composed of: 100 ml phosphate buffer, pH 7.4; 50 mg Fast Blue B Salt (Diazo Blue B Salt); 2.5 ml 1% solution of β-naphtyl acetate in 50% acetone. The phosphate buffer is composed of 70 ml 0.15 M Na_2HPO_4 and of 30 ml 0.15 M KH_2PO_4.

This solution is a slight modification of LAWRENCE's *et al.* (1960) original staining solution. Staining took from 30 min to 1 hr at laboratory temperature till esterase zones reached the required intensity. The staining solution was then discarded and the gel washed with distilled water. The coloured background was washed off and the zones fixed by means of a washing solution consisting of methanol, H_2O and CH_3COOH at a ratio of 5:5:1.

RESULTS AND DISCUSSION

Serum esterase of pigs may be observed in the gel in three regions. The first region migrates fastest in the electrophoretic field and it comprises two fractions with relatively low esterase activity. Their position corresponds to that of serum albumin. In the second region (Figs. 1 and 2) we can observe 4 fractions which form a configuration of certain types. Type A is the fastest in this region and has 3 fractions, two weak ones and 1 strong one. Type B has the slowest fractions and these are also 3 in number. The heterozygous manifestation of these types is represented by 4 fractions, two being stronger than the other two. These fractions are localized between the transferrin and post-albumin regions. By combined staining we have proved that these fractions are independent of transferrin or postalbumin fractions.

Esterase fractions in this region exhibit different activity so that with higher esterase activity the fractions diffuse and form blotches. Precise reading is then very difficult.

Family studies gave the following results:

The above mentioned results of phenotypical expressions of esterase in family studies suggest that esterase in this region is determined by a pair of codominant alleles (Es^A and Es^B). The distribution of phenotypes in the set of 136 non-related animals disclosed a frequency of:

$$q^A = 0.4817,$$
$$q^B = 0.5183.$$

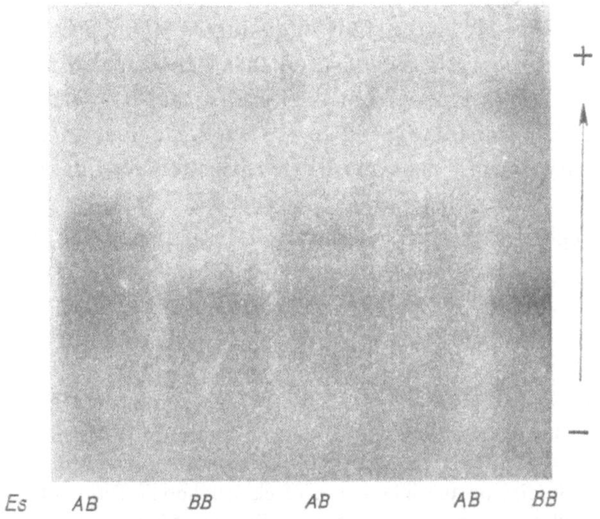

FIG. 1. Starch gel electrophoretic patterns of pig serum esterase (Es) in region II, using KRISTJANSSON's buffer system.

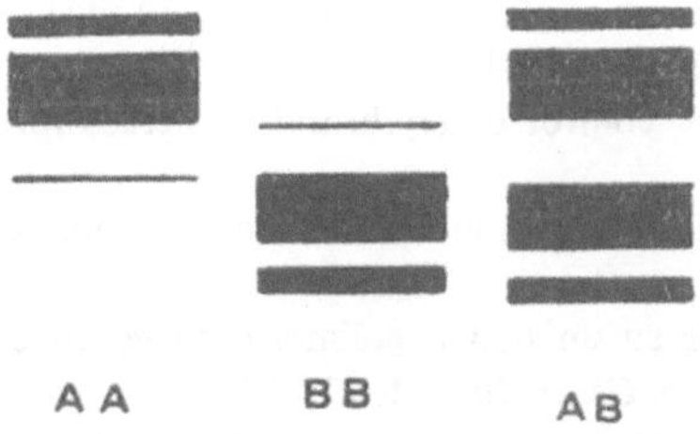

FIG. 2. Scheme of polymorphic phenotypes of pig serum esterase in region II.

Family study of I region of postalbumins in pig serum

Type of mating	No. of matings	No. of progeny	A		B		AB	
			obs.	exp.	obs.	exp.	obs.	exp.
AA×AA	2	8	8	8	—	—	—	—
AA×AB	1	4	3	2	—	—	1	2
AA×BB	1	6	—	—	—	—	6	6
AB×AB	5	22	7	5.5	3	5.5	12	11
AB×BB	4	17	—	—	5	8.5	12	8.5
BB×BB	1	4	—	—	4	4	—	—
Total	14	61	18	15.5	12	18	31	27.5

The distribution of phenotypes was in agreement with Hardy–Weinberg law.

The results obtained from the study of qualitative differences in esterases by means of starch gel electrophoresis demonstrate that the second esterase region is polymorphic. It may be presumed from the results that it is determined by two co-dominant alleles Es^A and Es^B. Owing to the small number of investigated animals, it will be necessary to carry out a more extensive study for final proof of our genetic hypothesis.

The third region with esterase activity is situated in the $S\alpha_2$ globulin region. There are always two fractions showing this electrophoretic mobility.

REFERENCES

AUGUSTINSSON, K. B. and OLSSON, B., 1961. Genetic control of arylesterase in the pig, *Hereditas*, **47**, 1–22.

GAHNE, B., 1966. Studies on the inheritance of electrophoretic forms of transferrins, albumins, pre-albumins and plasma esterases of horses, *Genetics*, **53**, 681–694.

GRUNDER, A. A., SARTORE, G. and STORMONT, C., 1965. Genetic variation in red cell esterases of rabbits, *Genetics*, **52**, 1345–1353.

KRISTJANSSON, F. K., 1963. Genetic control of two pre-albumins in pig, *Genetics*, **48**, 1059–1063.

LAWRENCE, S. H., MELNICK, P. S. and WEIMER, H.E., 1960. A species comparison of serum proteins and enzymes by starch gel electrophoresis, *Proc. Soc. Exp. Biol. Med.*, **105**, 572–575.

OKI, Y., TAKEDA, M. and NISHIDA, S., 1966. Genetic and physiological variations of esterases in mouse serum, *Nature*, **212**, 1390–1391.

PELZER, C. F., 1965. Genetic control of erythrocytic esterase forms in Mus musculus, *Genetics*, **52**, 819–828.

POULIK, M. D., 1957. Starch gel electrophoresis in a discontinuous system of buffers, *Nature*, **180**, 1477–1479.

SARTORE, G., 1966. Ricerche su un nuovo polimorfismo genetico riguardante una esterasi degli eritrociti bovini, *Atti Assoc. Genct. Ital.* **11**, 217–222.

SHAW, C. R., 1965. Electrophoretic variation in 8 enzymes, *Science*, **149**, 936–943.

SMITHIES, O., 1955. Zone electrophoresis in starch gels. Group variations in the serum protein of normal human adult, *Biochem. J.*, **61**, 629–641.

XITH EUROPEAN CONFERENCE ON ANIMAL BLOOD GROUPS AND BIOCHEMICAL POLYMORPHISM
Warsaw, July 2nd–6th, 1968

Results of the Third Pig Blood Grouping Comparison Test

H. DINKLAGE

Institute of Animal Husbandry and Animal Genetics, University of Göttingen, Göttingen, G.F.R.

The Third Pig Blood Grouping Comparison Test was organized by the Institute of Animal Husbandry and Animal Genetics of the University of Göttingen. Eighteen laboratories participated in this test, twelve of these investigated erythrocytes and serum, five studied only erythrocytes and one tested only serum.

The number of *blood group reagents* which have been reported totals nearly 130, of which 50 have been genetically systematized. Of the remaining 80, which are listed under the notation of the laboratory concerned, there are only four blood group reagents which are completely identical in at least two laboratories; all the other reaction patterns differ from one another. Judging from the test records, there is no laboratory possessing antisera for each of the 16 blood group systems listed. In endeavouring to come to a clear-cut and final coordination of the newly-determinable blood group factors, serum samples must be exchanged between interested laboratories. The exchange of serum samples could be performed in such a way that the laboratory possessing a test serum already identified receives about the same quantity of a test serum not yet identified. This would have the advantage that all the necessary investigations could be carried out in two laboratories independently of each other.

As a result of the great number of still unidentified blood group factors, it is to be expected that in the near future new and final genetic notations will be achieved. In order to avoid the possibility of two or more different blood group reagents being designated identically or that of identical test sera being designated differently, it would be of great advantage if each laboratory which participated in the last comparison test were kept informed of the introduction of new designations. In order to prevent confusion in nomenclature, it would seem advisable also to proceed in the same way in the case of a publication.

On the basis of the test records sent in and on the basis of data found in the literature, the identified blood group systems, factors and alleles are listed in Table 1. However, factors questioned by the participants in the Third Comparison Test or diverging completely from the previously accepted system of nomenclature are not been listed.

TABLE 1

Blood group systems in the pig

Blood group systems	Blood group factors	Alleles
A	A(Ac, Ap), O	A^A, a^o
S	A(Ac, Ap), O	S, s
B	Ba, Bb	B^a, B^b
C	Ca	C^a, C^-
D	Da	D^a, D^-
E	Ea, Eb, Ed, Ee, Ef, Eg, Eh, Ei	E^{bdg}, E^{edgh}, E^{aeg}, E^{efdh}, E^{bdf}, E^{aef}, E^{eg}, E^{aegi}
F	Fa, (Fb)	F^a, (F^b)
G	Ga, Gb	G^a, G^b
H	Ha, Hb, Hc	H^a, H^b, H^{ab}, H^c, H^-
I	Ia, Ib	I^a, I^b
J	Ja	J^a, J^-
K	Ka, Kb, Kc, Kd, Ke	K^{ace}, K^{ade}, K^{ac}, K^b, K^-
L	La, Lb, Lc, Ld, Lf, Lg, Lh, Li, Lj, Lk, Ll	L^{adhi}, L^{bcgi}, L^{bdfi}, L^{adhjk}, L^{adhjl}, L^{agi}
M	Ma, Mb, Mc, Md	M^a, M^b, M^c, M^d, M^{ab}, M^{bc}, M^{cd}, M^-
N	Na, Nb, Nc	N^a, N^b, N^{bc}
O	Oa, Ob	O^a, O^b

In addition, more than 70 non-systematized blood group reagents are recognized.

There is some overlapping in the notation of some of the blood group factors already identified. This is particularly true in the case of the C, H, K, L and M systems.

As has been mentioned in the comments on the Third Comparison Test, the Go_5–D/Gö monospecific antiserum always reacts when the allele E^{aeg} occurs in homozygous form. As this serum may be considered to be a dosage serum, no notation has been given to it. If no other suggestions are made, this factor will be given the notation E_j.

By means of systematic absorption trials it was found that the Go_{10}–D/Gö, Go_{11}–D/Gö and Go_{12}–D/Gö factors belong to the M system. It seems advisable to postpone a final notation for these newly determined factors until the relationship between them and the factors identified for the M system are elucidated. It may then turn out that the M system is similar in complexity to the E, K and L systems.

Apart from the increase in the number of newly-determinable blood group factors, there was also an increase in the number of *systems* detectable by means of *starch gel electrophoresis*. As in the case of blood group factors there is also some confusion in the notation of several systems and alleles. This is particularly true as far as the hemopexin and amylase systems are concerned (Table 2).

As a result of the rapid expansion of knowledge of blood components resulting from the use of the electrophoresis technique, the term "serum protein system" is no longer sufficient for our requirements. For example, genetically controlled systems have also been detected in red blood cells. As is shown in Table 2, the transcription "systems detectable by means of starch gel electrophoresis" is not satisfactory, because it is too detailed and laborious. It is suggested that the term "serum protein system" be maintained only for proteins which are found in the serum. Following the pattern of previously accepted nomenclature, e.g. "milk protein system" or "serum protein system", the term "red cell protein system" is suggested for the proteins found in the red cells.

TABLE 2

Systems detectable by means of starch gel electrophoresis

System name	Locus symbol	Gene symbols
Transferrin	Tf	A, B, C, D,
Pre-albumin	Pa (Pra)	A, B
Albumin	Alb_1	A, B, O
Hemopexin (hematin-binding β-globulin, haptoglobin)	Hp (Hpx, Hx, Hg)	O, 1, 1F (4), 2, 3, 3F
Amylase	Am	1, 2, 2F(2′), 3; or A, B, BF(B′), C
Ceruloplasmin	Cp	1, 2; or A, B
s α_2-globulin	S α_2	A, B, C
6-phosphogluconic dehydrogenase	6-PGD	A, B
Alkaline phosphatase	Akp	

Last but not least I should like to thank again all those who participated in this comparison test for their valuable cooperation and I sincerely hope that in our special field of blood group research and genetics, knowledge may be gained by further harmonious cooperation in the future.

CHAPTER 3

BLOOD GROUPS AND BIOCHEMICAL POLYMORPHISM IN POULTRY

XITH EUROPEAN CONFERENCE ON ANIMAL BLOOD GROUPS AND BIOCHEMICAL POLYMORPHISM
Warsaw, July 2nd–6th, 1968

The Influence of the Trace Elements Manganese, Cobalt and Iodine on Immunobiological Reactions in the Fowl

Lidia ERMENKOVA and K. ERMENKOV

Institute of Animal Breeding, Konstinbrod, G. Dimitrov Higher Institute of Agriculture, Sofia, Bulgaria

Numerous investigations in the last few years have been directed to elucidating the biological effect of trace elements in animals. It is now known that the trace elements Mn, Co and I, which activate or suppress some enzyme systems, hormones, and vitamins exert definite influence on metabolism and on a number of physiological processes (BERENSHGEIN, 1958; VOINAR, 1960; ROSENBAKH, 1954; DEVUYST *et al.*, 1956, 1960, 1961). Their role in the growth and development of the organism (MOMINOV, 1962), productivity (LEONOV *et al.*, 1963), haemopoietic processes (BERENSHGEIN, 1958; VOINAR, 1960; ROSENBAKH, 1954; DEVUYST *et al.*, 1956, 1960, 1961), tissue respiration (VOINAR, 1960; PEIVE, 1960; DEVUYST, 1956, 1960, 1961), etc. has been studied. There is a relative paucity of investigations concerning the influence of trace elements, Mn, Co and I included, on the build up of immunobiological reactions in the normal animal.

KOSOBRUHOV (1963) found that the feeding of Cu, Co and Mn to animals enhances the phagocytic activity of leukocytes. ERMENKOV's (1961) studies show that the amount of globulins in the blood serum, and of the gamma globulin fraction in particular, rises under the influence of the trace elements Co, Cu, Mn and I. The investigations of ALTUHOVA (1963) and AISINBUDAS *et al.* (1962) concerning the effect of some trace elements on the dynamics of titration variations are of interest. Of particular importance is the work of BABENKO *et al.* (1965) who has studied the effect of varying doses of Co, Cu, Mn and I on the build up of immune antibodies after the injection of *bacterium coli* and *bacterium dysenteriae Sonne* in rabbits.

The available literature shows a paucity of investigations about the influence of trace elements on antibody formation and reaction of the organism in normal immunogenesis. This research was undertaken for this reason and with a view to studying the influence of optimal Mn, Co and I doses an antibody formation, the reaction of the animal in normal immunogenesis and the possibility of maternal passage of antibodies from the hen to the eggs.

MATERIALS AND METHODS

Two series of experiments were conducted:

First series — with 25 laying hens of the New Hampshire breed, divided into 5 groups by liveweight and productivity: group I — control, group II — manganese (addition of 5.46 mg Mn per head per day to the basic ration), group III — cobalt (0.48 mg Co per head per day), group IV — iodine (2.28 mg I per head per day) and group V — mixed (5.46 mg Mn $\times$ 0.48 mg Co + 2.28 I per head per day).

Second series — with 15 cocks of the Leghorn breed, divided into 3 groups by liveweight: group I — control, group II — iodine (2.28 mg I per head per day), group III — mixed (received the same trace element doses as the hens of group V — (first series).

The birds in both experiments were similarly managed and received rations similar in composition.

In the course of 10 days prior to immunization and during immunization the birds in the experimental groups were treated in parallel with the specified quantities of trace elements.

Washed erythrocytes of cock donors of the same breed served as the antigen for immunization. The first injection of 2 mg erythrocytes was intramuscular and the following four — intravenous, every 2 days, with the same amount of erythrocyte suspension.

Washed erythrocytes but only from one cock donor of the same breed were also used as antigen in the immunization of cocks.

RESULTS

Occurrence, development and disappearance of antibodies in the blood serum and in the egg white

The data show that natural antibodies were not present in the experimental birds prior to immunization. Antibody formation began at almost the same time in all groups in both experiments — from the 6th to the 8th day after the first immunization — and gradually increased during succeeding days. The antibody titre reached 1:64 on the 14th day. The latent period of build up of antibodies under the conditions of our experiment almost completely agrees with the data of some other authors (Dixon *et al.*, 1954 and others). During the immunization (from the 1st to the 14th day) no special differences were found between the individual groups. Characteristic differences in the titre variations were observed on the 8th day after the last immunization, the highest titre being found in hens receiving a mixture of trace elements (Mn, Co and I) — from 1:1024 to 1:2048. Raised antibody titre was also found in hens of the iodine group — from 1:512 to 1:1024. No differences in the antibody titre were observed in the remaining groups (manganese and cobalt)

TABLE 1

Formation of erythroantibodies

Days \ Group	Control	Mn	Co	I	Mn, Co, I
			HENS		
			Prior to immunization		
	—	—	—	—	—
			During immunization		
8th	1:4–1:8	1:2–1:4	1:2–1:4	1:4–1:8	1:4–1:8
11th	1:16–1:32	1:8–1:16	1:16–1:32	1:16–1:32	1:16
14th	1:32–1:64	1:16–1:32	1:32–1:64	1:64	1:64
			After immunization		
8th	1:256–1:512	1:256–1:512	1:256–1:512	1:512–1:1024	1:1024–1:2048
15th	1:128–1:256	1:128–1:256	1:128–1:256	1:128–1:256	1:256–1:512
24th	1:64–1:128	1:16–1:64	1:32–1:64	1:64–1:128	1:128–1:256
30th	1:2	1:1	1:4–1:8	1:16–1:32	1:16
40th	—	—	—	1:2	1:2
			COCKS		
			Prior to immunization		
	—	—	—	—	—
			During immunization		
8th	1:4	—	—	1:2	1:8
11th	1:4–1:16	—	—	1:4–1:8	1:16
14th	1:8–1:32	—	—	1:8–1:32	1:64
			After immunization		
8th	1:32–1:128	—	—	1:16–1:128	1:64–1:512
14th	1:16–1:64	—	—	1:16–1:64	1:32–1:128
24th	1:4 –1:32	—	—	1:4 –1:16	1:8 –1:64
30th	1:4 –1:8	—	—	1:4 –1:8	1:8 –1:32
40th	1:1 –1:2	—	—	—	1:2 –1:8

compared with the control. Highest antibody titres were also observed in the cocks receiving trace element mixtures but these were lower than comparable titres of hens. This in all likelihood was due to the fact that the cocks had a single donor. In the succeeding observations the titre generally declined for all groups of both experiments. Hens, given a mixture of trace elements, maintained a relatively high antibody titre on the 15th and 24th day after the last injection. This retention of antibodies of raised titre for a prolonged period of time is of value in creating extra time in which to obtain immune sera from small animals and birds. The rapid drop of the antibody titre in the experimental hens on the 30th–40th day after the last erythrocyte injection might possibly be due to the comparatively short period of supplemental feeding of the hens with trace elements.

Our investigations showed that passage of antibody from the hen to its egg started earlier in the period of antibody formation (8th–10th day) was particularly evident in chickens receiving a supplemental mixture of trace elements and iodine, in which this process begins slightly earlier — on the 6th day of the first injection. Eggs, containing antibodies, increased at the end of immunization and after its termination, i.e. a definite regularity was observed between the amount of antibody present in the blood serum and their presence in the egg white.

Modifications in some blood components

Prior to the commencement of the experiment the number of erythrocytes found in the hens of the various groups was almost the same, with the exception of the manganese group, the numbers found in this group being comparatively fewer. The trace element supplement to the basic ration of the experimental group — 10 days before immunization and during the course of the immunization itself — resulted in increased erythrocyte numbers. The hemoglobin content, however, showed no particular change. The augmentation was greatest in hens of group V receiving the trace element mixture. The same tendencies were observed by RYBINA *et al.* (1962) in laying hens, fed Mn, Co and I. The numbers of erythrocytes and hemoglobin content decreased following the discontinuation of the trace element supplement to the ration. Increase in the erythrocyte count was also observed in hens of the control group as a result of immunization. The hemoglobin level, however, remained stable.

TABLE 2

Erythrocyte number, thousands

Days \ Group	Control	Mn	Co	I	Mn, Co, I
		Prior to trace element addition			
	2.750	2.270	2.510	2.610	2.700
		Prior to immunization			
	2.890	3.110	3.100	3.080	3.320
		During immunization			
8th	2.730	2.970	3.120	2.870	3.420
14th	3.020	3.260	3.560	3.460	3.850
		After immunization			
8th	2.940	3.220	3.210	3.210	3.970
15th	2.780	2.870	2.970	3.060	3.040
24th	2.790	2.940	3.120	2.880	3.040
30th	2.640	2.710	2.840	2.710	2.780
40th	2.290	2.290	2.430	2.450	2.560

Certain changes also took place in the quantity of blood glutathione under the influence of trace elements and immunization. Glutathione content was highest in hens of the iodine group and in the group receiving a trace element mixture. Glutathione increase occurred also in the hens of the control group. A glutathione decline, however, took place in all groups following cessation of injections and by the 40th day of the last injection its level was almost the initial one.

TABLE 3

Hemoglobin content

Days \ Group	Control	Mn	Co	I	Mn, Co, I
		Prior to trace element addition			
	54	54	54	54	56
		Prior to immunization			
	55	54	58	56	60
		During immunization			
8th	55	54	59	55	60
14th	54	56	59	56	60
		After immunization			
8th	54	56	59	55	60
15th	56	57	55	58	60
24th	56	54	57	55	57
30th	55	55	55	56	58
40th	55	55	51	54	60

The rise in glutathione shows that the build up of antibodies in an animal is attended by heightened oxidation–reduction processes. Existing data indicate that glutathione plays the role of a specific co-enzyme in the glyoxalase system and in some other systems. According to HANES *et al.* (1952) glutathione plays a special part in peptide synthesis. It may be surmised that in this case its increased content is connected with the build up of antibodies.

Data on the total blood serum protein showed that no special changes took place in individual groups during the experiment. In comparison with the control group, the group given a trace element mixture, had a slightly higher protein content at the end of immunization. Certain, though not very substantial changes, occurred in the amount of the globulin fractions. This increase was probably related to the heightened process of antibody formation in some of the groups. It might be assumed that all globulin fractions participate in the build up of antibodies. The ground for this assumption is the fact established by RAPP (1956), that the antireticular serum cytotoxic for the dog, obtained by rabbit immunization, is connected with the β-globulin fraction. This question will be the subject matter of our future investigations.

TABLE 4

Glutathione amount

Group / Days	Control	Mn	Co	I	Mn, Co, I
			Prior to immunization		
Total	50.03	50.96	55.87	57.71	57.71
Reduced	48.20	50.26	52.80	56.19	55.87
Oxidized	1.83	0.70	3.07	1.52	1.84
			During immunization		
8th					
Total	60.63	59.86	59.56	69.07	72.45
Reduced	58.33	56.39	58.02	63.47	68.77
Oxidized	2.30	3.47	1.54	5.60	3.68
14th					
Total	63.85	61.40	60.17	67.10	64.47
Reduced	62.32	59.55	58.75	64.95	62.93
Oxidized	1.54	1.85	1.42	2.15	1.54
			After immunization		
8th					
Total	54.03	52.80	54.64	52.49	66.92
Reduced	47.18	47.16	51.57	47.28	57.10
Oxidized	6.85	5.64	3.07	5.21	9.82
15th					
Total	52.80	53.43	50.06	51.88	63.85
Reduced	49.12	48.13	49.15	50.47	59.86
Oxidized	3.68	5.30	0.90	1.41	3.99
30th					
Total	49.12	57.10	54.03	49.11	62.63
Reduced	44.82	52.19	50.67	42.21	56.79
Oxidized	4.30	4.91	3.36	6.90	5.84
40th					
Total	53.42	52.19	51.57	49.10	57.71
Reduced	50.97	45.09	47.27	44.24	49.10
Oxidized	2.45	7.10	4.30	4.86	8.61

CONCLUSIONS

Our investigations on the influence of the trace elements Mn, Co and I on the build up of antibodies and the reaction of the animal during the period of immunization warrant the following inferences:

(1) The supplement of a trace elements mixture (5.46 mg Mn+0.46 mg Co+2.28 mg I per head per day) to the basic ration for cocks and laying hens exerts a rather positive effect on the formation of antibodies and on their retention in the animal for a certain period of time. The addition of iodine alone (2.28 mg per head

per day) to the feed also stimulates antibody formation but to a relatively lesser degree.

(2) Immunization of hens and the addition of trace elements to the food is attended by increase in the numbers of erythrocytes and in blood glutathione. Hemoglobin, however, shows no particular change in response to these stimuli.

(3) Immunization and the addition of trace elements to the food induces no essential changes in the total protein content and in the relative amount of protein fractions in the blood serum of hens.

REFERENCES

AISINBUDAS, L. B. *et al.*, 1962. O nekatorykh fiziologicheskikh metodakh otsenki effektivnosti podkormki mikroelementami (On some physiological methods of evaluating the effectiveness of trace elements supplement). Mikroelementi v zhivotnovodstve, 102–106.

ALTUHOVA, D., 1963. Izmenenie immunobiologicheskoy reaktivnosti organizma pod vliyaniem nekotorykh mikroelementov (Changes in the immunobiologic reactivity of the organism under the influence of some trace elements). Sb. Trudov IV Vses. soveshchanie po voprosam primenenia mikroelementov v sel'skom khozyaystve i meditşine, 621–624.

BABENKO, G. A. *et al.*, 1965. Vliyanie mikroelementov na immunobiologicheskie svoystva organizma (Effect of trace elements on the immunobiological properties of the organism). Mikroelementi v zhivotnovodstve i meditsine, 142–148.

BERENSHGEIN, F. F., 1958. Mikroelementi, ikh biologicheskaia rol' i znaçhenie dlia zhivotnovodstva (Trace elements, their biologic role and importance for animal husbandry). GIBSSR, Minsk.

BERLIN, N., 1951. Cobalt absorption and elimination, *Amer. J. Physiol.*, **164**, 1, 221.

DEVUYST, A. *et al.*, 1956. La manganèse en nutrition animale, sa teneur dans les tourteaux et dans les aliments de la ferme en campine, *Agricultura*, VII, 4. 543–590.

DEVUYST, A. *et al.*, 1960. La signification du cobalt et de la vitamine B_{12} en alimentation animal, *Agricultura*, VIII, 3, 509–551.

DEVUYST, A. *et al.*, 1961. La signification biologique de l'iode et son importance en nutrition animal, *Agriculture* IX, 2, 3, 407–449.

DIXON, F. F. *et al.*, 1954. Primary and specific anamnestic antibody response of rabbits to heterologous serum protein antigens, *J. Immunol.*, **72**, 2, 179.

ERMENKOV, K., 1961. Study of some physiological and biochemical indicators in calves in connection with the feeding of trace elements Co, Cu, Mn and I (In Bulgarian). -Nauč. trudove *Visš. selsko-stop. inst. G. Dimitrov, zootehn. fak.*, XI, 371–382.

HANES, C. S. *et al.*, 1952. Enzymic transpeptidation reactions involving glutamyl peptides and amino-acyl peptides, *Biochem. J.*, **51**, 25.

KOSOBRUHOV, A. N., 1963. Izmenenie immunobiologicheskikh svoystv organizma zhivotnykh pod vliyaniem mikroelementov medi, kobal'ta, margantsa (Changes in the immunobiological properties of the organism of animals under the influence of the trace elements copper, cobalt, manganese), Sb trudov IV Vses. soveshchanie po voprosam primenenia mikroelementov v sel'skom khozyaystve i meditsine, 503–504.

LEONOV, V. A. *et al.*, 1963. Obogashchenie produktov zhivotnovodstva iodom i kobaltom (Enrichment of products of animal origin with iodine and cobalt), Sb. trudov IV Vses. soveshchanie po voprosam primeneniya mikroelementov v sel'skom khozyaystve i meditsine, 499–502.

MOMINOV, L. S., 1962. K voprosu o zavisimosti rosta i razvitiya podsosnykh porosiat ot soderzhaniya mikroelementov v kormakh (On the problem of the dependence of gain and development of suckling pigs on the trace element content in the feed), Trudy novocherk. zootekhn. vet. inst., issue 14.

PEIVE, Ya. V., 1960. Mikroelementy i fermenty (Trace elements and ferments), Akad. Nauk Latv. SSR, Riga, 51–82.

RAPP, U. V., 1956. Ob ochistke i kontsentratsii ACS (On the purification and concentration of ACS), Tsitotoksiny v sovremennoy meditsine, Kiev, 270–274.

RYBINA, E. V. *et al.*, 1962. *Trudy nauchno-issled. Inst. zhivotnov. Uzbek. SSR*, issue 7.

VOINAR, A. I., 1960. Biologicheskaia rol' mikroelementov v organizme zhivotnykh i cheloveka (The biological role of trace elements in the organism of animals and man), GI Visshaya shkola, Moscow.

ROSENBÁKH, YA. YA., 1954. Vliyanie solei Co, Mn, Cu, Zn na iaitsenosnost' kur i inkubatsionnye kachestva iaits (Effect of the salts of Co, Mn, Cu, Zn on hen egg laying and incubation properties of the eggs), *Novosti AN Latv. SSR*, **4**, 55.

XITH EUROPEAN CONFERENCE ON ANIMAL BLOOD GROUPS AND BIOCHEMICAL POLYMORPHISM
Warsaw, July 2nd–6th, 1968

Abnormal Segregation for Blood Group Genes in Domestic Birds

A. PERRAMON and P. MERAT

with the technical collaboration of
J. L. MONVOISIN

Station Centrale de Génétique animale, C.N.R.Z. 78 Jouy-en-Josas, France

In two fowl strains, one of the Wyandotte, the other of the Rhode-Island breed three blood group loci have been identified (PERRAMON and MERAT, 1968).

In the Wyandotte, the first locus has three alleles (W^2, W^3, W^5) and the second two (W^7, W^8). At the third one at least two alleles are present (W^6, W^0 meaning the absence of the "6" antigen); its serological reactions are consistently weaker than those of the first two. Loci II and III are linked, with a recombination rate near 9%.

In the Rhode-Island strain, we identified one locus with 5 alleles (R^1, R^2, R^7, R^8, R^9), another with at least 2 alleles (R^4, "R^0"). The third one has at least three alleles (R^5, R^6, "R^0"). These last two loci are linked with a recombination rate of about 5%.

Having previously observed (MERAT, 1968) certain abnormal mendelian proportions in the segregation of loci with visible effect (colour or morphology), we examined also these proportions for the blood group genes described here.

A highly significant heterogeneity of segregation ratios between heterozygous dams was observed at loci W^6/W^0 and R^4/R^0 respectively for the two strains (W^6 representing an allele and W^0 its absence, and the same for R^4 and R^0) in crosses ♂ $W^0/W^0 \times$ ♀ W^6/W^0 and ♂ $R^0/R^0 \times$ ♀ R^4/R^0.

In the first case especially, the heterogeneity χ^2 between dams within sire amounts to 67.9 for 36 degrees of freedom ($P < 0.001$) for the total of the two years 1965 and 1966. A significant heterogeneity is found also at locus W^7/W^8 linked to the former ("between dams within sires" $\chi^2 = 76.5$ for 50 d.f., $P < 0.01$). In reciprocal crosses the sire being heterozygous, one does not find a significant heterogeneity.

On the same Wyandotte strain, in 1967, we could not type all birds for the W^6 antigen, but, for the linked gene W^7/W^8, mendelian proportions in the progeny of heterozygous dams are, in a similar way, more variable than expected ($\chi^2 = 166.1$ for 93 d.f., $P < 0.001$).

An occasional lack of reactivity of anti-W^6 and anti-R^4 sera is not likely, the overall proportion being normal, with an excess of the corresponding allele in some families. A difference of embryonic mortality associated with these loci would have to be of variable direction in different families; anyway, it is rendered even more unlikely by a comparison of hatching rates, which turn out to be as good for dams with an "abnormal" ratio as for the other ones (respectively 66.95% vs. 65.00%). This suggests an influence on mendelian ratios taking place before the female meiosis. Moreover, an excess of apparent "recombinants" with the linked locus, in both strains, appears in families with an "abnormal" segregation ratio. For the W^6/W^0 locus, the progeny of the corresponding dams comprises 36 "recombinants" on a total number of 150 (24.0%), compared to 32 on a total of 414 (7.7%) among the progeny of dams not significantly deviating from the 1:1 ratio. The difference between these two "recombinants" rates is significant at the 1 per 1000 level.

This seems to confirm that the explanation of the abnormal ratios can be found neither in a differential mortality, nor in an incomplete penetrance and suggests rather a "mitotic recombination"-type mechanism.

Seemingly analogous results were obtained on quails. In a family with the sire homozygous —/— for the absence of an antigen (corresponding to serum 3722, with a 1/64 titer) and the dam heterozygous +/— among 26 progeny, only 8 had this antigen (χ^2 amounting to 3.84, $P \simeq 0.05$). This would suggest a systematic lack of the relevant allele in this family. Moreover, the shortage seems to vary in time: the chicks being ranked according to the chronological laying order of the egg from which they were hatched, a test of the number of homogeneous sequences

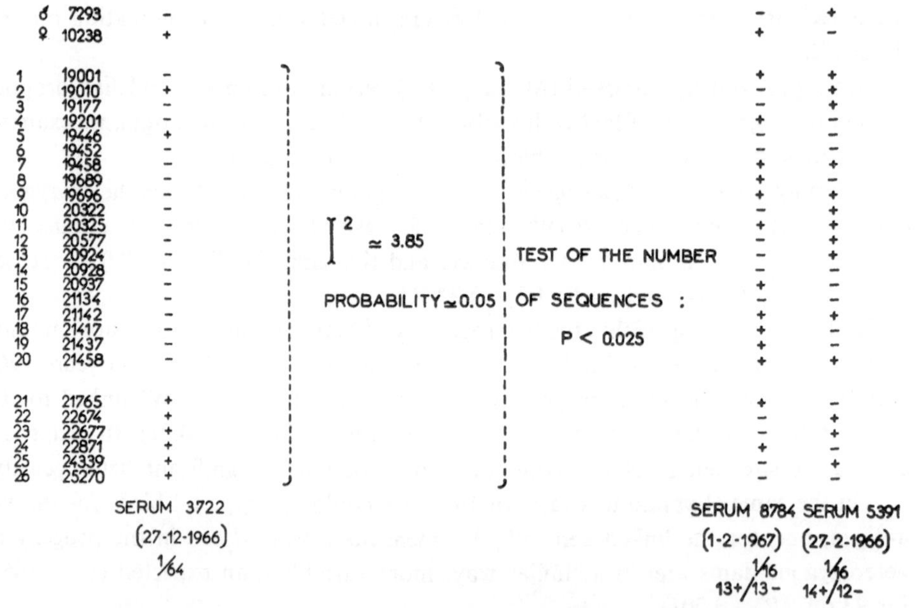

Abnormal segregation for an autosomal blood group gene in the Japanese quail.

for presence or absence of the antigen (MORICE and CHARTIER, 1954) reveals that this number is significantly inferior to the expectation ($P < 0.025$). As a comparison, in the same family, two other genes, for which, respectively, the dam and the sire were heterozygous, the other mate being recessive, gave the proportion of 13 progeny over a total of 26 and 14 over 26.

In the lack of another likely explanation, we are founded to suppose that non-random variation of the segregation ratio could take place at or before meiosis, for some dams.

We cannot, at the present time, discriminate between various hypotheses. One of them is preferential expulsion of an allele in the polar body, but it does not appear to be supported by our data on linked genes.

Another possibility should be "genic conversions", which we cannot prove in this sort of data but which was found in various species (e.g. RENNER, 1942; COE, 1959).

There is, finally, the possibility of mitotic crossing-over in the female germ line (STERN, 1936; PONTECORVO, 1958) or of any mechanism leading to mosaic gonads. This has been strongly suggested in higher vertebrates (GRÜNEBERG, 1966).

Such phenomena could possibly have some implications on the genetic evolution of a population, although MAYR (1963) states that long term evolutive consequences are probably unimportant.

REFERENCES

COE, E. H. JR., 1959. A regular and continuing conversion-type phenomenon at the B locus in maize, *Proc. Nat. Acad. Sci. U. S.*, **45**, 828–832.

GRÜNEBERG, H., 1966. The case for somatic crossing-over in the mouse. *Genet. Res.*, **7**, 58–75.

MAYR, E., 1963. Animal species and evolution, Harvard University Press.

MERAT, P., 1968. Hétérogénéité des proportions mendéliennes à plusieurs loci chez la poule lorsque la mère est hétérozygote, *Annls. Biol. Anim. Bioch. Biophys.*, **8**, 95–98.

MORICE, E. and CHARTIER, F., 1954. Méthodes statistiques, Imprimerie Nationale, Paris.

PERRAMON, A. and MERAT, P., 1968. Etude génétique des groupes sanguins dans deux populations de volailles. I. Transmission héréditaire des facteurs antigéniques, *Annls. Biol. Anim. Bioch. Biophys.* (In press).

PONTECORVO, G., 1958. Trends in genetic analysis, Columbia University Press, New-York, pp. x+145.

RENNER, O., 1942. Über den Erbgang des cruciata-merkmals der Oenotheren. III. Weitere Belge für somatische Konversionen, *Z. Vererb. Lehre*, **80**, 570–589.

STERN, C., 1936. Somatic crossing-over and segregation in *Drosophile melanogaster*, *Genetics*, **21**, 625–730.

XITH EUROPEAN CONFERENCE ON ANIMAL BLOOD GROUPS AND BIOCHEMICAL POLYMORPHISM
Warsaw, July 2nd–6th, 1968

Reaction Patterns of Monofactorial Reagents in Various Breeds and Lines of Chickens

M. PAPP

Blood Grouping Laboratory, University of Veterinary Science, Budapest, Hungary

SUMMARY. Investigations in various breeds and lines were carried out to try to solve the problem of the monofactorial reagents in chicken blood typing. Great attention was paid to the accidental non-specific cross-reactions.

The fact, that two or more identical reagents reacting against one and the same antigen, could be produced from various polivalent sera, proved to be the best evidence for the monospecificity. These sera or rather reagents, belonging to the A, E and B systems gave practically the same reaction patterns not only with the panel of control cells but also with those of birds belonging to various lines and breeds. These results — together with the encouraging data of the European comparative tests hitherto performed — suggest more intense continuation of international cooperation for selection of reference reagents in chickens.

INTRODUCTION

During studies carried out on different lines and breeds of chicken, it became clear, that an antigen which appeared simple within one line, might appear more complex in another one. In other words, reagents which show single allele specificity in one population may cross-react with several antigens of the same blood group system in another population (FANGUY *et al.*, 1961). These reactions are thought to be caused by similarities between the physical structure of antigens (FANGUY, 1961). According to another theory an antigenic product of an allele is considered as a particular antigen of high serological complexity (BRILES *et al.*, 1963); OOSTERLEE and BOUW (1966) also called the attention to the great difficulties of the selection of reference reagents.

The purpose of the present work was to investigate, whether monofactorial reagents can or cannot be used in various breeds and lines. In other words, is it really necessary to prepare separate antisera for every line of chicken?

METHODS AND MATERIALS

Our technique of producing reagents is the same as has been described by BRILES *et al.* (1950, 1962, 1963) and FANGUY (1961). All reagents were produced in the White Leghorn line L3 at Rákoshegy.

Absorptions were continued until the antisera proved to be undoubtedly monofactorial — at least according to the panel cells. In some cases, reagents giving the same reaction pattern could be produced from different recipients. Such antisera are supposed to react against one and the same antigen.

They are as follows: A8-979; A8-3450, A8-3927; E5-2596, E5-3212; E9-458, E9-3623; Bl-638, Bl-3481, Bl-3733, Bl-3916* and also X3-3450, X3-9026; X5-3398, X5-3493; X15-3528, X15-3454. (Number after the hyphen refers to the recipient from which the reagent was produced. Letter X designates an unknown system.) Accordingly, the 17 reagents react with 7 different antigens.

These identical reagents were very valuable from the point of view of the present study and only they were used in the undermentioned population test.

During the past year, on three occasions and in three state farms — Mosonmagyaróvár, Gödöllő, Rákoshegy — 50 Hungarian Yellow (only one line), 196 Rhode Island Red (50–50–96, three lines in three farms), 100 Sussex (50–50, two lines in one farm) and 223 White Leghorn (50–50–50–73, four lines in two farms) in all 569 birds were tested with the reagents mentioned above.

RESULTS

A8-3450 and A8-3927 reacted in parallel with the cells of each breed and line with few exceptions. A8-979 having been used only in the last test, gave fewer and rather weaker reactions than the other two. Obviously the test dilution which was suitable in the line L3, has not proved to be so in other lines. Reactions of E5-2596 and E5-3212 were throughout identical in each breed and line. E9-458 gave more reaction than E9-3623, chiefly in the S3, L1 and L5 lines, but it is to be noted, that the E9-458 has been stored more than 4 years, and thawed many times and it was bacteriologically contaminated. All the four B1 reacted very similarly, but the B1-3733 gave more reactions than the others in the line S3. It should be mentioned, however, that this reagent does not appear to have been perfectly absorbed in its reactions with the panel cells. There were still two unabsorbed weak reactions and it was used in the population test only for curiosity. The reactions of the two X5 reagents were exactly the same in each line. The few differences existing between the two X15 reagents were due to the fact, that the test dilution of X15-3428 was too high and thus it failed to react with some cells.

CONCLUSIONS

On the basis of the results obtained with identical monofactorial reagents, it is suggested, that antisera produced in a certain line can also be used in other breeds and lines.

* The letter designations of the above reagents were assigned on the basis of a comparison carried out with C. O. Briles' reagents.

The production of identical reagents on the basis of reaction with a panel of control cells — gives considerable guarantee of monospecificity.

Although it would be wrong to oversimplify the problem of monofactorial reagents and especially their usefulness in various breeds and lines, in my opinion we do not need to produce reagents for every line. If we rigorously respect the rules of preparing monofactorial sera, they will be reliable for every line or breed just like the reference reagents of other species.

REFERENCES

BRILES, W. E., 1962. Additional blood group systems in the chicken, *The New York Academy of Sciences*, **97**, 173–183.

BRILES, W. E., BRILES, C. O. and QUISENBERRY, J. H., 1963. The C and D blood group systems in the chicken, *Poultry Sci.*, **42**, 1096–1103.

BRILES, W. E., MCGIBBON, W. H. and IRWIN, M. R., 1950. On multiple alleles effecting cellular antigens in the chicken, *Genetics*, **35**, 633–652.

FANGUY, R. C., 1961. Blood typing techniques in poultry, The Agricultural and Mechanical College of Texas, Texas Agricultural Experiment Station, 1–23.

FANGUY, R. C., FERGUSON, T. M. and QUISENBERRY, J. H., 1961. The blood group spectrum of a non-inbred population as determined from cross-reaction with antisera produced in non-related populations, *Poultry Sci.*, **40**, 848–853.

GILMOUR, D. G., 1959. Segregation of genes determining red cell antigens at high levels of inbreeding in chickens, *Genetics*, **44**, 14–33.

OOSTERLEE, C. C. and BOUW, J., 1966. Basical aspects of chicken blood group research, Proc. 10th European Animal Blood Group Conference, 189–193.

DISCUSSION

K. HÁLA: You presented a paper on monofactorial reagents. Could you tell me how did you demonstrate that your reagents were monofactorial?

M. PAPP: I did it on the basis of a comparison carried out with C. O. Briles reagents 3 years ago. Besides my monofactorial reagents those which were supposed to be so were sent to W. E. Briles this year and according to him most of them proved to be monofactorial.

XITH EUROPEAN CONFERENCE ON ANIMAL BLOOD GROUPS AND BIOCHEMICAL POLYMORPHISM
Warsaw, July 2nd–6th, 1968

Complex Antigens of the A System in Chickens

Feldzžerita KNÍŽETOVÁ and K. HÁLA
Institute of Experimental Biology and Genetics, Czechoslovak Academy of Sciences, Prague, Czechoslovakia

SUMMARY. Segregation within the A locus and presence of 2 alleles — A13 and A14 — was detected within a highly inbred line of Iowa chickens. On the basis of genotype in the A system this line was divided into two sublines — IA and IB within which mutually exchanged skin grafts survived permanently. Specific antibodies were produced by reciprocal immunizations between individuals of these sublines. Analysis of these sera within a closed outbred WL population revealed the complexity of antigens A13 and A14. Typing of individual outbred families and absorption tests showed the presence of at least 2 antigenic factors within the A13 and of 4 antigenic factors within the A14 antigen. A different spectrum of anti-A antibodies obtained by immunization of recipients of the same blood group genotype with erythrocytes from the same donor or donors of identical genotype was found.

INTRODUCTION

The complexity of the B antigen in chickens is a known fact. Until recently analysis of A antigens has received little attention. Previously it was suggested that cross-reactions in the A system are not so apparent as in the B system. This suggests that the A system is less complex, although reports of the first workers indicate that 8 antigenic factors are present in the A system (BRILES *et al.*, 1950, 1952). In the experiments described here 2 antigens — A13 and A14, segregating in a highly inbred line of chickens, were analysed.

METHODS AND MATERIALS

The Iowa line is homogeneous for all histocompatibility loci as skin graft survival among all individuals is permanent (HÁLA *et al.*, 1966). On the basis of the genotype in the A system 2 sublines IA (A^{13}/A^{13}) and IB (A^{14}/A^{14}) were established. Specific antisera anti-13 and anti-14 were produced by reciprocal immunizations between individuals of these sublines. Antisera obtained were used for testing 2 other highly inbred lines C and W of chickens; members of these lines were shown to possess the genotype A^{14}/A^{14}. The sera were further analysed within a closed outbred population of White Leghorn chickens (WL).

RESULTS

Tests of specific reagents within inbred lines revealed the unitary nature of anti-A13 and anti-A14 antibodies (absence of anti-E antibodies) and the identity of the A14 allele (antigen) in all three lines. Further tests of the same reagents in a number of families within a closed outbred WL population, however, showed that the antisera in fact contain several antibody populations. Sera produced by absorption with erythrocytes from different animals of the outbred WL population can be regarded as monofactorial in this population. Results of testing and absorptions are given in the Table. It can be seen that antigen A13 consists of at least 2 antigenic factors and antigen A14 of at least 4 factors. The possibility is not excluded that further absorption of "monofactorial" sera with erythrocytes from a different population would allow detection of additional antigenic determinants. Further, attempts were made to determine the genotype of outbred individuals within the A locus (see Table). Seven alleles of the A locus were recognized. The genetic determination of A antigen must be verified in the offspring of selected parental pairs.

As with anti-B antibodies, a different antibody spectrum was observed in antisera prepared by immunization of recipients of the same genotype with erythrocytes from the same donor or donors of identical genotype (Table).

DISCUSSION AND CONCLUSIONS

The causes underlying the different reactivity of different recipients to the same antigenic stimulus have not yet been fully explained. The possibility of a different genetically determined antibody-forming capacity is not excluded. Thus some recipients react only against stronger antigenic factors, whereas recipients with higher reactivity also form antibodies against weaker factors. The different strength of antigenic factors within the A system was not analysed, but it was shown in the B system that the A and F factors are more immunogenic because specific antibodies directed against them are formed by all recipients (HÁLA and KNÍŽETOVÁ, 1966), and are present even in relatively weak antisera obtained after immunization with skin grafts.

We were successful in demonstrating a relatively considerable antigenic complexity of the A system and the resultant necessity of preparing monofactorial sera for the analysis of these complex antigens. The best way of producing such sera lies in the production of coisogenic lines, i.e. lines differing not only at a single allele, but at only one antigenic factor within this allele. This genetic approach is, however, lengthy. The immunologic approach, i.e. induction of polyvalent tolerance in individuals identical at the B system to a wide antigenic spectrum of A antigens, except for one antigenic factor to which the animal will be immunized after it is immunologically mature may be of some advantage.

Analysis of sera anti-A13 and anti-A14 by absorption of erythrocytes from outbred animals

Antiserum / Erythrocytes		Anti-A13					Anti-A14							Presence of antigenic factors on erythrocytes						Presumed genotypes
		208					343			340										
		n.a.	a.1591	n.a.	a.1511	a.1505	n.a.	a.1567	a.1589	n.a.	a.19	a.13	a.55	1	2	3	4	5	6	
♂	19	+	·	+	+	·	+	·	·	+	·	+	·	+	+	+	·	+		1,5/2,3
♀	13	+	·	+	+	·	+	·	·	+	·	·	·	+	+	+	·	·		1/2,3
F1	1503	+	·	+	+	·	+	·	·	+	·	·	·	+	+	+	·	·		1/2,3
	1505	+	·	+	+	·	+	·	·	+	·	+	·	+	+	+	·	+		1,5/2,3
	1511	+	·	+	·	·	+	·	·	+	·	·	·	·	+	+	·	·		2,3/2,3
♂	54	+	·	+	+	·	+	·	·	+	·	+	·	+	·	+	·	+	·	1/3,5
♀	52	+	·	+	+	·	·	·	·	+	+	+	·	+	+	·	+	·	+	1,2,6/4
F1	1567	·	·	·	·	·	+	·	·	+	+	+	·	·	·	+	+	+	·	3,5/4
	1569	+	·	+	+	·	+	·	·	+	·	+	·	+	+	+	·	+	+	1,2,6/3,5
	1573	+	·	+	+	·	·	·	·	+	·	+	·	+	+	·	·	·	+	1/1,2,6
	1575	·	·	·	·	·	+	·	·	+	+	+	·	·	·	+	+	+	·	3,5/4
♂	54	+	·	+	+	·	+	·	·	+	·	+	·	+	·	+	·	+	·	1/3,5
♀	55	·	·	+	·	·	+	·	·	+	+	+	·	·	+	+	+	+	+	2,3,5,6/4
F1	1581	+	·	+	+	·	+	·	·	+	·	+	·	+	+	+	·	+	+	1/2,3,5,6
	1583	·	·	·	·	·	+	·	·	+	+	+	·	·	·	+	+	+	·	3,5/4
	1585	+	·	+	+	·	+	·	·	+	·	+	·	+	+	+	·	+	+	1/2,3,5,6
♂	54	+	·	+	+	·	+	·	·	+	·	+	·	+	·	+	·	+		1/3,5
♀	72	+	·	+	+	·	+	·	·	+	·	·	·	+	+	+	·	·		1/2,3
F1	1589	·	·	+	·	·	+	·	·	+	·	+	·	·	+	+	·	+		2,3/3,5
	1591	+	·	+	+	·	+	·	·	+	·	+	·	+	·	+	·	+		1/3,5
	1593	·	·	+	·	·	+	·	·	+	·	+	·	·	+	+	·	+		2,3/3,5
	1595	+	·	+	+	·	+	·	·	+	·	·	·	+	+	+	·	·		1/2,3
♂	69	·	·	+	·	·	+	·	·	+	·	+	·	·	+	+	·	+		2,3/3,5
♀	14	·	·	+	·	·	+	·	·	+	+	+	·	·	+	+	+	·		2,3/4
F1	1597	·	·	+	·	·	+	·	·	+	·	·	·	·	+	+	·	·		2,3/2,3
	1599	·	·	·	·	·	+	·	·	+	+	+	·	·	·	+	+	+		2,3/4
	1701	·	·	·	·	·	+	·	·	+	+	+	·	·	·	+	+	+		2,3/4
	1703	·	·	+	·	·	+	·	·	+	·	·	·	·	+	+	·	·		2,3/2,3
	1705	·	·	+	·	·	+	·	·	+	·	·	·	·	+	+	·	·		2,3/2,3
Presence of antibodies in serum		1		1,2	1		3			3,4,5,6	4	4,5								

n.a. = non absorbed.
a = after absorption with.

REFERENCES

BRILES, W. E., BRILES, R. W. and IRWIN, M. R., 1952. Differences in specificity of the antigenic products of a series of alleles in the chicken, *Genetics*, **37**, 539.

BRILES, W. E., MCGIBBON, W. H. and IRWIN, M. R., 1950. On multiple alleles affecting cellular antigens in the chicken, *Genetics*, **35**, 633–652.

HÁLA, K., HAŠEK, M., HLOZÁNEK, I., KNÍŽETOVÁ, F. and MERVARTOVÁ, H., 1966. Syngeneic lines of chickens. II. Inbreeding and selection within the M, W and I lines and crosses between the C, M and W lines, *Folia biol.* (Praha), **12**, 407–422.

HÁLA, K., and KNÍŽETOVÁ, F., 1966. The analysis of complex B phenogroups in inbred lines of chickens, Xth European Conference on Animal Blood Groups and Biochemical Polymorphism, Paris.

XITH EUROPEAN CONFERENCE ON ANIMAL BLOOD GROUPS AND BIOCHEMICAL POLYMORPHISM
Warsaw, July 2nd–6th, 1968

Complex Antigens of the B System in Inbred Lines of Chickens

K. HÁLA and Feldzžerita KNÍŽETOVÁ

Institute of Experimental Biology and Genetics, Czechoslovak Academy of Sciences, Prague, Czechoslovakia

SUMMARY. Complex antigens of the B system were analysed in three inbred lines of chickens: C, I and W. Antisera were prepared by reciprocal immunization between individual animals from different sublines of the same line and between lines. We performed the analysis of complex antisera by absorption with erythrocytes of both inbred and outbred animals. Specificity was checked on animals of a backcross population. In 5 complex antigens from inbred sublines we found that at least 21 antigenic components were present of which some are common to several alleles and some are specific for a certain allele. We continued the analysis of these specific antigens in 12 families of full-sibs from outbred line WL and found them of further complexity.

INTRODUCTION

Complex antigens of the B system in chickens have been studied by various investigators (BRILES *et al.*, 1950; GILMOUR, 1959; OOSTERLEE and BOUW, 1966; HÁLA and KNÍŽETOVÁ, 1966). It was demonstrated that most of the B antigens are composed of a number of antigenic factors.

The aim of the present work was to determine the composition of B antigens and their similarity in inbred lines of chickens.

METHODS AND MATERIALS

Five sublines of inbred lines of chickens were used in the experiments. Antigen B1 is present in the subline CA, antigen B2 in the subline CC, antigen B9 in the subline WA, antigen B10 in the subline WB, and antigen B13 in the I line. Detailed characteristics of the sublines, their origin and the method of breeding were described in previous papers (HAŠEK *et al.*, 1966; HÁLA *et al.*, 1966).

Anti-B antisera were prepared by reciprocal immunizations between individuals from different sublines of the same line and between animals of a backcross population. All sera proved to be specific for B antigens. Complex antisera were analysed by typing erythrocytes of animals from other sublines and by absorption with these erythrocytes. Sera produced as monospecific within the 5 sublines, which reacted with a single antigenic factor, were further analysed on an outbred population of White Leghorns used in previous experiments (HÁLA and KNÍŽETOVÁ, 1966).

RESULTS

A total of 3 anti-B1, 9 anti-B2, 16 anti-B9 and 5 anti-B10 and 3 anti-B13 sera were produced and analysed. Each antiserum was tested against a panel of erythrocytes of known genotype. From reaction patterns and results obtained after absorption with the respective erythrocytes in this panel, the composition of antisera and the number of antibodies they contain was estimated.

Composition of complex antigens of B system

Subline	Alleles	Antigenic factors																				
		1	2	3	4	5	6	7	8	9	10	11	12	13	14	15	16	17	18	19	20	21
CA, CB	B1	.	.	.	.	5	6	7	8	9	.	.	12	13	14	.	.	.	.	.	.	21
CC, CD	B2	1	2	3	4	.	.	.	.	9	.	.	.	13	14	.	.	.	.	.	.	21
WA	B9	.	2	.	4	.	.	7	.	.	10	11	12	13	14	15	.	.	18	.	20	21
WB	B10	.	.	3	4	.	.	.	8	.	.	.	.	.	.	15	16	17	18	.	.	21
I	B13	.	.	.	4	.	6	.	.	.	.	11	12	13	.	15	.	17	.	19	20	.

It was found that individual complex antigens consist of 8 to 12 antigenic factors. Results are summarized in the Table. Twenty-one antigenic factors were proved to exist. Some of them are present in as many as 4 lines, others in only one line. The last mentioned factors are specific only for a certain B antigen and in fact for a certain line. Thus factor 5 is specific for CA subline, factor 1 for CC subline, factor 10 for WA subline, factor 16 for WB subline, and antigenic factor 19 for I line.

Antibodies to above mentioned antigenic factors specific for individual lines were further analysed on an outbred population of White Leghorn chickens. We found that complex antibodies were involved which could be divided by absorption. Antigenic factor 1 was found to be the most complex, because anti-1 antiserum contained 3 different specific antibody populations. Only in factor 19 no further antigenic components were detected.

DISCUSSION AND CONCLUSIONS

In contrast to the previous work (HÁLA and KNÍŽETOVÁ, 1966) in which we studied the number of antigenic factors composing antigen B1 and B2 within an outbred population of chickens, the present work was devoted to the analysis of common and distinct antigenic factors in 5 B antigens in inbred lines of chickens.

The same lines of chickens were used in the experiments of GILMOUR (1959). The lines have since then undergone further inbreeding, and, moreover, a new selection criterion has been applied—survival of skin grafts (KNÍŽETOVÁ *et al.*, 1966). Comparison with GILMOUR's results shows that our I line no longer segregates at B blood group system. It is still impossible to determine whether our designation of antigen B13 is synonymous with GILMOUR's designation of B_{R8} or B_{R9}. A similar situation is encountered with other antigens and the designation of antigenic factors.

We found that each antigen, which is present in the inbred lines under study, consists of 8 to 12 antigenic factors. This numbers does not seem to be finite. Results obtained indicate that some of the antigenic factors described are composed of further subunits. This complexity is additional evidence that the B system is the main histocompatibility system of the chicken.

REFERENCES

BRILES, W. E., MCGIBBON, W. H. and IRWIN, M. R., 1950. On multiple alleles affecting cellular antigens in the chicken, *Genetics*, **35**, 633.

GILMOUR, D. G., 1959. Segregation of genes determining red cell antigens at high levels of inbreeding in chickens, *Genetics*, **44**, 14–33.

OOSTERLEE, C. C. and BOUW, J., 1966. Basic aspects of chicken blood group research, Xth European Conference on Animal Blood Groups and Biochemical Polymorphism, Paris, p. 189.

HÁLA, K. and KNÍŽETOVÁ, F., 1966. The analysis of complex B phenogroups in inbred lines of chickens, Xth European Conference on Animal Blood Groups and Biochemical Polymorphism, Paris.

HÁLA, K., HAŠEK, M., HLOŽÁNEK, I., HORT, J., KNÍŽETOVÁ, F. and MERVARTOVÁ, H., 1966. Syngeneic lines of chickens. II. Inbreeding and selection within the M, W and I lines and crosses between the C, M and W lines, *Folia biol.* (Praha), **12**, 407–422.

HAŠEK, M., KNÍŽETOVÁ, F. and MERVARTOVÁ, H., 1966. Syngeneic lines of chickens. I. Inbreeding and selection by means of skin grafts and tests for erythrocyte antigens in C line chickens, *Folia biol.* (Praha), **12**, 335–342.

XITH EUROPEAN CONFERENCE ON ANIMAL BLOOD GROUPS AND BIOCHEMICAL POLYMORPHISM
Warsaw, July 2nd–6th, 1968

Detection of Alleles in Chicken Strains and Hybrids

C. C. OOSTERLEE and J. BOUW

Stichting Bloedgroepen Onderzoek, Landbouwhogeschool, Wageningen, The Netherlands

SUMMARY. The principle aim of studies of blood groups in the chicken is to detect the genes controlling such systems and to use these genes as genetic markers. Development in this field of research and collaboration between laboratories is limited as a result of difficulty in production of specific reagents for animals belonging to different strains of chicken.

In an attempt to avoid this problem the authors have set up reagent-antigen specificity tables for the blood group systems A, E and B. In this report results will be presented indicating the usefulness of such reagent-antigen specificity tables for chicken blood typing and as an aid to international collaboration in the field.

INTRODUCTION

The complexity of the blood groups which are controlled by the allelic series of genes of the blood group systems A, E and B in the chicken and resulting cross-reactivity of the reagents make it extremely difficult for workers in this field to reach agreement about the antigenic specificity of the blood groups and to establish an uniform nomenclature.

McDermid and Oosterlee (1966) reported the results of some comparison tests for antisera and discussed existing problems. The close linkage between the loci controlling the A and E systems interferes with the production of specific reagents to detect the antigens of these systems. This often results in reagents with combined anti-A and anti-E specificities. The comparison of antisera produced within strains to detect antigens of the B system shows characteristic cross-reactivity when these antisera are tested against cells of other strains of chicken.

Oosterlee and Bouw (1966) therefore suggested to produce monofactorial reagents. However, absorptions which limit the specificity of the antisera usually reduce the titre of the antisera, as a result of which it is hardly or even not possible to use such reagents for studies in which a large number of birds is involved.

Briles (1960 and 1966) described the value of reagent-antigen specificity tables using sera detecting a variety of factors which were controlled by one locus.

* This investigation was supported by a grant from the Netherlands Organization for the Advancement of Pure Research (Z.W.O.).

On the basis of the approach of BRILES the authors have constructed reagent-antigen specificity tables for the systems A, E and B. These tables were used for the detection of alleles in the Dutch experimental strains and hybrids and in birds of the inbred lines of the Institute of Experimental Biology and Genetics, Prague.

This report considers the value of these tables and possibilities of using them internationally.

MATERIAL AND METHODS

The birds used in this study belong to the experimental flock of the Institute for Poultry Breeding of the Agricultural University, Wageningen. This flock is composed of a white Leghorn population and one of medium heavy birds. Both populations are crossbreds of two commercial hybrid lines. Each of the commercial hybrid lines has been developed from crosses of up to 3 different strains. For the production of antisera 9 original strains are available.

The antisera produced were tested with a panel of reference cells from birds of the original strains and compared to reagents originating from the DeKalb (Illinois) and Thornbers (Halifax) laboratories. The antisera contained antibodies for antigens of the A, E, B, C, D and L systems. Antisera against antigens controlled by the A and E locus are considered to be reagents when they are reproducible within strains and when they can be used to detect alleles in different strains of chicken. B locus antisera are accepted as reagents when they clearly trace allelic antigens within strains.

The test procedure has been described by MCDERMID and OOSTERLEE (1966).

RESULTS

The reagent-antigen specificity tables have been constructed on the basis of family studies and tests of more than 3000 hybrids of the described populations.

The reagent-antigen specificity tables for the A, E and B locus are presented in the Tables 1, 2 and 3.

The tables for the A and E locus determine the genotypes as far as possible. The B locus table is based upon the reaction pattern of 5 antisera (reagents Ba, Bb, Bc, Bd and Be). These sera are stimulated by blood groups common in the original strains, are easy to produce and have a marked specificity. Each of these 5 antisera has been used as a basic reagent. The blood groups which give reactions to such a basic reagent are subdivided into related groups by means of the reaction pattern with other reagents, resulting in the blood groups Ba_1, Ba_2, Ba_3, etc. This table presents only the reaction pattern of the supposed homozygous birds. The B locus nomenclature is schematic and preliminary and has to be considered as a first approach.

TABLE 1

A locus reagent-antigen specificity table

Genotypes \ Reagents	A_1	A_2	A_3	A_4^*	A_5^*
A_1/A_1	+	+			
A_1/A_2	+	+			
A_1/A_3	+	+	+	+	
A_1/A_4	+	+		+	
A_1/A_5	+	+		+	+
A_2/A_2		+			
A_2/A_3		+	+	+	
A_2/A_4		+		+	
A_2/A_5		+		+	+
A_3/A_3			+	+	
A_3/A_4			+	+	
A_3/A_5			+	+	+
A_4/A_4				+	
A_4/A_5				+	+
A_5/A_5				+	+

* See text.

The nomenclature for the A and E locus is adapted as much as possible to the accepted international designations and is partly taken over from BRILES (1966). The antigens are indicated as A_1, A_2, A_3, A_4 (formerly A_6) and A_5, resp. E_1, E_2, E_3, E_4, E_5 and E_6. The antisera with the designation of the donor antigen which has stimulated the antibodies. The E_6 reagent has not been produced in the Dutch strains and was kindly offered by the DeKalb laboratories.

The antigen A_5 is absent in the White Leghorn strains studied. The other A locus antigens have been traced in the White Leghorn populations and the medium heavy birds.

The E_4-antigen has been traced in the medium heavy birds, E_6 only in the White Leghorn populations.

The B locus table presents the reaction pattern of 21 allelic antigens of supposed homozygous birds of our own experimental animals and four of the Prague Institute.

By using B locus reagent-antigen specificity tables for heterozygous birds (not published) on about 10% of the tested 3000 hybrids, only one of the two alternative blood groups could be detected. In the White Leghorns nine alleles could be differentiated; in the medium heavy birds twelve.

TABLE 2

E locus reagent-antigen specificity table

Reagents / Genotypes	E_1	E_2	E_3	E_4	E_5	E_6
E_1/E_1	+					
E_1/E_2	+	+				
E_1/E_3	+		+			
E_1/E_4	+		+	+		
E_1/E_5	+				+	
E_1/E_6	+				+	+
E_2/E_2	+	+				
E_2/E_3	+	+	+			
E_2/E_4	+	+	+	+		
E_2/E_5	+	+			+	
E_2/E_6	+	+			+	+
E_3/E_3	+		+			
E_3/E_4	+		+	+		
E_3/E_5	+		+		+	
E_3/E_6	+		+		+	+
E_4/E_4	+		+	+		
E_4/E_5	+		+	+	+	
E_4/E_6*						
E_5/E_5					+	
E_5/E_6					+	+
E_6/E_6					+	+

* See text.

DISCUSSION AND CONCLUSIONS

The A locus table is, presently the simplest so far. Even in this table it can be seen that several antigens react with 2 or more reagents. For instance, the antigen A_3 reacts with the A_4 reagent. Anti A_5 (which is an experimental serum) in some strains parallels the A_4 pattern, except for A_3 cross-reactivity. However, more detailed family studies must confirm specificity at the A locus. The same can be stated for the anti E_3 and E_4. It is likely that further studies will lead to a larger number of blood groups in the E system. BRILES (1966) indicated the existence of nine E locus alleles. Also this locus shows a marked cross-reactivity pattern. One of the first

TABLE 3

B locus reagent-antigen specificity table

Reagents / Alleles	B_a	B_b	B_c	B_d	B_e	B_f	B_g	B_h	B_i	B_j	B_k	B_l	B_m
B_{a1}	+												
B_{a2}	+	+											
B_{a3}*	+	+				+							
B_{a4}*	+			+									
B_{a5}*	+			+	+	+							
B_{a6}	+					+							
B_{b1}		+											
B_{b2}		+	+										
B_{b3}		+		+		+							
B_{c1}			+			+	+						
B_{c2}			+						+				
B_{c3}			+			+			+				
B_{c4}			+			+	+		+				
B_{c5}			+			+			+	+			
B_{d1}				+		+							
B_{d2}				+		+	+						
B_{d3}				+		+		+					
B_{d4}*				+								+	
B_{e1}					+						+		
B_{e2}					+	+					+	+	
B_{e3}					+							+	
B_f						+							
B_l												+	
B_m													+

* Foreign alleles, not traced in our own strains, see text.

criteria for a reagent-antigen specificity table is that the reagents contain only antibodies against the antigens controlled by one locus.

In the introduction it has been indicated that this is rather difficult for the A and E locus, because of the close linkage between these loci. When different strains with a variable A–E linkage pattern are introduced, however, it becomes easier to make adequate donor–recipient combinations for the production of pure anti-A and anti-E sera. In cattle blood typing an antiserum is called a reagent when it detects one antigenic factor. In this field of research the ideas are changing, however. FIORENTINI and BOUW (1970) will demonstrate at this Conference that related but nevertheless distinctly different antigens can stimulate antibodies which show similar patterns of reactions.

On the basis of the cross-reactivity pattern it is also likely that the antisera for the A and E locus cannot be considered to trace monofactorial antigens. No study has yet been performed to solve the question of whether the cross-reactivity patterns in the A and E loci are due to antigenic complexes with more factors or whether this type of reactivity is based on subtypes.

Another basic aspect of a reagent-antigen specificity table is the necessary variety in specificities between the several reagents belonging to one table. Some years ago it was supposed that the A locus controlled two alleles A_2 and A_4. On the basis of studies of BRILES (1960) and comparison tests performed by MCDERMID and OOSTERLEE (1966), it is obvious that more alleles are involved in the A and E loci, a fact also demonstrated by these tables. The tables clearly demonstrate that some genotypes for the A and E system cannot be detected directly from the blood typing test. However, when population genetic studies are carried out giving frequencies of A_2 and A_4, estimates for frequencies of the other genes can be made. The B locus table indicates the existence of a large number of antigenic factors. It is likely that the related blood groups Ba_1, Ba_2, Ba_3, Bb_1, Bb_2, Bb_3, etc. have only some antigenic factors in common and not all such factors.

The test results with the cells of homozygous birds from the Prague Institute indicate that the table can also be useful for other strains, although the genotyping of heterozygous birds will be more difficult, so that family studies will be necessary.

How far internationally accepted reagent-antigen specificity tables can be introduced will partly depend on the results of the comparison tests as these are performed by the chicken section of the E.S.A.B.R. It is quite possible, however, that it will be difficult to reach agreement about the specificity of antisera which are produced in different strains of birds. The value of a reagent-antigen specificity table depends on the reaction pattern and the variety of the introduced antisera. Therefore the exchange of antisera between research workers has to be considered as a means of testing antigens of different strains with the same antisera.

From a practical point of view this approach is possible since our experience has shown that a reagent can be reproduced within a strain. Already a relatively small amount of the antisera is enough to genotype a control panel of about 50 cells representing several lines of chickens. On basis of this exchange of sera, agreement about the specificity of the various allelic antigens can be established.

ACKNOWLEDGEMENTS

The authors are indebted to Miss R. van GINKEL who very accurately performed a large part of the technical work of this study.

REFERENCES

BOREL, J. F., 1964. Recherches immuno-génétiques sur des substances spécifiques du groupes chez la poule et sur la utilization comme marqueurs des gènes dans l'élevage, Juris-Verlag, Zürich, 66–75.

BRILES, W. E., 1960. Blood groups in chicken, their nature and utilization, *World's Poult. Sci. J.*, **16**, 223–242.

BRILES, W. E., 1966. Personal communication.

FIORENTINI, A. and BOUW, J., 1969. Specificities of antibodies detecting antigenic substances on cattle red blood cells, XIth European Conference on Animal Blood Groups and Biochemical Polymorphism, Warsaw, pp. 117–122.

MCDERMID, E. M. and OOSTERLEE, C. C., 1966. Results of the first and second European comparison tests for blood typing of chicken, Proceedings of the Xth Eur. Conf. on Animal Blood Groups and Biochemical Polymorphism, Paris.

OOSTERLEE, C. C. and BOUW, J., 1966. Basical aspects of chicken blood group research, Xth European Conference on Animal Blood Groups and Biochemical Polymorphisms, Paris, pp. 189–193.

XITH EUROPEAN CONFERENCE ON ANIMAL BLOOD GROUPS AND BIOCHEMICAL POLYMORPHISM
Warsaw, July 2nd–6th, 1968

Blood Groups and Biochemical Polymorphism in Different Populations of Chickens

E. PETROVSKÝ, J. MÁCHA and Jitka SOUTOROVÁ

Department of Genetics, University of Agriculture, Brno, Czechoslovakia
Laboratory of Animals Physiology and Genetics, Czechoslovak Academy of Sciences, Liběchov, Czechoslovakia

SUMMARY. The frequencies of individuals with agglutinogen Vh (genotypes Vh/Vh and Vh/vh) and with agglutinogen Hi (genotypes Hi/Hi and Hi/hi) in several breeds of chickens are presented. The preliminary results referring to phenotype and genotype frequencies at the G3 locus of strains sensitive and resistant to coccidiosis are described. The question of the manifestation of the Hi agglutinogen and its dependence on the intensity of egg production is discussed.

INTRODUCTION

In the past few years we have witnessed a keen interest in the immunogenetics of chickens that has led to the detection of a number of immunological and biological polymorphism. The full utilization of the first discovery, the blood group systems, such as A, B, etc., for genetical analyses of populations is impeded by their immunological complexness. However, a number of simpler systems partly of blood and partly of non-cellular substances are known which, on the whole, can be used without limitation. Here belong the erythrocytic systems Vh (GILMOUR, 1959) and Hi (SCHEINBERG and RECKEL, 1962), the serum systems Alb (MCINDOE, 1962), Tf (OGDEN *et al.*, 1962), Ap (WILCOX, 1963), Es (CSUKA and PETROVSKÝ, 1968) the egg white systems Ov (LUSH, 1964), G2 and G3 (LUSH, 1961; BAKER and MANWELL, 1962). Until recently it seemed that the manifestation of all genetic systems in chickens is relatively equally stable as is the case in mammals, this rule being confirmed by the exceptional Hi system. This view has been somewhat shaken by our findings that the polymorphic esterase system also shows changes dependent on physiological processes. Therefore it is possible to divide all the chicken markers which have been discovered hitherto and which will be discovered in the future into 2 groups, i.e. systems influenced by physiological processes (e.g. Hi and Es systems) and systems, whose manifestation do not depend on the individual's physiological state (e.g. blood groups and polymorphic systems of proteins).

At the beginning of systematic immunogenetical analysis of chicken populations we first of all learned how to use, reliably, the systems of both above mentioned groups by repeated testing of chickens of the same origin (PETROVSKÝ *et al.*, 1968a). After this we tackled testing of populations of different origin and production specialization. For the very limited content of this contribution we cannot present the frequencies of all the tested systems and populations, but select from both groups of systems only the erythrocytic ones, Vh and Hi, for different breeds and strains and their influence on selection, and the biochemical systems G2 and G3 in strains sensitive and resistant to coccidiosis.

MATERIALS AND METHODS

For testing the erythrocytic systems we used blood samples obtained from chickens at breeding farms in Czechoslovakia and the German Democratic Republic towards the end of the first year of egg production. An exception to this was the White Cornish and White Plymouth hens from Czechoslovakia, whose blood was taken at the beginning of egg production.

The strains sensitive and resistant to coccidiosis were maintained at the Veterinary Faculty, University of Agriculture, Brno.

For detection of the Vh agglutinogen we used phospholipids, e.g. Sachs–Witebski antigen, produced by the Institute of Sera and Vaccines, Prague, in optimum dilution established beforehand.

For detection of the Hi agglutinogen an extract from common vetch seeds was prepared by mixing coarsely ground seeds with physiological saline in suitable dilution (1:2 and 1:5).

The polymorphic egg white protein systems G2 and G3 were established with the aid of starch-gel electrophoresis with KRISTJANSSON's buffer (1963).

RESULTS

Erythrocytic Vh and Hi Systems in Different Chicken Strains and Breeds and in Selection

Table 1 shows the frequencies of Vh and Hi positive individuals of genotypes Vh/Vh, Vh/vh or Hi/Hi and Hi/hi in several strains of different breeds. Between the breeds there exist significant differences in both systems. In various strains of the same breed we sometimes also found significant differences.

Chickens of the meat type displayed, in most cases, a low frequency of Vh positive individuals, which manifests itself in a low Vh allele frequency, established through calculation. An exception was one strain only of the Cornish breed from Czechoslovakia. On the other hand, in the laying type the frequency of Vh positive

TABLE 1

Frequency of Vh and Hi positive individuals in different chicken populations

Breed-strain	Location	Utility type	Number of chickens	Vh positive, % (Vh/Vh/,Vh/vh)	Hi positive, % (Hi/Hi,Hi/hi)
W. Plymouth-2	Ger.	m.	95	5.3	75.8
W. Plymouth-1	Ger.	m.	95	12.6	56.8
W. Cornish-2	Ger.	m.	102	24.5	71.6
W. Plymouth-15	Cz. x	m.	302	25.8	20.2
W. Plumberg	Ger.	m.	166	27.1	30.7
N. Hampshire	Ger.	m.	132	40.1	28.8
N. Hampshire	Cz.	m.-l.	46	52.2	47.8
Rhode Island	Cz.	m.-l.	36	66.6	36.1
W. Cornish-13	Cz. x	m.	306	71.9	13.7
W. Leghorn 4 strains	Cz.	l.	2345	88.3	27.6
W. Leghorn 4 strains	Ger.	l.	699	93,6	5.7
W. Leghorn 4 strains	Cz.	l.			
A-1966			264	81–96	43–84
B-1966			603	81–98	15–31
A-1967			953	88–95	18–62
B-1967			1437	82–88	4–37

Note. A, Chickens excluded from mating; B, chickens selected for further mating; x, tested at beginning of egg production; Ger., German Democratic Republic; Cz., Czechoslovakia; m., meat; l., laying.

chickens was relatively high in all strains tested. In populations for combined use — breeds RIR and NH — the frequency was intermediate.

A similar situation was also encountered in the Hi system. Here, too, Cornish and Plymouth chickens of the meat type about 15-month old exhibited a characteristic — in this case high-frequency of Hi agglutinogen, while in White Leghorn chickens of the laying type a preponderance of the other group was to be seen, however, there was greater overlapping of frequencies than was the case in the Vh system.

At the end of the control period the more useful chickens of all populations were selected for further mating while the worse chickens were separated and removed. In four strains of Leghorns we observed over a two year period changes in the frequencies of the Vh and Hi systems, caused by this selection (Table 1).

In the Vh system no difference between the two groups of chickens were found in 2 strains in the first year of observation; in 1 strain the more productive chickens exhibited a higher, while in the remaining one a lower frequency of Vh agglutinogen. A somewhat different result was obtained in the second year, where 1 strain displayed no difference, while in the remaining 3 strains the more productive layer-hens exhibited a significantly higher frequency of Vh agglutinogen.

In the Hi system the chickens selected for further mating displayed always a markedly lower frequency of Hi agglutinogen than the group of eliminated chickens did.

Systems G2 and G3 in strains sensitive and resistant to coccidiosis

In the strains sensitive and resistant to coccidiosis there was no difference in the G_2 locus — in both only the G_2 B phenotype was found. On the other hand, in the G_3 locus a difference was observed, the resistant strain possessed 2.6 times more heterozygotes than the sensitive strain (Table 2).

TABLE 2

Phenotype and gene frequencies at G_3 locus in strains sensitive and resistant to coccidiosis

Strain	Number of chickens	Phenotype frequency pc/%			Allele frequency	
		A	AB	B	A	B
Sensitive	54	47 87.04	7 12.96	— —	0.935 ± 0.024	0.065 ± 0.024
Resistant	68	45 66.18	23 33.82	— —	0.831 ± 0.032	0.169 ± 0.032

DISCUSSION AND CONCLUSIONS

It was found that the frequency of the Vh system groups differed markedly in populations of different production specialization. An exception was the Cornish breed from Czechoslovakia. The most acceptable explanation of this fact is to be found in the different origin of the breeds of varying production specialization. It may be expected that tests of other populations will not give such an unambiguous picture, for practically all production populations arise from very complicated hybrids.

We hold the view that the interbreed differences in the Hi system have a physiological rather than genetical background. It is probable that breeds of the laying type display a low frequency of Hi agglutinogen because intensive egg production does not lead to its better manifestation as suggested by SCHEINBERG and RECKEL, 1962, but, on the contrary, is not favourable for this manifestation. Therefore, in populations of the meat type with lower egg production we find a high frequency of Hi positive chickens. In favour of this view the less productive White Leghorn chickens not included in further mating display a markedly higher frequency of this antigen than the chickens for further breeding so that they do not differ from the chickens of heavier breeds.

Therefore, such systems as Hi, whose manifestation depends on the intensity of the physiological processes, cannot be regarded as immunogenetic markers, but rather as physiological markers. Details about the change of frequency in the Hi system in selection are described elsewhere (PETROVSKÝ *et al.*, 1968b).

ACKNOWLEDGEMENTS

The authors wish to express their sincere thanks to Prof. K. LÖHLE and Dr. D. MULSOW of the Institute of Poultry Breeding, Humbolt University, Berlin, Prof. B. KLIMEŠ of the Veterinary Faculty in Brno and to Dr. OREL of the Moravian Museum in Brno, collaborating with us in obtaining certain results.

REFERENCES

BAKER, C. M. A. and MANWELL, C., 1962. Molecular genetics of avian proteins. I. The egg white proteins of the domestic fowl, *Brit. Poult. Sci.*, **3**, 161.

CSUKA, J. and PETROVSKÝ, E., 1968. Study of polymorphism of esterase of chicken egg white and blood serum, *Fol. Biol.*, **XIV**, 165.

GILMOUR, D. G., 1959. Blood groups in chickens, 6th Int. Blood Group Congress, Munich.

KRISTJANSSON, F. K., 1963. Genetic control of two prealbumins in pigs, *Genetics*, **48**, 1059.

LUSH, I. E., 1961. Genetic polymorphisms in the egg albumen proteins of the domestic fowl, *Nature*, **189**, 981.

LUSH, I. E., 1964. Egg albumen polymorphism in the fowl: the albumin locus, *Genet. Res.*, **5**, 257.

MCINDOE, W. N., 1962. Occurrence of two plasma albumins in the domestic fowl, *Nature*, **195**, 353.

OGDEN, A. L., MORTON, J. R., GILMOUR, D. G. and MCDERMID, E. M., 1962. Inherited variants in the transferrins and conalbumins of the chicken, *Nature*, **195**, 1026.

PETROVSKÝ, E., MÁCHA, J. and SOUTOROVÁ, J., 1968a. Genetika imunologických a fysiologických vlastností drůbeže. I. Využití imunogenetických znaků k analyse populací slepic, *Živoč. vyr.* (In press).

PETROVSKÝ, E., MÁCHA, J. and SOUTOROVÁ, J., 1968b. V. Změny fonetypových frekvencí v erytrocytárním Hi systému slepic při selekci na ekonomicky významné vlastnosti, *Acta Univ. Agric.*, (Brno) (In press).

SCHEINBERG, S. L. and RECKEL, R. P., 1962. Studies on the "Hi" agglutinogen in chicken, *Annals N.Y. Acad. Sci.*, 194.

WILCOX, F. H., 1963. Genetic control of serum alkaline phosphatase in the chicken, *J. Exp. Zool.* **152**, 195.

DISCUSSION

C. C. OOSTERLEE: Did you test the offspring of the selected birds on the Hi antigen?

E. PETROVSKÝ: All WL strains were tested in two generations, but not the offspring of the birds selected on the Hi antigen.

K. HÁLA: This was a very interesting work and I am sure that it will be possible to study the relationship between blood groups and various diseases in a not too distant future. Certain promising results begin to appear in mice.

E. PETROVSKÝ: We must test more animals before we can evaluate the differences between sensitive and resistant strains of chickens.

Blood Groups in the Goose

S. LOSONCZY
Bloodgrouping Laboratory, University of Veterinary Science, Budapest, Hungary

INTRODUCTION

Blood group research in poultry was started by HADDA and ROSENTHAL in 1913. Amongst poultry chicken blood group systems are the best characterized and international comparison of chicken blood grouping reagents has been started recently.

IRWIN (1956) thoroughly studied the blood group characteristics and inheritance in various pigeon species and MUNKÁCSI (1959) and PODLIACHOUK *et al.* (1964) reported on duck blood group characteristics and LAW *et al.* (1965) on those of turkeys, but to date nothing has been published on goose blood groups.

MATERIAL AND METHOD

One–two year old geese were used for the production of blood group antibodies. The number of individuals examined is indicated in the result section. A 30% suspension of washed red blood cells was used for intravenous isoimmunization and the presence of antibodies was detected by means of haemagglutination.

RESULTS

Immunizations were carried out in Rhein, Landes, Rhein×Landes, Chinese swan and Hungarian breeds. There was no significant difference in the antibody producing ability of the different breeds. Table 1 indicates the percentage distribution of the animals treated at one time with different amounts of blood and responsing by production of antibody of characteristic titre on the 8th day after the injection.

The classification into the titre categories 0; 1:1–1:16, and 1:16–1:512 is reasonable, because sometimes raw sera giving reaction at a dilution of 1:16 may also yield reagents.

The animals immunized with initial doses 3 and 5 ml of blood, received 4 further injections at 2-day intervals but there was no change in the titre of antibody

TABLE 1

Number of individuals examined	Quantity of blood injected, ml	No antibody detectable, %	Titre value between 1 : 1–1 : 16, %	Titre value between 1 : 32–1 : 512,%
15	3	55	20	25
15	5	87	—	13
20	8	50	30	20
130	40	43	22	35

produced. Reimmunization after an interval of one month did not cause titres to rise to any degree although antibodies could be revealed in each animal.

Time of appearance of antibodies and titre formation

Experiments were carried out in 30 animals in such a way that one half of them received 3 ml blood every second day the rest of them received 40 ml in one dose.

Antibodies could be detected as soon as the 3rd day after the first injection, and the titre rose appreciably by the 4–6th day. From the 8–9th day onwards a slow decrease was detectable. In animals injected with 40 ml of blood, antibodies appeared sooner and the average titre level was higher by one dilution, than in those receiving several smaller doses. Bleeding is best performed between the 5th and 8th day.

Bleeding

Depending on the body weight (4.5–7 kg) 160–180 ml blood can be taken from the wing wein without any serious injury. The use of a vacuum bottle or record syringe is essential, otherwise the above amount of blood cannot be obtained and that which is collected hemolyses. On the basis of our own experience the quantity of blood which may be taken from the wing vein by the above technique is equal to that which can be collected by cutting the carotid artery.

Quantity of absorbent blood (ratio of red blood cells to serum)

When using 0.05 ml of serum and 0.025 ml of 2% red cell suspension for the haemagglutination test, reactions designated by + indicate that all antigens are combined with antibody. When using double the dilution and the serum gives weak (0+) or negative reaction, it can be easily calculated that 1 ml serum can be cleared of unwanted antibody with the blood quantity indicated in Table 2.

Absorption time

Two and a half hours proved to be the best length of time for absorption. Trial absorptions can be performed in one or two phases at room temperature of 23–25°C. If no haemagglutination appears within an hour, the absorption can be made in one phase, otherwise the quantity of the absorbing blood should be divided into

TABLE 2

Antibody titre in the serum	Quantity of absorbent red blood cells, ml
T	0.02
2	0.04
4	0.08
8	0.16
16	0.32
32	0.64
64	1.28
128	2.56
256	5.12
512	10.24
1024	20.48
2048	40.96

equal portions and 2×75 min trial absorptions carried out. Naturally occurring antibodies could not be detected in the goose.

Direct agglutinins

To date goose blood group research has revealed only the presence of direct agglutinins obtained by isoimmunization. Twenty-one monospecific sera have been produced from the complex sera obtained by isoimmunization. These are as follows: anti-A, -B, -C, -D, -E, -F, -G, -H, -J, -K, -L, -M, -N, -O, -P, -Q, -R, -S, -T, -U, -V. These designations are only temporary ones. Monospecific sera are designated by the letters of the alphabet in the order of detection, thus the designations do not refer to any relation amongst each other. The inheritance of antigens, determined by these of monospecific sera was studied in family material. The classification of the antigenic factors into genetic systems requires further investigations to be carried out.

REFERENCES

FISCHBEIN, M., 1913. Isoagglutination in man and lower animals, *J. Infectious Diseases,* **44**, 86–93.

HADDA, S. and ROSENTHAL, F., 1913. Studien über den Einfluss der Hämolysine auf die in Kultur lebendes Gewebe ausserhalb des Organismus, *Z. Immun. Forsch.*, **16**, 524–548.

IRWIN, M. R., 1956. Blood grouping and its utilization in animal breeding, Proc. VII. Int. Congr. Anim. Husb., Sect. 2, 7–32.

LAW, G. R. J., MILLER, W. J., ASMUNDSON, V. S. and STORMONT, C., 1965. Blood groups of turkeys, *Genetics*, **51**, 253–261.

MUNKÁCSI, F., 1959. Magyar Tudományos Akadémia Agrártud, *Osztályának Közleményei*, **3–4**, 452–455.

PODLIACHONK, L., BÖSINGER, E. and BENOIT, J., 1964. Sur les groupes sanguines de canards domestiques, *C. R. Sc. Acad. Paris*, **256**, 2435–2437.

XITH EUROPEAN CONFERENCE ON ANIMAL BLOOD GROUPS AND BIOCHEMICAL POLYMORPHISM
Warsaw, July 2nd–6th, 1968

Leukocyte Antigens in Chickens*

D. O. SCHMID and P. THEIN

Institute of Animal Blood Group and Resistance Research, Livestock Breeding Research Organization, Munich, G.F.R.

SUMMARY. Specific antileukocyte sera have been obtained by immunization of chickens with spleen, bone marrow and blood from other chickens. Methods of immunization, isolation of leukocytes, preparation of antisera and detection of antibodies are described. By means of these specific antileukocyte reagents we succeeded in detecting specific leukocyte antigens.

INTRODUCTION

The leukocyte antigens are cell substances, which cause the formation of antibodies and which can be detected like the red cell antigens by serological methods in an antigen-antibody reaction.

We intended to produce isologous leukocyte-antisera in chickens and to demonstrate their mode of action in order to test for genetically determined markers. Because of the lymphocytic blood picture of the chicken we speak of antileukocytic sera, a separation of granular leukocytes from lymphocytes in the cell suspension utilized for the immunization not being possible.

MATERIAL AND METHODS

Immunization

We immunized three groups of 67 cross-bred Sussex × New Hampshire hens. The first group containing 41 animals was immunized with a 25% spleen cell suspension. The second group composed of 6 animals immunized with a 25% bone marrow suspension and the third group of 20 animals immunized with blood leukocytes. One chicken was immunized with spleen cells and a leukocyte suspension from different donors. Each immunization series was composed of 8 intramuscular injections per animal. We immunized twice a week with 2.0 ml antigen. The cell suspen-

* This research work has been carried out with the assistance of the Deutsche Forschungsgemeinschaft.

sion of the blood-leukocytes was adjusted to 1,600,000 cells per injection. In contrast to the first and third experimental series, in the second experimental series, in which we used bone marrow as antigen, each donor was used to supply inoculum for one recipient. Recipient chickens of the spleen cell series were re-immunized one month after completion of the initial immunization and were given 6 intra muscular injections of 3.0 ml of a 25% cell suspension. All immunizations were done without adjuvant. Seven to 10 days after the end of the immunization we collected the plasma.

Detection of antibodies

In order to isolate leukocytes a lot of different methods have been described. Hemolytic methods and filtration methods were not appropriate for the collection of a useful leukocyte fraction, so we applied the differential centrifugation of whole blood with addition of dextran described by DAUSSET (1956). This method gave very good results. In isolation of leukocytes for antigen we used only female blood because in contrast to males females have 45% more white blood cells. The collection of blood has to be done very carefully. We stabilized 10.0 ml whole blood with 2.0 ml of 2% EDTA-solution. Citrate anticoagulants reduce the leukocyte content. We always used glass tubes instead of plastic tubes, because in our first experiments we found them much better. Immediately after the collection of blood we added 2.0 ml of a 5% dextran-solution to each tube. The blood was sedimented over night at 4°C as the inclined position of the tube recommended by DAUSSET does not contribute towards acceleration of sedimentation with chicken blood. We had the same results with Poviet-dextran having a molecular weight of 160,000 (Amstel Brewery Amsterdam) and with dextran of Pharmacia-Uppsala having a molecular weight of 250,000.

After spontaneous sedimentation the plasma was discarded and the layer of white blood cells above the red cells was carefully pipetted off and stored. The red cell sediment was centrifuged for 8 min at 1000 rpm. The leukocytes still present formed a layer and were pipetted off and centrifuged together with the previously separated leukocytes, for 15 minutes at 2000 rpm. Although the leukocytes which are pipetted off are still contaminated with a small quantity of erythrocytes the isolation can be regarded as terminated. In order to exclude plasma-reactions, leukocytes which are used as antigens in antigen-antibody tests are washed in saline. For the storage of the test leukocytes isologous plasma is recommended. The storage must be carried out 4°C and should not exceed 24 hr. Detection of antibodies was by means of the agglutination technique. Agglutination was performed in glass tubes of 7 mm diameter. The test contained one drop of antiserum and one drop of a 20% cell suspension mixed by shaking. After an incubation time of 90 min at 37°C interrupted twice by shaking at intervals of 30 min one drop of the test was pipetted onto a slide and was read microscopically. The strength of agglutination was judged according to the criteria of van Loghem from negative to ++++

according to the number of agglutinated cells and the size of the aggregates. A macroscopic reading of the reactions on plates was not possible. Each test was controlled by suspending the test leukocytes in saline and in normal chicken plasma.

RESULTS

Heterologous antilymphocyte sera have been prepared during recent years in different animal species. To date such heteroimmune sera have been prepared in mouse, rat, guinea pig and dog. Immunization with chicken leukocytes has been tried before. WARNER (1964) stated, that an intradermal isoimmunization causes antibody formation. This was proved histologically by the existence of highly inflammatory exudates with perivascular, lymphocytic infiltration. BOREL (1964) obtained a serum with a titre of 1/128 after isoimmunization of four chickens with white cell suspension in Freund's adjuvant. However it turned out, that this serum was not an antileukocyte-serum but an anti-erythrocyte serum of the specificity N-1. TERASAKI *et al.* (1959) detected leukocyte agglutinins after the immunization of chickens with spleen cells, and therefore we also began to immunize with the spleen of healthy chickens. After the successful termination of this immunization series we immunized with bone marrow cells and leukocytes, the collection of which from the blood causes more difficulties in chickens than in mammals.

Human medicine utilizes in experiments the immunosuppressive effect of lymphocytic antisera in order to protect grafts. Another possible application lies in the treatment of reactions of incompatibility after blood transfussion and pregnancies and possibly in the diagnosis and therapy of the autoimmune diseases in humans and animals. The investigation of genetically determined structures on chicken leukocytes induced us to produce and purify specific leukocyte antisera. The immunosuppressive effect of these sera and their uses have still to be examined.

By means of isoimmune chicken anti-leukocyte antisera we succeeded in detecting antigens on white cells of chickens by means of the agglutination test. The leukocytotoxic reaction, a complement fixation test, was, possibly because of in appropriate complement, not successful. As these antigens can not be regarded as chemically characterized compounds they can only be defined by means of antibody formation. The research on leukocyte antigens focuses on the differentiation of the antibodies in question. Natural antibodies to white blood cells of chicken have never been observed in constrast to those which exist to red blood cells in many animals.

With the exception of 3 reagents shows reaction patterns which do not differ from those of the red cell antigens A_7, B_{13} and B_{38}, we can say, that the antigens determined by us can be regarded as group-antigens detected only on leukocytes. This statement was confirmed very recently by extensive comparison tests in MC DERMID's laboratory at Oswestry, United Kingdom. I want to thank Mr. MCDERMID once more for a most effective cooperation.

Observations in human leukocyte serology, indicating that completely identical leukocyte agglutination patterns are not detected have been essentially confirmed for the chicken. We did however detect 3 completely identical agglutination patterns with 2 and in 1 case 3 immune antisera. Reproduction of serological results was good. To date we have not observed the agglutination negative, absorption positive, phenomenon described by VAN ROOD and coworkers. In contrast to work in human lymphocyte serology we do not feel satisfied with a statistically significant uniformity of agglutination reactions as a criterion for the identity of two antibodies defining the same antigen and the characterization of antigens and antibodies by computer analysis. We feel, that the serology of leukocytes ought not to deviate from the principle rules of general blood group serology. DAUSSET stated only recently, that with regard to the 40 described leukocyte antigens in human beings no two agglutinating or cytotoxic leukocyte antibodies are known, which would be likely to give identical reaction against a sufficiently numerous leukocyte panel.

Actually we detect leukocyte antigens in chickens by means of oligo or multispecific phenogroup reagents. We have produced 90 different reagents with which 58 different reaction patterns can be determined. These reagents we call as preliminary measure L1–L58. Absorptions proved, that we are dealing with specific reactions and that the reaction patterns can be divided into different phenogroups.

Analogous complexity of antigen structure of chicken red cells and the antigen structure of chicken white cells is apparent. We cannot exclude, that several genetic systems participate in the control of these antigens.

Family studies have shown, that the inheritance of leukocyte antigens is autosomal and codominant.

REFERENCES

BOREL, J. F., 1964. Recherches immunogénetiques sur les substances specifiques de groupes chez la poule et sur leur utilisation comme marquers de gènes dans l'élevage, Inaug. Diss. ETH Zürich Nr. 3485.

DAUSSET, J., 1956. Immunohematologie biologique et clinique, Flammarion, Paris.

ROOD, J. J. VAN, 1962. Leukocyte grouping, Thesis, Leiden.

ROOD, J. J. VAN, 1963. Leukocyte grouping, *J. Clin. Invest.*, **42**, 9.

SCHIERMAN, L. W. and NORDSKOG, A. W., 1962. Relationship of erythrocyte to leukocyte antigens in chickens, *Science*, **137**, 620.

TERASAKI, P. I., CANNON, J. A. and LONGMIRE, W. P., 1959. Antibody response in homografts. I. Technique of lymphoagglutination and detection of lymphoagglutinins upon spleen injection, *Proc. Soc. Exp. Biol.*, **102**, 280.

THEIN P. and SCHMID, D. O., 1968. Über isolage Antileukocytenseren vom Huhn, *Zeitschrift für Immunitätsforschung*, **135**, 388.

WARNER, N. L., 1964. Immunological reactions produced by the local injections of adult fowl leukocytes in chickens, *Austral. J. Exp. Biol. Med. Sciences*, **42**, **417**.

XITH EUROPEAN CONFERENCE ON ANIMAL BLOOD GROUPS AND BIOCHEMICAL POLYMORPHISM
Warsaw, July 2nd–6th, 1968

Gel Diffusion Studies on Polymorphic Serum Proteins in the Chicken

E. M. McDERMID

Charles Salt Research Centre, The Robert Jones and Agnes Hunt Orthopaedic Hospital, Oswestry, Shropshire, Great Britain

SUMMARY. Iso-immune antisera to chicken blood serum proteins have been stimulated by prolonged immunization using a variety of methods. Results of testing all individuals of 4 lines of chicken in one generation with all available antisera are given. Several reaction patterns of antisera are demonstrated. The relationship between these immune antibodies and the naturally occurring antibodies of other workers is discussed.

INTRODUCTION

SKALBA (1966) has shown that both the gamma-globulins and beta-globulins of chicken blood serum are polymorphic. Using iso-immune antisera she was able to identify 4 antigens on gamma-globulins and 8 antigens on beta-globulins. The expression of these antigens develops slowly in young chickens and can be identified on proteins of the yolk of egg. DAVID *et al.* (1965) and PETROVSKÝ *et al.* (1966) have both described a system of serum groups in the chicken identified by naturally occurring antibodies. PETROVSKÝ found some birds to possess the antigen, others the antibody and others to possess neither. This antigen is carried on the albumen. DAVID also found an isoprecipitin requiring high salt concentration to produce reaction and was able to stimulate iso-immune antibodies giving positive and negative reactions in some lines of chicken tested. SKALBA found no similarity between PETROVSKÝ's natural system and her iso-immune systems.

MATERIALS AND METHODS

The four lines of chicken used in these studies have been established at Oswestry from importations made from Thornber control flocks. These flocks bred by a method designed to maintain a constant genetic variability from generation to generation were established in 1959 from the basic breeding flocks used in the blood grouping studies. Breeding at Oswestry has been carried out so as to maintain the antigens of the B blood group system within the 4 populations. The lines are described in Table 1.

The same immunization procedure was used irrespective of the inoculum except when alum precipitated gamma-globulin was used as antigen. The schedule adopted was that of weekly inoculation of one millilitre of solution by intraperitoneal, intravenous and intramuscular routes in regular rotation. Dosage was adjusted according to antibody titre and was that just sufficient to give visible reaction in the inoculum. Test samples were obtained monthly, immediately prior to immunization which was continued for at least 6 months and where possible over 10 months.

TABLE 1

Description of 4 lines of chicken maintained at Oswestry

Breed	Name of line	Antigens of B blood group system
1. White Leghorn	Red	1, 2, 11, 12, 13 and 14
2. White Leghorn	Green	19 and 21
3. Light Sussex	Blue	35, 36, 37 and 38
4. Rhode Island Red X New Hampshire	Gold	6, 7, 8, 9 and 10

The inocula used were whole serum, alum precipitated gamma-globulin, agglutinated Salmonella bacteria (S. London O), agglutinated proteus bacteria (Proteus OX 2), agglutinated pooled sheep red cells and immune precipitates of bovine serum albumin — anti B.S.A. complexes.

The alum precipitated gamma-globulin was prepared by the method of PROOM (1943) as quoted by DUNSFORD and GRANT (1959). The antisera to S. London O, Proteus OX 2, sheep red cells and B.S.A. were all prepared in chickens by the same

TABLE 2

Immunization procedure used to prepare inocula antibodies

Substance	Dose	Immunization day						
		1	8	16	23	26	29	33–37
Gamma-globulin	50% Suspension	Intra-peritoneal (I.P.)	Intra-muscular (I.M.)	Intra-venous (I.V.)	I.P.	I.M.	I.V.	Bleed
S. London O.	Normal concentration							
Proteus OX 2	Normal concentration							
Sheep r.b.c.	50% Suspension							
B.S.A.	2% Suspension							

Note: S. London O and Proteus OX 2 were obtained from Burroughs Wellcome Ltd., London as 15× concentrated suspensions. They were diluted in 15 ml isotonic sodium chloride to give normal concentration of bacteria.

procedure as was used to stimulate antibodies to alum precipitated gamma-globulin (see Table 2). The sheep red cells and B.S.A. were not emulsified in Freund's complete adjuvant, but the other antigens were.

The antisera raised to all these inocula were tested by a variation of the Ouchterlony agar gel precipitation methods. 1% solutions of Ionagar No. 2 (OXO Limited, London) with 1% sodium azide as bacteriostat and either 1% or 8% sodium chloride were poured to a depth of approximately 2 mm into 9 cm diameter Petri dishes. The wells were made according to a template with well diameter 1.5 cm and distance between wells 3 mm by means of a Harshaw disposable Pasteur capillary pipette (Harshaw Chemicals Limited, Daventry, Northamptonshire) 5¾ in. (14 cm) long attached to a water suction line. Normally 6 or 8 peripheral wells are arranged around the central well and by using this micro-technique (lambda) quantities of reactants are required and approximately 10 templates can be set up on the same Petri dish. Sera are introduced into the wells by capillary attraction from samples held in micro-haematocrit tubes (Hawksley & Sons Ltd., Lancing, Sussex). Precipitations are allowed to develop in a moist chamber at room temperature, 15–20°C, for 10–14 days and are controlled by adding a rabbit anti-chicken antiglobulin antisera, known to produce a precipitate with chicken serum within 24 hr by this method, to one or more of the peripheral wells on the dish. They are stained with a 1% solution of Napthalene black (Amido black 10 B — George T. Gurr, London) in water, after having been washed in several changes of saline over a period of 2 days. This staining can reveal precipitation lines not otherwise seen.

RESULTS

Preliminary results indicated that there were 35 sera which potentially contained precipitating antibodies. The results of testing these 35 sera against approximately 160 sera from the 4 lines of chicken are given in Table 3.

TABLE 3

Distribution of reaction of precipitating antisera

No. reaction	1 Line out of 4	2 Lines out of 4	3 Lines out of 4	All 4 lines	Total
8	9(4.3)	9(6)	6(4)	3	27/35

The four lines can occur together in 15 combinations, 4 single lines, 6 pairs of lines, 4 triplets of lines and 1 group of all 4 lines. Of those antisera which react with antigens of more than one line, 6 react with sera of Blue and Gold lines and 4 with sera of Blue, Red and Green lines.

Table 4 gives details of the success rate obtained with each of the different methods of stimulation.

TABLE 4

Comparison of different methods of stimulating iso-precipitins

Inoculum	No. Immunized	No. yielding antisera	% success
1. Alum precipitated gamma globulin	14	1	7
2. Proteus OX 2	24	9	37
3. Salmonella London O	16	2	12
4. Sheep r.b.c.	16	1	6
5. Bovine serum albumin	24	10	42
6. Whole serum	9	1	11
Total	103	24	23

This evidence suggests that the classically accepted method of raising antisera against serum proteins — that of using aggregates coated with antibody — is for chickens the one most likely to be successful. However, it seems that bovine serum albumin is at least as good a carrier particle as bacteria and that bacterial species vary amongst themselves as suitable carriers. It is worth noting that in my previous attempts to raise iso-precipitating antisera (McDERMID, 1964) I used *Salmonella mansoni* as carrier bacterium and found the system more successful in stimulating precipitins than any of these quoted in Table 4. The precipitins raised with B.S.A. aggregates appear to have a more restricted reactivity than those raised with Proteus agglutinates. DAVID *et al.* (1965) used Brucella agglutinates as inoculum and raised antisera.

TABLE 5

Reaction patterns obtained using 27 precipitating antisera

Line	No. of different patterns
Blue	3
Gold	6
Red	3
Green	3

Table 5 shows the least number of reaction patterns obtained in each line using all 27 iso-precipitins for identification. In each line some sera do not react with any of the antisera and so probably more antigens remain to be detected.

SKALBA suggests that there are at least 2 genetic loci controlling antigen of chicken blood serum identified by iso-precipitins and that each of these possess a series of multiple alleles. My results seem to support the idea of multiple alleles controlling the antigens identified.

DISCUSSION

No one appears to have been successful in reproducing antibodies of the specificites obtained by SKALBA. My earlier antisera, although reactive by her techniques did not give similar reaction patterns in her flocks. PETROVSKÝ *et al.* (1968) find isoimmune antisera to give results exactly like those obtained with his naturally occurring antibodies which are usually found in laying hens. DAVID *et al.* (1965) found naturally occurring antibodies in 4 birds only. Out of 27 antisera apparently containing precipitins in this study, 24 birds were definitely immunized with one or other of 6 inocula used and 3 were not (see Tables 3 and 4). Of these 3 birds, 2 were donors of Proteus antisera and therefore have anti-Proteus antibody in their sera; one was a donor of whole serum and was not known to have circulating antibody in its serum. This bird might therefore possess naturally occurring antibodies like those of PETROVSKÝ and DAVID.

I have recently had the opportunity of comparing some of PETROVSKÝ's antisera with my own and some of them do give reaction patterns like those of my antisera against selected sera from my flocks.

Evidence has recently been produced that naturally occurring antierythrocyte antibodies occur in the Fayoumi breed of chicken (MANSO, *et al.* 1968). The chickens used in these experiments and in the one I describe here have not been given any prophylactic immunization as is common practice in both America and Europe with chicken flocks, so if the "natural" antibodies found have arisen as a result of bacterial stimulation, then it was from normal exposure to bacteria and not as a result of deliberate immunization. It does, however, seem necessary to investigate the possible role of bacterial infection as a "revealing" mechanism for chicken isoprecipitins, perhaps because of enhanced immunoglobulin production in response to bacteria. Eight out of 12 Proteus agglutinate recipients and 9 out of 11 B.S.A. aggregate recipients show antibody to the respective carrier particle 155 to 165 days after the initial injection of the coated particles. Thus, response to immunization may not necessarily have been to the aggregate but possibly to the carrier alone.

ACKNOWLEDGEMENTS

I wish to acknowledge the generous invitation issued by the Polish Academy of Sciences and the receipt of a research travel grant issued by the Welcome Trust. This work was also aided by a grant from the Medical Research Council. Without these I would have been unable to attend the conference. I wish also to acknowledge the technical assistance of Mrs. J. BANCROFT and the help of Mrs. MURIEL JACKSON in the preparation of the manuscript.

REFERENCES

SKALBA, D., 1966. Antigenic differences of hen serum proteins detected by anti-allotypic immune sera, Polymorphismes Biochimiques des Animaux, Xth European Conference on Animal Blood Groups and Biochemical Polymorphisms, Institut National de la Recherche Agronomique, Paris, pp. 477–480.

DAVID, C. S., KAEBERLE, M. L. and NORDSKOG, A. W., 1965. Gamma-globulin allotypes in chickens, *Immunogenetics Letter*, **4**, 95–97.

PETROVSKÝ, E., MACHA, J., MÜLLEROVÁ, Z. and BEZECNA, O., 1966. Blood and serum groups in chickens, Polymorphismes Biochimiques des Animaux, Xth European Conference on Animal Blood Groups and Biochemical Polymorphisms, Institut National de la Recherche Agronomique, Paris, pp. 207–212.

PROOM, H., 1943. The preparation of precipitating sera for the identification of animal series, *J. Path. Bact.*, **55**, 419.

DUNSFORD, F. I. N. and GRANT, J., 1959. The anti-globulin (Coomb's) test in laboratory practice, Oliver and Boyd, Edinburgh, p. 20.

McDERMID, E. M., 1964. Immunogenetics of the chicken — a review, *Vox Sang.*, **9**, 249–267.

PETROVSKÝ, E., 1968. Personal communication.

MANSO, F., FAVRET, E. A. and LIFSCHITZ, E., 1968. A preliminary report on the occurrence of natural iso-agglutination in fowl, *Immunogenetics Letter*, **5**, 132–133.

DISCUSSION

R. L. SPOONER: Did I not catch your routes of immunization?

E. M. McDERMID: All the three normal methods I/P, I/M and I/V were used and I alternated these.

R. L. SPOONER: Did you find any differences between the different routes?

E. M. McDERMID: No, because in one immunization schedule all three routes are used.

R. L. SPOONER: This is interesting because I have found a very greatly increased response using adjuvant by the intravenous route.

E. M. McDERMID: I adopted this procedure because my previous experience in chickens was in the preparation of anti-red-cell sera. For instance, intramuscular is every bit as good as any other route. But to produce a response to bacterial human lipo-proteins a combination of the three routes seems to be the most effective.

M. KAMINSKA: I do not have any personal experience with chicken allotypes but I worked on allotypes in ducks some years ago. This paper has not been published yet. We obtained two precipitating systems, one in IgG and the other in IgM globulins. In both cases the immunization was performed by using different routes of injection (several groups of rabbits immunized simultaneously): intradermal with Freund adjuvant, intramuscular and intravenous. The production of antibodies was exactly similar in the three groups and we had quite a high levels of antibodies.

XITH EUROPEAN CONFERENCE ON ANIMAL BLOOD GROUPS AND BIOCHEMICAL POLYMORPHISM
Warsaw, July 2nd–6th, 1968

Proteins of the Seminal Fluid from the *Vas deferens* of Cocks: their Polymorphism and Relation to Serum Proteins

A. STRATIL

Laboratory of Physiology and Genetics of Animals, Czechoslovak Academy of Sciences, Liběchov, Czechoslovakia

SUMMARY. Proteins of the seminal fluid from the *Vas deferens* of cocks were studied by means of starch gel electrophoresis and immunoelectrophoresis. The presence of "transferrins" (Tf_{SP}) and "albumins" (Alb_{SP}) was proved. Both proteins are immunologically related to the corresponding proteins of the blood serum. The existence of Tf_{SP} polymorphism was confirmed which is analogous to the polymorphism of serum transferrins. Heterogeneity of Tf_{SP} is discussed. Alb_{SP} is polymorphic too, and the phenotypes correspond to the albumin phenotypes of the serum.

A number of other proteins, which call for more profound study, was detected in the seminal fluid of cocks.

INTRODUCTION

Cock semen consists of the content of *Vas deferens*. "Transparent fluid" is either not present or, if it is, its amount is variable (LAKE, 1966).

NISHIYAMA (1957) proved electrophoretically the presence of the same protein fractions in "transparent fluid" as in the serum; the only difference was in their quantitative distribution. OGDEN (1962) found, by means of starch gel electrophoresis, in seminal plasma of roosters a protein which conformed to serum transferrin but differed from it in its pattern. Genetic variants of this protein corresponded to those of the serum transferrins and the conalbumins of egg white. Similar results were obtained by STRATIL (1968a) who suggested that an analogous situation also existed in "albumins". More detailed results concerning these problems are presented in this report.

MATERIALS AND METHODS

Cocks (*Gallus gallus*) of different breeds (White Leghorn, Partridge Leghorn, Light Sussex, White Cornish) and their crosses were used for investigations. The animals were 7–24 months old. After they were slaughtered their semen was retrieved from *Vas deferens* and the fluid from the testes. Contamination with blood could not be avoided in the collection of testicular fluid. Spermatozoa were removed by centrifugation, seminal fluid (SP) and testicular fluid were used either instantly or they were stored at $-15°C$. Several months' storage did not affect the resulting separation (except of the fractions with esterase activity).

Starch gel electrophoresis was carried out in alkaline and acid buffers according to previously published procedures (STRATIL, 1967b; 1967c; 1968a; 1968b).

Fractions with esterase activity were detected according to the method described by KÚBEK (1970).

Microimmunoelectrophoresis was carried out by essentially the same procedure as described in detail by BRUMMERSTEDT–HANSEN (1967). The antibodies used were the same as in the previous work (STRATIL, 1967a): anti-non-laying hen serum, anti-egg yolk serum and anti-egg white serum.

Nomenclature of polymorphic proteins

To designate products of one locus in different body fluids we use the symbol of the locus and the abbreviation of the body fluid in the form of a subscripts; e.g. serum transferrins — Tf_S, egg yolk transferrins — Tf_{EY}, egg white conalbumins — Tf_{EW}, seminal plasma "transferrins" — Tf_{SP}; serum albumins — Alb_S, egg yolk albumins — Alb_{EY}, seminal plasma "albumins" — Alb_{SP} (see BAKER *et al.*, 1968). For designation of genetic variants of transferrins and albumins we use the letters A, B and C.

RESULTS

Electrophoretic separation of SP proteins

Figure 1*a* shows the separation of seminal fluid proteins from the *Vas deferens* (seminal plasma — SP) after starch gel electrophoresis performed in alkaline buffer. The whole pattern resembles, to a certain extent, that of serum proteins. Characteristic is the presence of the "albumin" fraction (Alb_{SP}) and the "transferrin" fractions (Tf_{SP}). In the region between the borate line and the Alb_{SP} fraction there are several zones, the best revealed of them is fraction Pa 1, migrating a little slower than the borate line. This fraction is characteristic for SP — it is not present neither in testicular fluid nor in the serum. It is difficult to characterize the other fractions of this region because they are so weak. Their number, localization and intensity differ in individual patterns. In the region between Alb_{SP} and Tf_{SP} there is a number of other weak fractions (at least 6) which may also be found in testicular fluid where they exhibit much stronger intensity (see Fig. 1*a*). Of special interest is the presence of zones with strong esterase activity; in Fig. 1*b* they are designated Es 1–Es 5. None of these esterases could be detected in the serum.

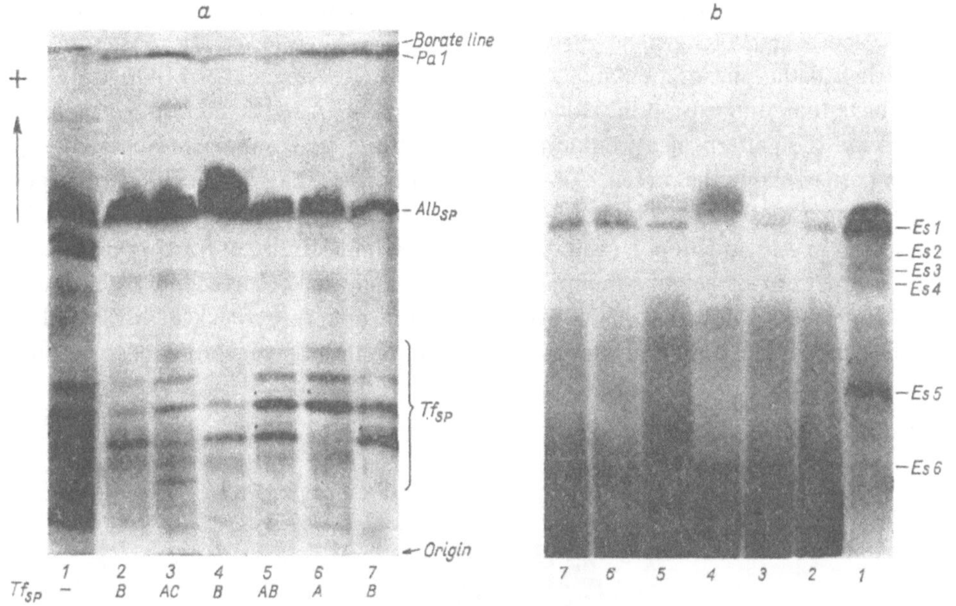

FIG. 1. Starch gel electrophoresis of seminal fluid from the *Vas deferens* and testicular fluid of cocks (alkaline buffer system). *a* — Amido Black staining. Note polymorphism and heterogeneity of Tf$_{SP}$. *b* — Gel after staining for esterase activity. Sample 1 — testicular fluid, samples 2–7 — seminal fluid. Sample 4 was stored 3 months at −15°C.

Another distinct region is Tf$_{SP}$ which is represented by a larger number of fractions (see further). Between Tf$_{SP}$ and the origin there are some other zones.

Anodically from the origin, far beyond Tf$_{SP}$ there is a diffuse area with distinct esterase activity which differs in intensity from esterase activities of the testicular fluid (Fig. 1*b*) and the serum. Only the position designated as Es 6 on the photo shows the fraction with esterase activity which is stronger in the serum than in testicular fluid or seminal plasma.

When using acid buffer, the majority of fractions of SP with strong esterase activity are localized cathodically from the origin, while in testicular fluid there is only one very weak esterase fraction, and there is none in the serum.

Polymorphism of SP proteins

"Transferrins" (Tf$_{SP}$). Some Tf$_{SP}$ phenotypes are shown in Fig. 1*a*. Since this polymorphism has already been described, we include only a brief note. It is known that Tf$_{SP}$ corresponds to Tf$_S$ and Tf$_{EW}$ phenotypes, homozygotes having three fractions. We have succeeded in proving the presence of a further weak fraction which is localized anodically from the slowest zone and is marked with a dot in some patterns on the photo. This fraction occurs in all phenotypes, but in some patterns with low intensity of fractions it is not visible.

"*Albumins*" (*Alb*$_{SP}$). Alb$_{SP}$ observed are shown in Fig. 2. Genetic variants of Alb$_{SP}$ correspond to genetic variants of Alb$_S$. We have tested 81 roosters, 16 of which had the phenotype Alb$_{SP}$ AB, 45 Alb$_{SP}$ B, 15 Alb$_{SP}$ BC, 5 Alb$_{SP}$ AC and we have found no exception. However, the pattern of Alb$_{SP}$ is interesting. Figure 3*a* presents the pattern of a homozygote Alb$_{SP}$ B and its scanning in reflected light. The pattern shows the presence of 4 fractions, fraction 1 having the same or a slightly lower intensity than fraction 3. An analogous picture appears also in Alb$_{EY}$ (Fig. 3*b*), whereas in Alb$_S$ fraction 1 exhibits much lower intensity than fraction 3 (Fig. 3*c*). In the pattern of serum albumins, fraction 1 may be observed only when using a stronger concentration of the sample for electrophoresis. Peak 4 of SP and egg yolk may be found in all phenotypes but photographic documentation of correspond-

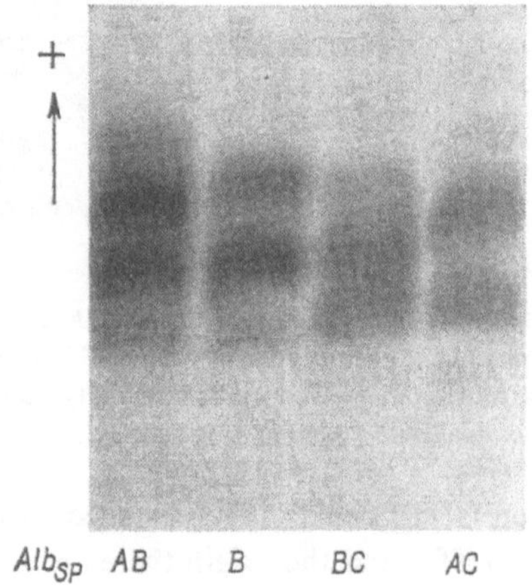

FIG. 2. "Albumin" phenotypes of seminal fluid from the *Vas deferens* of cocks (Alb$_{SP}$). Acid buffer system.

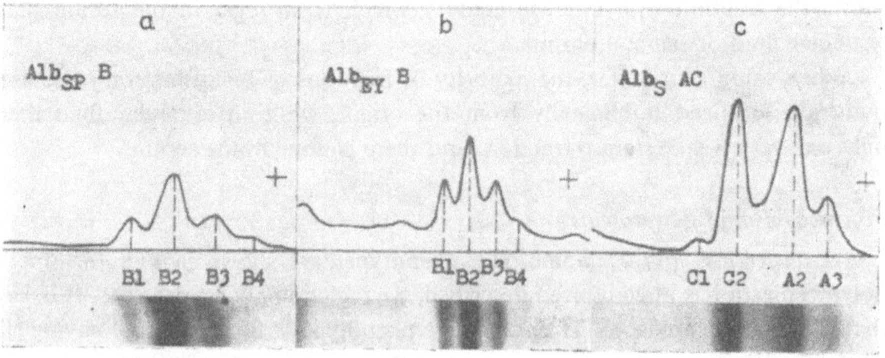

FIG. 3. Electrophoretic separation of albumins: *a* — seminal fluid, *b* — egg yolk, and *c* — serum, in acid buffer. Patterns were scanned in reflected light after staining with Amido Black. The peaks, designated A, B and C respectively belong to the cerresponding genetic variants.

ing fraction is difficult because of the low intensity. All fractions (1–4) in all fluids (SP, egg yolk, serum) change their mobility in dependence on genetic variation, so that it may be presumed that they are all part of the albumin variant.

Immunoelectrophoresis of SP

Figure 4 shows the pattern of immunoelectrophoretic precipitation of SP. Using anti-non-laying hen serum we detected two strong arcs, the anodically localized one being "albumin" and the line localized cathodically representing "transferrin". In

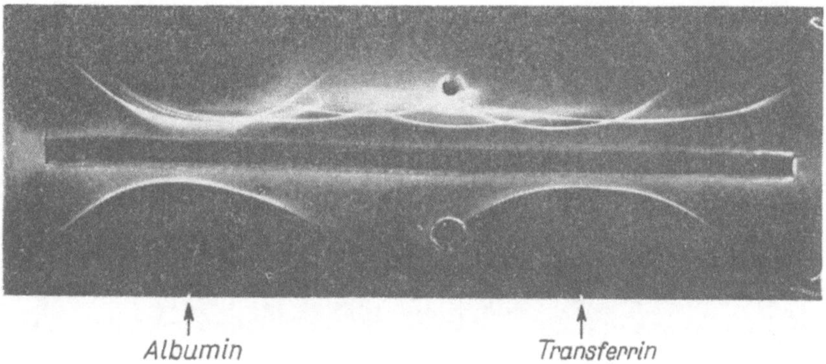

FIG. 4. Immunoelectrophoresis of cock serum (upper hole) and seminal fluid (lower hole). Anti-non-laying hen serum was used as antibody.

48 samples we found only these two arcs; in the remaining 14 tested samples there were, besides the two mentioned, also some other very weak lines in the α_1, α_2 or β_1 regions. The relation of intensity between Alb_{SP}/Tf_{SP} lines was not constant in all investigated samples.

To demonstrate the similarity of Alb_{SP} and Alb_S, and of Tf_{SP} and Tf_S we have carried out combined diffusion experiments (acc. to CLAUSEN and HEREMANS, 1960). In both, albumins and transferrins fusion occurred, which shows that Alb_{SP} are immunologically related to Alb_S, and Tf_{SP} to Tf_S (Fig. 5). The same results were obtained when using anti-egg yolk serum. Combined diffusion also showed immunological relationship between Tf_{SP} and egg white conalbumins when using anti-egg white serum.

DISCUSSION

The results obtained from the starch gel electrophoretic studies of genetic polymorphism and immunoelectrophoretic investigations furnish adequate evidence that "transferrins" and "albumins" of the seminal fluid from the *Vas deferens* of roosters are genetically determined in the same manner as the same proteins in the serum.

The considerable heterogeneity of Tf_{SP} is evidently caused by the non-protein part of the protein molecule (sialic acid). This hypothesis is supported by the results from the study of transferrins and conalbumins in hens (WILLIAMS, 1962; 1966) and from the research on iron binding proteins of human cerebrospinal fluid (see SCHULTZE and HEREMANS, 1966). The presence of a further weak fraction in the Tf_{SP} pattern will apparently be an analogy to the heterogeneity of conalbumins in egg white.

The similarity of Alb_{SP} and egg yolk albumin patterns is very interesting, and so is the difference between Alb_{SP} and serum albumin pattern. The cause of the

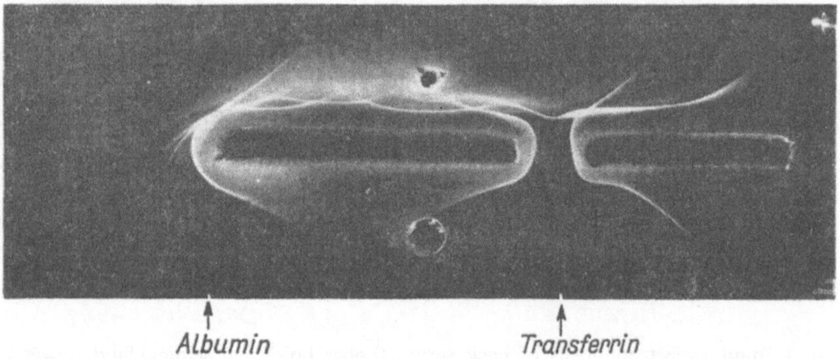

FIG. 5. Combined diffusion, showing fusion of Alb_{SP} with Alb_{S}, and Tf_{SP} with Tf_{S}. At the top, cock serum; below, seminal fluid. Antibody: anti-non-laying hen serum.

considerable heterogeneity of Alb_{SP} (and also Alb'_{BY}) and their phenotypical difference from Alb_{S} has not yet been disclosed. It is probable, however, that this heterogeneity is caused by the binding of extraneous substances by albumin and/or polymer formation.

The numerous protein fractions in SP of cocks, detected by means of starch gel electrophoresis, show that this fluid is, with respect to proteins, just as complex as the serum. Worth mentioning, besides Alb_{SP} and Tf_{SP} is the presence of the prealbumin fraction (Pa 1) which has no analogy in the testicular fluid or in the serum, and the fractions exhibiting esterase activity. Es 1–Es 5 were found to have much higher intensity in the testicular fluid than in SP, whereas Es 6 is the predominant esterase in the serum. On the other hand there are fractions with esterase activity in SP (when using acid buffer) which have not been detected in the testicular fluid or the serum.

Our results show that some SP proteins of roosters are specific for the *Vas deferens,* while others have their analogy in the testicular fluid. Proteins found in SP which showed analogy to those of the serum (Tf_{SP}, Alb_{SP}) differed from them in some features, however.

REFERENCES

BAKER, C. M. A., CROIZIER, G., STRATIL, A. and MANWELL, C., 1968. Proposal on nomenclature of chicken polymorphic proteins. (In preparation).

BRUMMERSTEDT-HANSEN, E., 1967. The serum proteins of the pig. An immunoelectrophoretic study, Munksgaard, Copenhagen, pp. 169.

CLAUSEN, J. and HEREMANS, J., 1960. An immunologic and chemical study of the similarities between mouse and human serum proteins, *J. Immunol.*, **84**, 128–134.

KÚBEK, A., 1970. Electrophoretical study of the esterases in pig serum, XIth European Conference on Animal Blood Groups and Biochemical Polymorphism, Warsaw, pp. 355–358.

LAKE, P. E., 1966. Physiology and biochemistry of poultry semen. In: Advances in reproductive physiology. Edited by A. MCLAREN, Logos Press, London, Vol. I, pp. 93–123.

NISHIYAMA, H., 1957. On the characteristics of the transparent fluid. II. An electrophoretic study of proteins of the transparent fluid, *J. Fac. Agric. Kyushu Univ.*, **11**, 63–68.

OGDEN, A. L., 1962. Expression of the transferrin gene in the serum, egg white and seminal fluid proteins of the chicken, Mimeographed Report on the Eighth European Conference on Animal Blood Group Research. Ljubljana.

SCHULTZE, H. E. end HEREMANS, J. F., 1966. Molecular biology of human proteins with special reference to plasma proteins. Vol. 1. Nature and Metabolism of Extracellular Proteins, Elsevier Publishing Company, Amsterdam–London–New York, pp. 732–745.

STRATIL, A., 1967a. Relationship between chicken serum, egg yolk, and egg white proteins, Royal Veterinary and Agricultural College Yearbook 1967, Royal Veterinary and Agricultural College, Copenhagen, pp. 57–76.

STRATIL, A., 1967b. The effect of iron addition to avian egg white on the behaviour of conalbumin fractions in starch gel electrophoresis, *Comp. Biochem. Physiol.*, **22**, 227–233.

STRATIL, A., 1967c. Genetic and non-genetic differences in chicken egg white ovalbumins, *Folia biol.* (Praha); **13**, 476–478.

STRATIL, A., 1968a. Transferrin and albumin loci in chickens, *Gallus gallus* L, *Comp. Biochem. Physiol.*, **24**, 113–121.

STRATIL, A., 1968b. Genetická, fysiologická a chemická variabilita proteinů slepic, Thesis (In Czech). Liběchov, pp. 153.

WILLIAMS, J., 1962. A comparison of conalbumin and transferrin in the domestic fowl, *Biochem. J.*, **83**, 355–364.

WILLIAMS, J., 1966. The sites of attachment of carbohydrate to conalbumin and transferrin in the hen. In: Protides of the Biological Fluids. Vol. 14. Elsevier Publishing Company, Amsterdam, pp. 65–70.

DISCUSSION

I. E. LUSH: Do you find any trace of genetic variation in the esterases of seminal plasma?

A. STRATIL: We have observed some differences in the mobility of esterase fraction E_s1 in individual samples, but at present there is not sufficient evidence if it is genetic polymorphism or not.

M. KAMINSKI: How do you reveal the esterases — what substrate and azo-coupler?

A. STRATIL: B-naphtylacetate and Diazo-Blue-B salt were used for revelation of the esterase activity.

M. KAMIŃSKI: Do you incubate the gel in alkaline buffer before revelation?

A. STRATIL: All gels were incubated in alkaline buffer before revelation.

XITH EUROPEAN CONFERENCE ON ANIMAL BLOOD GROUPS AND BIOCHEMICAL POLYMORPHISM
Warsaw, July 2nd–6th, 1968

Polymorphism of Chicken Serum Transferrin

F. S. VYSHINSKY and V. I. MURAVIEV
All-Union Research and Technological Poultry Institute, U.S.S.R.

SUMMARY. Sera from various chicken breeds were subjected to disc electrophoresis. Transferrin variations of five types were found: A, B, AB, BC and AC. As a result of the analysis of these types it was concluded that transferrin types in chickens are controlled by a triallelic system of genes at a single locus with six corresponding phenotypes: TfA, TfB and TfC in homozygous form and TfAB, TfBC and TfAC in heterozygous form. The triallelic hypothesis was tested by the genetic analysis of offspring from matings of tested parents. The existence of transferrin subtypes, in particular two subtypes of TfB are suggested.

INTRODUCTION

The study of genetical polymorphism of tissue proteins and tissues and animal protein substrates is of great value both scientifically and practically. The problems raised are new and complicated and it is not to be expected that they will be solved easily.

The initial aim of the work described was to study transferrin polymorphism in the serum of various chicken breeds and to see if whether such polymorphism has value which can be exploited practically.

The work of OGDEN *et al.* (1962) is unique in the literature of transferrin polymorphism in chickens in providing genetic analysis of the polymorphism reported. The authors used starch gel electrophoresis of chicken serum and observed three inherited transferrin types (a, b, ab) controlled by two alleles at a single locus. These transferrin types are inherited as simple codominant characters.

METHODS AND MATERIALS

ORNSTEIN and DAVIS (1962), DAVIS (1964), RAUSCH *et al.* (1965a,b), SAFONOV and SAFONOVA (1964), DEBABOV and REBENTISCH (1966) and others in their works describe the advantage of protein electrophoresis in polyacrylamide gel. RAUSCH *et al.* (1965b) show that the transferrin type picture obtained as a result of disc electrophoresis in polyacrylamide gel is completely similar to that obtained by using electrophoresis in a starch gel. For these reasons we used disc electrophoresis according to DAVIS (1964) with the following modification of our own.

The apparatus allows the simultaneous electrophoresis of twenty samples. Cylindrical 260×300×80 mm cups are used as buffer reservoirs. Twenty evenly spaced holders for gel containers are situated on the bottom of the upper reservoir 105 mm from the centre. Spiral platinum electrodes are held by special lids in the reservoir centres. Gel containers are made of cylindrical 83 mm glass tubes with an inside diametre of 5.5 mm. 1.4 ml small-pore division gel was used. The sample of serum for dispersion was about 400 γ protein. With 1000 ml of buffer and 2.5 mA current per container electrophoresis was carried for 2 hr at 4°C. Electrophoregrams were stained using acid blue and black dye.

Blood serum of hens and cocks from two populations of the Russian White breed, Leghorn "K", Zagorsk Salmon, Light Sussex and Cornish "M" breeds were studied. Studies were also carried out on blood serum of embryos (incubator by-products) and chicks.

RESULTS AND DISCUSSION

As previously reported (VYSHINSKY, 1968) the transferrin zone of the electrophoregrams shows from 2 to 4 bands in the different types. The bands are designated by letters a, b, c and d. Band a migrates the fastest and band d the slowest.

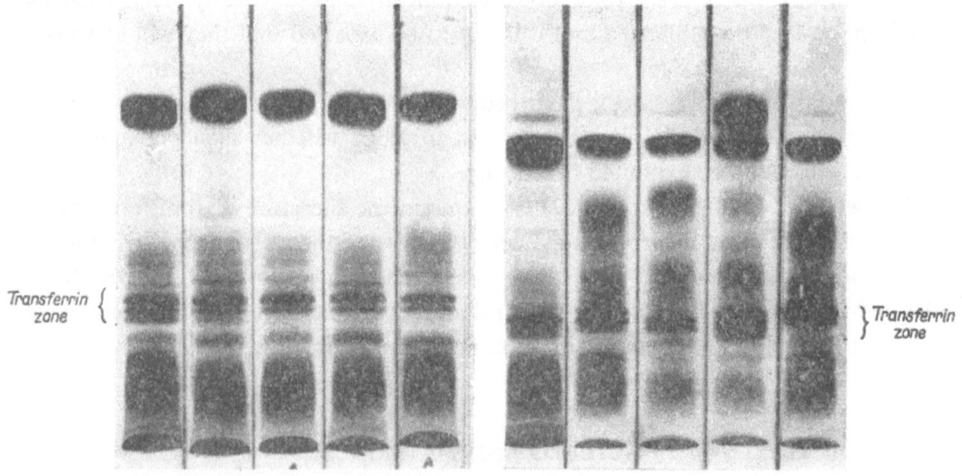

Electrophoregrams of chicken sera.

In the Russian Whites and Zagorsk Salmon 2 variations were found in each breed. In Sussex 5 variations were found each with all the four bands and in Leghorns and Cornish one variation was found (Figure).

Based on analysis of the variations observed it can be suggested that chicken transferrin types are controlled by a triallelic system of gene at one locus with six corresponding phenotypes and not by a diallel gene with but three corresponding phenotypes as proposed by OGDEN *et al.* (1962).

Chicken serum transferrin types

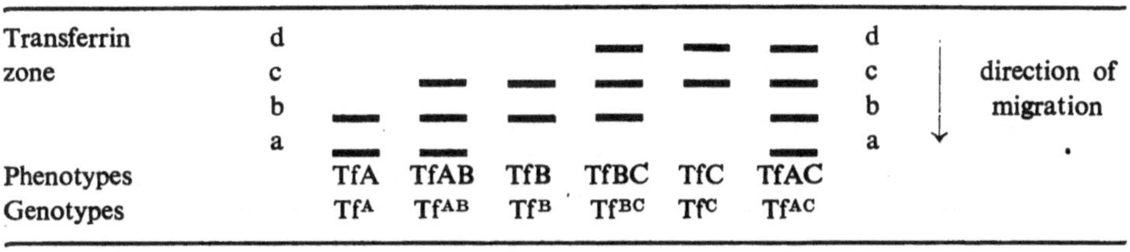

Obviously three different band pairs correspond to the three homozygote phenotypes viz: bands a and b to TfA, bands b and c to TfB and bands c and d to TfC; in heterozygotes the bands abc, bcd and abcd correspond to phenotypes TfAB, TfBC and TfAC as is shown in the scheme.

Of six suggested phenotypes we have as yet found five: TfA, TfB, TfAB, TfBC, TfAC the distribution of which are given in Table 1.

TABLE 1

Transferrin phenotypes in various breed chickens

Breed	No. tested	Phenotypes					
		TfA	TfAB	TfB	TfBC	TfC	TfAC
Russian Whites hens	190	—	—	175	15	—	—
cocks	96	—	—	93	3	—	—
Leghorn "K" hens	67	—	—	67	—	—	—
cocks	33	—	—	33	—	—	—
Zagorsk Salmon hens	70	—	—	67	3	—	—
Light Sussex hens	60	—	6	52	2	—	—
cocks	5	—	1	4	—	—	—
Embryos and chicks	70	3	21	45	—	—	1
Cornishes "M" hens	100	—	—	100	—	—	—
cocks	40	—	—	40	—	—	—
Total	731	3	28	676	23	—	1

The results of blood analysis of offspring from matings of tested parents are inconclusive due to small numbers of offspring and lack of certain mating types. But they provide some evidence for the hypothesis of the triallelic system of genes at the transferrin locus and codominant inheritance of these genes so they are given in Table 2.

Simultaneously the phenotype TfB was found to have two variants or subtypes

TABLE 2

Segregation of transferrin types among offspring of various matings

Matings type	No. of matings	Transferrin types and offspring number				Total No. of offsprings
		A	AB	B	AC	
AB×AB	3	3	8	4	—	15
AB×B	5	—	3	7	—	10
AB×BC	2	—	1	3	1	5
B×AB	3	—	9	10	—	19
B×B	9	—	—	23	—	23
Total	22	3	21	47	1	72

according to the intensity of band expression. In one of them the band c stains intensely and is wide. This variant is temporarily designated TfB^1. In another subtype, TfB^2 the band is thin and weak. Subtypes of other phenotypes may exist.

REFERENCES

DAVIS, B. J., 1964. Disc electrophoresis — II. Method and application to human serum proteins, *Annals of the New York Academy of Sciences*, **121**, 404–427.

DEBABOV, W. G. and REBENTISH, B. A., 1966. Fractionation of histones by electrophoresis in polyacryl-amide gel, *Biochemistry Acad. Sci. U.S.S.R.*, **31**, 943–947.

VYSHINSKY, F. C. O geneticheskom polymorphizme transferrina syvorotky krovy kur razlichnyh porod (On the genetic polymorphism of various breed chicken serum transferrin), *Sbornik rabot molodykh uchenykh VNITIP*,* **XI** (In press).

OGDEN, A. L. *et al.*, 1962. Inherited variants in the transferrins and conalbumins of the chicken, *Nature*, **195**, 1026–1028.

ORNSTEIN, L. and DAVIS B. J., 1962. Disc electrophoresis, Preprinted by Distillation Products Ind. N.Y.

RAUSCH, W. H. *et al.*, 1965a. The determination of bovine transferrin types by disc electrophoresis, *J. Dairy Sciences*, **48**, 6, 720.

RAUSCH, W. H. *et al.*, 1965b. Inheritance of bovine transferrin types as determined by disc electrophoresis, *J. Dairy Science*, **48**, 7, 990.

SAFONOV, V. I. and SAFONOVA, M. P., 1964. Metod electroforeza belkov rasteny v synteticheskoi srede — poliakrilamidnom gele (The method of plant protein electrophoresis in the synthetic medium of poliacrylamid gel), *Fiziologia rastenii*, **11**, 147.

* VNITIP — All-Union Research and Technological Poultry Institute.

XITH EUROPEAN CONFERENCE ON ANIMAL BLOOD GROUPS AND BIOCHEMICAL POLYMORPHISM
Warsaw, July 2nd–6th, 1968

Taxonomical Value and False Individual Variations of Esterases in Avian Sera: a Survey of Bio-, Histo- and Immunochemical Characters

Marie KAMINSKI

Laboratoire d'Enzymologie du C.N.R.S., Gif-sur-Yvette, France

SUMMARY. Five lipoesterases and 2 non-lipoesterases were identified in avian sera; lipoesterases being mainly function-specific while the non-lipoesterases were species-specific.

INTRODUCTION

Taxonomical investigations have generally been carried out on structural proteins, more likely to differ from one species to another being less involved in essential physiological events.

However, as the part of the molecule containing the active site might be controlled by genes other than those controlling the "inert" part, it is conceivable that functional proteins could also undergo a phylogenetical differentiation. It is therefore of great interest to investigate such components, provided that appropriate methods are used for analysis. Numerous attempts were made to detect analogous enzymes in different animals and to define a species by its enzyme pattern.

The esterases for example, were studied either by measuring activities of fractions separated by a physicochemical method or by histochemical revelation after starch gel electrophoresis. Unfortunately, the first method proved unsuitable for discriminating between isodynamic enzymes and the second tends to favour the demonstration of minute heterogeneities of each component but can not positively establish relationships between them. Therefore it is an advantage to characterize enzymes, in addition to their catalytic and physicochemical properties, by a third parameter such as antigenicity.

The immunohistochemical method permits differentiation of isodynamic enzymes, which are different proteins, from families of isoenzymes, which are multiple molecular forms of the same enzyme; the method can also demonstrate relationship between proenzymes, active enzymes and inert derivatives.

During our previous studies on avian serum esterases (KAMINSKI, 1964, 1966, 1967) other causes of misinterpretation became evident, related to the lipid moiety present in many esterases. These enzymes, besides their hydrolytic function, may serve to fix temporarily and/or to transport lipids. The lipid content of the complex may therefore be modified and cause significant alterations of physicochemical properties, reflected mainly by changes of electrophoretic mobility or elution from an ion exchanger.

In the present paper species specificity of avian esterases in whole serum, and DEAE-cellulose fractions of 5 species of Gallinaceae, 2 of Anatidae and 2 of Columbidae were compared in homo- and heterologous reactions.

MATERIALS AND METHODS

Duplicate immunoelectrophoretic plates were revealed with β-naphthyl acetate coupled with Diazo Blue B, and with Indoxyl acetate. Other substrates or dye-couplers were used occasionally. The second plate was generally stained with Lipid Crimson subsequently to esterase revelation. The DEAE-cellulose fractionation was performed by elution with an ionic strength gradient of a phosphate buffer, pH 6. The α- and β-lipoproteins were isolated by ultracentrifugation-flotation in KCl gradient; the delipidization was performed by digestion with a lipase or by ethanol-ether treatment. The full details will be presented elsewhere (KAMINSKI and PIROELLE; KAMINSKI and AYRAULT).

RESULTS

In homologous reaction significant differences are observed between sera from Anatidae and Columbidae, containing large amounts of α_1-nonlipoesterase and sera from Gallinaceae containing mainly a β_1-nonlipoesterase; the first was identified as an acetylesterase and the second as cholinesterase. These components did not show individual variations and were stable upon storage. The lipoesterases were present in all species and showed marked variations between samples. The β- or low-density lipoproteins were partially precipitated by agar, and the reaction consisted in a long trail bordered by elongated arcs of immunological precipitate. The electrophoretic location varied between origin and α_0-region; no consistent correlation with either species or hereditary characters could be demonstrated. At any rate, these components are antigenically different from other lipoesterases, even though in a given sample the electrophoretic location was the same and the enzymatic properties similar. Treatment with lipase did not significantly change their electrophoretic behaviour.

The α- or high-density lipoproteins, likewise present in all species, were even more troublesome to identify without use of immunological criterion. The relation of electrophoretic location and duration of storage has already been reported (KAMINSKI, 1966); it was subsequently shown that digestion with lipase reproduces variations of mobility even though delipidization was not complete. Several agents, mainly detergents have similar effects. The esterase and antigenic properties are maintained, but quantitatively reduced. A chemical treatment releasing lipids produces a protein migrating as an α_2-protein, still immunologically and enzymatically active. A progressive transformation of a lipoprotein migrating faster than albumin into an α_2-component was also obtained by treatment with a cationic detergent. The preserved immunological individuality permitted its identification despite the presence of another α_2-esterase. The physicochemical heterogeneity of high-density lipoproteins was further reflected by their elution behaviour: in low ionic strength the faster fraction alone is obtained; the slower one is eluted together with serum albumin.

In starch gel differences between slowly and fast-migrating α-lipoesterases are much less pronounced; however, lipase treatment produces an acceleration of migration and detergent a slowing. In the pattern obtained, the fastest band corresponds to another esterase, having the same biochemical characters, which can lead to confusion in interpretation of results, especially for family studies. This esterase was identified by immunoelectrophoretic analysis.

In heterologous reaction, DEAE cellulose fractions from sera of different species were compared by immunodiffusion or immunoelectrophoresis tests with different antisera; esterases belonging to the group of lipoesterases showed little or no species specificity and were hardly different between families; conversely, the non-lipoesterases proved strictly family and species specific.

DISCUSSION

Besides the fact that the number of different components is far larger in our study than in those previously reported, the use of different methods and the rarity of such studies in birds prevents precise comparison with results of other authors. The phylogenetical signification of antigenic similarity between esterases of different species and families as compared to marked differences between other body components is that probably function had a dominant effect on the evolution of molecular structure.

Another interesting point is the difference in cross-reactivity between lipo- and non-lipoesterases. Do the non-lipoesterases which are species specific, represent a class of esterases restricted to birds or do they represent a less evolved level in the phylogenetical line?

REFERENCES

KAMINSKI, M., 1964. Esterases in avian sera: species specific pattern and individual variations, *Experientia*, **20**, 286.

KAMINSKI, M., 1966. Immunoelectrophoretic examination of avian serum esterases, *Nature*, **209**, 723.

KAMINSKI, M., 1967. Unpublished data.

Serum Esterase Polymorphism in Chickens

J. CSUKA and E. PETROVSKÝ

Department of Genetics, University of Agriculture, Brno, Czechoslovakia
Laboratory of Animal Physiology and Genetics, Liběchov, Czechoslovakia

SUMMARY. The manifestation of polymorphic plasma esterase fractions in chickens is influenced by physiological processes. In the period before reaching sexual maturity the active esterase phenotypes (Es A, Es B, and Es AB) change into the Es 0 type in some chickens. In cocks such a change does not occur. In younger populations and in hens with a high level of egg production the active phenotypes disappear earlier and in a larger number of birds than in less productive populations.

INTRODUCTION

Genetic polymorphism of esterases has been observed in a number of animal species, such as insects, fish, laboratory and farm animals. As a rule, three phenotypes are involved, controlled by two co-dominant alleles; in mice only, four phenotypes have been found, controlled by two dominant and one recessive allele (RANDERSON, 1965).

Besides genetic esterase variability, physiological changes have also been described. TANAKA and NAKAJO (1959) observed a varying level of cholinesterase in the diencephalon of hens in the period between ovulation and egg-laying. ALLEN and HUNTER (1960) described changes which involved the complete disappearance of esterase fractions in the mouse epididymis after castration. SHAW (1965) discovered the dependence of esterase in the mouse liver on sexual hormones: male adults have one fraction more than females or juvenile males. Changes in the cholinesterase level in rats after the application of ACTH were also found by VACCAREZZA and WILSON (1964). In mice, the expression of the electrophoretic zone C of the blood serum increases during growth, is strongest in gravid females, and weak during lactation (OKI *et al.*, 1966).

In birds this enzyme has been studied by several investigators. KAMINSKI and JEANNE-ROSE (1964) did not demonstrate polymorphism in several species of birds, including the chicken. On the other hand, BOREL (1964) reported that in chickens genetic variability occurred in the form of the presence or absence of one esterase fraction (types Es+ and Es−). CSUKA and PETROVSKÝ (1968) described four types: Es 0 (without esterase activity), Es A (a rapidly migrating fraction), Es B (a more slowly migrating fraction), and Es AB (with both fractions).

BAKER and MANWELL in their study of esterase polymorphism in pheasants did not observe any physiological changes due to age, sex, or egg production. On the other hand, CSUKA and PETROVSKÝ (1968) found that the manifestation of esterase fractions changes in chickens in connection with the beginning of egg production.

The present paper contains a more detailed description of these physiological changes in the polymorphic esterase system.

METHODS AND MATERIAL

For the study of physiological changes in the polymorphic system of esterases we used a total of 133 hens and cocks. Thirty-five cocks and hens were from the inbred strain A of breed W, 24 cocks and hens from the inbred strain R of breed RIR, 32 individuals were hybrids produced by mating cocks of the inbred strain A with hens of strain R (group 9), while the rest were hybrids of cocks of strain A and semi-inbred hens of the breed RIR (group 10).

Hens and cocks were sampled at approximately monthly intervals (a total of 11 samples) from 5 months of age. Blood was taken into heparin to allow collection of plasma.

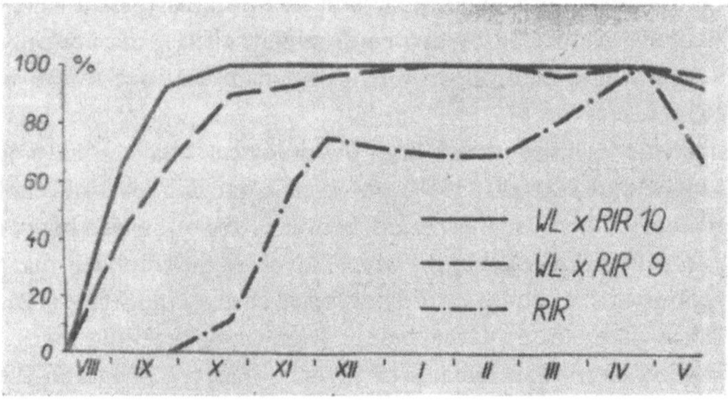

FIG. 1. Ontogenetic changes in Es 0 frequency.

Esterase variation was studied with the aid of horizontal electrophoresis in 13% starch gel from hydrolyzed starch. We used POULIK's (1957) discontinuous system modified according to ASHTON and BRADEN (1961). The voltage drop was 5–6 V/cm and the gel was water cooled.

Esterase was detected by the method of LAVRENC *et al.* (1960).

RESULTS

At the age of 5 months individuals studied displayed the following phenotypes: all the individuals of breed LW — type Es 0, all the individuals of breed RIR — type Es B. The latter was also exhibited by all cocks and hens from group 9, while in group 10 there were 37 individuals of type Es B and 3 hens of type Es 0.

In succeeding tests we found that no change in type occurred in the hens and the cocks of breed W, i.e. in the original phenotype Es 0, or in the cocks of breed RIR and the hybrids. On the other hand, in all hybrid hens and hens of breed RIR the type Es B sooner or later changed into type Es 0.

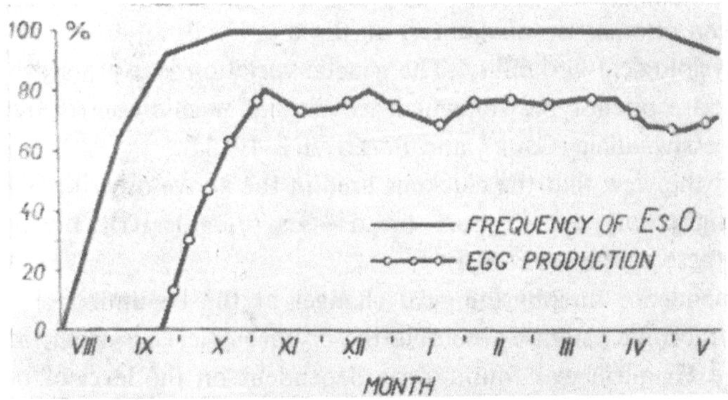

FIG. 2. Ontogenetic changes in Es 0 frequency and egg production (WL × RIR 10).

Figure 1 illustrates the changes in the frequency of type Es 0 in strain R and both groups of hybrids. Inactivation of the "active" type Es B in the hybrids started at the beginning of September, while in the individuals of breed RIR this did not occur until the end of October. Complete inactivation was recorded in the April measurements, while in May we noticed that in several laying hens the original — Es B-type manifested itself again.

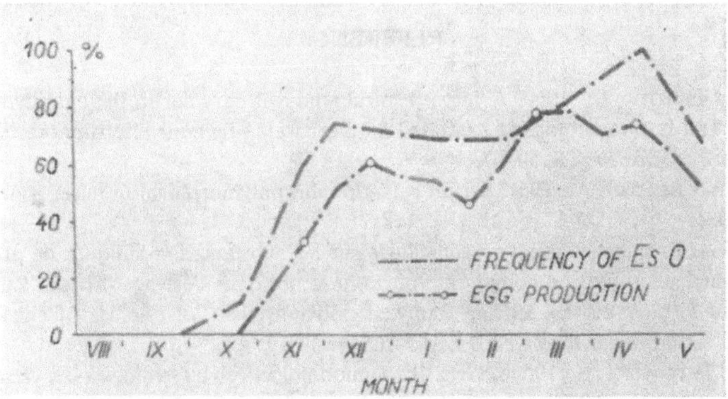

FIG. 3. Ontogenetic changes in Es 0 frequency and egg production (RIR 7).

The following Figures demonstrate the time course of these changes and its relationship to the beginning of egg production. In the group of hybrids — 9 and 10 — the change of active-type esterase started at about 40–45 days after the beginning of egg production (Fig. 2), while in the hens of breed RIR where both the change of esterase and the beginning of egg production occurred much later, this time difference was only 15–20 days (Fig. 3) and the course of the changes displayed a character different from that in the hybrids.

DISCUSSION AND CONCLUSIONS

In chicken esterase we encounter, on the one hand, a genetic, and on the other hand, a physiological variability. The genetic variation is expressed by the existence of four esterase phenotypes for which we assume genetic control by two dominant and one recessive allele (CSUKA and PETROVSKÝ, 1968a).

We hold the view that the chickens used in the above described experiments had the following genotypes: Leghorn breed — Es 0/Es 0, RIR breed — Es B/Es B, hybrids of these breeds — Es B/Es 0.

The dependence on physiological changes at the beginning of egg production makes polymorphic esterase resemble the erythrocytic Hi system in which the activity of the Hi allele was found to be dependent on the level of oestrogenic hormones (SHEINBERG and RECKEL, 1962, and other authors).

The fact that in hens with a high rate of egg production a change from the active phenotypes to a type without esterase activity occurs, reflects itself in management practice of selection of more productive layer-hens of the Es 0 type, which is associated with a significantly higher egg production (CSUKA and PETROVSKÝ, 1968a,b). These differences, of course, have a purely physiological background, and therefore, we rank polymorphic esterase among "physiological" markers.

REFERENCES

ALLEN, J. and HUNTER, R., 1960. A histochemical study of enzymes in the epididymis of normal, castrated, and hormone replaced castrated mice separated by zone electrophoresis in starch gel, *J. Histochem. Cytochem.*, **8**, 50–57.

ASHTON, G. and BRADEN, A., 1961. Serum beta-globulin polymorphism in mice, *Aust. J. Biol. Sci.*, **14**, 248–253.

BOREL, J., 1964. Recherches immuno-génétiques sur les substances spécifiques de groupes chez la poule et sur leur utilisation comme marquers de génes dans l'élevage, Thesis, Zürich.

CSUKA, J. and PETROVSKÝ, E., 1968a. Study of polymorphism of esterase of chicken egg white and blood serum, *Folia Biologica*, **14**, 165–168.

CSUKA, J. and PETROVSKÝ, E., 1968b. Genetika immunologických a fysiologických vlastností drůbeže. IV. Zmeny fenotypových frekvencií sérové esterázy slepic při selekci na ekonomicky významné vlasností, *Acta Univ. Agric.* (Brno) (In press).

KAMINSKI, M. and JEANNE-ROSE, M., 1964. Esterases in avian sera: Species specific patterns and individual variation, *Experientia*, **20**, 286–289.

LAVRENC, S., MELNICK, P. and WEIMER, H., 1960. A species comparison of serum proteins and enzymes by starch gel electrophoresis, *Proc. Soc. Exp. Biol. Med.*, **105**, 572–575.

OKI, Y., TAKEDA, M. and NISHIDA, S., 1966. Genetic and physiological variations of esterases in mouse serum, *Nature*, **212**, 1390–1391.

POULIK, M., 1957. Starch gel electrophoresis in a discontinous system of buffer, *Nature*, **180**, 1477–1479.

RANDERSON, S., 1965. Erythrocyte esterase forms controlled by multiples alleles in the deer mouse, *Genetics*, **52**, 999–1005.

SHAW, C. R., 1965. Electrophoretic variation in enzymes, *Science*, **149**, 936–943.

SHEINBERG, S. L, and RECKEL, R, P., 1962. Studies on the "HY" agglutinogen in chicken, *Ann. N. Y. Acad. Sci.*, **97**, 194–204.

TANAKA, K. and NAKAJO, S., 1959. Cholinesterase in the diencephalon of the hen in relation to egg laying, *Poultry Sci.*, **38**, 991–995.

VACCAREZZA, J. and WILSON, J., 1964. The relationship between corticosterone administration and cholinesterase activity in rats, *Experientia*, **20**, 425–426.

DISCUSSION

J. E. LUSH: Do you mean that there are two alleles or three?

E. PETROVSKÝ: We suppose there are three alleles, 2 dominant and 1 recessive.

CHAPTER 4

BLOOD GROUPS AND BIOCHEMICAL POLYMORPHISM IN HORSE

XITH EUROPEAN CONFERENCE ON ANIMAL BLOOD GROUPS AND BIOCHEMICAL POLYMORPHISM
Warsaw, July 2nd–6th, 1968

Preliminary Investigations on Blood Groups in Horses

Research Note

S. SIUDZIŃSKI, A. KACZMAREK and J. ZWOLIŃSKI

College of Agriculture in Poznań, Department of Animal Husbandry, Poznań, Poland

SUMMARY. The test sera anti-A, anti-C and anti-H were obtained. By iso-immunization of 9 horses we succeeded in obtaining immune test sera: anti-A, anti-D, anti-H, anti-K and anti-$PO_3 = Sw_{10}$.

The studies on blood groups in horses in Poland, were initiated by HIRSZFELD and PRZESMYCKI (1921), and carried on later by WOYCIECHOWSKA and LILLE–SZYSZKOWICZ (1960), KOWNACKI and SZENIAWSKA (1960) and WADOWSKI (1964).

The first investigations in our faculty, carried out in cooperation with the Pasteur Institute in Paris, were reported on in 1963 under the title: Study on blood groups in some breeds of horses bred in Poland (PODLIACHOUK *et al.*, 1963).

In 1964, we began to study blood groups in horses making iso- and hetero-immunizations and working on normal iso-agglutinins, and so obtained the test sera anti-A, anti-C and anti-H.

It was only last year that we started more systematic studies, utilizing our previous iso-immunizations. Donors and recipients were chosen from among 16 tested horses, on the results of a cross-test of their blood. Four horses were immunized with blood cells, previously washed in saline solution, by injecting 20 ml portions 6 times at 5-day intervals. This treatment resulted in formation of agglutinating antibodies in 3 horses of titre 1:8 to 1:128. The sera obtained were absorbed by conventional methods and monovalent immune sera were elaborated. Two of them were identical, differing only in titre. These sera were identified using blood samples from 54 horses, whose blood factors had been determined at the Pasteur Institute in Paris. The sera obtained appeared to be anti-H and anti-D (two sera).

Further iso-immunizations were carried out and choice of donors and recipients was based on the known blood antigenic structure of the experimental horses. In this series of immunizations we obtained the immune serum anti-K which was produced by injecting blood cells from two donors. In one case, whilst attempting to produce the serum anti-E by iso-immunization, we obtained an unknown serum that we designed Po_3. In the Comparison Test 1968 we found that it was identical with the Swedish Sw_{10}.

In our investigations we twice observed the phenomenon of the increase in the titre of normal antibody as a result of immunization, although the donors did not possess the blood cell antigens to cause such reaction. In the case of the normal serum anti-A the titre increased from 1:8 to 1:128 after immunization, and in that of the normal serum anti-D its titre increased from 1:4 to 1:64.

At present we are in possession of the following sera

Normal sera	Immune sera	
anti-A	anti-A	anti-K
anti-C	—	anti-Po_3 = Sw_{10}
—	anti-D	
anti-H	anti-H	

From a total number of 11 horses subjected to iso-immunization only in 6 horses were there positive reactions strong enough to allow production of monovalent test sera.

It should be mentioned that our observations show that iso-immunizations have had no harmful effect on the health of the experimental animals.

We are continuing our work to obtain more test sera.

REFERENCES

HIRSZFELD, L. and PRZESMYCKI, F., 1921. Badania nad aglutynacją normalną. IV. O izoaglutynacji u koni (Study on normal agglutination. IV. Iso-agglutination in horses), *Przegląd Epidemiolologiczny*, I, VI.

KOWNACKI, M. and SZENIAWSKA, D., 1960. Badania nad otrzymaniem izoaglutynin odpornościowych u koni (Studies on obtaining immune iso-agglutinins in horses), *Roczn. Nauk Roln.*, **76-B-1.**

PODLIACHOUK, L., KACZMAREK, A. and ZWOLIŃSKI, J., 1963. Badania nad grupami krwi niektórych ras koni hodowanych w Polsce (Study on blood groups in some breeds of horses bred in Poland), *Roczn. Nauk Roln.*, **82-B-4.**

WADOWSKI, S., 1964. Koń lidzbarski i jego grupy krwi (Lidzbark horse and its blood groups), Olsztyn.

WOYCIECHOWSKA, St. and LILLE-SZYSZKOWICZ, J., 1960. Badania nad grupami krwi różnych ras koni w zespołach zakażonych wirusem zakaźnego ronienia klaczy Dimocka oraz w zespołach wolnych od zakażenia (Studies on blood groups in horses of various breeds, in areas infected with Dimock's virus of infectious abortion in mares and in infection-free areas), *Roczn. Nauk Roln.*, **69-E-4.**

XITH EUROPEAN CONFERENCE ON ANIMAL BLOOD GROUPS AND BIOCHEMICAL POLYMORPHISM
Warsaw, July 2nd–6th, 1968

Blood Groups in Horses. Genetic Study

Luba PODLIACHOUK
with the technical assistance of
R. BEAUD
Pasteur Institute, Paris, France

SUMMARY. Genetic and mathematical studies enable us to state that the 18 horse blood factors belong to at least 8 genetic systems.

The A system contains 6 blood factors: A_1, F, H_1, H_2, I and O. The J system contains 4 blood factors: E, G, J_1 and J_2. The D system contains 2 blood factors: D_1 and D_2.

The blood factors C, K, Fr_4 and Fr_5 are bi-allelic.

A relationship exists between the blood factors Fr_1 and Fr_3.

The genes that determine the horse's blood factors are located at at least 8 loci.

INTRODUCTION

During our serological studies of horse blood groups we used as blood typing reagents either natural iso- or hetero-antibodies or immune ones. The antibodies may be either agglutinating or hemolysing, or both. The source of complement is fresh guinea pig or rabbit serum suitably absorbed.

So far we are in possession of 18 reference sera: anti-A_1, C, D_1, D_2, E, F, G, H_1, H_2, I, J_1, J_2, K, O, Fr_1, Fr_3, Fr_4 and Fr_5, which define 18 specific equine blood factors. Nine of them are identical with those described by STORMONT and SUZUKI (1964), who identified 16 horse blood factors belonging to 8 blood group systems, 4 of which are multi-allelic.

RESULTS

With our blood typing reagents, we have determined the blood groups of about 4500 horses. A certain number of them were horses of pure breed, others had a well defined pedigree.

Our genetic observations lead to hypotheses of relationships between genes which may take various forms: multiple alleles, a pair of alleles more or less closely linked on the same chromosome, or genes independent of each another. These hypotheses have been subjected to statistical study. The results of the mathematical study of inheritance of phenotypes in cross-breedings from known parents were not inconsistent with the suggested genetic hypothesis that every equine blood factor is determined by a pair of alleles with the presence of the allele dominant to its absence (PODLIACHOUK, 1958).

TABLE 1

χ^2 $n=1$	Probability	Total
0 –0.0158	1 –0.9	12
0.0158–0.0642	0.9–0.8	9
0.0642–0.148	0.8–0.7	11
0.148–0.455	0.7–0.5	22
0.455–1.074	0.5–0.3	28
1.074–1.642	0.3–0.2	13
1.642–2.706	0.2–0.1	4
2.706–3.841	0.1–0.05	11
3.841–5.412	0.05–0.02	4
5.412–6.635	0.02–0.01	2
6.635–10.827	0.01–0.001	8
>10.827	<0.001	12

In order to ascertain whether there is a linkage between the genes or whether their transmission is independent, we applied the chi-square test (χ^2). The chi-square has been determined for all the blood factors, grouping them by pairs, in a group of 300 unrelated horses taken at random at the Garches, Annexe of the Pasteur Institute.

TABLE 2

Probability	Antigen
0.05–0.02	AC CD EG J_2O
0.02–0.01	IH_1 GFr_3 Fr_3Fr_5
0.01–0.001	AH_2 FH_1 FH_2 IC IO DJ_1 H_1O H_2O
<0.001	AF AI AH_1 AO FI FO EJ_1 EJ_2 GJ_2 J_1J_2 H_1H_2 Fr_1Fr_3

The number of possible paired combinations of 17 factors is as follows: $\frac{17\times 16}{2} = 136$ (the blood factor D_2 having been excluded). The degree of freedom chosen by us was 0.05, which is that generally used in biological studies. The results of the chi-square test and the corresponding limits of probability are shown in Table 1.

The value of the chi-square test for 110 pairs of blood factors was not significant, the observed variations being merely accidental. Out of 26 significant results 4 were between the limits of probability 0.05 and 0.02; 2 were between 0.02 and 0.01; 8 were between 0.01 and 0.001 and 12 of them were below a probability of 0.001. The 6 significant results corresponding to probabilities between 0.05 and 0.02, and between 0.02 and 0.01 deviate only slightly from the chosen degree of freedom 0.05. It is difficult to assert that these results are due to connection between the genes. This applies mainly to the pairs A_1C, CD_1, J_2O, and Fr_3Fr_5 (Table 2).

In our previous papers we have pointed out that there is linkage between the blood factors A_1, F and I (PODLIACHOUK, 1957, 1958) and between J_1J_2 and G (PODLIACHOUK *et al.*, 1966). In the present study we observed linkage of the blood factors H_1H_2 and O with A_1, F and I; and also of E with J_1J_2 and G. The blood factor I is a sub-group of A_1. One observes 3 phenotypes A_1+I+, A_1+I−, A_1−I−, but never A_1−I+. Likewise I is a subgroup of F. The blood factor O is a subgroup of A_1 and F. H_1 is a subgroup of H_2. Thus the A system comprises 6 blood factors: A_1, F, I, H_1, H_2 and O. In the J system that contained 3 blood factors J_1, J_2 and G, a fourth must be included, the blood factor E. The D system contains 2 blood factors: D_1 and D_2, the first being a subgroup of the second. The blood factors C, K, Fr_4 and Fr_5 are independent of each another and all other blood factors. Thus they belong respectively to the systems C, K, Fr_4 and Fr_5. Each one is determined by a pair of alleles with the presence of the allele dominant to is absence. Therefore the systems C, K, Fr_4 and Fr_5 are bi-allelic. The blood factors Fr_1 and Fr_3 are not included in these 7 systems, and a relationship exists between them (Table 3).

TABLE 3

Blood groups in horses

Genetic systems	A	C	D	J	K		Fr_4	Fr_5
Blood factors	A_1 F I H_1 H_2O	C	D_1 D_2	E G J_1 J_2	K	Fr_1 Fr_3	Fr_4	Fr_5

CONCLUSION

The 18 horse blood factors belong to at least 8 genetic systems. The genes that determine horse blood factors are located at at least 8 loci.

ACKNOWLEDGEMENTS

My thanks to Dr. NICOL, Dr. GIRARD and Dr. CORVAZIER of the Garches annexe of Pasteur Institute, for kindly supplying most of blood samples and to Dr. LABERT for assistance with the isoimmunization program.

REFERENCES

PODLIACHOUK, L., 1957. Thèses de Doctorat ès-Sciences.

PODLIACHOUK, L., 1958. Les groupes sanguins des équidés, *Ann. Inst. Pasteur*, **95**, 7–22.

PODLIACHOUK, L., SALERNO, A. and LABERT, D., 1966. Les groupes sanguins du cheval salernitain, *Ann. Inst. Pasteur*, **110**, 208–211.

STORMONT, C. and SUZUKI, Y., 1964. Genetic systems of blood groups in horses, *Genetics*, **50**, 915–992

XITH EUROPEAN CONFERENCE ON ANIMAL BLOOD GROUPS AND BIOCHEMICAL POLYMORPHISM
Warsaw, July 2nd–6th, 1968

Blood Group Factors and Erythrocytic Protein Polymorphism in Swedish Horses

K. SANDBERG

Department of Animal Breeding, Agricultural College, Uppsala, Sweden

SUMMARY. A number of 17 horse blood typing reagents have been prepared. The inheritance of the corresponding factors A_1, A′, C, D_1, E, H_1, J_2, K, P_1, P′, Q_1, Sw12, U_1, U_2, Sw9, Sw10 and Sw14 was investigated on family material from two native Swedish horse breeds. The factors were assigned to 8 blood group systems, one of which appeared to be "new".

Starch gel electrophoresis revealed a genetic polymorphism in the enzyme carbonic anhydrase from horse erythrocytes. Family data were consistent with the interpretation that the carbonic anhydrase phenotypes were controlled by 5 codominant autosomal alleles, designated CA^F, CA^I, CA^L, CA^O and CA^S.

INTRODUCTION

At least 20 horse blood group factors have been described in literature (see PODLIACHOUK, 1957; STORMONT *et al.*, 1964). Some of these blood groups are controlled by series of multiple alleles (STORMONT and SUZUKI, 1964).

Electrophoretic variants of the erythrocyte carbonic anhydrase (CA) have been reported in man (SHAW *et al.*, 1962), in domestic cattle and bison (SARTORE *et al.*, 1967) and in sheep (TUCKER *et al.*, 1967).

The present report will briefly describe 17 horse blood group reagents prepared in Sweden, one new blood group locus and one extensive series of multiple alleles. A genetic polymorphism in the horse erythrocyte CA will also be demonstrated. A preliminary report on this polymorphism is published elsewhere (SANDBERG, 1968).

MATERIALS AND METHODS

The blood typing reagents were produced by intramuscular injections of horse blood into 50 horses, 34 rabbits and 10 cattle. The blood typing tests were set up in one saline agglutination section and one lytic section employing absorbed rabbit complement.

The nomenclature proposed by PODLIACHOUK and HESSELHOLT (1962), SUZUKI and STORMONT (1962) and STORMONT *et al.* (1964) will be used. Reagents which have not parallelled already known reagents or are proved to be related (subgroups) to such reagents in comparison tests are given preliminary designations (e.g. Sw9).

In the starch gel electrophoresis a slight modification of the discontinous buffer system (pH 8.5) described by GAHNE (1966) was used. Hemolysates prepared by freezing and thawing fresh, washed horse erythrocytes were inserted into the gel slot by pieces of filter paper (Whatman 3 MM). The gels were stained in a solution of amido black and nigrosin. Carbonic anhydrase activity was demonstrated according to WALDEYER and HÄUSLER (1959) and HYYPPÄ *et al.* (1966).

The family material used in the inheritance study belonged mainly to the two native Swedish breeds the North-Swedish horse and the Gotland pony.

RESULTS AND DISCUSSION

Source and nature of the Swedish blood-typing reagents are shown in Table 1. The genetic study will here be concentrated on factors of special interest. All factors except U_1 were inherited as dominant traits (Table 2). Absorbtions showed that U_1

TABLE 1

Source and nature of Swedish horse blood-typing reagents

Reagent	Nature of antibody	Source
A_1	Hemolysin (agglutinin)	Isoimmune
A′	,, (,,)	Heteroimmune (rabbit)
C	,, (,,)	Isoimmune
D_1	Agglutinin	Heteroimmune (cattle)
E	,,	Isoimmune
H_1	Hemolysin (agglutinin)	,,
J_2	Agglutinin	,,
K	,,	,,
P_1	Hemolysin	,,
P′	,, (agglutinin)	,,
		Heteroimmune (rabbit)
Q_1	,,	Isoimmune
Sw12	,,	,,
U_1	,,	,,
U_2	,,	,,
Sw9	Agglutinin	,,
Sw10	,,	,,
Sw14	,,	,,

and U_2 were related to each other as if they were linear subgroups. In the 12 cases where two U_1 negative parents gave a U_1 positive progeny, both parents were U_2 positive. This observation indicates that the U_1 reagent may detect homozygosity

TABLE 2

The inheritance of the individual blood factors in 252 matings

Blood factor	Mating types of parents: + × +		+ × —		— × —	
	No. of offspring with (+) and without (—) the factor					
	+	—	+	—	+	—
A_1	158	27	39	24	0	4
A′	20	7	75	60	0	90
C	147	10	60	26	0	9
D_1	1	4	18	20	0	209
E	19	4	63	54	0	112
H_1	7	5	28	28	0	184
J_2	99	12	70	47	0	24
K	24	4	43	55	0	116
P_1	62	9	84	52	0	45
P′	7	0	21	33	0	191
Q_1	1	0	12	7	0	232
Sw12	11	2	67	47	0	125
U_1	1	0	6	42	12	191
U_2	56	14	56	49	0	77
Sw9	0	2	38	34	0	178
Sw10	63	8	73	56	0	52
Sw14	195	11	32	12	0	2

for the U_2 factor. This possibility is also supported by the fact that 26 matings between one U_1 positive (U_2 genotype +/+) and one U_2 negative (U_2 genotype —/—) parent, all gave U_2 positive (U_2 genotype +/—) progeny (Table 3). The family material is too small to decide whether all individuals homozygous for the U_2-factor are U_1-positive.

The "new" factors Sw10 and Sw14 seemed to be controlled by allelic genes. None of about 800 bloods tested locked both factors. It was also observed that

TABLE 3

The inheritance of the U_2 factor as revealed by the reagents U_1 and U_2 under the assumption that the U_1 reagent detects the homozygotes +/+ for the U_2 factor

Mating type	Offspring: +/+ (U_1)	+/—	—/—
—/—×—/—	—	—	76
+/—×—/—	—	34	45
+/—×+/—	12	21	14
+/+(U_1)×—/—	—	26	—
+/+(U_1)×+/—	6	17	—
+/+(U_1)×+/+(U_1)	1	—	—

Sw10-positive bloods were either weakly reactive or negative with Sw14 antibodies, while Sw10-negative bloods were always strongly reactive with Sw14 antibodies. Additional evidence supporting the theory of a closed system is provided in Table 4. The similarities with the cattle F-V system as described by STORMONT (1951) are

TABLE 4

The inheritance of the horse blood factors Sw10 and Sw14 as alternatives in 252 matings

Mating type	Offspring		
	Sw10	Sw10 Sw14	Sw14
Sw10 × Sw10	2	—	—
Sw10 × Sw10, Sw14	12	16	—
Sw10 × Sw14	—	16	—
Sw10, Sw14 × Sw10, Sw14	11	22	8
Sw10, Sw14 × Sw14	—	57	56
Sw14 × Sw14	—	—	52

obvious. However, the horse system comprises additional factors. Absorbtion tests showed that the factors E and Sw10 are related as linear subgroups and accordingly the factors E, Sw10 and Sw14 all appear to belong to one system. In addition the family data indicated that these factors belonged to the D-system, which previously (STORMONT and SUZUKI, 1964) has shown to contain the factors D and J. The evidence for this is too extensive to be presented here. Six alleles (d^{Sw14,J_2}, d^{Sw14}, $d^{E,Sw10}$, d^{Sw10}, d^{Sw14,D_1}, d^{Sw10,J_2}) were found in this system.

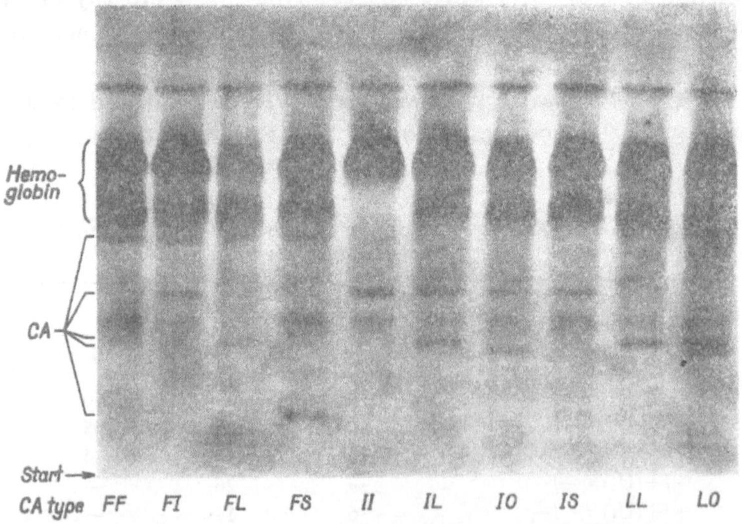

A photograph of a starch gel with 10 different types of carbonic anhydrase.

TABLE 5
Inheritance of CA phenotypes in 236 matings

Mating type	No. of offspring of phenotype							
	II	IL	FI	IO	LL	FF	FS	IS
II×II	112							
II×IL	13	7						
II×FL		5	2					
II×FI	29		39					
II×LO				1				
IL×LO					1			
IL×IL		1						
II×LL		2						
FI×FI	2		5			4		
FI×IS			1				2	1
II ×IS	2							1
FI×IL	1							
FI×FS						1		
II ×FS			1					1
IL×FL		1						
FF×II			1					

The family data also supported the findings of STORMONT and SUZUKI (1964) that A_1, A' and H_1 belong to one system, P_1 and P' to one and Q_1 and Sw12 (possibly R subgroup) to one system. Each of the factors C, K and Sw9 seemed to belong to one-factor systems.

Carbonic anhydrase

Starch gel electrophoresis of hemolysates from the above family material revealed the existence of 10 CA phenotypes (Figure). The different bands were named F, I, L, O and S in order of decreasing electrophoretic mobility. Data on the inheritance of the phenotypes are summarized in Table 5. The observed types of the offspring were consistent with the hypothesis that the CA phenotypes are controlled by five codominant autosomal alleles designated CA^F, CA^I, CA^L, CA^O and CA^S. A considerable quantitative variation was observed in the staining intensity of the enzyme between individuals.

REFERENCES

GAHNE, B., 1966. Studies on the inheritance of electrophoretic forms of transferrins, albumins, prealbumins and plasma esterases of horses, *Genetics*, **53**, 681–694.

HYYPPÄ, M., KORHONEN, L. K. and KORHONEN, E., 1966. Electrophoretic separation of carbonic anhydrase and naphthyl esterase activities, *Ann. Med. Exp. Fenn.*, **44**, 63–66.

PODLIACHOUK, L., 1957. Les antigènes de groupes sanguins des équidés et leur transmission héréditaire, Thése, Paris.

PODLIACHOUK, L. and HESSELHOLT, M., 1962. Les groupes sanguins des équidés. Les sérums de référence, *Immunogenetics Letter*, **7**, 69–71.

SANDBERG, K., 1968. Genetic polymorphism in carbonic anhydrase from horse erythrocytes, *Hereditas*. (In press).

SARTORE, G., STORMONT, C. and GRUNDER, A. A., 1967. Electrophoretic forms of esterase and carbonic anhydrase in the red cells of cattle and bison, *Genetics*. (In press).

SHAW, G. R., SYNER, F. N. and TASHIAN, R. E., 1962. New genetically determined form of erythrocyte esterase in man, *Science*, **138**, 31–32.

STORMONT, C., 1951. The F-V and Z systems of bovine blood groups, *Genetics*, **37**, 39–48.

STORMONT, C. and SUZUKI, Y., 1964. Genetic systems of blood groups in horses, *Genetics*, **50**, 915–929.

STORMONT, C., SUZUKI, Y. and RHODE, E. A., 1964. Serology of horse blood groups, *Cornell Vet.*, **54**, 439–452.

SUZUKI, Y. and STORMONT, C., 1962. A comparison of equine blood-typing reagents, *Immunogenetics Letter*, **2**, 138.

TUCKER, E. M., SUZUKI, Y. and STORMONT, C., 1967. Three new phenotypic systems in the blood of sheep, *Vox Sang.*, **13**, 246–262.

WALDEYER, A. and HÄUSLER, G., 1959. Histochemische Studien über die Carboanhydraseaktivität, der Samenwege und ihrer Anhangsdrüschen von *Mus rattus*, *Acta Biol. Germ.*, **2**, 568–589.

XITH EUROPEAN CONFERENCE ON ANIMAL BLOOD GROUPS AND BIOCHEMICAL POLYMORPHISM
Warsaw, July 2nd–6th, 1968

Blood Group and Serum Type Studies in Basuto Ponies

D. R. OSTERHOFF*, D. O. SCHMID** and I. S. WARD-COX*

* *Department of Zootechnology, Faculty of Veterinary Science, Onderstepoort, South Africa*
** *Institute of Animal Blood Group and Resistance Research, Livestock Breeding Research Organization, Munich, G.F.R.*

SUMMARY. The Basuto pony is perhaps the only equine in Southern Africa which has characteristics so typical that it could be regarded as an indigenous breed.

Remarkably low frequencies of blood factors A_2, C and I, and relatively high frequencies of H and K point to the uniqueness of this breed. A relatively high frequency of the transferrin allele Tf^H and a very low frequency of Tf^D add to the assumption that a specific horse breed has been formed in Lesotho.

No haemoglobin polymorphism could be found. Albumin and esterase types were also investigated and here, evidence was brought forward that a relationship exists with the Boer horse, being a cross between the original Cape horse and English Thoroughbreds imported later.

INTRODUCTION

Breed differences regarding equine blood groups have been extensively reported by SCHMID (1967). The serum polymorphic types of equidae have been examined by KAMINSKI (1964) and by OSTERHOFF and WARD-COX (1967). Concurrent investigations of differences have been performed on Norwegian Döle and Fjord breeds by BRAEND (1964), the American Thoroughbred and Shetland ponies by STORMONT *et al.* (1965), Icelandic horses by HESSELHOLT (1966) and Salernitana horses by GAHNE (1965). In this paper the results of the mentioned investigations are compared with those of the South African Basuto pony.

The Basuto pony is not originally an indigenous African breed. The horse (*Equus caballus*) was unknown in South Africa until the middle of the 17th century, although his ultimate ancestor was flourishing on the North African coast. The opening of the sea routes to the East by the Portuguese and Dutch prepared the way, and in 1653 the first horses of Arab and Barb blood from Java were landed in the Cape by the Dutch East India Company. These horses became the foundation of the Cape horse, which was the direct ancestor of the Basuto pony.

This breed had its origin in Lesotho during the early 18th century when the

Basuto tribes conducted border raids on the migratory farmers and colonial troops. By 1830 the type had developed into the Basuto pony and by 1870 the whole Basuto nation was mounted. Due to the rugged terrain and extreme climatic conditions of Lesotho, the pony developed into a sturdy, sure-footed, hardy animals capable of carrying 190 lb for a distance of 70 miles in one day under the worst possible conditions.

Although Basuto ponies were much used in the Boer War, the breed had already begun to decline, the descent becoming more rapid in the beginning of this century. Since then other stock has been introduced to increase the numbers but isolated groups of true Basutos have been preserved in studs and distributed to selected farmers for breeding purposes.

METHODS AND MATERIAL

Altogether 104 samples were collected from Basuto ponies at the Nooitgedacht Research Station in the Ermelo district of the Transvaal and from various farms in that district.

Standard blood typing techniques (SCHMID, 1967) were used. Transferrin determinations were carried out using the procedure of KRISTJANSSON (1963) and making use of the nomenclature as outlined by BRAEND and STORMONT (1964). Haemoglobins were typed according to the technique of BUSCHMANN (1963). The separation of esterases and albumine was accomplished using the bridge buffer of KRISTJANSSON (1963) and the gel buffer of ASHTON and LAMPKIN (1965). Several modifications as outlined by OSTERHOFF and WARD-COX (1967) were applied.

RESULTS

Blood groups

For immunogenetic characterization 72 Basuto ponies from South Africa were bloodtyped with 28 horse bloodgroup reagents. Table 1 shows the phenotype frequencies.

Remarkably low frequencies were found for A_2, C_1 and J_1, high frequencies were observed for H and K in comparison to breeds investigated earlier (SCHMID, 1967).

Transferrins

The analysis of the transferrin gene frequencies in the various South African breeds together with those foreign investigations are presented in Table 2. It can be seen that a large deviation exists between the Basuto pony and the local and foreign breeds with respect to the genes Tf^D and Tf^H the former being far less and

TABLE 1

Frequency of blood factors in Basuto ponies

A_1	0.459	K	0.194	Mü-18	0.333
A_2	0.777	L	0.013	Mü-19	0.055
C_1	0.125	V_1	0.055	Mü-20	0.027
C_2	0.222	V_2	0.373	Mü-21	0.291
O	0.750	Mü-3	0.041	Mü-22	0.305
D	0.027	Mü-5	0.068	Mü-23	0.106
E	0.055	Mü-9	0.346	Mü-24	0.097
H	0.346	Mü-13	0.001		
J_1	0.000	Mü-15	0.430		
J_2	0.000	Mü-16	0.459		
J_3	0.194	Mü-17	0.055		

the latter being far more than what is normally realized. The frequency of the Tf^R gene is the highest of all the local breeds and the Tf^M gene is absent. These decided differences suggest the existence of a definite breed.

Haemoglobins

As in the case of Thoroughbreds, no polymorphism was revealed in this system, all the animals being of the type A_1A_2, using the nomenclature of BRAEND and EFREMOV (1965).

Albumins

From Table 2 it can be seen that the Basuto albumin gene frequencies fall between those of the Arabs and Thoroughbreds, and coincide with the frequencies pertaining to the "Common horse", which is often called the Boer horse, also developed from the original Cape horse (OSTERHOFF and WARD-COX, 1967).

Esterases

The only comparison with foreign breeds available to date is that of the Salernitana horses of Italy (GAHNE, 1965). The allele Es^I (GAHNE, 1965) was lacking in every South African breed so far investigated. It is, however, possible that our technique does not allow a differentiation between the I and S bands. The allele Es^O makes an appearance in the Basuto pony and has been found occasionally in horses sampled at random in South Africa (OSTERHOFF and WARD-COX, 1967). As this allele is also present in the Salernitana horses, it would be of interest to compare these results with other breeds so as to arrive at more conclusive results.

TABLE 2

Frequency of transferrin, albumin and esterase alleles in local and foreign horse breeds

Alleles	Basutoes (S.A.) (n = 104)	Arabian (S.A.) (n = 45)	Thoroughbreds (S.A.) (n = 54)	Döle (Norway) (n = 220)	Fjord (Norway) (n = 104)	Thoroughbreds (U.S.A.) (n = 150)	Shetland ponies (U.S.A.) (n = 273)	Icelandic horses (n = 925)	Salernitana (Italy) (n = 147)	"Common horses" (S.A.) (n = 235)
Tf^D	.062	.300	.167	.002	.149	.267	.172	.200	.410	.319
Tf^F	.485	.477	.648	.234	.620	.563	.460	.270	.390	.478
Tf^H	.221	.056	.009	.075	.000	.027	.026	.070	.030	.048
Tf^M	.000	.000	.000	.000	.000	.000	.031	.010	.000	.000
Tf^O	.046	.167	.046	.021	.039	.090	.108	.250	.160	.091
Tf^R	.186	.000	.130	.668	.192	.053	.203	.200	.010	.028
Alb^A	.529	.620	.278	.580	.350	.214	.387	—	.340	.479
Alb^B	.471	.380	.722	.420	.650	.786	.613	—	.660	.521
Es^F	.143	.000	.037	—	—	—	—	—	.120	.135
Es^S	.797	.978	.926	—	—	—	—	—	.060	.865
Es^O	.060	.000	.000	—	—	—	—	—	.090	.000
Es^I	.000	.000	.000	—	—	—	—	—	.730	.000
Es_1^X	.000	.022	.000	—	—	—	—	—	—	.000
Es_2^X	.000	.000	.037	—	—	—	—	—	—	.000

DISCUSSION AND CONCLUSIONS

The ponies are selected today for the typical Arab nose and forehead with its fine features, tremendously powerful shoulders and forequarters and small hooves. It would thus be logical to seek its genetic background in the data pertaining to the Arab, the Thoroughbred and perhaps other oriental horses. As the latter has not been available, this aspect is purely speculation. As the numbers are too small to warrant any investigation into the possibility of them being in genetic equilibrium, any relationship will have to be found in the data at present available.

The differences that exist in the blood group and transferrin systems suggest the existence of a peculiar breed. The albumin and esterase types indicate a relationship to "Common horses" (Table 2) which can be regarded as a parallel branch of the Basuto pony also originating from the Cape horse. The appearance of the Es^O allele is at present an unexplained phenomenon although it may be possible that this is inherited from an Oriental breed, as no other breed so far investigated in South Africa reveals this allele.

REFERENCES

ASHTON, G. C. and LAMPKIN, G. H., 1965. Serum albumin and transferrin polymorphism in East African cattle, *Nature*, **205**, 209.

BRAEND, M., 1964 Serum types of Norwegian horses, *Nord. vet. med.*, **16**, 363.

BRAEND, M. and EFREMOV, G., 1965. Haemoglobins, haptoglobins and albumins of horses, Proc. 9th Eur. Conf. Anim. Blood Groups, 253.

BRAEND, M. and STORMONT, C., 1964. Studies on haemoglobin and transferrin types of horses, *Nord. vet. med.*, **16**, 31.

BUSCHMANN, H., 1963. Die Bedeutung der Serumtypenbestimmung für die forensische Veterinärmedizin, *Zentr. Bl. vet. med.*, **B. X**, 49.

GAHNE, B., 1965. Studies on the inheritance of electrophoretic forms of transferrins, albumins, pre-albumins and plasma esterases of horses, *Genetics*, **53**, 681.

HESSELHOLT, M., 1966. Studies on blood and serum types of the Icelandic horses, *Acta Vet. Scand.*, **7**, 207.

KAMINSKI, M., 1964. Serum proteins in equidae: species, race and individual differences, Proc. 9th Eur. Conf. Anim. Blood Groups, 245.

KRISTJANSSON, F. K., 1963. Genetic control of two prealbumins in pigs, *Genetics*, 48, 1059.

OSTERHOFF, D. R. and WARD-COX, I. S., 1967. Preliminary horse breed comparison with regard to haemoglobin and serum type polymorphism, Proc. S., Afr. Soc. Anim. Prod., 6 (In press).

SCHMID, D. O., 1967. Erforschung der Blutgruppen bei Rind, Pferd und Huhn, Habil. Schrift, München.

STORMONT, C., SUZUKI, Y., and RENDEL, J., 1965. Application of blood typing and protein tests in horses, Proc. 9th Eur. Conf. Anim. Blood Groups, 221.

XITH EUROPEAN CONFERENCE ON ANIMAL BLOOD GROUPS AND BIOCHEMICAL POLYMORPHISM
Warsaw, July 2nd–6th, 1968

Chimerism in Horses

M. VANDEPLASSCHE and Luba PODLIACHOUK

with the technical assistance of
R. BEAUD

Ruksuniversiteit, Gent, Belgium
Institut Pasteur, Paris, France

SUMMARY. Observations made on chorial membranes from about 50 cases of dizygotic twin pregnancy in the mare have shown that nearly all cases are bicornual and that invagination of the chorial sacs develops in about 75%. Tissular fusion was observed from the fifth month on in about 40% of cases. Macroscopic vascular anastomosis between both foetal placentae was shown in 5 out of 27 cases. Blood groups in the parents and foals of 10 cases of twin-pregnancy have been determined, using 18 reference-antibodies. Identical blood groups were found in both foals from 4 cases of twin gestation, indicating chimerism and blood-anastomosis between both chorions; blood chimerism was probable in 2 other cases in which one foal only could be examined. Macroscopic fusion between both chorial sacs may occur without the establishment of vascular anastomosis, and blood-chimerism has been shown in some cases where both chorial sacs seemed to be free, thus indicating the occurrence of rather microscopical blood anastomosis. Different blood groups occurred in three cases of twin-pregnancy. Thus there was altogether a surprisingly high incidence of vascular anastomosis between the chorial sacs in cases of bicornual twin-gestation.

In two cases of still-born and in three cases of living bisexual twins of one to two years of age, the genital apparatus of the female co-twin was found to be normal. These observations point to a much lower incidence of freemartin-deformation than in cattle.

INTRODUCTION

LILLIE (1916) observed a vascular anastomosis between the chorions of embryos of dizygotic bovine twins. OWEN (1945) established that these twins are of the same blood group, but that their blood is composed of two different red cells populations. Ninety per cent of dizygotic bovine twins are chimeric. Chimerism or mozaicism of blood can be revealed by partial serological reactions between certain reference antibodies and the corresponding erythrocytic antigens. In men, twelve cases of chimeric twins have been published (VAN der HART and VAN LOGHEM, 1967). Chimerism has also been noticed in marmosets (BENIRCHKE and BROWNHILL, 1962), in sheep 5 cases (STORMONT *et al.*, 1953; MOORE and ROWSON, 1958; SLEE, 1963); in chicken one case (BILLINGHAM *et al.*, 1956) and in mink 2 cases (RAPACZ *et al.*, 1964).

A chimeric cow which has a male co-twin is always sterile. This phenomenon is called freemartinism (LILLIE, 1916). No freemartin has been observed in human chimeric twins, nor in marmosets, nor in mink.

In the horses, one case of anastomosis between chorial blood vessels in twin gestation at 7 months has been shown by KELLER (1934). Some hypoplasia of the ovaries was observed in this case. In another case at term, severe ovarian hypoplasia combined with undifferentiated genital ducts was observed (FREUDENBERG, 1960). Marked invagination of both chorial sacs may exist at fourth months of pregnancy (VANDEPLASSCHE, 1957); real fusion between both chorial sacs is rather frequent (VANDEPLASSCHE *et al.*, 1965).

OBSERVATIONS

Observations made on chorial membranes from about 50 cases of dizygotic twin pregnancy in the mare have shown that nearly all cases are bicornual and that invagination of the chorial sacs develops in about 75%.

Ten cases of twin gestation examined in detail were dizygotic (proved by sex, type, coat and hoof-characteristics). We examined their blood groups as well as those of their parents. The blood groups were determined by use of 18 blood typing reagents: anti-A_1, C, D_1, D_2, E, F, G, H_1, H_2, I, J_1, J_2, K, O, Fr_1 (agglutinogenic), Fr_3, Fr_4 and Fr_5 (hemolytic) (PODLIACHOUK, 1966).

In five cases (Nos. 1, 10, 11, 12 and 13) the blood groups of the twins were identical. In two cases (Nos. 7 and 9) there was only one twin alive (Table). In three cases (Nos. 2, 5 and 6) the blood groups of the twins were different.

A marked degree of invagination of the chorial sacs was noted in a bicornual and in a monocornual gestation at 5 months of pregnancy, a fair chorial adherence, but no fusion. A clear fusion of the chorial sacs was found to have developed in two cases of mummified foetus which were at about the 6th month of gestation. The incidence of real fusion is about 40% in advanced pregnancy or at term. Macroscopic fusion as a rule is confirmed in histological examination by complete disappearance of the chorial epithelial layers in some areas. A macroscopically recognizable anastomosis between the blood circulation of both chorions was shown (by the naked eye or by injections of a stained solution) in 5 out of 27 cases examined.

However blood chimerism occurred in 4 out of 5 cases with identical blood group (Table). Blood chimerism was probable in 2 other cases in which one foal only could be examined. The anastomosis in some cases may be rather microscopical (cases Nos. 11 and 12). On the other hand strong tissular fusion may occur without establishment of eithei blood vessel anastomosis or blood chimerism (case No. 6). The absence of anastomosis and blood chimerism in case No. 6 was confirmed by

Blood chimerism in the horse

Case No.	Blood group of twins	Reference serum	Titre of erythrocytis	
			parents	twins
10	$A_1\ C\ \overset{*}{D}_2\ F\ I\ J_2\ \overset{*}{Fr}_1\ \overset{*}{Fr}_3\ \overset{*}{Fr}_5$	anti-D_2	1/16	1/4
		anti-Fr_1	1/1024	1/64
		anti-Fr_3	1/32	1/1
		anti-Fr_5	++++	++
11	$D_1\ D_2\ Fr_1\ \overset{*}{Fr}_3$	anti-Fr_3	1/64	1/8
12	$A_1\ C\ D_2\ F\ I\ Fr_1\ Fr_3\ \overset{*}{Fr}_5$	anti-Fr_5	++++	+
13	$A_1\ C\ F\ I\ Fr_1\ \overset{*}{Fr}_3$	anti-Fr_3	++++	+
	One twin alive			
7	$\overset{*}{A}_1\ \overset{*}{D}_1\ \overset{*}{D}_2\ \overset{*}{F}\ \overset{*}{I}\ \overset{*}{J}_2\ Fr_1\ \overset{*}{Fr}_3\ Fr_5$			
9	$\overset{*}{A}_1\ C\ \overset{*}{D}_1\ D_2\ \overset{*}{F}\ \overset{*}{I}\ Fr_1\ Fr_3\ Fr_5$			

* Weak reaction.

the male karyotype in cell culture, in addition to the different blood groups of the twins.

In cases Nos. 7 and 8, respectively at 10 and 10.5 months of gestation, the genital apparatus of the female co-twin of a male was macroscopically normal: in case No. 7, the histological picture of the ovary was similar to that of a single foal-foetus. In all three female co-twin foals of males (Nos. 1, 11, 12) examined respectively at the age of 20 months, 9 months and 2 years, the genital apparatus was found to be completely normal. Thus in the mare, in contrast to the cow-deformation in the sense of freemartinism is not the rule in the case of bisexual twin-pregnancy and chorial blood vessel anastomosis.

REFERENCES

BENIRCHKE, K. and BROWNHILL, L. E., 1962. Further observations on marrow chimerism in marmosets, *Cytogen.*, **2**, 331.

BILLINGHAM, R. E., BRENT, L. and MEDAWAR, P. B., 1956. Quantitative studies on tissue transplantation immunity. III. Actively acquired tolerance, *Phil. Trans.*, **239**, 357–414.

FREUDENBERG, F., 1960. Intersexuelle Genitalmissbildungen beim Stutfohlen eines zweigeschlechtlichen Zwillingspaares, *Dtsch. tierärztl. Wschr.*, **67**, 214–216.

KELLER, K., 1934. Plazentargefässanastomose bei Pferdezwillingen verschiedenen Geschlechts, *Z. Tierz. u. Zuchtbiol.*, **30**, 241–253.

LILLIE, F. R., 1916. The theory of the free-martin, *Science*, **43**, 611–613.

MOORE, N. W. and ROWSON L. E. A., 1958. Freemartins in sheep, *Nature*, **182**, 1754–1755.

OWEN, R. D., 1945. Immunogenetic consequences of vascular anastomoses between bovine twins, *Science*, **102**, 400–401.

PODLIACHOUK, L., 1966. Groupes sanguins et sériques des équidés, *Rev. Path. Comp.*, T3-3-776, 185–187.

RAPACZ, J., SHACKELFORD, R. M. and JAKÓBIEC, J., 1964. Blood groups in the mink, Proc. IXth Conf. Eur. An. Bl. Gr. Res., Prague, 211–215.

SLEE, J., 1963. Immunologic tolerance between litter-maters in sheep, *Nature*, **200**, 654–656.

STORMONT, C., WEIR, W. C. and LANE, L. L., 1953. Erythrocyte mosaicis in a pair of sheep twins, *Science*, **118**, 695–696.

VAN DER HART, MIA, VAN LOGHEM, J. J., 1967. Blood group chimerism, *Vox Sang.*, **12**, 161–172.

VANDEPLASSCHE, M., 1957. The normal and abnormal presentation, position and posture of the foal-foetus during gestation and at parturition, Mededelingen, Veeartsenijschool, Gent, 1, No. 2, 68.

VANDEPLASSCHE, M., SPINCEMAILLE, J., HERMAN, J. and BOUTERS, R., 1965. Die Zwillingträchtigkeit bei der Stute, *Dtsch. tierärztl. Wschr.*, **72**, 541–548.

DISCUSSION

D. R. OSTERHOFF: What are the possibilities of using blood groups in horses for the differentiation between monozygous and dizygous twins? — We had a pair of albino horse twins, 1.5 years old, and wanted to make sure that the members of the pair were identical. All blood groups and serum types gave identical results and we thought that we actually had a pair of identical albino twins, which would appear at a frequency of 1 in 240,000 births. But skin transplantations performed on three places of the body showed that the twins were in actual fact dizygous. The transplants were rejected after 3 weeks.

L. PODLIACHOUK: In principle utilization of blood groups for the differentiation between mono- and dizygous horse twins gives a supplementary criterion to such criteria like sex, type, bristle-colour and hoof-prints.

XITH EUROPEAN CONFERENCE ON ANIMAL BLOOD GROUPS AND BIOCHEMICAL POLYMORPHISM
Warsaw, July 2nd–6th, 1968

Erythrocytic Antigens on Horses Spermatozoa

Luba PODLIACHOUK and V. DIKOV
with the technical assistance of
R. BEAUD

Pasteur Institute, Paris, France
Institute of Biology and Pathology of Animal Production and Non-infectious Diseases, Sofia, Bulgaria

SUMMARY. Using inhibition, elution and mixed agglutination tests we were unable to establish the presence of 17 equine blood factors on the spermatozoa of 13 Bulgarian stallions.

INTRODUCTION

Numerous studies have ascribed to the antigenic characteristics of spermatozoa an important role in the immunology of sterility: if a species possesses naturally occurring isoantibodies in the uterine cavity, reaction with the blood factors present on the spermatozoa may interfere with the fertilization (BRATANOV, 1966; KRIEG and EYQUEM, 1964). So far research on blood factors on spermatozoa has been carried out chiefly in man.

The extension of blood group studies to the animal kingdom lead to the research on blood factors on spermatozoa. The first investigations carried out on cattle by DOCTON *et al.* (1952) revealed the presence of blood factors on spermatozoa. On the contrary, SCHMID *et al.* (1964) did not find any cross-reaction between bovine spermatozoa (carefully washed) or seminal plasma and blood factors. MATOUŠEK (1964) likewise obtained negative results when searching for 30 bovine blood factors on spermatozoa. One bovine blood factor only (J) exists in a soluble form in serum, saliva and secretions (seminal plasma included) (STONE, 1962). According to SCHMID *et al.* (1964) the J antigen exists in seminal plasma and on spermatozoa, but MATOUŠEK (1964) thinks that it is found only in seminal plasma.

Very few studies have been carried out on other species. None of 42 known blood factors has been revealed on the spermatozoa of boars (MATOUŠEK *et al.*, 1966). Search for 12 cock blood factors on spermatozoa gave negative results (PAZDERKA and KNÍŽETOVÁ, 1968).

Our previous studies on horse blood groups and soluble blood factors in serum, saliva and organ extracts induced us to carry out such studies on this species.

MATERIALS AND METHODS

Sperm was collected from 13 Bulgarian stallions belonging to the Arabian, English and Eastern Bulgarian races. It was sent by air to Paris and preserved frozen at −20°C. The number of spermatozoa varied between 600,000 and 800,000 per mm^3.

The 17 blood typing reagents used for the study of horse blood groups are agglutinins (anti-A_1, C, D_1, D_2, E, F, G, H_1, H_2, I, J_1, J_2, K, O, Fr_1) or haemolysins (anti-Fr_3, Fr_4 and Fr_5).

The test red cells were obtained from horses of the Garches Annexe of the Pasteur Institute.

For the research into the presence of equine erythrocyte antigens on spermatozoa of horses, 4 methods were used:

(1) *Inhibition test*: all test sera were diluted to a titer of about 1/16.

(2) *Elution test.*

(3) *Mixed agglutination test* as described by COOMBS and BEDFORD (1955).

(4) *Immunofluorescence technique*: the indirect test was used.

Determination of horse blood groups was as described in a previous paper (PODLIACHOUK, 1957).

RESULTS

The blood groups of the 13 stallions that supplied the sperm were determined with 17 blood typing reagents.

The presence of antigen A_1 was detected 13 times; F — 12 times; C and Fr_4 — 11; I — 10; Fr_1 — 7; Fr_3 — 6; Fr_5 and J_2 — 5; J_1 — 3; G and O — 2; H_1, H_2 and K — once only; D was absent.

(1) *Inhibition test*: 4 samples of spermatozoa were studied with all 17 typing reagents; the other 9 only with the test sera corresponding to blood factors present on the red cells of the stallion. We did not observe notable inhibition of the activity of the antibody after contact with spermatozoa.

(2) *Elution test*: the elution test was carried out with 7 test sera of high titer: anti-A_1, C, H_2, J_2, K, O and Fr_5. Two samples of spermatozoa were examined with these 7 test sera, independently of the blood group; the other 11 were examined only with antibodies corresponding to the blood factors that were present. We did not observe decrease in activity of the antibodies in the eluates.

(3) *Mixed agglutination test*: One sample of spermatozoa was examined with all the test sera, the others with antibodies corresponding only to the blood factors which were present on the stallion's red cells. All results were negative. As a positive control we utilized an anti-C reagent of human origin, which coated C+ spermatozoa and also reacted with C− spermatozoa because of anti-species specificity.

After elimination of the anti-species heteroantibody (by absorbing it with a mixture of C negative horse red cells) the mixed agglutination test became negative.

(4) *Immunofluorescence test*: interpretation was not possible because of autofluorescence of both coated and uncoated spermatozoa.

DISCUSSION AND CONCLUSIONS

WEIL and RODENBURG (1962) postulate that the antigenic material of human and rabbit spermatozoa originates from the liquid constituents of the sperm. These antigens fix themselves on the spermatozoa during their transit along the male genital tract.

In men EDWARDS *et al.* (1964) found A and B antigens on spermatozoa of secretors only. Recently the presence of human A and B antigens was also detected on the spermatozoa of non secretors (KRIEG, 1967; POPIVANOV *et al.*, 1967).

In the animal kingdom, the soluble antigen J of cattle was detected by SCHMID *et al.* (1964) in the seminal plasma and on the spermatozoa, while MATOUŠEK (1964) found it only in the seminal plasma. In certain pigs the A antigen is present in a soluble form in the serum and saliva (PODLIACHOUK and EYQUEM, 1956), while the Na antigen is found in the serum (MATOUŠEK and SCHRÖFFEL, 1966). These 2 antigens are present in ovarian follicule fluid of sows and in the epididymal fluid of the boar, while seminal vesicle fluid contains only weak concentration of Na. These antigens, A and Na, have not been found on the spermatozoa of pigs (MATOUŠEK *et al.*, 1966).

In the horse, we were unable to establish the presence of any of the 17 equine blood factors on the spermatozoa. This fact is in agreement with the absence of these antigens in soluble form in serum, saliva and extracts of organs (PODLIACHOUK and WRÓBLEWSKI, 1958). It is also in agreement with the absence of blood factors on the spermatozoa on other species examined: cattle (except the presence of the soluble J factor) pig and cock.

REFERENCES

BRATANOV, K., (Ed.), 1966. Problemi na razmnojavaneto pri jivotnite, Sofia, pp. 301.

COOMBS, R. R. A. and BEDFORD, D., 1955. The A and B antigens on human platelets demonstrated by means of mixed erythrocyte platelet agglutination, *Vox Sang.*, **5**, 111–115.

DOCTON, F. L., FERGUSON, L. C., LAZEAR, E. and ELY, F., 1952. The antigenicity of bovine spermatozoa, *J. Dairy. Sci.*, **35**, 706–709.

EDWARDS, R. G., FERGUSON, L. C. and COOMBS, R. R. A., 1964. Blood group antigens on human spermatozoa, *J. Reprod. Fert.*, **7**, 153–161.

KRIEG, H. and EYQUEM A., 1964. Immunologie und Sterilität, *Gynäk, Rdsch.*, **1**, 243–264.

KRIEG, H., Sept. 1967, Blood group substances and infertility, International Symposium on Immunology of Spermatozoa, Varna.

MATOUŠEK, J., 1964. Antigenic characteristics of spermatozoa from bulls, rams and boars. I. Erythrocytic antigens in bulls spermatozoa, *J. Reprod. Fertil.*, **8**, 1–3.

MATOUŠEK, J., DOSTAL, J. and FULKA, J., 1966. Antigeniticity and polymorphism of the seminal vesicle fluid in boars, Xth Eur. Conf. Animal Blood Groups, Paris, 523–531.

MATOUŠEK, J. and SCHRÖFFEL, J., 1966. Blood group substance A, Na and serum polymorphic characters in ovarian follicle fluid of sows, *Immunogenetics Letter*, **4**, 182–185.

PAZDERKA, V. and KNÍŽETOVÁ, F., 1968. The presence of erythrocytic antigens on cock spermatozoa, *Immunogenetics Letter*, **5**, 123–127.

PODLIACHOUK, L., 1957. Thèses de Doctorat ès-Sciences, Paris.

PODLIACHOUK, L., and EYQUEM, A., 1956. Les antigènes érythrocytaires et plasmatiques des porcs, *Ann. Inst. Pasteur*, **91**, 751–758.

PODLIACHOUK, L. and WRÓBLEWSKI, A., 1958. Recherche des substances des groupes sanguins chez les chevaux et les mules, *Ann. Inst. Pasteur*, **94**, 748–752.

POPIVANOV, R., SHTARKALEV, I., EVREV, T. and ANANIEV, T., Sept. 1967. Blood group antigens in human spermatozoa of non secretors, International Symposium on Immunology of Spermatozoa, Varna.

SCHMID, D. O., CONNEALLY, P. M. and STONE, W. H., 1964. Blood group antigens on bull spermatozoa, *Journal of Animal Sci.*, **23**, 4, 1198–1199.

STONE, W. H., 1962. The J substance of cattle, *N. Y. Acad. Sci.*, **97**, 269–280.

WEIL, A. J. and RODENBURG, J., 1962. The seminal vesicle as the source of the spermatozoa. Coating antigens of seminal plasma, *Proc. Soc. Exp. Med. Biol.*, **109**, 567–572.

XITH EUROPEAN CONFERENCE ON ANIMAL BLOOD GROUPS AND BIOCHEMICAL POLYMORPHISM
Warsaw, July 2nd–6th, 1968

Serum Esterases in Equidae

Marie KAMINSKI and Luba PODLIACHOUK
with the technical collaboration of
F. PIGACHE

Laboratoire d'Enzymologie, C.N.R.S., Gif-sur-Yvette, France
Institut Pasteur, Paris, France

SUMMARY. Frequencies of 5 esterases occurring in sera from 5 light, 3 draft breeds and 2 races of primitive horses were compared: 4 combinations of esterases were observed representing polymorphism or variations in intensity and mobility of 3 main components resulting in a total of 33 phenotypes. At least 2 esterases are genetically controlled.

INTRODUCTION

Several plasma esterases in Equidae have been reported using histochemical characterization after starch gel electrophoresis: KAMINSKI and GAJOS (1964) demonstrated interspecies difference between horse and donkey, the latter lacking a fast-migrating esterase; OKI *et al.* (1964) described inter-breed differences in horses and reported a genetically controlled variant of cholinesterase; GAHNE (1966) found 6 phenotypes in a light breed and proposed a hypothesis for their inheritance.

In the present work, inter-species differences were confirmed; inter-breed differences were extended to several breeds and genetical determination was demonstrated in family studies.

MATERIALS AND METHODS

318 samples (Table 1) of horse serum, and 50 samples of donkey and mule sera were examined either fresh or stored for periods from a few months up to 6 years.

Horizontal starch gel electrophoresis was performed using the discontinuous buffer system; esterases were characterized by incubation of gel slices in following substrates: α- or β-naphthyl acetates, α-naphthyl butyrate, naphtol AS-D acetate. Indoxyl acetate, β-carbonaphthoxycholine iodide. The dye-couplers were: Naphthanil Diazo Blue B, Fast Blue BB or RR, Fast Red B or Fast Violet B, Garnet GBC. The inhibitors used were: DFP (diisopropylfluorophosphate) prostigmine, pCMB (parachloromercuribenzoate).

TABLE 1

Horse esterases

Breed/race	No. of samples	E1	E2	E3	E4	E5
Light						
Arab	16	—	16	—	16	16
Thoroughbred	24	2	24	—	18	24
French military	30	—	30	—	30	30
Salernitana	28	—	28	—	6	28
Poznań	20	—	20	—	14	20
	118	2	118	0	84	118
Draft						
Lidzbark	40	—	40	1	13	38
Boulonnais	77	—	77	3	64	77
Bavarian	58	—	d 10	—	m 9	m 3
	175	0	175	4	94	148
Primitive						
Tarpan	22	—	22	—	19	22
Fjording	3	—	3	—	3	3
	25	0	25	0	22	25
Total	318	2	318	4	200	271

d — double, m — multiple.

RESULTS

Esterases in horse serum

Figure 1 summarizes the main zones observed: E_1 is located close to the lipoprotein band; because of different conditions of revelation the precise correlation was not ascertained. However, as the frequency of E_1 is very low compared to that of lipoprotein, they are rather different components. E_1 is revealed by either α- or β-naphthyl acetates with Fast Blue BB. The E_2 band is probably a doublet; it is preferentially revealed by indoxyl acetate or β-carbonaphthoxycholine and inhibited by DFP and prostigmine. E_3 is a well-defined single band, hydrolyzing indoxyl acetate, resistant to DFP. E_4 was the most heterogeneous zone observed: it occurred as a smudge-like zone, extending over the region of β- and α-globulins, with more intense spots varying in electrophoretic mobility, leaving usually behind them a faint trail of activity; occasionally, 5 to 7 fine striations were observed, also varying in mobility; usually, however, a well-defined zone of rather fast mobility was detected. The color obtained using β-naphthyl acetate and Diazo Blue B differs for E_4 from other zones revealed by the same reaction: it is blue instead of purple. E_4 is never revealed by indoxyl acetate nor by β-carbonaphthoxycholine, is inhibited by prostigmine and resistant to DFP. The fastest zone, E_5, is sometimes slightly heterogeneous, but is the most intensely stained when revealed by naphthyl acetates or Naphthol AS, very weakly by indoxyl acetate; β-carbonaphthoxycholine is never hydrolyzed. It is resistant to prostigmine and inhibited by DFP.

Esterase patterns of Equidae

As previously reported (KAMINSKI and GAJOS, 1964) the donkey lacks the fast esterase, E_5 (Fig. 2). Fresh sera of different origin, as well as those stored for several years were similar in this respect. E_4 was present in all donkey samples, and was usually seen as a doublet. Mule sera examined seemed homogeneous and resembled the main pattern of horses.

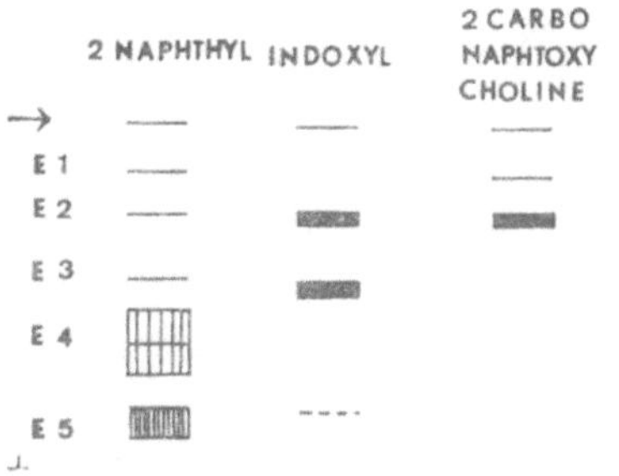

FIG. 1. General esterase diagram in horses.

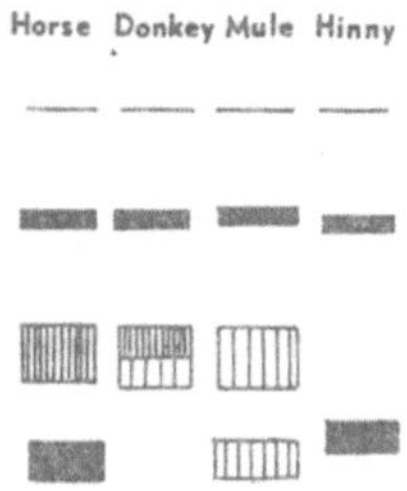

FIG. 2. Esterases in Equidae.

TABLE 2

Main esterase patterns in horses

				No.
Light	$E2_+$			—
	$E2_+$	$E4_+$		—
	$E2_+$		$E5_+$	34
	$E2_+$	$E4_+$	$E5_+$	84
				118
Draft	$E2_+^*$			10
	$E2_+$	$E4_+^{**}$		17
	$E2_+$		$E5_+$	71
	$E2_+$	$E4_+$	$E5_+^{***}$	77
				175
Primitive	$E2_+$		$E5_+$	3
	$E2_+$	$E4_+$	$E5_+$	22
				25

* None in Boulonnais. ** Only in German. *** None in German.

Esterase patterns in different races and breeds of horses

Figure 3, Tables 1 and 2 summarize our findings: very low incidence of E_1 and E_3; presence of E_2 in all samples and high frequency of E_5. The majority of breeds appeared relatively homogeneous especially the light horses, in spite of differences observed between them: thus, 100% Arab and French military horses presented the pattern E_2, E_4, E_5, while 80% Salernitan lacked E_4. It should, however, be emphasized that the diversity of phenotypes, as it appears in Fig. 3, is often attenuated by the weakness of less frequent bands: for example, E_4 present in 20% Salernitan breed is a faint trail instead of a clear-cut band seen in other breeds.

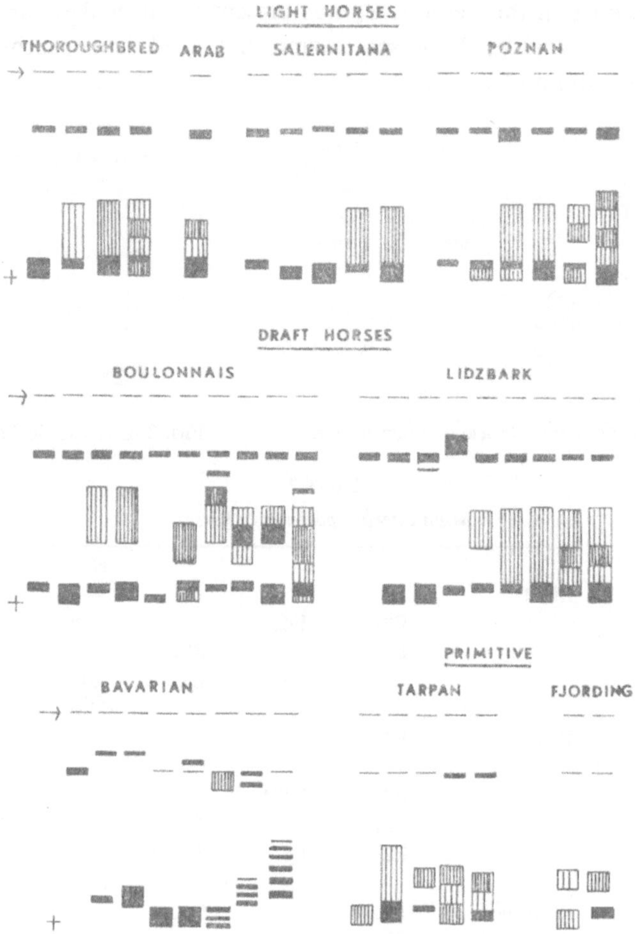

FIG. 3. Esterase phenotypes in different breeds of horses.

Bavarian horses differed strikingly from other breeds by frequent absence of E_5 and polymorphism of E_2, E_4 and E_5. E_2, usually the most uniform band, was either a doublet, or extremaly faint or much slower than in other horses. The hereditary transmission of E_4 and E_5 was studied in 21 families; the 2 sires had the simple pattern E_2, E_5, both bands being strong and non polymorphic. 13 dams presented either an E_4' composed of 5 or 7 striations with absence of E_5, or a simple E_4 with absence of E_5, or absence of both E_4 and E_5. The offsprings of these 13 dams presented in about 50% the striated or absent E_4 and absence of E_5; none of the offsprings of other dams showed these features.

DISCUSSION

As pointed out previously, light horses exhibit rather homogeneous patterns, with, however, some interbreed differences; the most frequently observed pattern E_2, E_5 seems equivalent to that of OKI *et al.* (1964). The esterases reported as variable by these authors correspond in mobility to our E_1 and E_3, or to E_5 (GAHNE, 1966), with slight differences in specificity. On the other hand, no variability was described corresponding to our zone E_4: consequently the number of phenotypes is higher than previously reported.

It is important to stress that presence of several zones in a zymogram, reacting with the same substrates and inhibitors does not necessarily implicate that they are multiple forms of one enzyme: they may well be different proteins, as can easily be demonstrated by comparing their antigenic properties in immunoelectrophoretic analysis. On the contrary, the multiple bands formed by E_4 in our study correspond the most probably to groups of isoenzymes.

Family studies in Boulonnais and Salernitan breeds did not produce any new evidence concerning the inheritance of E_2, E_4, or E_5, the samples examined having the same esterase patterns.

ACKNOWLEDGEMENTS

The autors are indebted to Institut Pasteur, Annexe Garches, Fédération Nationale Chevaline, France, Dr. JASIOROWSKI (PAN, Warszawa), Drs. ZWOLINSKI and KACZMAREK (Poznań), WADOWSKI (Olsztyn), Poland, Dr. SALERNO, Italy, Dr. Schmid, G.F.R. for providing sera used in this study.

REFERENCES

GAHNE, B., 1966. Studies on the inheritance of electrophoretic forms of transferrins, albumins prealbumins and plasma esterases of horses, *Genetics*, **53**, 681.

KAMINSKI, M. and GAJOS, E., 1964. Comparative examination of carboxylic esterases in sera of horse, donkey and their hybrids, *Nature*, **201**, 716.

OKI, Y., OLIVER, W. T. and FUNNELL, H. S., 1964. Multiple forms of cholinesterase in horse plasma, *Nature*, **201, 605**.

XITH EUROPEAN CONFERENCE ON ANIMAL BLOOD GROUPS AND BIOCHEMICAL POLYMORPHISM
Warsaw, July 2nd-6th, 1968

Transferrin and Albumin Polymorphism in the Polish "Tarpan" Breed

Krystyna TOMASZEWSKA-GUSZKIEWICZ and M. ŻURKOWSKI
Institute of Experimental Animal Breeding, Polish Academy of Sciences, Warsam Poland

SUMMARY. In 114 Tarpans examined 11 transferrin phenotypes were found: DD, FF, RR, DF, DH, DO, DR, FH, FO, FR, HO, controlled by 5 alleles the frequency of which are: $q_D = 0.2105$, $q_F = 0.6228$, $q_H = 0.0438$, $q_O = 0.0658$ and $q_R = 0.0570$. On the basis of testing 76 animals it was established that three albumins phenotypes occur controlled by two alleles Alb^F and Alb^S of frequency $q_F = 0.4589$ and $q_S = 0.5410$.

INTRODUCTION

The name "Tarpan" applied now to a small population of primitive horses which show certain features of wild horses in really a courtesy name arising from the fact that these horses are derived from the forest Tarpan — *Equus caballus gmaeliny, Ant. forma silvatica Vet.*

The Tarpans are an interesting material for research work because: (1) they are the nearest as regards origin to the wild horses which lived in Europe, (2) they are a small and isolated population in which a strong selection is carried out in order to maintain the initial type.

The researches carried out by BRAEND (1964), BRAEND and STORMONT (1964), GAHNE (1966), HESSELHOLT (1966) and PODLIACHOUK *et al.* (1966) on transferrin polymorphism in horses, showed their great variety enabling distinction of breeds or types of horses.

MATERIAL AND METHODS

The study was carried out on 114 horses coming from the elite Tarpan studs in Popielno and Jeżewice.

Transferrin were determined in all these horses, the albumins in only 73 animals.

Electrophoretic analyses of sera were carried out in gels made with hydrolysed commercial starch. The buffer used for preparing the gels was a tris-citric acid buffer pH = 7.35, and the electrolyte buffer was a borate buffer pH = 8.5. The time of electrophoresis was 3–4 hours completed when the brown line was at a distance 12 cm from the insert line. Staining was done with amido-black 10B in mixture methanol–water–glacetic acetic acid (5:5:1).

These are the modified conditions of electrophoresis (TOMASZEWSKA-GUSZKIEWICZ, 1968), to give good separation of transferrins and albumins on the same gel.

RESULTS

Transferrins

In the test population of Tarpan we found 11 transferrins phenotypes, controlled by 5 autosomal codominant alleles: Tf^D, Tf^F, Tf^H, Tf^O and Tf^R.

Table 1 shows the result of testing for Hardy–Weinberg equilibrium. No significant deviation between the observed and expected values was found.

Table 2 shows the frequency of transferrin phenotypes and Table 3 gives the frequency of transferrin alleles in the tested population of Tarpan.

TABLE 1

The observed and expected distributions of transferrin phenotypes among 114 Tarpans

	DD	FF	RR	DF	DH	DO	DR	FH	FO	FR	HO
Observed	1	38	1	39	2	4	1	7	10	10	1
Expected	5	44	0.6	29	2.1	3.1	2.7	6.2	9.3	8.1	0.6

$\chi^2 = 10$ 10 d.f. $p < 0.05$

TABLE 2

Frequency of transferrins phenotypes in Tarpans

In percents										
DD	FF	RR	DF	DH	DO	DR	FH	FO	FR	HO
0.8	33.3	0.8	34.2	1.7	3.5	0.8	6.1	8.7	8.7	0.8

TABLE 3

Frequency of transferrin alleles in Tarpans

Tf^D	Tf^F	Tf^H	Tf^O	Tf^R
0.2105	0.6228	0.0438	0.0658	0.0570

The degree of variability of transferrins in Tarpans is rather small when 5 genes combine to give only 11 phenotypes with a distinct predominance of two phenotypes Tf-FF and Tf-DF. These phenotypes are present in 76% of examined animals. This may result from the fact that two foundation dams of female lines (Hanczora and

Liszka) which play an important role in the breeding of Tarpans, have the phenotypes Tf-FF and Tf-DF.

It is characteristic that whilst the frequency of the Tf^D allele is quite high (Tf^D 0.2105), only one animal was observed of homozygous type Tf D/D, and 46 animals carrying the Tf-D allele were heterozygous.

Albumins

It was found that similar to other breeds of horse 3 albumin phenotypes were present: FF, FS and SS. The frequency of the alleles were: $q_F = 0.4589$ and $q_S = 0.5410$.

REFERENCES

BRAEND, M., 1964. Serum types of Norwegian horses, *Nord. Vet.-Med.*, **16**, 363.

BRAEND, M. and STORMONT, C., 1964. Studies on hemoglobin and transferrin types of horses, *Nord. Vet.-Med.*, **16**, 31.

GAHNE, B., 1966. Studies on the inheritance of electrophoretic forms of transferrins, albumins, prealbumins and plasma esterases of horses, *Genetics*, **53**, 4, 681.

HESSELHOLT, M., 1966. A study of blood groups and serum types of the Icelandic horse, Proc. Xth European Conference on Animal Blood Groups and Biochemical Polymorphisms, Paris, 1966.

PODLIACHOUK, L., MADEYSKA, A. and PEGUINES, D., 1966. Etude immunogenetique du cheval polonais, Proc. Xth European Conference on Animal Blood Groups and Biochemical Polymorphisms, Paris, 1966.

TOMASZEWSKA-GUSZKIEWICZ, K., 1968. The separation of the transferrin and albumin types using the same starch gel. (In press).

XITH EUROPEAN CONFERENCE ON ANIMAL BLOOD GROUPS AND BIOCHEMICAL POLYMORPHISM
Warsaw, July 2nd–6th, 1968

Serum Transferrin and Haemoglobin Polymorphism in Lipizzaner Horses

W. SCHLEGER and P. SOOS

Blood Group Laboratory of the Department of Animal Breeding and Genetics at the Veterinary School Vienna, Austria

SUMMARY. Transferrin was investigated in 456 Lipizzaner horses from Austria, Hungary and Czechoslovakia. The three populations have been separate for about 30 years. Considerable difference with regard to transferrin distribution was found in the three populations, and the causes thereof were discussed.

Haemoglobin determinations on 293 Austrian and Czechoslovakian Lipizzaner horses failed to show differences between the populations. The mode of inheritance was revealed by examination of 161 direct pedigrees.

The probability of exclusion of wrong pedigree by use of the Tf system = Austria 32.68%, Hungary 31.51% and Czechoslovakia 25.65%. The combined probability both for transferrin and haemoglobin amounts to about 40%.

The Lipizzaner breed originated around the year 1580. It was developed from breeding animals of Arabian and North African descent which were brought from Spain, to the stud in Lipizza. The imported animals showed, according to history, the so-called "Spanish gait" and had a "proud" appearance. Various stallions were imported from the various European countries so that at some time seven strains existed. Even though their names are still used in designating breeding animals there is little justification today to consider in reality as separate strains. The breeding of the Lipizzaner Horses was mainly accomplished by the original Lipizza stud, and also the Hungarian studs Mesöhegyes, Fogaros and Babolna. In addition to these, studs in Czechoslovakia — Kladrub and Topolcianky — and since 1953 the Austrian Piber stud have been responsible for the breeding work.

It should be pointed out that until 1914 breeding animals were exchanged between all the studs so that the Lipizzaner breed could be considered as one large population with very little differentiation. After 1918 the exchange of breeding animals was severely reduced. In particular Piber closed its population more and more to imports from the other studs. After 1935 all the other studs also terminated the exchanges.

It appeared to be of interest to investigate whether the separation of the original population into three sub-populations was reflected in the transferrin frequency differences. Soos supplied the results of transferrin determinations of 163 Hungarian Lipizzaner horses, and Dozent Dr. LABIK, chairman of the Department of Animal Breeding of the Veterinary School in Brno supplied the blood samples of the Czechoslovakian Lipizzaners and sent them to Vienna where transferrin and haemoglobin types were determined. The laboratory in Vienna, therefore, investigated blood samples from 172 Austrian and 122 Czechoslovakian Lipizzaner horses.

In 1965 and 1966 we investigated the procedure and were able to reproduce the results of STORMONT, BRAEND, GRAETZER and HESSELHOLT. We could observe, in 300 sera from horses, 15 different transferrin phenotypes. We never observed, however, an M-band. In 1966 and 1967 we investigated the sera from the Lipizzaner horses from the Topolcianky and Piber studs. From both studs all stud stallions, stud mares and their progeny from 3 years were investigated. Analogous material was investigated by Soos in Hungary.

		No.	Tf^D	Tf^F	Tf^H	Tf^O	Tf^R	χ^2		D.P.O.
Austria	♂	8	0.19	0.25	0.12	0.44	—			
	♀	39	0.22	0.20	0.06	0.41	0.11			
	N	125	0.22	0.20	0.07	0.47	0.04			
Total		172	0.22	0.20	0.07	0.45	0.06	14.231	$0.50<p<0.30$	114
Hungary	♂	11	0.13	0.37	0.32	0.13	0.05			
	♀	52	0.04	0.57	0.21	0.17	0.01			
	N	100	0.11	0.40	0.28	0.21	—			
Total		163	0.09	0.45	0.26	0.19	0.01	6.566	$0.97<p<0.50$	107
Czechoslovakia	♂	6	0.08	0.42	—	0.42	0.08			
	♀	23	0.06	0.30	—	0.46	0.18			
	N	92	0.07	0.33	0.02	0.46	0.12			
Total		121	0.07	0.33	0.01	0.46	0.13	19.904	$0.30<p<0.10$	47
All		456	0.13	0.33	0.12	0.36	0.06	22.413	$0.10<p<0.05$	268

From the Table the differences between the three Lipizzaner populations in respect to the transferrin distribution is evident. The Table shows the gene frequencies of the stud stallions, the dams and the progeny within each population, the number of direct parent offspring relationships as well as the results for the total population. Each population is in the Hardy–Weinberg equilibrium; however, the total population shows a deviation from this genetic equilibrium which is almost statistically significant. This is caused by the differences between the three sub-populations which in turn are the result of a separation going back more than 30 years. The frequency of the individual transferrin genes shows considerable differences in the three populations. For example, Tf^H in Czechoslovakia, Austria, Hungary shows a relation of 1:7:26, the Tf^R one of 13:6:1. A similar observation was reported by MILLER

(1966) fom two herds of Longhorn cattle in the U.S., which had been separated for 35 years. He also observed a significant difference between the two herds, which was, however, considerably smaller than ours herd A : herd B for transferrin Tf^A 1:2, for Tf^D 7:5 and for Tf^E 3:7. These deviations from the equilibrium for the whole population can be explained by genetic drift in small populations. However, the three Lipizzaner populations are bred for different purposes, in Austria for the classic school, the stallions are used in the Spanische Hofreitschule, in Hungary for riding, and in Czechoslovakia for working.

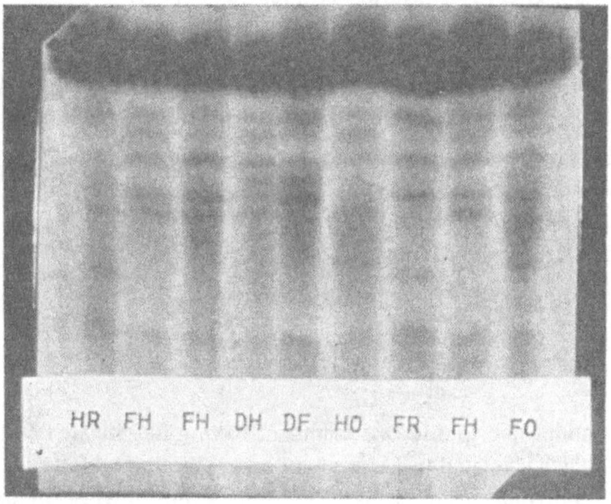

We would like to point out that the investigations on a total of 465 animals involved 268 which are related as parent and offspring. In nine cases we had to exclude the pedigree as given by the stud, in all of these the sire was given incorrectly (3.36%).

The determination of the haemoglobin type was done on the Austrian and Czechoslovakian animals (239) only. Three haemoglobin types could be observed and the inheritance of these was found to be according to the theory suggested by SCHMID. In all three types the haemoglobin band is distinct and well defined. Homozygous Hb^A/Hb^A phenotypes show the slower moving band very distinctly, heterozygous Hb^A/Hb^a types show this band more weakly, whilst the Hb^a/Hb^a types show no slow moving band at all. The gene frequencies for Hb^A, Hb^a are alike in both the Austrian and the Czechoslovakian population, Hb^A 0.87, Hb^a 0.13. There is no difference, therefore, between the two populations with regard to the haemoglobins. Out of the 293 animals 220 showed type Ab^A/Ab^A, 70 the type Hb^A/Hb^a, and three the type Hb^a/Hb^a.

The heterogeneity of the populations with regard to the Tf types permits the use of the transferrin types for clarifications of pedigrees. According to the formula of GAHNE (1961) for a 3-allele-system, the probability of exclusion of wrong pedigree by the transferrin system in horses (6 alleles) was computed as $P_{Tf/Austria}$ 32.68, $P_{Tf/Hungary}$ 31.51 and $P_{Tf/CSSR}$ 25.65. The probability of exclusion of wrong paternity by haemoglobin amounts to 10.02%. Therefore the combined probability of exclusion of wrong pedigree is equal in Austrian Lipizzaners to 39.41, and in Czechoslovakian to 33.1.

REFERENCES

BRAEND, M. and STORMONT, C., 1964. *Nord. Vet.-Med.*, **16**, 31–37.
GAHNE, B., 1961. *Anim. Prod.*, **3**, 135.
GAHNE, B., 1966. *Genetics*, **53**, 681.
GRAETZER, M. A., HESSELHOLT, M., MOUSTGAARD, J. and THYMANN, M., 1964. Proc. 9th Europ. Anim. Blood Group Conf., 179–193.
HESSELHOLT, M., 1966. Proc. 10th Europ. Anim. Blood Group Conf., 325.
MILLER, W. J., 1966. *Genetics*, **54**, 391.
SCHMID, D. O., 1965. *Z. Immunforsch.*, **128**, 499.
STORMONT, C. and SUZUKI, Y., 1964. *Genetics*, **50**, 915.

DISCUSSION

J. SCHRÖFFEL: I should like to add one comment having investigated 54 samples of Old Kladruby horses, a breed which is related to Lipizzaner horses. We have found all common alleles with he exception of Tf^M. Polymorphism seems to exist in serum amylase of the investigated breed. We have found 3 fenotypes which correspond perhaps to two allelic systems.

CHAPTER 5

BLOOD GROUPS AND BIOCHEMICAL POLYMORPHISM IN SHEEP AND GOAT

XITH EUROPEAN CONFERENCE ON ANIMAL BLOOD GROUPS AND BIOCHEMICAL POLYMORPHISM
Warsaw, July 2nd–6th, 1968

M Antigen and Potassium Types in Lambs

Elisabeth M. TUCKER
A.R.C. Institute of Animal Physiology, Babraham, Cambridge, Great Britain

SUMMARY. Red cells from sheep foetuses and newborn lambs of the genotype Mm and MM react only weakly or not at all with the M reagent and yet have a high red cell potassium concentration. The red cells become fully M-positive at about the same time that adult red cell potassium levels are established and foetal haemoglobin has disappeared, that is, 40–50 days after birth.

INTRODUCTION

RASMUSEN and HALL (1966 a, b) reported that there was an association between the potassium types and the antigen M on the red cells of adult sheep. All high potassium (HK type) red cells were M-positive whereas low potassium (LK type) red cells were either M-positive or M-negative. Sheep red cells which were M-negative were always LK type.

On the basis of these results RASMUSEN and HALL put forward the hypothesis that sheep homozygous for M (MM) were HK type ($ka^h ka^h$) while sheep homozygous for m (mm) were homozygous for the gene for LK ($Ka^L Ka^L$). The heterozygous LK type sheep ($Ka^L ka^L$) were also heterozygous for M (Mm).

All lambs are born with red cells containing high potassium concentrations and adult potassium levels are not reached until about 50 days after birth (DRURY and TUCKER, 1963). It was of interest therefore to determine if a lamb undergoes any changes in respect of the M antigen during development. The present paper reports on an attempt to correlate the changes in red cell potassium concentration with the M antigen activity and the disappearance of foetal haemoglobin which occurred in lambs after birth. Reticulocyte counts were also made because it is known that the presence of reticulocytes can raise the potassium concentration to high levels (DRURY and TUCKER, 1963).

MATERIALS AND METHODS

Sheep. Clun Forest sheep kept at pasture under normal husbandry conditions were used and blood samples were collected by jugular venepuncture into heparin.

"Layering" of red cells. This method (DRURY and TUCKER, 1963) was used to separate the red cell samples into 4 fractions of different mean ages. After centrifugation in a polythene tube the washed packed red cells were sliced into 4 fractions. Layer 1 from the top of the column contained the youngest and layer 4 from the bottom of the column the oldest red cells. Layers 2 and 3 were the intermediate fractions.

Potassium measurements. These were carried out in a flame photometer (DRURY and TUCKER, 1963). The potassium concentration was related to the haemoglobin concentration of the sample and the results expressed as m-equivalents of potassium per litre of haemoglobin solution containing 116 g of haemoglobin. (This standard was chosen because normal sheep blood has a haemoglobin concentration of about 11.6 g/100 ml.)

Electrophoresis. The separation of adult and foetal haemoglobin was carried out by starch gel electrophoresis at pH 6.5 (Phosphate buffer, DRURY and TUCKER, 1963) and at pH 8.9 (Tris-EDTA-borate buffer after GAHNE, RENDEL and VENGE, 1960).

Reticulocyte counts were made from blood films stained with brilliant cresyl blue (DACIE, 1950).

Serology. The red cells were tested against doubling-up dilutions of M reagent using rabbit complement (absorbed sheep red cells). The test was read after 2 hr at 32°C. The degree of haemolysis in each tube was recorded as 0–5, and the haemolysis score was obtained by adding up the results for each dilution. The control adult M-positive red cells had a score of 21 at the 2 hr reading. A check reading was also done after 4 hr.

RESULTS

Foetal red cells. Red cells from 12 foetuses aged between 71 and 140 gestational days were tested for M antigen and potassium concentration. All of the samples had high potassium concentrations and most of them showed no lysis with the M reagent at the 2 hr reading. In one or two samples there was a trace of lysis at high concentrations of anti-M and this became more obvious after prolonged incubation. Six of the foetuses had HK type dams and 5 of them LK type M-positive dams. Only one foetus had an M-negative mother, so that the majority of the foetuses must have been of the M-genotype.

The foetal red cells were also tested against a panel of 8 blood typing reagents and several of the samples reacted with A, B′, N′, U and D reagents, thus indicating that these blood factors are present on the red cells of sheep foetuses.

Newborn lambs. Blood samples were taken from 5 lambs of the genotype Ka^L ka^h Mm and 4 of the genotype ka^h ka^h MM at birth and at regular intervals thereafter. The red cells were separated into the 4 population fractions (see Methods) and each fraction tested for M antigen and potassium concentration and also for the presence of reticulocytes and foetal haemoglobin.

Figure 1 shows the mean results for the 5 Ka^L ka^h Mm lambs. At birth the red cells had high potassium concentrations and only reacted very poorly if at all with anti-M. As the lambs grew, the potassium concentration fell sequentially in the 4 population fractions and at the same time the red cells reacted more completely with the M reagent.

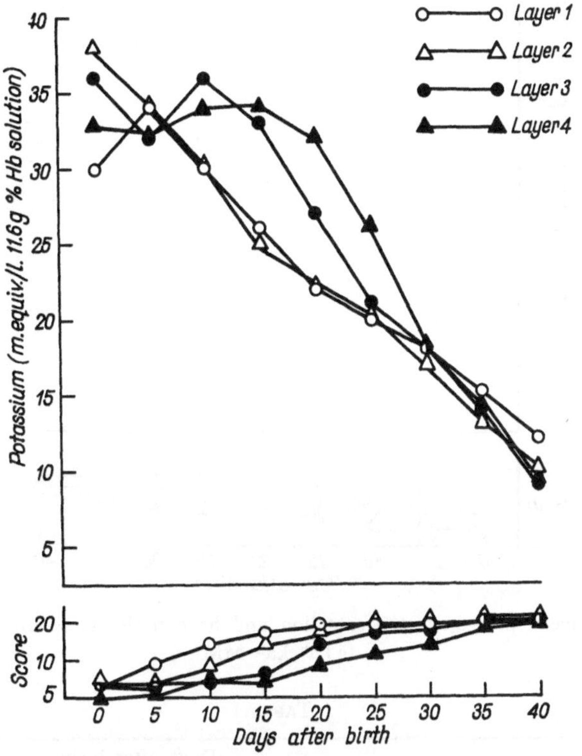

FIG. 1. The mean potassium concentration and haemolysis score for lamb red cells (5 Ka^L ka^h Mm).

A similar result was obtained with the 4 ka^h ka^h MM lambs (Fig. 2) but since they were KH type the potassium levels only fell slightly.

Red cells from newborn Ka^L Ka^L mm lambs were also tested and were M-negative and had high potassium concentrations which fell to normal adult levels at the usual 40–50 days after birth.

Reticulocyte counts. Table 1 shows the mean counts for 2 of the Ka^L ka^h Mm and 4 of the ka^h ka^h MM lambs. Reticulocytes were always present in layer 1, as would be expected in these rapidly growing lambs. They were also sometimes found in layer 2, rarely in layer 3, and never in layer 4.

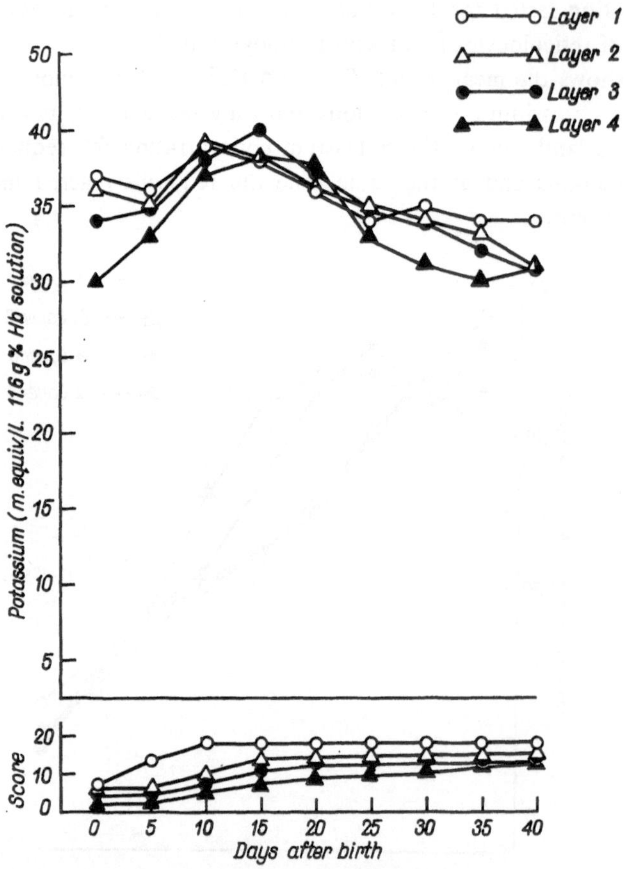

FIG. 2. The mean potassium concentration and haemolysis score for lamb red cells (4 ka^h ka^h MM).

TABLE 1

Genotype	Layer	Days after birth							
		0	7	14	21	28	35	42	49
Mm	1	4.0	4.1	7.9	17.0	9.0	3.0	0.8	0.9
	2	0.8	0	0.5	3.4	1.2	0.6	0	0.2
	3	0.2	0	0.1	0.3	0	0	0	0
	4	0	0	0	0	0	0	0	0
MM	1	7.1	7.0	5.4	2.9	1.6	1.2	0.7	0
	2	0.3	0.1	0	0.1	0	0.1	0	0
	3	0	0	0	0	0	0	0	0
	4	0	0	0	0	0	0	0	0

Reticulocyte counts (%) in 4 layers of lamb red cells at intervals after birth. The Mm results are the mean for 2 lambs and the MM the mean for 4 lambs. The differences between the 2 genotypes are not significant because of the great individual variation encountered.

Electrophoresis. Adult haemoglobin first appeared in layer 1 of the samples and then appeared sequentially in all 4 layers as the foetal haemoglobin correspondingly disappeared.

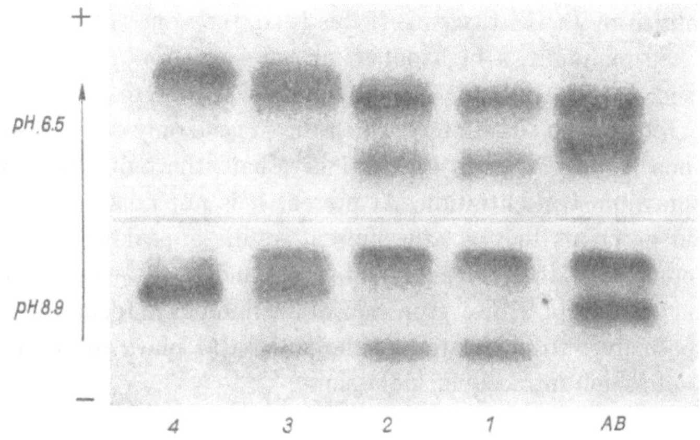

FIG. 3. Starch gel showing the haemoglobin in the 4 layer of red cells from a 14 day old lamb. Reading right to left: AB — control; layer 1 — HbA+C; layer 2 — HbA+C; layer 3 — HbA+F+C; layer 4 — HbF.

Figure 3 shows a typical haemoglobin pattern in one of the lambs 14 days after birth. This lamb was of adult type A haemoglobin, and at this stage also had same C haemoglobin present as well as foetal. In all the lambs tested, traces of foetal haemoglobin were still present in layer 4 on the 35th day, but had disappeared by the 45th day. The disappearance of foetal haemoglobin therefore closely followed the development of adult potassium type and of the M antigen in the red cells.

DISCUSSION

The results showed that the development of the M antigen on the red cells closely followed that of adult potassium types. The new red cells which came into the circulation after birth in the Mm lambs had progressively less potassium and gave a progressively higher haemolysis score with anti-M. The fact that the red cells in layer 1 were giving adult type haemolysis scores before adult potassium types were reached can be explained by the presence of reticulocytes in this layer. In layer 4 which had no reticulocytes, the increase in M antigen activity closely paralleled the development of adult potassium types. The finding that foetal haemoglobin disappeared at about the same time that adult potassium and full M antigen activity appeared strongly suggests that all these changes were caused by a gradual replacement of a foetal population of red cells by an adult one.

RASMUSEN and HALL (1966a, b) tested many sheep and found no exception to their rule of association between M and potassium type. On this evidence they concluded that the association was not due to close linkage, and suggested that the gene controlling potassium types also controls the M antigen. In other words they postulated that an antigen involved in the molecular mechanism controlling potassium levels is identified by the M reagents. If this is so, then one still has to explain how a foetal red cell maintains a high potassium concentration and yet has no detectable M antigen. It is possible that it is the presence of the m gene which is important. One can envisage a situation in which the m gene only exerts its influence after birth and when it does so, even in Mm individuals, the red cells cannot maintain their high potassium concentration. At present it is not possible to do more than speculate, but as far as they go, the present results appear to support RASMUSEN and HALL's contention that the potassium types and M antigen system are closely associated in adult sheep. However in view of the findings in foetal red cells it would seem to be premature to postulate that the antigen M plays any part in the actual maintenance of a high intracellular potassium.

ACKNOWLEDGEMENTS

I am much indebted to Sir ALAN DRURY, F.R.S., for many helpful discussions. I am also grateful to Professor B. A. RASMUSEN who kindly supplied the original M reagent. This enabled us to type the Babraham sheep and thus prepare our own anti-M which was used for all these tests. I also thank Mr. L. KILGOUR, Mrs. V. A. HERBERT and Miss F. A. THOMPSON, for valuable technical assistance.

REFERENCES

DACIE, J. V., 1950. Practical Haematology, Churchill, London, pp. 22–25.

DRURY, A. N. and TUCKER, E. M., 1963. Red cell volume, potassium and haemoglobin changes in lambs, *Res. Vet. Sci.*, **4**, 568–579.

GAHNE, B., RENDEL, J. and VENGE, O., 1960. Inheritance of β-globulins in serum and milk from cattle, *Nature* (Lond.), **186**, 907–908.

RASMUSEN, B. A. and HALL, J. G., 1966a. Association between potassium concentration and serological type of sheep red blood cells, *Science* (Wash.), **151**, 1551–1552.

RASMUSEN, B. A. and HALL, J. G., 1966b. Xth European Conference on Animal Blood Groups and Biochemical Polymorphisms, Institut National de la Recherche Agronomique, Paris, pp. 453–457.

XITH EUROPEAN CONFERENCE ON ANIMAL BLOOD GROUPS AND BIOCHEMICAL POLYMORPHISM
Warsaw, July 2nd–6th, 1968

Studies on Blood Groups in Sheep

Czesława LIPECKA

Department of Animal Husbandry, College of Agriculture, Lublin, Poland

SUMMARY. Studies were carried out on the Long-Wooled Polish breed. Twenty-three antisera were obtained as a result of a series of immunizations and 30 re-immunizations. From these 23 sera, 15 were separated after use in hemolytic tests and after absorption as sera identifying single antigenic factors of sheep blood. These sera were given identifying letters according to the author's own nomenclature. Sheep flocks were tested with these sera and the frequency of antigen occurrence was established. Studies of the inheritance of antigens to aid further classification were made on 761 sheep, including 9 rams, 238 ewes and 514 of their offspring. An analysis of the inheritance within particular animals (sire–dam progeny) showed that the sera which had been tested belonged to 4 systems. Seventy-five percent of the sera were classified as belonging to the B-system.

INTRODUCTION

Studies on blood groups in sheep were performed by BIAŁOSUKNIA and KĄCZKOWSKI (1924). Later ANDERSEN (1938), YCAS (1949), STORMONT (1951), RENDEL (1957), TUCKER (1965) and others carried out experiments on the serology and genetics of the RO-system in sheep and its similarity to the human AB0-system and bovine J-system. Recent investigations have shown that it is possible to obtain hemolysins by immunization of sheep. Using hemolysins, it is much easier to show individual differences in the blood of sheep than by use of natural antibodies. YCAS (1949) showed the presence of 9 antigens identifying individual antigenic factors of sheep blood. RASMUSEN (1958) discovered the XZ-system in the blood of sheep, its genetic locus being different from that of the RO-system. RASMUSEN (1960), RASMUSEN *et al.* (1960) discovered five more systems in sheep blood: A, B, C, D and M. Altogether 7 systems have now been discovered in blood groups of sheep. Studies on blood groups in sheep, however, are less advanced than in other animals. The objective of this work was to obtain as many sera as possible, to identify individual antigenic factors, and to characterize a flock of sheep by means of these sera. The inheritance of these antigens was also studied.

MATERIALS AND METHODS

Studies were carried out on the Long-Wooled Polish breed of sheep. A series of 75 immunizations and 30 re-immunizations was made in 220 sheep. Immunizations were made by different methods and at different times. Twenty-three antisera with the titre varying from 1:64 to 1:2048 were obtained as the result of these immunizations and re-immunizations. The sera were separated by means of hemolytic and absorption tests, using 1.5% suspension of erythrocytes in 0.92% physiological solution. Lyophylised sera of the guinea pig were used as complement. Estimation of the titre of each serum was done on red cell samples obtained from 30 sheep selected at random. This material was used for absorption tests. All absorption tests were repeated three times: the first time for 30 min and the other two for 1 hr each (30 min in the refrigerator and 30 min at room temperature). Fifteen of the 23 sera were monovalent, and were additionally checked for monovalency with red cells samples from 70 sheep.

RESULTS

The analysis of the results showed that older sheep were the best material for the production of isoimmune sera, and that the re-immunization of sheep was the best way of obtaining such sera. Sixteen out of the 23 sera were obtained by re-immunization, that is 70% ($\chi^2 = 23.66$). The optimal time between immunization and re-immunization was 3–4 months. It was shown that the method of the immuniza-

TABLE 1

The frequency of occurrence of blood factors in sheep

Blood factor	Parents		Offspring=514
	rams=9	ewes=238	
A	0.556	0.525	0.581
B	0.333	0.303	0.313
C	0.556	0.513	0.572
D	0.556	0.596	0.564
E	0.556	0.163	0.313
F	0.222	0.247	0.247
G	0.222	0.210	0.239
H	0.111	0.211	0.152
I	0.333	0.235	0.296
K	0.222	0.181	0.169
L	0.222	0.193	0.230
M	0.444	0.408	0.479
N	0.222	0.283	0.239
O	0.333	0.324	0.334
P	0.111	0.176	0.188

tion (intravenous or intramuscular) and the quantity of the injected blood had no influence on the production of antibodies of high titre. Intramuscular injection was much easier to perform than intravenous, and 5 ml was considered to be sufficient quantity of blood. Fifteen monovalent sera were obtained after several absorptions and hemolytic tests. They have been described in previous papers and identified using the author's own letter nomenclature (1963, 1968). The frequency of the occurrence of the antigens in the flock of sheep studied and in their offspring was estimated using these sera (Table 1). The highest frequency of occurrence was shown by the antigens A, C and D, both in the flock and in the offspring. Altogether over 50% individuals possessed one of these three antigens. Antigens B, M and O had lower and E, K, P the lowest frequency. The frequency of antigens in the offspring was almost the same as that in the parents. The only exception was antigen E ($\chi^2 = 16.56$). Knowledge of antigenic factors within species or breed can be used for practical purposes, provided that the mechanism of inheritance of the factors is known. Accordingly, 9 rams, 238 ewes and 514 of their offspring were studied using the 15 monovalent sera. Analysing 382 different families (all except two) it was

TABLE 2

Comparison of the theoretical and actual frequency of antigens in the blood of offspring

Blood factor	Actual	Theoretical	χ^2
A	299	301.3	3.073
B	161	152.8	3.113
C	294	311.4	3.382
D	290	299.6	1.251
E	161	182.6	6.496
F	127	129.7	1.928
G	123	113.6	5.721
H	83	81.5	3.338
I	151	151.9	0.709
K	87	84.2	0.214
L	118	111.8	1.318
M	246	240.7	0.967
N	123	126.9	2.026
O	172	170.5	0.199
P	97	79.0	9.145

possible to show that none of the antigens which were absent in the blood of parents occurred in the blood of the offspring. The theoretical and actual frequency of the occurrence of the antigens in the offspring are presented in Table 2. There were great differences between the theoretical and actual frequency of the antigens E, G and P. It is not possible to explain this phenomenon without further studies. Preliminary results show that all sera obtained can be classified into 4 blood systems (A, B, C, D): two factors in system C, eleven in system B, one in system A, and one in D. This report can be considered as a preliminary to further studies, both of a theoretical and practical character.

REFERENCES

ANDERSEN, E., 1938. Untersuchungen über die Blutgruppeneigenschaften der Schafe, *Zeitschr. Rassenphysiol.*, **10**, 88–103.

BIAŁOSUKNIA, W. and KĄCZKOWSKI, B., 1924. Studies on the blood groups in sheep, *Com. R. Soc. Biol.*, **XC**, 1196.

LIPECKA, C., 1963. Studies on the blood groups in sheep, *Agricultural Sc. Annals*, **76**, B-3, 629–643.

LIPECKA, C., 1968. Occurrence of the antigenic factors in Long-Wooled Polish breed of sheep. *Annals Univ. Mariae Curie-Skłodowska*, Lublin. (In press).

RASMUSEN, B. A., 1958. Blood groups in sheep. I. The XZ system, *Genetics*, **43**, 6, 814–821.

RASMUSEN, B. A., 1960. Blood groups in the sheep. II. The B system, *Genetics*, **45**, 10, 1295–1417,

RASMUSEN, B. A., STORMONT, C. and SUZUKI, Y., 1960. Blood groups in sheep III. The A, C, D and M systems, *Genetics*, **45**, 12, 1595–1603.

RENDEL, J., 1957. Further studies in some antigenic characters of sheep blood determined by epistatic action of genes, *Acta Agr. Scand.*, **VIII**, 2, 224–259.

STORMONT, C., 1951. An example of a recessive blood group in sheep, *Genetics*, **36**, 577–578.

TUCKER, E. M., 1965. Further observations on the blood group in sheep, *Vox Sang.*, **10**, 195–205.

YCAS, M., 1949. Studies of the development of a normal antibody and cellular antigens in the blood of sheep, *Journal of Immun.*, **61**, 327–427.

DISCUSSION

E. M. TUCKER: Is your D factor an agglutinogen?

C. LIPECKA: Yes.

I. G. HALL: Please would you tell us the titres of your reagents.

C. LIPECKA: 1/4 to 1/16.

XITH EUROPEAN CONFERENCE ON ANIMAL BLOOD GROUPS AND BIOCHEMICAL POLYMORPHISM
Warsaw, July 2nd–6th, 1968

Some Blood Group Antigens in Polish Merino Sheep

Wanda PORĘBSKA

Department of Sheep Breeding, College of Agriculture, Cracow, Poland

SUMMARY. This was our introduction to sheep blood group studies. The test sera were not available at the beginning of this study and the experimental material was of unknown composition in respect to blood group antigens.

The reagents were prepared by isoimmunization. In the informal comparison tests arranged by Dr. RASMUSEN this year the test sera were identified as reacting with A, B′, C, M, R factors.

The two flocks of Polish Merino sheep were tested by these reagents and their gene frequencies were calculated. No gene equilibrium was found in either flocks and only phenotype frequencies were estimated. The latter were similar in both flock, but it differed considerably in comparison with those for other breeds quoted in the literature.

INTRODUCTION

This was the introductory study on the sheep blood groups in our Laboratory. This study is a part of a research programme we are to continue under the supervision of Prof. Dr. St. JEŁOWICKI.

The purpose of the work was to prepare reagents and then by means of them to carry out inheritance studies.

METHODS AND MATERIALS

The Longwool sheep (Crossbreed — local Bocheńska × Kent) were used as an experimental material for the preparation of test sera. In further inheritance studies Polish Merino sheep were used. The test sera were prepared by isoimmunization. The blood samples and the techniques used in performing the blood-typing hemolytic tests were those described by RASMUSEN (1958), the complement used was guinea pig serum.

The two flocks of Polish Merino sheep from the Ram Testing Stations at the Experimental Stations of Czechnica and Mełno attached to the Institute of Zootechnics were tested by the reagents produced. The blood samples were taken from unselected young sheep of 10 months of age and their parents in total 417 animals, comprising 136 families from 12 rams.

RESULTS AND DISCUSSION

The 18 ewes (one year and a half old) were immunized and reimmunized with their mothers red cells. It was expected that the serological differences between donor and recipient would be relatively smaller, and the preparation of the reagents would be easier. In three series of immunizations ten sheep produced isoimmune antisera, and two of them antisera containing one-antibody population only (anti-C and anti-R) which are relatively easy to produce. A total of 9 reagents were prepared. To identify these reagents a comparison test with 49 reagents used for blood typing of cattle was carried out. Only 7 of this number reacted with erythrocytes of 30 sheep blood samples giving segregation of the experimental material (the test sera J_1, J_2, S_1, R, F, B, O_1) and only two J_1, J_2 proved to be identical with one of the reagents prepared in this study. It is probable therefore that this test serum is the anti-R reagent of the sheep blood group system R–O. It is also unlikely that any of the reagents prepared would belong to the X–Z blood group system for as yet X and Z reagents have not been produced by isoimmunizations (RASMUSEN, 1962). The blood samples of 136 families (male, female and progeny) were tested with the produced test sera. The inheritance studies indicate that these reagents identify the antigens of six blood group systems. The confirmation of the assignment of several examined antigens to different blood group systems was obtained by the informal comparison test arranged by Dr. RASMUSEN this year. The reagents prepared in this study identified following antigens: A, B′, C, M, F, (the anti-M serum reacted with fewer cells than the test serum anti-M of Dr. RASMUSEN).

A further two reagents Pl/B and Pl/F have not been identified as reacting in any known blood group systems. The inheritance studies indicate that the Pl/B reagent probably reacts with antigens of B blood systems. The two other reagents Pl/H and Pl/I probably react in the A system, but behave differently to the test serum anti-A of Dr. RASMUSEN. For the identified antigens, except the antigen B′ belonging to the very complex system D, recessive gene frequencies (*q*) in Polish Merino sheep were estimated.

TABLE 1

Recessive gene frequencies and Chi-square test for offspring and matings in Polish Merino sheep in Ram Testing Stations of Czechnica and Mełno

Ram Testing Station	Czechnica			Mełno		
		Chi-square test			Chi-square test	
Antigen	*q*	matings	offspring	*q*	matings	offspring
A	0.2757	0.2400	6,3412*	0.4848	6.6413*	13.8415**
C	0.7253	0.2592	2.9914	0.6603	0.5452	2.4823
M	0.9879	0.00	2.5123	0.9260	0.5701	4.2328*
R	0.5620	0.3493	1.7220	0.6554	3.4671	5.6041*

* $P < 0.05$.
** $P < 0.01$.

The fact that antigen A in the Czechnica flock, and antigens A, M, R in Mełno flock were not in equilibrium does not allow any conclusions about values of gene frequencies to be drawn. Therefore the phenotype frequency of the examined antigenes was estimated.

Neither of the investigated Polish Merino sheep flocks differ greatly in the phenotype frequency. But the Polish Merino sheep did differ considerably from other breeds mentioned in Table 2. The phenotype frequency of the antigen A in Polish Merino sheep was similar to that of Rambouillets, Shropshires and Suffolks. The phenotype frequency of the antigen C was close to that of the Shropshires, Suffolks and Corriedales. The frequency of the M-allele was similar to that of the same allele in Rambouillets, Targhees, Columbias and Corriedales. The phenotype frequency of antigen R in Polish Merino sheep was markedly more frequent than that of the antigen R in other breeds.

TABLE 2

Breed differences in the phenotype frequencies

Breed or flock	Number of animals	Locus and alleles									
		A		C			M			R	
		A	a	C	C_x	c	M	M_x	m	R	r^0
Polish Merino sheep											
Czechnica flock	194	67.5	32.5	42.3	—*	—	4.6	—	—	62.4	37.6
Mełno flock	212	67.5	32.5	50.0	—	—	9.0	—	—	61.3	39.7
Rambouillet**	195	66	34	1	99	0	1	0	99	11	89
Targhee**	192	40	60	1	87	12	7	3	90	33	67
Columbia**	177	22	78	2	88	10	11	1	88	25	75
Shropshire***	75	65.3	34.7	89.3	2.7	8.0	57.3	0	42.7	82.7	17.3
Suffolk***	23	65.2	34.8	0	95.7	4.3	78.3	0	21.7	21.7	78.3
Corriedale***	8	50.0	50.0	0	100.0	0	0	0	100.0	50.0	50.0

* Not tested.

** After W. D. STANSFIELD *et al.* (1964).

*** After B. A. RASMUSEN *et al.* (1960).

This large variability of blood group antigen frequency in several sheep breeds points to a great genetic diversity which is chiefly the outcome of different origin of the breeds.

ACKNOWLEDGEMENTS

I wish to express my appreciation to Dr. LIDIA DOLA for carrying out the comparison test with the reagents used for blood typing of cattle, as well as to Dr. A. ŻARNECKI for his assistance in selecting statistical methods and to Mrs. ANNA FREY-ŻARNECKA and Miss ALICJA GRABOŃ for their invaluable assistance in the laboratory studies.

REFERENCES

Ching Chun Li, 1961. Human genetics, principles and methods, McGraw-Hill Book Company, Inc., New York, 28–35.

Rasmusen, B. A., 1958. Blood groups in sheep. I. The X–Z system, *Genetics*, **43**, 814–821.

Rasmusen, B. A., 1960. Blood groups in sheep. II. The B system, *Genetics*, **45**, 1405–1417.

Rasmusen, B. A., 1962. Blood groups in sheep, *Ann. N. Y. Acad. Sci.*, **97**, 306–319.

Rasmusen, B. A., Stormont, C. and Suzuki, Y., 1960. Blood groups in sheep. III. The A, C, D and M stystems, *Genetics*, **45**, 1595–1603.

Stansfield, W. D., Bradford, G. E., Stormont, C. and Blackwell, R. L., 1964. Blood groups and their associations with production and reproduction in sheep, *Genetics*, **50**, 1357–1367.

DISCUSSION

E. M. Tucker: Did I understand you to say that you use guinea pig complement?

W. Porębska: Yes.

E. M. Tucker: At what dilution?

W. Porębska: 1 in 15, unabsorbed.

I. G. Hall: I notice your anti-M is like Dr. Rasmusen's. His was very powerful — about 1/1000. May I ask the titre of your anti-M′?

W. Porębska: 1/16.

XITH EUROPEAN CONFERENCE ON ANIMAL BLOOD GROUPS AND BIOCHEMICAL POLYMORPHISM
Warsaw, July 2nd–6th, 1968

Preliminary Determination of Transferrin Types in Polish Merino Sheep

H. BALBIERZ and Maria NIKOŁAJCZUK

Laboratory of Immunopathology, Department of Obstetrics and Gynecology; College of Agriculture in Wrocław, Poland

Encouraging observations and the detection of some relationships whether small or not, between transferrin types and some production characters in domestic cattle (ASHTON and FERGUSON, 1963; OSTERHOFF, 1967) have induced us to extend such investigations to sheep (NASRAT and OOSTERLEE, 1965; OOSTERLEE and BOUW, 1966).

Until the 10th ESABR Conference (Paris, 1966) the studies on this subject had been hampered by the lack of uniform nomenclature of transferrin types in sheep, which in turn led to failure to obtain comparable results.

OUR OWN INVESTIGATIONS

There have been no transferrin studies on sheep in Poland. The purpose of the present paper is to present results of such studies in Polish Merino sheep and to determine the percentage distribution of individual types in a selected flock.

Material

The sera examined orginated from 340 Polish Merino sheep of a single flock bred in the Wrocław region.

Methods

Electrophoretic separation was made on starch gel with discontinuous buffer system under the following conditions:

(a) starch, hydrolysed in our laboratory;

(b) electrode-vessel buffer (A): 1.2 LiOH; 11.8 boric/1 litre;

(c) buffer (B): 8.67 tris; 1.33 citric acid/1 liter;

(d) gel buffer: 5.5 volume of buffer (B) and 1 volume of buffer (A);

(e) separation: 5.2 V/cm for 14 hr.

After separation, the gel was cut horizontally into layers about 1.5 mm thick.

Amido Black 10B was used for staining. The distance of migration from the origin to the cathodic edge of albumin spots was about 12 cm.

After completion of the first phase of electrophoretic studies, sera with similar migration rate were divided into groups and a further electrophoretic separation was performed under the same conditions in the presence of reference sera*.

RESULTS AND DISCUSSION

The flock under examination showed all the transferrin alleles as described and classified in 1966 by the ESABR Commission (OOSTERLEE and BOUW, 1966), with the exception of the allele determining the type TfP with the slowest migration rate. We have detected also a hitherto unidentified band situated between Tf-D and Tf-E; arbitrarily referred to as Tf-E_1. The type was found to occur three times in heterozygous combination, twice with Tf-M and once with Tf-B.

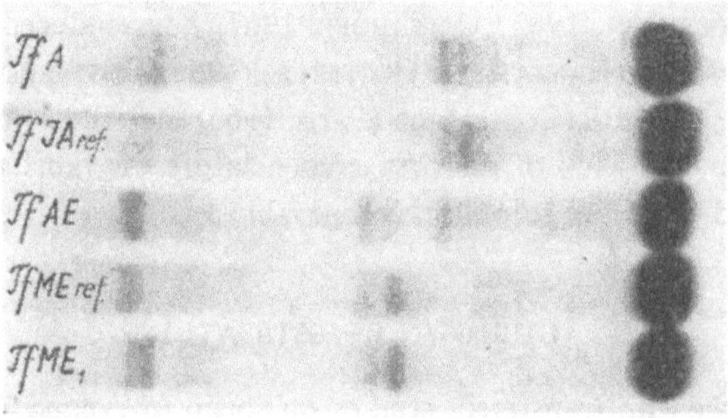

Transferrin phenotype	Number of sheep	Transferrin phenotype	Number of sheep	Transferrin phenotype	Number of sheep
IA	2	GG	3	MM	34
AA	33	GC	1	MD	21
AG	7	GM	4	ME	5
AB	2	BM	1	ME_1	2
AC	67	BD	1	DD	17
AM	29	BE	1	DE	10
AD	21	BE_1	1	EE	1
AE	24	CC	29		
		CM	18		
		CD	3		
		CE	3		

Total : 340 sheep

* Reference sera were obtained through the courtesy of Dr. BOUW and Dr. OOSTERLEE from Wageningen.

Most sheep (223) carried transferrins in heterozygous combination (Figure); the transferrins which occurred in homozygous state were: A, G, C, M, D, E.

Quantitative distribution of individual transferrin types in the flock of sheep examined is given in the Table.

It must be admitted that with the technique used for separation even with the use of reference sera there was still difficulty in distinguishing allele A from allele G and allele M from allele D.

REFERENCES

Ashton, G. C. and Ferguson, K. A., 1963. Serum transferrin polymorphism in Merino sheep, *Genetical Research*. **4**, 240.

Nasrat, G. E. and Oosterlee, C. C., 1965. Transferrin types in Dutch and Egyptian sheep, *Tijdschr. Diergeneesk*, **90**, afl. 12.

Oosterlee, C. C. and Bouw, J., 1966. Nomenclature of transferrin types in sheep, Wageningen (Typewritten).

Osterhoff, D. R., 1967. Transferrin type adaptability in cattle, Polymorphismes Biochimiques des Animaux, Institut National de la Recherche Agronomique, Paris.

XITH EUROPEAN CONFERENCE ON ANIMAL BLOOD GROUPS AND BIOCHEMICAL POLYMORPHISM
Warsaw, July 2nd–6th, 1968

On the Excess of BM Transferrin Genotype in Sheep

L. FÉSÜS

Bloodgrouping Laboratory, University of Veterinary Medicine, Budapest, Hungary

SUMMARY. Starch-gel electrophoresis described by ASHTON (1965) was applied, with slight modifications (FÉSÜS, 1967a) to determine the transferrin types of 431 Hungarian Merino, 213 Hungarian Black Racka, 69 Hungarian White Racka, 255 Cigaja, 139 Karakul and 70 Ile de France sheep (FÉSÜS, 1967b). On the basis of the gene frequencies the expected numbers of the different transferrin types were determined and the differences between the expected and observed number of each transferrin type were also calculated.

In the first part of this study the electrophoretic differences between the Tf M and Tf D bands were not considered and all Tf BM and Tf BD animals were taken to be Tf BD. In five of the six breeds examined the number of the Tf BD types was higher than expected.

In the second part of the study the zone pairs controlled by the alleles Tf^M and Tf^D were considered separately. In five of the six breeds studied the number of the Tf BM type was higher than expected and these differences were significant at the 5% level in four breeds (Hungarian Merino, Hungarian Black Racka, Hungarian White Racka and Karakul).

INTRODUCTION

Sixteen sheep transferrin alleles have been described by different authors: Tf^I Tf^A, Tf^G, $Tf^{B'}$, Tf^B, Tf^C ($Tf^{C_{Hungary}}$), Tf^M, Tf^D, Tf^U, Tf^N ($Tf^{N_{Hungary}}$), Tf^Q, Tf^E, Tf^R, Tf^V and Tf^P (ASHTON, 1958; ASHTON and FERGUSON, 1963; KHATTAB *et al.*, 1964; FÉSÜS, 1967a, b; OSMAN, 1967; FÉSÜS, and ORBÁNYI 1968).

Studying the genetics of the transferrins in different sheep breeds EFREMOV and BRAEND (1964), KHATTAB *et al.* (1964), COOPER (1966), KING and FECHTER (1966), NI *et al.* (1967), and OSMAN (1967) found differences between the expected and observed numbers of Tf BD type. All of these authors found an excess of this type. The extent of deviations of the expected and observed numbers of this genotype were different and in certain cases even significant.

These authors did not distinguish between the zones controlled by the alleles Tf^M and Tf^D, respectively.

The present study was carried out to reveal whether lack of distinction between alleles Tf^M and Tf^D was responsible for the excess of the Tf BD type shown in previous studies.

Breeds	Without separation of Tf M and Tf D				With separation of Tf M and Tf D							
	Tf BD				Tf BM				Tf BD			
	expected	observed	difference	95% confidence limits	expected	observed	difference	95% confidence limits	expected	observed	difference	95% confidence limits
Hungarian Merino (431)	44.44	53.00	+8.56	40.44–48.44*	25.54	31.00	+5.46	22.54–28.54*	20.84	22.00	+1.16	17.84–23.84
Hungarian Black Racka (213)	28.10	37.00	+8.90	23.10–33.10*	15.73	25.00	+9.27	11.73–19.73*	11.36	12.00	+0.64	8.36–14.36
Hungarian White Racka (69)	13.60	18.00	+4.40	6.60–20.60	6.79	11.00	+4.21	3.79–9.79*	6.27	7.00	+0.73	3.27–9.27
Cigaja (255)	51.41	53.00	+1.59	46.41–56.41	25.71	30.00	+4.29	20.71–30.71	25.71	23.00	−2.71	20.71–30.71
Karakul (139)	6.91	7.00	+0.09	3.91–9.91	0.38	3.00	+2.62	0.18–0.58*	3.01	4.00	+0.99	1.01–5.01
Ile de France (70)	2.57	2.00	−0.57	−0.43–5.57	1.34	1.00	−0.34	−0.56–3.34	1.22	1.00	−0.22	−0.78–3.22

* Significant at the 5% level.

MATERIAL AND METHODS

Starch-gel electrophoresis described by ASHTON (1965) was applied with slight modifications (FÉSÜS, 1967a) to determine the transferrin types of 431 Hungarian Merino, 213 Hungarian Black Racka, 69 Hungarian White Racka, 255 Cigaja, 139 Karakul and 70 Ile de France sheep (FÉSÜS, 1967b). The separation of the zone pairs controlled by the alleles Tf^A and Tf^G, Tf^M and Tf^D, Tf^Q and Tf^E was very clear. On the basis of the gene frequencies the expected numbers of the different transferrin types were determined and the differences between the expected and observed number of each transferrin type were also calculated.

RESULTS AND DISCUSSION

The Table shows the expected and observed number of Tf BM and Tf BD genotypes and also the differences between them.

In the first part of this study the electrophoretic differences between the Tf M and Tf D bands were not considered and all Tf BM and Tf BD animals were taken to be Tf BD. In five of the six breeds examined, the number of the Tf BD types was higher than expected. The differences between the expected and observed number was significant at the 5% level in two breeds (Hungarian Merino and Hungarian Black Racka; Table).

In the second part of this study the zones controlled by the alleles Tf^M and Tf^D were considered separately. The differences between the expected and observed number of both the Tf BM and Tf BD types were determined to ascertain whether the allele Tf^M or the Tf^D was responsible for the excess of the Tf BD type observed in the first part of this study. In five of the six breeds studied, the number of the Tf BM type was higher than expected and these differences were significant at the 5% level in four breeds (Hungarian Merino, Hungarian Black Racka, Hungarian White Racka and Karakul; Table).

REFERENCES

ASHTON, G. C., 1958. Further beta-globulin phenotypes in sheep, *Nature*, **182**, 1101–1102.

ASHTON, G. C., 1965. Serum transferrin D alleles in Australian cattle, *Au. J. Biol. Sci.*, **18**, 665–670.

ASHTON, G. C. and FERGUSON, K. A., 1963. Serum transferrins in merino sheep, *Genet. Res.*, **4**, 240–247.

COOPER, D. W., 1966. Some results of genetical studies on the transferrin variants of the Australian merino, Polymorphismes Biochimiques des Animaux, Paris, 301–305.

EFREMOV, G. and BRAEND, M., 1964. Haemoglobins, transferrins and albumins of sheep and goats, Blood Groups of Animals, Prague, 313–317.

FÉSÜS, L., 1967a. A new sheep transferrin allele: Tf^I, *Acta Vet. Acad. Sci. Hung.*, **17**, 95–98.

FÉSÜS, L., 1967b. Transferrin alleles in some sheep breeds in Hungary, *Acta Vet. Acad. Sci. Hung.*, **17**, 433–438.

Fésüs, L. and Orbányi, I., 1968. On the occurrence of the alleles $Tf^{N_{Hungary}}$, Tf^{U} and Tf^{V} in sheep, *Acta. Vet. Acad. Sci. Hung.* (In press).

Khattab, A. G. H., Watson, J. H. and Axford, F. R. E., 1964. Associations between serum transferrin polymorphism and disturbed seggregation ratios in Wales Mountain sheep, *Animal Prod.*, **6**, 207–213.

King, P. and Fechter, H., 1966. Transferrin polymorphism in South African sheep breeds, Polymorphismes Biochimiques des Animaux, Paris, 1966, 307–312.

Ni, G. V., Egorov, E. A, and Ris, M. A., 1967. Transferrin types and their inheritance in Karakul sheep (Tipy transferinov i ikh nasledovanie u karakul'skikh ovec), *Dokl. VASHNIL* (Moscow), **2**, 32–34.

Osman, H. El S., 1967. Serum transferrin polymorphism in the desert sheep of the Sudan, *Nature*, **215**, 162–163.

XITH EUROPEAN CONFERENCE ON ANIMAL BLOOD GROUPS AND BIOCHEMICAL POLYMORPHISM
Warsaw, July 2nd–6th, 1968

Inherited Variations in the Prealbumins of Sheep Serum*

G. EFREMOV, B. VASKOV and R. HRISOHO

Department of Physiology and Biochemistry, Faculty of Agriculture, and Department of Internal Medicine, Faculty of Medicine, Skopje, Yugoslavia

A number of inherited variations of serum proteins in the sheep have been described (ASHTON, 1958; RENDEL and STORMONT, 1964; EFREMOV and BRAEND, 1965; LEE, 1966; TUCKER *et al.*, 1966, 1967). Polymorphism of prealbumins has been reported to occur in some domestic animals (KRISTJANSSON, 1963; BRAEND and EFREMOV, 1965; GAHNE, 1966; BRAEND, 1967) detected by using starch gel electrophoresis. The present report describes a starch gel electrophoresis procedure for fractionation of sheep serum prealbumin and presents results on the genetic control of the prealbumins of sheep serum.

MATERIALS AND METHODS

Samples from adult sheep of the native Macedonian breeds, Sharplanina and Ovchepole, and from the imported French Merino de L'est breed were investigated. From these 430, 256 and 117, respectively, were examined. For genetical studies samples from sheep of 4 different flocks of Ovchepole breed were used. In each of the flocks 1–2 rams of known prealbumin types were used to fertilize approximately 40 sheep. A total of 7 rams, 152 sheep and 155 lambs were sampled. Samples from the lambs were taken 40–60 days after lambing.

All blood samples were taken into heparin. Plasma samples were freshly frozen and kept at −25°C until used.

The technique used was horizontal starch gel electrophoresis based upon POULIK's (1957) discontinuous system as modified by KRISTJANSSON (1963). The cathode vessel contained Kristjansson buffer at double concentration. Gels were made from 22 g starch, 8.7 ml A and 2.3 ml B in 250 ml distilled water, pH 4.7. The plasma samples were inserted in the gel on double Whatman 3 MM paper. The starting

* This work was supported by the Institute of Animal Breeding Development at Skopje in the Socialist Republic of Macedonia, and by the Federal Foundation for the Scientific Investigation, Belgrad (No. 1876/2-1966).

voltage was 200 V and increase to 350 V was made when the boundary line was 0.2 cm before the sample insertion line. The cathode bridge was 2 cm from the insertion line and the run was stopped when boundary line had moved 5 cm beyond the insertion line. Another buffer giving a gel of pH 6.5 (10 ml A and 6.5 ml B in 250 ml distilled water) was also used. The conditions were as the first buffer except that the boundary line was allowed to migrate 11 cm beyond the insertion line.

Isolation of the seromucoids was made by WINZLER's method (1955). Neuraminidase treatment was performed as described by GAHNE (1967).

RESULTS AND DISCUSSION

The resolution of the serum proteins which is achieved in acid buffer pH 4.7 is illustrated in Fig. 1. Migration of the proteins is in both directions mostly close to the insert line. The prealbumins migrate toward the anode as bands well sepa-

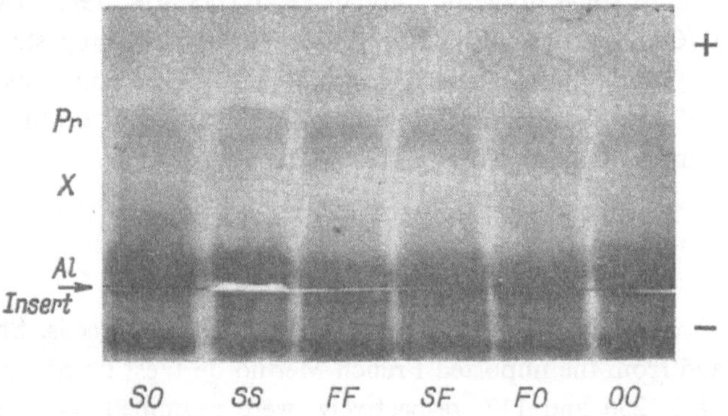

FIG. 1. A photograph of the result of a starch gel pH 4.7 with six sheep plasma samples representing six different prealbumin types.

rated from the albumin, and usually appear as one main diffuse band and one or two other thinner bands with faster rates of migration, or alternatively there are no bands at all. If the time of running is prolonged and the pH of the gel slightly decreased (4.6) then the main band separates into up to 6 fainter bands (Fig. 2). When the pH of the gel is 6.5 the prealbumins also separate from the albumin but the resolution is not complete. The protein fractions separated at pH 4.7 and 6.5 were compared by using two dimensional starch gel electrophoresis. First the proteins were separated at pH 4.7, and a piece of this gel with separated proteins was placed in the insertion line of a gel of pH 6.5 (Fig. 3). The reverse procedure was also adopted. The genetically variable prealbumin zone demonstrated at pH 4.7 migrated in the prealbumin area at pH 6.5, and the same was found when the first separation was made in a gel of pH 6.5 and the second in a gel of pH 4.7. These

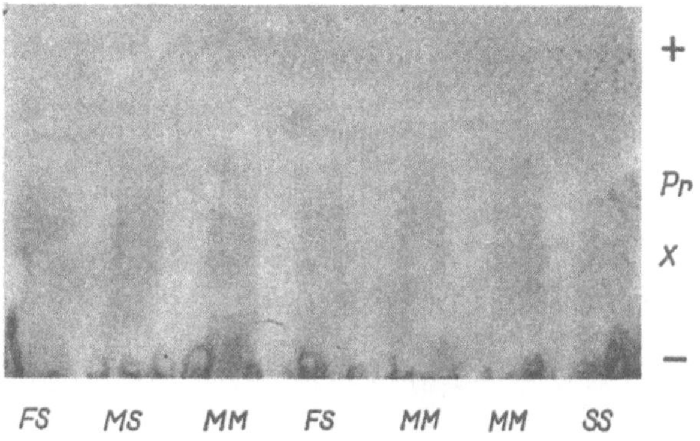

Fig. 2. Fractionation of prealbumins in a starch gel pH 4.6.

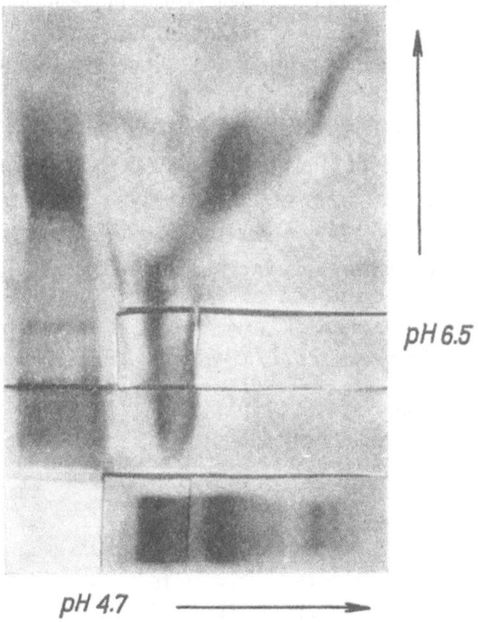

Fig. 3. Two-dimensional starch gel electrophoresis of sheep plasma of Pr type SS.

analyses showed that all of the protein fractions indicated in Fig. 1 are prealbumins.

The fastest migrating bands designated as Pr F and Pr S were found to show a regular variation among the plasma samples. Six different prealbumin phenotypes were observed (Figs. 1 and 4). Some sera contained only Pr S (phenotypes SS and S0), some Pr F (phenotypes FF and F0), some both Pr S and Pr F (phenotype SF), and other contained neither Pr S nor Pr F (phenotype 00). Using gels of both pH level and some other variations of the described electrophoretic conditions, we

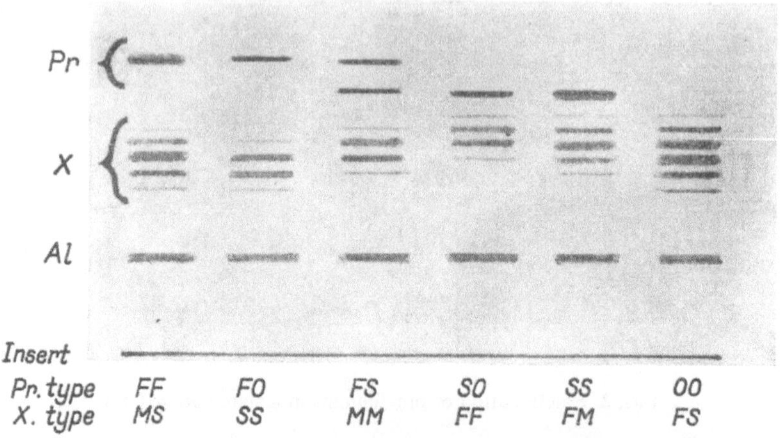

FIG. 4. A diagram of two protein systems in prealbumin region.

did not observe any protein in the phenotype 00 except the main prealbumin zone which was also present in the other phenotypes. An allele which has no detectable product has been found by KRISTJANSSON (1966) in the serum albumin of pigs.

Data on the inheritance of the prealbumin (Pr) phenotypes are summarized in Table 1. Three of the 7 sires were of the phenotype S0, two of the phenotype SS and two of the phenotype 00. The observed types of the offspring were in all cases, except six, in accordance with the hypothesis that the prealbumin phenotypes are

TABLE 1

Inheritance of prealbumin phenotypes of 155 offspring having sire and dam of known type

Kinds of matings		No.	No.	No. of offspring of phenotypes						Total
sire	dam	of sires	of mating	SS	S0	00	FF	F0	FS	No.
SS	S0	2	6	3	3	—	—	—	—	6
	F0		2	—	1	—	—	—	1	2
	FS		2	—	—	—	—	—	2	2
	00		15	—	16	—	—	—	—	16
S0	SS	3	27	11	13	3*	—	—	—	27
	S0		24	5	12	7	—	—	—	24
	00		34	—	15	19	—	—	—	34
	FF		4	—	1*	1*	1*	—	1	4
	F0		3	—	—	1	—	1	1	3
	FS		4	1	1	—	—	2	1	5
00	00	2	7	—	—	7	—	—	—	7
	SS		11	—	12	—	—	—	—	12
	S0		8	—	3	5	—	—	—	8
	FS		5	—	2	—	—	3	—	5
Total		7	152	20	79	43	1	6	6	155

* Not expected on the basis of the genetic theory proposed.

determined by three alleles, designated Pr^S, Pr^F and Pr^0. The alleles Pr^S and Pr^F control the synthesis of Pr S and Pr F whereas the Pr^0 has no detectable product under our analytical conditions. The six exceptional cases were probably caused by interchange of samples or by illegitimacy. If we exclude them from further consideration there is agreement of the data with the above hypothesis only those progeny phenotypes expected from the hypothesis being observed.

Distributions of observed and expected prealbumin phenotypes and gene frequencies are presented in Table 2. In the two native, Sharplanina and Ovchepole, breeds the Pr^0 allele has the highest frequency (0.532 and 0.515 respectively), and the Pr^F allele the lowest (0.077 and 0.049), the Pr^S being intermediate (0.390 and 0.435). Contrary to this, the imported Merino de L'est breed has the Pr^F allele in highest frequency (0.401). For the native Sharplanina breed there is small deviation between observed and expected numbers of sheep of phenotypes SS, S0, 00 and F0, and for Ovchepole breed of phenotypes S0, 00 and F0. This might be explainable on the basis of differences in the thickness and intensity of staining of the S band in the serum of some sheep of phenotypes SS and S0 which it difficult to type SS and S0, the possible presence of very small quantities of F producing a band of very weak intensity in some plasma samples which would then be wrongly typed as 00 instead of F0. The most probable explanation for the variations of the thickness of the bands are individual differences in the total content of the serum prealbumin. For the imported Merino de L'est breed there is agreement between observed and expected numbers.

TABLE 2

Observed and expected distribution of prealbumin phenotypes and gene frequencies

Breed		Phenotypes						Total	Gene frequency
		SS	S0	00	FF	F0	FS		
Sharplanina	obs.	73	163	132	4	31	27		$Pr^S = 0.390$
	exp.	65.4	178.4	121.7	2.55	35.3	25.8	430	$Pr^0 = 0.532$
									$Pr^F = 0.077$
Ovchepole	obs.	50	111	72	2	9	12		$Pr^S = 0.435$
	exp.	48.5	114.7	67.9	0.6	12.9	10.9	256	$Pr^0 = 0.515$
									$Pr^F = 0.049$
Merino de L'est	obs.	12	20	10	19	26	30	117	$Pr^S = 0.316$
	exp.	11.7	20.9	9.4	18.8	26.6	29.6		$Pr^0 = 0.283$
									$Pr^F = 0.401$

Analyses of serum collected from the same animals over a period of a year did not show any change either in the individual phenotype or in the proportions of the bands in SF phenotype.

The main prealbumin zone also showed individual variations and it is assumed that these represent another protein system (Figs. 2 and 4). This system has been given the tentative name X. Six different phenotypes were observed and the designations FF, MM, SS, FM, FS and MS were used. The phenotypes FF, SS and MM show different migration rates, and are characterized by the presence of two distinct bands which are accompanied by two faint bands, one in front of the main bands and the other behind. In the phenotypes FM and MS five bands are visible while in the phenotype FS six bands can be observed. These results suggest genetic control by three codominant alleles, X^F, X^M and X^S.

The chemical nature of the prealbumins detected in this study is still an open question. FAGERHOL and BRAEND (1966) found that the fastest migrating protein in the prealbumin area of human serum is thyroxine binding prealbumin. Two other proteins in the same area were identified as orosomucoid and protease inhibitor. Preliminary biochemical investigations of the starch gel after the electrophoresis

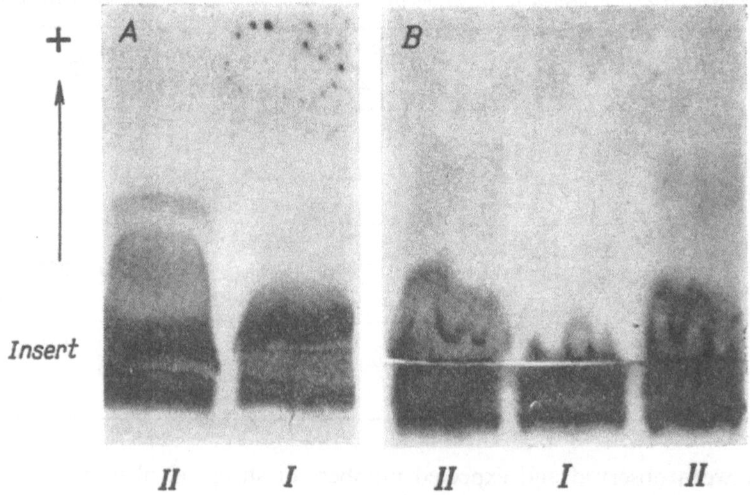

FIG. 5. Starch gel showing the effect of neuraminidase treatment. A — gel buffer pH 4.7; B — gel buffer pH 4.6 prolonged run. I — treated; II — control.

indicates that all fractions in the prealbumin region are glycoproteins, i.e. seromucoids, as all give positive reactions with PAS and DOT. Starch gel electrophoresis of perchloric acid extract of sheep serum shows the presence of all bands which are seen in untreated serum.

Treatment with neuraminidase caused disappearance of all the fractions in the prealbumin region (Fig. 5). Possible explanations are that treated fractions either did not enter the anodal side of the gel or that they migrated in the same position as the albumin. On the basis of these results it may be concluded that the two protein systems in the prealbumin region of sheep serum contain sialic acid but the sialic acid is not the cause of the genetic differences between prealbumins fractions.

CONCLUSIONS

Starch gel electrophoretic conditions which yield fraction of prealbumins of sheep serum have been described. Two protein systems in the prealbumin region have been found. The faster migrating system designated Pr showed six different phenotypes, and the results obtained suggest genetic control by three codominant alleles. The alleles Pr^S and Pr^F control the synthesis of Pr S and Pr F, and the allele Pr^0 has no detectable product under the analytical condition used.

The slower migrating system, designated X, was also found to show six different phenotypes, and the data obtained support the hypothesis that they are determined by three alleles, X^F, X^M and X^S. Preliminary biochemical investigations showed that all proteins in the prealbumin region of sheep serum are glycoproteins, i.e. seromucoids.

REFERENCES

ASHTON, G. C., 1958. Further beta-globulin phenotypes in sheep, *Nature*, **182**, 1101.

BRAEND, M., 1967. Variation of horse prealbumins in acidic starch gels, *Acta Vet. Scand.*, **8**, 193.

BRAEND, M. and EFREMOV, G., 1965. Haemoglobins, haptoglobins and albumins of horses, Proc. 9th Europ. Anim. Blood Group Conf., Prague, p. 253.

EFREMOV, G. and BRAEND, M., 1965. Haemoglobins, transferrins and albumins of sheep and goats, Proc. 9th Europ. Anim. Blood Group Conf., Prague, p. 313.

FAGERHOL, M. K. and BRAEND, M., 1966. Classification of human serum prealbumins after starch gel electrophoresis, *Acta Path. et Microbiol. Scandinav.*, **68**, 434.

GAHNE, B., 1966. Studies on the inheritance of electrophoretic forms of transferrins, albumins, prealbumins and plasma esterases of horses, *Genetics*, **53**, 681.

GAHNE, B., 1967. Alkaline phosphatase izoenzymes in serum seminal plasma and tissues of cattle, *Hereditas*, **57**, 100.

KRISTJANSSON, F. K., 1963. Genetic control of two pre-albumins in pigs. *Genetics*, **48**, 1059.

KRISTJANSSON, F. K., 1966. Fractionation of serum albumin and genetic control of two albumin fractions in pigs, *Genetics*, **52**, 627.

LEE, R. M., 1966. Genetic control of the hydrolysis of aromatic esters by sheep plasma A-esterase, *Genet. Res.* **7**, 373.

POULIK, M. D., 1957. Starch gel electrophoresis in a discontinuous system of buffers, *Nature*, **180**, 1477.

RENDEL, J. and STORMONT, C., 1964. Variants of ovine alkaline serum phosphatases and their associations with the R–O blood groups, *Proc. Soc. Exptl. Biol. Med.*, **115**, 853.

TUCKER, E. M., SUZUKI, Y. and STORMONT, C., 1966. Genetic variation in an esterase of sheep serum (Abstr.), *Feder. Proc.* **25**, 612.

TUCKER, E. M., SUZUKI, Y. and STORMONT, C., 1967. Three new phenotypic systems in the blood of sheep, *Vox Sang.*, **13**, 246.

WINZLER, R. J., 1955. In: Methods in Biochem. Analysis, Intersc. Publ., New York. Vol. II, p. 279.

XITH EUROPEAN CONFERENCE ON ANIMAL BLOOD GROUPS AND BIOCHEMICAL POLYMORPHISM
Warsaw, July 2nd–6th, 1968

Studies on the Serological Constitution of Goats

S. SUZUKI and S. WATANABE

Institute of Animal Serology, Tokyo University of Agriculture, Tokyo, Japan

SUMMARY. Four cellular antigenic factors, Ch_1, Ch_2, Ch_3 and Ch_4, classified by isoimmune antisera were shown to be genetically controlled by independent dominant genes. A serum beta-globulin of the goat was identified as transferrin by labelling with Fe^{59}. The serum transferrin was classified into three phenotypes genetically controlled by two autosomal alleles, Tf^{I} and Tf^{II}. And also, serum albumin of goats divided into three phenotypes, Al AA, Al BB and Al AB, with different mobilities in agar gel electrophoresis, was shown to be genetically controlled by two autosomal alleles, Al^{A} and Al^{B}.

Two isozymes, Apf and Aps of alkaline phosphatase were recognized in sera of goats. The Apf zone is genetically controlled by the genes Ap^{f} and Ap^{0}, Ap^{f} being dominant to Ap^{0}. Aps zone is transferred to the kid by the milk of the dam. The sera of goats were classified into two groups; the Apf group having the Apf zone and Ap0 group not having the Apf zone.

Large differences in frequencies of cellular antigens, serum transferrin types, serum albumin types and serum alkaline phosphatase types were observed among breeds of goats.

INTRODUCTION

Studies on the blood groups of goats were reported by WANG (1950) and SUZUKI *et al.* (1962) Two cellular agglutinogens G_1 and G_2 and two lysinogens Y_1 and Y_2 were found on the blood cells of goats by SUZUKI *et al.*, (1960) stimulating four blood group specific reagents in rabbit antisera. In addition, research has been carried out by many investigators on the biochemical polymorphisms of animals by using the starch gel electrophoresis technique introduced by SMITHIES (1955). The vari able beta globulins of goats were classified into three types, AA, AB and BB, by ASHTON *et al.* (1958) To investigate the serological constitution of goats the experiments reported were carried out on the blood groups and polymorphism of serum protein and isozymes in goats.

MATERIALS AND METHODS

The reagents to classify the blood groups were prepared from isoimmune hemolysins. Vertical starch gel electrophoresis with a discontinuous buffer system was performed to classify the serum transferrin types. The classifying of serum albumin was carried out by horizontal agar gel electrophoresis in a cold room (5–8°C). The buffer solution containing 8.5 g barbital sodium per 450 ml of distilled water was adjusted to pH 9.0 with 1 N HCl.

The experiments on the alkaline phosphatase isozymes in serum used agar gel prepared with 100 ml of barbital-HCl solution (pH = 8.4, $\mu = 0.05$) containing 0.7 g of J-agar and 0.7 g of PVP. To detect the alkaline phosphatase isozymes, a reaction mixture was prepared with 100 mg of Na_2-beta naphthyl phosphate, 100 mg of naphthanil diazo blue B, five drops of 10% $MgCl_2$ solution and 20 ml of barbital-HCl buffer solution (pH = 9.0, $\mu = 0.1$).

RESULTS AND DISCUSSION

Different reaction patterns were observed among the reactions of isoimmune antisera for blood cells of individual goats. Absorption tests of the antisera were performed to isolate the specific antibodies. Four hemolytic reagents, Ch_1, Ch_2, Ch_3 and Ch_4, differing from anti-G_1 and anti-G_2 were produced from the isoimmune antisera. Four cellular antigenic factors corresponding to these four reagents were genetically controlled by independent dominant genes, respectively. Blood sera of 1,979 goats were observed to be divided into different patterns according to the migration of components in beta-globulin. The fast migrating zones (a and b) were observed in the first pattern called the 1st type (I), the slow (c and d) in the third pattern, a, b, c and d components appeared to be three components in the second pattern, because b and c components were piled on each other. The variable proteins were identified as transferrins by labelling with Fe^{59}. From the distribution of the offspring of five matings among the three phenotypes of transferrin in goats and from the fact the II type goat was considered to be a heterozygote between the I and the III type goat, it was suggested that transferrin types in goats are controlled by two allelomorphs Tf^{I} and Tf^{II}, with three genotypes Tf^{I}/Tf^{I}, Tf^{II}/Tf^{II} and Tf^{I}/Tf^{II}, and three phenotypes I, II and III type. In electrophoretograms of unrefined agar gel two different migrating zones were recognized in albumin fractions. The individuals having the A zone were called Al AA type, the B zone Al BB type, and the individuals having two albumin zones were called Al AB type and considered to be the heterozygote between Al AA and Al BB types. From the comparison of results of the frequencies of serum albumin phenotypes in adult and juvenile animals in 686 goats, no variations of serum albumin could be detected as they grew older. Investigation of the distribution of the offspring of 6 matings among the three phenotypes of serum albumin in goats was carried out. The observed values were similar to the expected ones. It is suggested that the albumin types in goats are controlled by two allelomorphs, Al^{A} and Al^{B}, which produce three genotypes, Al^{A}/Al^{A}, Al^{B}/Al^{B} and Al^{A}/Al^{B}, and three phenotypes, Al AA, Al BB and Al AB.

In goat sera agar gel electrophoretograms two zones of alkaline phosphatase were recognized on the anode side. The faster migrating zone was called the f zone and the slower migrating zone the s zone of the alkaline phosphatase isozyme in goat serum. There were significant differences in the frequencies of s zone between adult and juvenile goats. The alkaline phosphatase zone in milk and the s zone in goat serum was recognized to have the same mobility. Therefore, it was realized that the s zone in goat serum is transferred to the offspring through the milk of the dams. So, goats were classified into two groups, Apf type having the f zone, and Ap0 type lacking the f zone.

From the results obtained in offspring of mating tests it is assumed that the Apf type is genetically controlled by a gene which is dominant to the gene controlling the Ap0 type which lacks the Apf zone. Two alleles, the Ap^f gene which produces the Apf zone and the Ap^0 gene which causes lack of the Apf zone or produces another zone were hypothesized. The genotypes of serum alkaline phosphatase type in goats are designated Ap^f/Ap^f, Ap^f/Ap^0 and Ap^0/Ap^0.

REFERENCES

ASHTON, G. C. *et al.*, 1958. Beta-globulin polymorphism in cattle, sheep and goats, *Nature*, **182**, 945–946.

SMITHIES, O., 1955. Zone electrophoresis in starch gel, group variations in serum proteins of normal human adults, *Biochem. J.*, **61**, 629–641.

SUZUKI, S. *et al.*, 1960. Studies on the goat blood groups and inheritance mode, *Z. Tierzücht. Züchtungsbiol.*, **74**, 236–247.

SUZUKI, Y. *et al.*, 1962. The J systems of goats, *Immunogenetics Letter*, **2**, 47–48.

WANG, K., 1950. On the new blood groups, Z_1 and Z_2, of goats by immune agglutinins, *Jap. J. Genetics*, **25**, 47.

XITH EUROPEAN CONFERENCE ON ANIMAL BLOOD GROUPS AND BIOCHEMICAL POLYMORPHISM
Warsaw, July 2nd–6th, 1968

Researches on Protein Polymorphism in a Goat Population of South Italy

A. SALERNO, N. MONTEMURRO and A. L'AFFLITTO
Animal Production Institute, University of Naples, Naples, Italy

SUMMARY. In this paper the distribution of the transferrins and albumins in a goat population bred in Lucania (South Italy), was studied. The results obtained in relation to the source of genetical variation are discussed.

INTRODUCTION

The β-globulins have been studied in many domestic animals in order to point out differences between various species and among individuals of the same species.

ASHTON and MCDOUGALL (1958), EFREMOV and BRAEND (1964) and WATANABE *et al.* (1965) carried out researches on goat transferrin polymorphism. EFREMOV and BRAEND (1964) studied also goat albumin polymorphism, finding the same phenotype in all animals examined.

MATERIALS AND METHODS

Plasma samples from a total 100 goats, bred in Lucania (South Italy) and belonging to three flocks, were examined.

Horizontal electrophoresis was used for the separation of both the transferrins and albumins.

Transferrins

The block gel-buffer solution (pH 8.6) contained 1.73 g Tris (hydroxymethylaminomethane), 0.85 g citric acid, and 1000 ml distilled water. The vessel buffer solution (pH 8.1) was prepared by dissolving 18.55 g boric acid, 4 g NaOH in a final volume of one liter. The gel was prepared as follows: 28 g hydrolysed starch were dissolved in 160 ml of the gel buffer and added to 40 ml of the vessel buffer solution. The quantities quoted were sufficient to prepare two blocks of starch gel 22 cm × 11 cm × 6 mm in which nine plasma samples could be run. The mixture was swirled by hand in a 1.5 l. conical flask over a Bunsen flame. With increasing temperature the viscosity first increased and then decreased as the temperature approached 90°C. The liquid was cleared of gas bubbles by connecting the flask to a suction pump at 70 cm Hg negative pressure for 25 sec. The hot liquid was poured into trays and stored in the refrigerator for 2 hr at about 4°C to set. Electric current of 50 mA was transmitted for 1.5 hr through the starch gel. After electrophoresis the starch gel was sliced longitudinally and stained in amido black 10B for 5 min. And then it was washed with methanol: glacial acetic acid: water (450:95:450).

Albumins

The different albumin types were demonstrated by using a sodium acetate EDTA buffer (pH 5.4). The composition of the buffer in the cathode vessel was 17.01 g Na-acetate and 2.48 g EDTA in a final volume of one liter. Some of the heat effects were avoided by halving the buffer concentration in the anode vessel. The gel buffer was a four fold dilution of cathode buffer. The starch concentration in the gels was 12%. Electric current of 75 mA was applied for 7 hr. After electrophoresis the starch gel was sliced, and stained in amido black and washed with methanol, glacial acetic acid and water.

RESULTS

Transferrin showed two bands A and B. The A band had a slower migration while the B band was faster. The apperance of the transferrins on starch gel (Fig. 1)

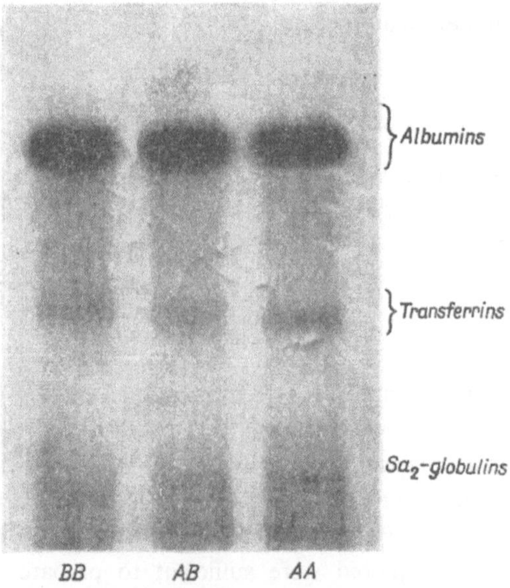

FIG. 1. Electrophoregram of three different phenotypes of transferrin in goat plasma by starch gel electrophoresis.

and the distribution of the phenotypes can be explained by a genetic theory of two codominant alleles: Tf^A and Tf^B. The distribution of the phenotypes and allele frequency is shown in the Table.

Distribution of transferrin and albumin types in a goat population bred in Lucania (South Italy)

Number of goats		Transferrin types			Allele frequency	
		AA	AB	BB	Tf^A	Tf^B
100	obs.	68	31	1	0.835	0.165
	exp.	69.7	27.5	2.8		
	$\chi^2 = 2.8010$; $P > 0.050$; f.d. = 1					
		Albumin types			Allele frequency	
		FF	FS	SS	Al^F	Al^S
100	obs.	3	64	33	0.34	0.65
	exp.	11.56	44.25	42.20		
	$\chi^2 = 17.1587$; $P < 0.0005$; f.d. = 1					

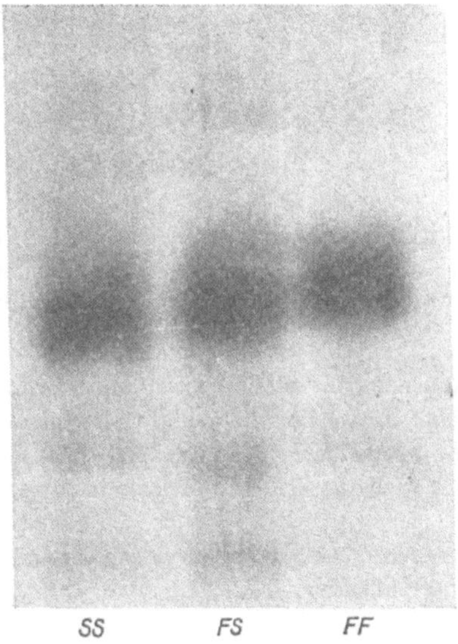

FIG. 2. Goat plasma albumin phenotypes.

Three albumin phenotypes were found: FF, FS and SS (Fig. 2), the distribution of which is shown in the Table. FF phenotype was observed in three animals only. The migration of the three phenotypes suggests that the two banded one is heterozygous in accordance with a genetic theory of two codominant alleles: Al^F and Al^S, as shown by the position and colour intensity of the bands.

DISCUSSION AND CONSIDERATIONS

Using De Finetti's equation ($E = 2\sqrt{DR}$), to examine the phenotype distribution of the transferrin and albumin loci, there appears to be an excess of heterozygotes over homozygotes.

The genetic equilibrium test for Tf and Al systems showed deviation from binomial distribution, for the second only.

In conclusion it can be said that in relation to the small size of the population examined, there is still a fair source of genetic variation since there are more heterozygotes of the transferrin and albumin loci than homozygotes. This is in conformity with De Finetti's equation.

These observations must be regarded as preliminary, and there remains much opportunity for further research.

REFERENCES

ASHTON, G. C., and MCDOUGALL, E. I., 1958. *Nature*, **182**, 945.
EFREMOV, G. and BRAEND, M., 1964. Proceedings of the 9th European Animal Blood Group Conference, Prague, 313.
WATANABE, S., NOZAWA, K. and SUZUKI, S., 1965. *Proceedings of the Japan Academy*, **41**, 326.

DISCUSSION

J. G. HALL: Reports on goat transferrins from several parts of the world agree that only three phenotypes are involved. In sheep however there are twenty or more. We regard sheep and goats as very similar phylogenetically. Is there any explanation for this difference?

A. SALERNO: Thank you, Dr. Hall, for your interesting remark. The history of ovine phylogenesis (sheep and goat) is quite complicated: the speciation of these animals seems to have appeared during the late Miocene. What do we know about transferrins evolution during those far away times? Furthemore, there are great differences in the social and sexual organization between goat and sheep life, in spite of their similarity. The sub-division in sub-specific ethnical groups is greater in sheep than in goats.

It is my opinion that these reasons might somehow explain numerical differences of the phenotypes at Tf locus between goats and sheep.

CHAPTER 6

BLOOD GROUPS AND BIOCHEMICAL POLYMORPHISM IN FISH

XITH EUROPEAN CONFERENCE ON ANIMAL BLOOD GROUPS AND BIOCHEMICAL POLYMORPHISM
Warsaw, July 2nd–6th, 1968

Inbreeding in Fresh Water Fish Populations Using Transferrins and Esterases as Markers

R. S. MALECHA and G. C. ASHTON
Department of Genetics, University of Hawaii, Honolulu, Hawaii, U.S.A.

SUMMARY. A species of fresh water fish, *Tilapia macrochir* was examined for biochemical polymorphism, and transferrin and esterase polymorphism were found. Phenotype frequencies were examined on several occasions from three different regions of a reservoir. No difference in phenotype distribution between areas was found in either system. However, a consistent excess of presumed homozygotes in both the transferrin and esterase systems was obtained, suggesting population structuring. Estimates of the inbreeding coefficient from both systems were consistent and averaged 0.42.

INTRODUCTION

The purpose of this investigation was to examine the population genetics of a fresh water fish, *Tilapia macrochir*, in a reservoir in central Oahu, using protein and enzyme polymorphisms. It was hoped that evidence for the mechanisms maintaining these polymorphisms might be found.

MATERIALS AND METHODS

Tilapia macrochir is a successful colonizing species, introduced into Hawaii from the Belgian Congo in the early 1950's. In 1958, 888 fish produced from the original 52 imported were released into the Wahiawa reservoir, which is a U-shaped reservoir 4 miles long and 200 yd wide at its widest point. Between August 1966 and March 1967 three areas of the reservoir were sampled on several days, but not on the same days. The areas sampled were coded AB and YZ for the upper reaches of both arms, and M for a region at the base of the "U".

The fish were examined for polymorphisms by starch gel electrophoresis and transferrin and ali-esterase polymorphisms were found. Transferrin polymorphism was routinely typed by ^{59}Fe autoradiography, because of the complexity of the serum patterns. Esterases were determined after incubation with α-naphthylacetate and detection of the released α-napthol with a mixture of Fast Red TR and Fast Blue RR.

RESULTS

Six transferrin phenotypes were found each showing either one or two zones, typical of the pattern expected for a system of three codominant alleles, Tf^A, Tf^B, and Tf^C each producing a single zone. Three esterase phenotypes were found, producing phenotypes typical of the pattern expected for a system of two codominant alleles, Es^F and Es^S each producing a single zone on starch gel.

The distribution of transferrin phenotypes between days within each of the three areas did not differ significantly, nor did the distributions between the pooled data for each area (Table 1). The distribution of esterase phenotypes between sampling days in each of the three areas was significantly different; area AB, $P < 0.005$; area M, $P < 0.01$ and area YZ, $P < 0.005$. However, comparing the total number of fish caught in each area there was no significant discrepancy in distribution of phenotypes between the three areas (Table 1).

TABLE 1

Transferrin and esterase phenotypes: Observed/expected distributions, pooled for days in each area

Phenotype	Area		
	AB	M	YZ
Tf A	20/16.0	34/30.3	8/15.7
Tf AB	13/12.4	27/23.5	8/12.2
Tf AC	2/ 2.6	4/ 4.9	4/ 2.5
Tf B	17/23.2	41/44.0	32/22.9
Tf BC	12/ 9.5	14/18.1	11/ 9.4
Tf C	3/ 3.4	7/ 6.4	3/ 3.3
Es F	52/49.6	84/90.0	56/52.5
Es Fs	22/19.9	36/36.0	19/21.0
Es S	12/16.5	36/30.0	16/17.0

On the assumption that the transferrin system is controlled by three and the esterase system by two codominant alleles, gene frequencies and the expected distribution of genotypes were calculated. It immediately became apparent that nearly every collection of fish showed an excess of homozygotes and a deficiency of heterozygotes, compared with expectation, for both the transferrin and esterase systems (Table 2). The consistency of this excess was checked by WOOLF's (1955) procedure. This essentially is a method for checking heterogeneity between 2×2 contingency tables, and for obtaining a weighted pooled estimate of any deviation from the expectation of unity. In the present application the ratio tested was: homozygotes observed/homozygotes expected — heterozygotes observed/heterozygotes expected. A ratio greater than unity indicates an excess of homozygotes and a deficiency of heterozygotes compared with expectation. The results obtained were as follows:

TABLE 2

Distribution of transferrin and esterase presumed homozygotes and heterozygotes, observed/expected

Area	Day	Transferrins homoz.	Transferrins heteroz.	Esterases homoz.	Esterases heteroz.
AB	1	9/ 7.5	11/ 12.5	14/ 12.5	6/ 7.5
	2	—	—	16/ 14.3	6/ 7.7
	3	19/ 11.9	11/ 18.1	18/ 17.5	10/ 10.5
	4	8/ 6.7	9/ 10.3	16/ 8.5	0/ 7.5
M	1	—	—	15/ 13.3	8/ 9.7
	2	14/ 8.0	6/ 12.0	13/ 11.6	6/ 7.4
	3	8/ 5.8	8/ 10.2	11/ 7.5	4/ 7.5
	4	17/ 10.2	8/ 14.8	19/ 17.4	2/ 3.6
	5	8/ 4.4	3/ 6.6	10/ 8.7	2/ 3.3
	6	35/ 23.1	20/ 31.9	52/ 33.1	14/ 32.9
YZ	1	11/ 8.1	9/ 11.9	26/ 20.3	9/ 14.7
	2	12/ 8.6	3/ 6.4	8/ 8.2	2/ 1.8
	3	—	—	4/ 5.8	6/ 4.2
	4	—	—	5/ 2.6	0/ 2.4
	5	20/ 14.5	11/ 16.5	29/ 22.6	2/ 8.4
Total		165/107.6	95/152.4	256/192.0	77/141.0

Note: Due to technical difficulties some fish were not analyzed for transferrins, and not all fish were typed for both systems.

Comparison	Deviation of ratio χ^2 (ld.f.)	Heterogeneity χ^2
Transferrin (11 samples from three areas)	17.2, $P < 0.01$	4.7, N.S. (10 d.f.)
Esterases (15 samples from three areas)	15.2, $P < 0.01$	12.4, N.S. (14. d.f.)
Transferrins and esterases from above pooled	32.4, $P < 0.01$	17.2, N.S. (25 d.f.)

DISCUSSION AND CONCLUSIONS

The possible reasons for obtaining a consistent excess of homozygotes and a deficiency of heterozygotes are: (1) misinterpretation of the phenotypes genetically, (2) selection for homozygotes, (3) selection against heterozygotes, and (4) inbreeding.

The first three reasons seem unlikely because of the consistency of results for both transferrins and esterases; it is hard to think of selection affecting two independent loci to the same extent. The fourth reason seems most likely, and implies that in Wahiawa reservoir the fish travel, and presumably breed in small schools.

This would explain the heterogeneity between samples found for the esterase system (the failure to find similar heterogeneity in the transferrins could be due to the greater number of phenotypes and hence the larger number of degrees of freedom).

The inbreeding coefficients for each area and system were calculated from the relationship (Li, 1955),

$$F = 1 - \frac{\text{Prop. of heterozygotes observed}}{\text{Prop. of heterozygotes expected}}$$

The values obtained were as follows:

Source	F
Tf, area AB	0.3453
Tf, area M	0.4071
Tf, area YZ	0.3475
Tf, all samples	0.3891
Es, area AB	0.3472
Es, area M	0.4902
Es, area YZ	0.4824
Es, all samples	0.4574

The mean coefficient of inbreeding seems to be about 42%, which is surprisingly high. Estimates from a shallow half acre reservoir containing *T. melanopleura* also gave consistent estimates for F from both transferrins and esterases, averaging 57%.

It is clear that protein and enzyme polymorphisms are powerful tools for investigating population structure. However data such as those presented here must be accepted with caution until it has been established beyond doubt that the mode of inheritance postulated is correct.

REFERENCES

Li, C. C., 1955. Population Genetics, University of Chicago Press, Chicago, Illinois, pp. 366.

Woolf, B., 1955. On estimating the relation between blood group and disease, *Eugenics*, **19**, 251-253.

DISCUSSION

I. E. Lush: If there were inbreeding would you not expect to find differences between the sampling areas?

G. C. Ashton: There were non-significant differences in gene frequencies between the areas. The "areas" were simply different regions of the reservoir, and were not isolated. Presumably, some interchange between populations in the different areas does occur.

N. P. Wilkins: Did you find any differences in environmental factors in the three areas sampled?

G. C. Ashton: Not in any of the factors we measured.

J. S. Alabaster: Is there any information on the movements of the fish in the system you have been studying?

G. C. Ashton: No.

XITH EUROPEAN CONFERENCE ON ANIMAL BLOOD GROUPS AND BIOCHEMICAL POLYMORPHISM
Warsaw, July 2nd–6th, 1968

Ontogenic Changes of Serum Transferrins in Plaice

Wilhelmina de LIGNY
Netherlands Institute for Fishery Investigations Ijmiiden, The Netherlands

SUMMARY. The inheritance of serum transferrins in plaice was investigated by means of breeding experiments. Using starch gel electrophoresis, unexpected types were found in the offspring during the first year of life. Changes into expected patterns were observed in some of the animals sampled again at a later stage. An explanation, involving developmental changes in the structure of the transferrin molecules, superimposed on their inherited characteristics, is suggested.

INTRODUCTION

Polymorphism of serum transferrins in plaice (*Pleuronectes platessa*) has been described previously (de LIGNY, 1967). Using starch gel electrophoresis over 15 varieties of single transferrin bands were found, successive bands occurring at close regular intervals. In each serum either one or two bands were present. Although population genetical data suggested that codominant multiple alleles may control the three groups in which they, on base of their frequency distribution, can be divided, no conclusive evidence was obtained with regard to the nature of the individual bands.

Breeding experiments that had been started in collaboration with Statens Biologiskie Stasjon Flødevigen, Norway, in order to investigate the genetics of red cell antigens in plaice, have in addition revealed some information on the inheritance of the individual transferrin bands. These results will be reported here.

METHODS AND MATERIALS

Blood samples were obtained from the adult fish by cardiac puncture and from the offspring by severing the caudal artery and inserting a capillary pipet. A minimum lenght of 6 cm was found to be required in order to obtain enough blood. Part of the animals could be investigated at an age of 8 months, while the remaining mostly had reached or exceeded this size at 12 months. Maximum size at both times of sampling was 12 cm.

Transferrins were separated by means of starch gel electrophoresis, according to the method described before (de LIGNY, 1966). Neuraminidase treatment was carried out according to KRISTJANSSON and CIPERA (1963), using Sigma type V, 2 mg dissolved in 0.2 ml 1 M acetate buffer, pH 5.5, containing 0.4 M $CaCl_2$, and added at a ratio of one drop to two drops of serum. For labeling with ^{59}Fe $^{59}FeCl_3$, activity 40 micro-Curie/ml, was added at a ratio of 1 to 2.

Offspring of four matings between two males and two females was investigated. Direct comparison of the parental transferrins with those of the offspring was possible in the case of one parent (male No. 1). The other parents had died before the offspring was sampled. Sera of other plaice that survived throughout the experiment were used as intermediary in comparing the transferrin types in those parents and their offspring. In the case of female No. 2 the position of one band remained uncertain.

RESULTS

Results are shown diagrammatically below (Figure). In the majority of the offspring of the four matings two transferrin bands were found, each corresponding with one band of each parent. In 15 animals from matings Nos. 12 and 17, however, one band unlike the parental bands was found, while 8 animals from mating Nos. 12, 17 and 18 possessed three bands.

In the cases where one band unlike the bands of the parents was observed, a band, or occasionally two, corresponding to those of the mother was present. The unexpected band appeared in 14 of the animals in a position similar to a weak third band occasionally observed in the serum of the father, male No. 1, in particular when this had been stored in the refrigerator during a few days. Two of the animals, No. 1315 and No. 1312 from mating No. 12, possessing a band unlike the parents, could be sampled again at a later stage of development, 8 months after the first sampling, when they had increased in lenght to 14 and 16 cm respectively. It was observed that the unlike band had disappeared during this time, while one of the paternal bands had come into the pattern. One of the animals possessing three bands, No. 1313 from mating No. 12, also was sampled again at this time, having reached a length of 15.5 cm. Only two bands were found, each corresponding to one band of each parent.

It may be noticed that in the changes observed, a band with higher electrophoretic mobility replaced the "unlike" band, or was left in the case where an additional band disappeared.

During the course of population studies involving deep frozen storage of plaice sera, it had been noticed that in some sera production of additional bands occurred when the storage at −20°C was continued for more than 3 months. In these cases one new band was found with lower electrophoretic mobility than the original transferrin. Additional bands, migrating behind the major bands were also occasionally

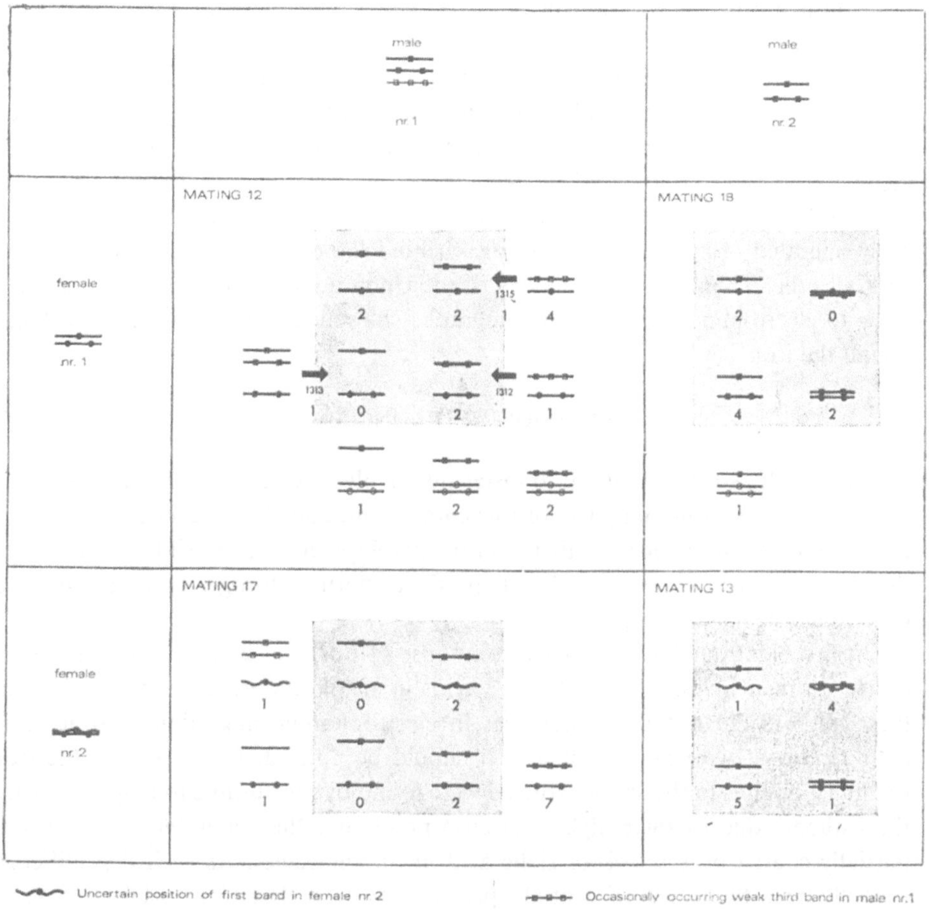

Diagram of serum transferrin types in parents and offspring of four matings. In shaded squares: types expected. Numbers indicate the number of individuals of each type found.

found in samples taken from plaice, brought into the aquarium, and run immediately afterwards. These phenomena suggest the existence of derivative forms of the transferrins.

Studies on the structure of mammalian transferrins have established that they are glycoproteins, their carbohydrate moiety containing a number of sialic acid residues. PARKER and BEARN (1962) have shown that the electrophoretic mobility of human transferrins decreases stepwise upon removal of sialic acid molecules under influence of the enzyme neuraminidase. They also found that, while the single transferrin of adults contains four of such residues, in cord blood additional iron-binding bands were present, coinciding with the bands produced in adults by neuraminidase action, suggesting that "at a particular stage of embryonic development an enzyme mechanism begins of function in the liver, which is capable of adding sialic acid to transferrin".

In order to check if the normal and "abnormal" transferrin bands observed in plaice could represent molecules containing different amounts of sialic acid, serum samples from adult plaice were treated with neuraminidase. When the samples after incubation during 12 hr at 37°C were subjected to starch gel electrophoresis, and run parallel with untreated samples, presence of additional bands was observed. They migrated at a comparable distance behind the original bands as was the case with all other "abnormal" bands described. Upon addition of ^{59}Fe to the samples, prior to electrophoresis, the autoradiographs showed that the newly formed bands bound the iron.

DISCUSSION AND CONCLUSIONS

Our results suggest that the transferrins of plaice contain sialic acid, and that derivative forms, containing less of this component, and showing a decrease in electrophoretic mobility, occur either, during development, as possible precursors of the final product, or as degradation products during storage, or occasionally as metabolic sidelines during adult life.

At first sight our findings differ from those of BOFFA *et al.* (1967) on transferrins in fish. In their investigations of transferrins in an Elasmobranch and a Cyclostome these authors did not find differences in electrophoretic migration of transferrins after treatment with neuraminidase. It should be noticed, however, that the iron-binding globulins in the species studied were found by the authors to migrate towards the cathode, due to their high isoelectric point that they consider to be at least partially caused by the lack of sialic acid in the molecules. As such they differ essentially from the transferrins of all species of Teleosts described so far (FINE *et al.*, 1965; CREYSSEL *et al.*, 1966; MØLLER and NAEVDAL, 1966; KOEHN and JOHNSON, 1967; BARRETT and TSUYUKI, 1967), that in a pH range from 8.0–9,0 all migrate to the anode.

We would like to conclude that our results indicate that the transferrins of the Teleost Pleuronectes platessa resemble those of mammalia in containing sialic acid. The unexpected types found in the offspring during the first year of life and the subsequent changes observed in some of the animals may represent the occurrence of transferrins containing less sialic acid then the corresponding molecules from the adults.

This phenomenon obscures interpretation of the numerical relationships of the transferrin types in the offspring. It is felt that the experiments in spite of this provide evidence that the individual transferrin bands are inherited, developmental processes however influencing their structure during the first year of life.

Details of the mode of inheritance may result from further breeding experiments under way.

ACKNOWLEDGEMENT

Breeding experiments and part of the investigations described were carried out at Statens Biologiske Stasjon Flødevigen, Norway. I thank its director Dr. G. DANNEVIG and his collaborators for their cooperation and hospitality.

REFERENCES

BARRETT, I. and TSUYUKI, H., 1967. Serum transferrin polymorphism in some Scombroid fishes, *Copeia*, 551–557.

BOFFA, G. A., FAURE, A., GOT, R., DRILHON, A. and FINE, J. M., 1967. Sur les caractères physico-chimiques des transferrines sériques de la lamproie (Cyclostome) et de la roussette (Sélachien), *Protides of the Biological Fluids*, **14**, 97–102, Elsevier, Amsterdam.

CREYSSEL, R., RICHARD, G. and SILBERZAHN, P., 1966. Transferrin variants in carp serum, *Nature* (Lond.), **212**, 1362 only.

FINE, J. M., DRILHON, A., BOFFA, G. A. and AMOUCH, P., 1965. Les types de transferrines chez certains poissons migrateure, *Protides of the Biological Fluids*, **12**, 165–168, Elsevier, Amsterdam.

KOEHN, R. K. and JOHNSON, D. W., 1967. Serum transferrin and serum esterase polymorphism in an introduced population of the bigmouth buffalo fish, *Ictiobus cyprinellus*, *Copeia*, 805–808.

KRISTJANSSON, F. K. and CIPERA, J. D., 1963. The effect of sialidase on pig transferrins, *Canad. J. Biochem. Physiol.*, **41**, 2523–2528.

DE LIGNY, W., 1966. Polymorphism of serum transferrins in plaice, *Polymorphismes biochimiques des Animaux*, 373–378. Inst. Natn. Rech. Agron., Paris.

MØLLER, D. and NAEVDAL, G., 1966. Transferrin polymorphism in fishes, *Polymorphismes biochimiques des Animaux*, 367–372. Inst. Natn. Rech. Agron., Paris.

PARKER, W. C. and BEARN, A. G., 1962. Studies on the transferrins of adult serum, cord serum, and cerebrospinal fluid, *J. Exp. Med.*, **115**, 83–106.

DISCUSSION

N. WILKINS: Did you find that the abnormal transferrins bind more (or less) radioactive iron than the normal transferrin bands?

W. de LINGY: We found in the neuraminidase treated adult sample that the new formed bands, even if showing up very weakly in the protein stain, bound the radioiron very strongly compared with the original transferrin bands.

I. LUSH: Did you find more fish showing these signs of immaturity in the 8-months sample than in the 12-month sample?

W. de LIGNY: At both dates of sampling we selected fish of similar length and between these we did not notice differences in the occurrence of the immature types. The number of animals was small, however.

XITH EUROPEAN CONFERENCE ON ANIMAL BLOOD GROUPS AND BIOCHEMICAL POLYMORPHISM
Warsaw, July 2nd–6th, 1968

Cod Transferrins and Genetic Isolates

A. JAMIESON

M.A.F.F., Fisheries Laboratory, Lowestoft, Great Britain

SUMMARY. The transferrins in the cod, *Gadus morhua* L. indicate genetic differences between a number of distinct stocks in different regions of the North Atlantic. The identities of the stocks are maintained by obscure isolating mechanisms which operate in the absence of any obvious physical barrier between some cod stocks spawning in relatively close proximity.

INTRODUCTION

The cod, *Gadus morhua* L. is found throughout the North Atlantic ocean. Varieties of cod found in different regions can be distinguished by examining the number of vertebrae and the number of rays in the second dorsal fin. The mean numbers of vertebrae from 51.5 at Rockall, west Scotland, to 55.5 near Cape Bauld, Newfoundland. The mean numbers of fin rays range from 17.2 near Kaliningrad to 20.9 at Sukkertoppen, west Greenland (SCHMIDT, 1930). In recent years cod stock differences have been demonstrated using erythrocyte antigens (MØLLER, 1966b, 1967) and haemoglobins (FRYDENBERG *et al.*, 1965; SICK, 1965a and b). Similarly cod transferrins have been used at Norway (MØLLER, 1966a, 1968), Greenland (JAMIESON, 1967) and Faroe (JAMIESON and JONES, 1967).

The present paper extends the cod transferrin data to most of the oceanic range of this species.

MATERIAL

Blood samples were collected from cod caught on research vessel cruises and also on chartered and normal commercial fishing trips. The total number of fish examined exceeds 4000, and their lenghts ranged between 20 and 130 cm. They were caught at different places and seasons between November 1965 and June 1968.

METHOD

Batches of cod blood samples were collected into citrated vacutainers (Beckton, Dickinson and Company, 2 ml tubes numbered 3273W) and placed on ice in thermos containers during transit by air, as necessary, to the testing laboratory. The great majority of the tests were made on unfrozen sera less than 6 days after bleeding. Where this strict time-scale was not feasible, adequate results were obtained using sera which had been centrifuged, separated and frozen at sea.

The electrophoretic runs were made at pH 9.0 in gels containing starch plus agar in a large refrigerator. The system was essentially the same as that used in similar work at Bergen (MØLLER, 1966b). Each unknown serum was matched directly in parallel against a fresh sample of a control serum from a familiar heterozygote. Repeated tests were used to resolve doubtful results, and the identities of rare alleles were confirmed in tests using the most appropriate controls for each.

NOTATION

All of the cod transferrins tested to date can be described adequately using a series of seven co-dominant alleles: Tf^{A1}, Tf^{A}, Tf^{B}, Tf^{C1}, Tf^{C}, Tf^{D}, and Tf^{E}. Although this interpretation is widely supported by population data, it cannot be tested experimentally in the absence of family data.

Examples of four cod transferrin alleles, Tf^{A}, Tf^{B}, Tf^{C} and Tf^{D}, were present in control sera from the Havforskningsinstitutt, Bergen. Those gave equivalent results at the Fisheries Laboratory, Lowestoft. Three additional alleles, Tf^{A1}, Tf^{C1} and Tf^{E}, have yet to be compared by exchange of material, but their relative electrophoretic mobilities indicate that they refer to the bands which MØLLER and NAEVDAL (1967) supposed to be new cod transferrins.

RESULTS

Cod show seven serum transferrins which are attributed to a series of alleles giving 28 possible genotypes. Examples of 22 of these genotypes have been observed among 4000 cod tested at Lowestoft. The six rare genotypes which remain postulated are Tf^{A1}/Tf^{A1}, Tf^{A1}/Tf^{A}, Tf^{A1}/Tf^{C1}, Tf^{A1}/Tf^{E}, Tf^{A}/Tf^{E}, and Tf^{E}/Tf^{E}.

This implied genetic interpretation of cod transferrin type is supported by the genetic equilibrium shown in the widespread examples of population data, but the numerical details of the many genotypes are not presented here. Instead, the allele frequency statistics in the Table present a condensed outline of the results.

DISCUSSION

The cod transferrin data may be applied in the formation of opinions about the genetic affinities, structures and migrations of cod within and between the various region. Cod populations from all of the major regions were distinctive, and in some regions there were useful genetic differences between sub-samples.

The frequencies of transferrins in cod sera tested at Lowestoft

Region	Chart ref. approx.	N*	Transferrin allele frequencies						
			Tf^{A1}	Tf^{A}	Tf^{B}	Tf^{C1}	Tf^{C}	Tf^{D}	Tf^{E}
Canada									
Grand Rivière	49°N 46°W	102	—	0.015	0.427	0.270	0.142	0.142	0.005
Halifax	44°N 63°W	100	—	0.005	0.425	0.180	0.175	0.180	0.035
N. Grand Banks	49°N 51°W	86	—	—	0.494	0.314	0.029	0.134	0.029
S.E. ,, ,,	45°N 49°W	29	—	—	0.552	0.328	0.069	0.052	—
E. ,, ,,	48°N 48°W	174	—	0.003	0.376	0.331	0.138	0.126	0.026
Flemish Cap	47°N 45°W	120	—	0.058	0.192	0.283	0.304	0.154	0.008
Greenland									
Disko Bay	70°N 55°W	138	—	0.130	0.243	0.040	0.504	0.080	0.004
Store Hellefiske	68°N 54°W	180	—	0.097	0.239	0.067	0.508	0.083	0.006
Fyllas Bank	65°N 54°W	195	0.008	0.192	0.192	0.039	0.526	0.044	—
Cape Farewell	60°N 44°W	198	—	0.139	0.149	0.033	0.616	0.063	—
Iceland									
Selvogsbanki	63°N 21°W	100	—	0.100	0.190	0.020	0.615	0.075	—
Reykjanes 1967	64°N 23°W	52	—	0.067	0.125	0.010	0.721	0.077	—
,, 1968	64°N 23°W	60	—	0.042	0.325	0.033	0.517	0.083	—
Strandagrunn	67°N 22°W	106	—	0.094	0.094	0.014	0.698	0.099	—
Eyjafjördur	66°N 19°W	56	—	0.134	0.194	—	0.598	0.071	—
Langanes	66°N 14°W	65	—	0.100	0.100	0.008	0.708	0.077	0.008
Faroe									
Plateau	62°N 07°W	281	—	0.064	0.137	0.007	0.749	0.043	—
Bank	61°N 08°W	125	—	0.079	0.070	0.020	0.772	0.063	—
W. Scotland									
N. Rona	59°N 06°W	189	—	0.027	0.114	0.005	0.833	0.021	—
N. Minch	58°N 06°W	29	—	0.103	0.069	—	0.776	0.052	—
Bara Head	57°N 07°W	11	—	—	0.136	—	0.864	—	—
Galloway	55°N 05°W	13	—	0.039	0.077	—	0.885	—	—
Channel									
Plymouth	50°N 04°W	34	—	0.015	0.177	—	0.794	0.015	—
Folkstone	51°N 01°E	101	—	0.020	0.188	—	0.743	0.049	—
North Sea									
Smith's Knoll	53°N 03°E	39	—	0.051	0.090	0.013	0.833	0.013	—
Lowestoft 1966–67	53°N 02°E	97	—	0.010	0.134	0.010	0.835	0.010	—
,, 1967–68	53°N 02°E	155	—	0.013	0.165	—	0.816	0.007	—
Silver Pits	54°N 03°E	84	—	—	0.042	0.018	0.899	0.042	—
W. Mud Hole	54°N 03°E	79	—	—	0.120	0.006	0.835	0.038	—
Skate Hole	54°N 02°E	21	—	0.024	0.071	—	0.881	0.024	—
Flamborough	54°N 00°	69	—	0.007	0.051	—	0.928	0.014	—
Clay Deep	55°N 04°E	16	—	0.031	0.094	—	0.875	—	—
Dogger	55°N 05°E	11	—	—	0.273	—	0.727	—	—
,,	56°N 02°E	15	—	0.067	—	0.033	0.900	—	—
Aberdeen	57°N 02°W	27	—	0.019	0.093	—	0.852	0.037	—

(contd.)

Region	Chart ref. apporox.	N*	Transferrin allele frequencies						
			Tf^{A1}	Tf^{A}	Tf^{B}	Tf^{C1}	Tf^{C}	Tf^{D}	Tf^{E}
Belt Sea									
Sejerø Bay	56°N 11°E	95	—	0.032	0.184	0.005	0.753	0.021	0.005
Southern Baltic									
S. Trelleborg	55°N 13°E	51	—	0.029	0.226	—	0.696	0.049	—
S.W. Ystad	55°N 14°E	91	—	0.016	0.258	0.011	0.681	0.033	—
W. Bornholm	55°N 14°E	182	—	0.008	0.247	0.014	0.714	0.016	—
N. ,,	55°N 15°E	60	—	—	0.300	—	0.700	—	—
S. ,,	55°N 15°E	109	—	—	0.260	0.014	0.711	0.014	—
S. Öland Point	56°N 16°E	52	—	—	0.317	—	0.673	0.010	—
Central Baltic									
S. Gotland	56°N 19°E	50	—	—	0.350	0.020	0.620	0.010	—
S.E. ,,	57°N 19°E	75	—	0.013	0.207	—	0.773	0.007	—

* N—number of fishes.

Canada. The cod population at Canada differ distinctly from those found elsewhere. The allele Tf^{C} showed consistently lower frequencies at Canada than in any other batch of cod. The allele Tf^{C1} was common at Canada and rare elsewhere; this was partly true also for Tf^{B} and Tf^{D}. With four alleles moderately well represented, this locus promises to be very useful in discriminating the genetic subdivisions among cod at Canada.

Greenland. The cod population at Cape Farewell differs genetically from those at west Greenland. The rare allele Tf^{A1}, which appeared once in each of three new genotypes, occurred at Fyllas Bank only.

Iceland. Until recently the transferrins in cod tested at Iceland appeared uniformly intermediate in frequencies, in keeping with their place on the Atlantic chart, apart from one exception; this was a northern fjord stock which showed genetic isolation. However, in the spring of 1968 a batch of spawning cod caught at Reykjanes showed transferrin frequencies akin to those observed at west Greenland. This work is being continued in conjunction with colleagues in Reykjavik.

Faroes. The distinction between the two cod stocks at Faroe has been described (JAMIESON and JONES, 1967).

West of Scotland. The Hebridean cod transferrin alleles closely resemble those found in the North Sea cod.

Channel. The cod in the English Channel were represented by two similar samples, differing distinctly from the cod in the North Sea.

The North Sea. The transferrins give no evidence for the genetic sub-division of the stock of cod in the North Sea. The apparent homogeneity of this population is characterized by the extremely high frequency of Tf^{C}.

Norway. MØLLER has reported a marked genetic difference between the Arctic cod and the coastal cod at Norway, and also indications of lesser genetic variation within both of the major Norwegian cod populations (MØLLER, 1966a, 1968).

Belt Sea. The Danish cod are probably a unit stock; their transferrin alleles are mainly Tf^C and Tf^B, both being intermediate in frequency between those of the North Sea and the Baltic populations.

Baltic. The Central and Eastern Baltic regions are tabulated separately but their transferrin values do not appear to distinguish the two stocks east and west of Bornholm, known to have very different gene frequencies at the locus for haemoglobin.

CONCLUSION

Preliminary population studies recognizing serum transferrin types in cod indicate that these are controlled by a series of seven co-dominant alleles which occur with distinctive frequencies in different regions of the North Atlantic. The frequencies are not necessarily influenced to any appreciable extent by selection in their present environments. The regional cod stocks maintain characteristic transferrin groupings, although distinct stocks may spawn in close proximity at Lofoten, Faroe and south-west Iceland, and possibly at Newfoundland and Greenland also. The evidence suggests that this potentially mobile species, whose larvae are often carried involuntarily in surface currents, has some highly developed mechanism ensuring the genetic isolation and continuity of the contemporary breeding units or stocks.

ACKNOWLEDGEMENT

I gratefully acknowledge the receipt of cod blood samples from Dr. P. DANDO, the Laboratory, Plymouth; Director J. JÓNSSON, Reykjavik; Dr. J. TØNNFS NIELSEN, Aarhus University; Dr. PAUL ODENSE, Fisheries Research Board of Canada; Dr. OTTERLIND, Fisheries Research Institute, Lysekil; also Mr. B. C. BEDFORD, Mr. M. J. HOLDEN and other members of the Lowestoft Laboratory staff.

The tests were assiduously carried out by Mr. D. F. SIMPSON and Mr. D. THOMPSON.

REFERENCES

FRYDENBERG, O., MØLLER, D., NAEVDAL, C. and SICK, K., 1965. Haemoglobin polymorphism in Norwegian cod populations, *Hereditas*, **53**, 257–271.

JAMIESON, A., 1967. New genotypes in cod Greenland, *Nature* (Lond.), **215**, 661–662.

JAMIESON, A. and JONES, B. W., 1967. Two races of cod at Faroe, *Heredity*, **22/4**; 610–612.

MØLLER, D., 1966a. Genetic differences of cod in the Lofoten area, *Nature* (Lond.), **212**, 824.

MØLLER, D., 1966b. Polymorphism of serum transferrin in cod, *Fisk. Dir. Skr. Ser. Havunders.*, **14**, 51.

MØLLER, D., 1967. Red blood cell antigens in cod, *Sarsia*, 29, 413–430.

MØLLER, D., 1968. Genetic diversity in spawning cod along the Norwegian coast, *Hereditas*, **60**, 1–32.

MØLLER, D. and NAEVDAL, C., 1967. Transferrin polymorphism in fishes. In Polymorphismes biochimiques des animaux, Xth European conference on animal blood groups and biochemical polymorphisms, Paris, 1966, pp. 367–372.

SCHMIDT, J., 1930. Racial investigations. X. The Atlantic cod (*Gadus callarias*) and local races of the same, *C. F. Lab.* (Carlsberg), **18**, 1–72.

SICK, K., 1965a. Haemoglobin polymorphism of cod in the Baltic and the Danish Belt Sea, *Hereditas*, **54**, 19–48.

SICK, K., 1965b. Haemoglobin polymorhpism of cod in the North Sea and the North Atlantic Ocean, *Hereditas*, **54**, 49–73.

XITH EUROPEAN CONFERENCE ON ANIMAL BLOOD GROUPS AND BIOCHEMICAL POLYMORPHISM
Warsaw, July 2nd-6th, 1968

Lactate Dehydrogenase Isoenzymes in Atlantic Salmon (*Salmo salar* L.)

N. P. WILKINS

Department of Agriculture and Fisheries for Scotland, Marine Laboratory, Aberdeen, Great Britain

SUMMARY. The lactate dehydrogenase isoenzymes in various tissues of Atlantic salmon (*Salmo salar* L.) were studied by electrophoresis in starch gel. Tissue specific isoenzyme patterns were observed, and those of the heart, red and white skeletal muscles are discussed in detail.

The occurrence of four distinct sub-units is postulated to explain the major LDH isoenzymes observed.

INTRODUCTION

The enzyme lactate dehydrogenase (LDH) is a tetrameric molecule, exhibiting, in many vertebrate species, five isoenzymic forms. These represent the tetramers resulting from the assembly of two genetically distinct sub-units — generally termed the H and M chains — in all possible combinations of four (CAHN *et al.*, 1962). Additional sub-units, either synthesised independently of the H and M sub-units in specific tissues, or representing allelomorphic mutants of one or both of the H and M sub-units, account for the larger number of isoenzymes observed in other species.

This report describes the multiple LDH isoenzymes in the Atlantic salmon *Salmo salar* and discusses their sub-unit structure.

MATERIALS AND METHODS

Extracts of the liver, kidney, spleen, heart and skeletal muscle of adult salmon caught in the Davis Strait at West Greenland, and in a number of Scottish rivers were analysed by electrophoresis in starch gel (SMITHIES, 1959), employing the buffer system described by MARKERT and FAULHABER (1965).

Extracts were prepared by homogenizing small (1–2 g) aliquots of tissue with two volumes (W/V) of distilled water, buffer or 0.25 M sucrose. The resultant homogenates were centrifuged at 30,000 g for 30 min in a refrigerated centrifuge. All procedures were carried out at 2–3°C. Zones of enzyme activity in the gels were visualized by the reduction of nitro blue tetrazolium in the usual manner (MARKERT and FAULHABER, 1965) with lithium lactate as substrate.

RESULTS

The isoenzyme patterns observed in heart muscle, and in red and white skeletal muscle, are illustrated in Fig. 1. Five fractions of relatively high anodal mobility (fractions 1–5) are observed in extracts of heart muscle. Of these, the slower fractions 4 and 5 predominate. No significant differences are observed when extracts are prepared with distilled water, buffer or 0.25 M sucrose.

In white skeletal muscle five major fractions of lower anodal mobility (fractions 6–10) are observed. Five fractions with mobilities similar to those of the heart muscle are also present, but only in very low concentration.

The enzymic activity of fractions 1–5, but not of fractions 6–10, is inhibited to a great extent in the presence of 15×10^{-3} M pyruvate.

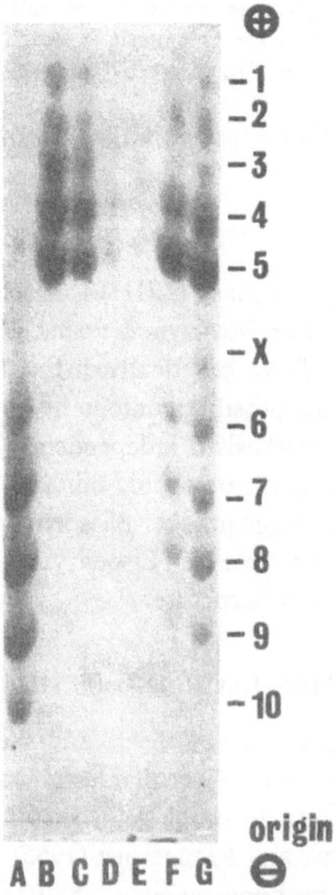

FIG. 1. Electrophoregram of lactate dehydrogenase isoenzymes in heart and skeletal muscle of the salmon. A — Distilled water extract of white skeletal muscle; B — Distilled water extract of heart muscle; C — 0.25 M sucrose extract of heart muscle; D — Insoluble residue of B, re-extracted with distilled water; E — Insoluble residue of C, re-extracted with distilled water; F — Distilled water extract of red superficial muscle; G — Sucrose extract of red superficial muscle.

In extracts of the red superficial skeletal muscle, five major isoenzymes with mobilities identical to those of heart muscle predominate; five slower fractions similar to those observed in white skeletal muscle are also present, but in lower concentration. One further fraction, marked "X" in Figs. 1 and 2 is sometimes evident in the extracts of red and white skeletal muscle, but never in that of the heart.

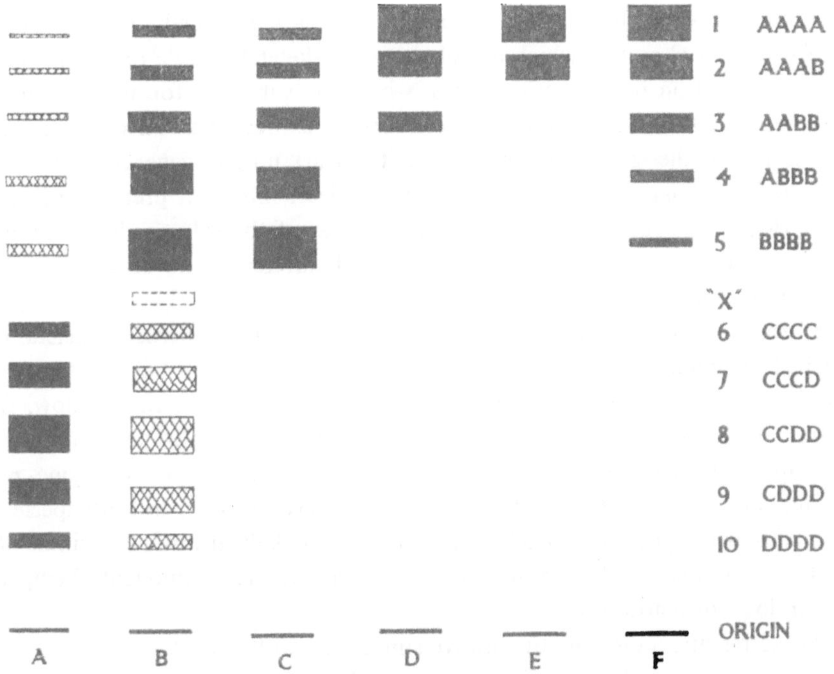

FIG. 2. Diagrammatic representation of the LDH isoenzymes in various tissues of the salmon, together with their postulated sub-unit structure. A — White skeletal muscle; B — Red superficial skeletal muscle; C — Heart muscle; D — Liver; E — Spleen; F — Kidney.

The isoenzyme patterns observed in the various tissue extracts are illustrated diagrammatically in Fig. 2, which also indicates the proposed sub-unit structure of the different fractions.

No intra-specific variations in the isoenzyme patterns of the heart or of the liver were observed in fifty individuals analysed.

DISCUSSION

The occurrence of two pairs of sub-units, i.e. of 4 genetically distinct polypeptide chains, is postulated to explain the existence of 10 major LDH isoenzymes in the salmon. One pair of sub-units — termed A and B — is expressed in the LDH phenotypes of heart, liver, spleen and kidney. In white skeletal muscle, both pairs of subunits are synthesized, the second pair—termed C and D sub-units—being predominant. Interactions between the sub-units of the different pairs, e.g. A–D interactions, have not been observed *in vivo*, nor have they been successfully produced by freezing and thawing *in vitro*. The possibility that fraction "X", observed in some muscle extracts, represents such a hybrid interaction cannot, however, be confidently excluded.

Sub-units A and B, which occur in all the tissues examined, are more ubiquitous than sub-units C and D which are observed in skeletal muscle only. This contrasts with the situation in most vertebrates, where the sub-units forming the major isoenzyme system in skeletal muscle are normally expressed in all other tissues also. In species synthesizing only two sub-units, heart muscle generally exhibits a preponderance of one sub-unit type (H) and skeletal muscle a preponderance of the other (M). The heart and white skeletal muscles of the salmon, however, differ in that separate isoenzyme systems, composed of different pairs of sub-units, are expressed in each of these tissues.

In this context, also, the isoenzyme pattern of the red superficial skeletal muscle, in which the major isoenzyme system resembles that of the heart, is of interest. A similar situation is observed in the mackerel *Scomber scomber*: a single, slow, fraction is present in white skeletal muscle, and a single fast fraction in heart muscle; in the red superficial muscle both fractions occur with the heart-type isoenzyme predominating, and no sub-unit interaction being observed *in vivo*. In both species, then, the major isoenzyme system of the red superficial skeletal muscle is similar to that of the heart muscle, the normal white skeletal muscle components being present only in low concentration.

These results, indicating qualitative similarities in heart and red muscle, complement and augment those of KAPLAN *et al.* (1960), who showed that in both trout and mackerel the reaction rate of red muscle LDH resembles that of heart muscle more closely than that of white skeletal muscle. There is, however, no histological or embryological evidence to support the suggestion of KAPLAN *et al.* (1960) that this resemblance may indicate a common origin of heart and red superficial skeletal muscle. It is more probable that the similarity reflects the functional adaptation of red superficial muscle to slow prolonged activity which is the prime function of this tissue in fish (BONE, 1966).

REFERENCES

BONE, Q., 1966. On the function of the two types of myotomal muscle fibre in elasmobranch fish, *J. Mar. Biol. Ass. U.K.*, **46**, 321–349.

CAHN, R. D., KAPLAN, N. O., LEVINE, L. and ZWILLING, E., 1962. Nature and development of lactic dehydrogenases, *Science*, **136**, 962–969.

KAPLAN, N. O., CIOTTI, MARGARET, HAMOLSKY, M. and BIEBER, R., 1960. Molecular heterogeneity and evolution of enzymes, *Science*, **131**, 392–397.

MARKERT, C. L. and FAULHABER, ILSE, 1965. Lactate dehydrogenase isozyme patterns in fish, *J. Exp. Zool.*, **159**, 319–332.

SMITHIES, O., 1959. An improved procedure for starch gel electrophoresis: further variations in the serum proteins of normal individuals, *Biochem. J.*, **71**, 585–587.

DISCUSSION

I. E. LUSH: Do you have any comment to make on the theory that tetraploidy may account for the complexity of the LDH of Salmon-like species?

N. P. WILKINS: The salmonids are an extremely complex group, both biochemically and otherwise. I would prefer not to discuss this complexity without adequate preparation and more time then is at our disposal today.

J. HYLDGAARD-JENSEN: Electrophoretic separation of LDH from Salmon tissues. Owing to the abundance of bands and the presence of several systems it may be useful to use the electrophoresis technique as the agar gel electrophoresis. Have you tried this?

N. P. WILKINS: Yes, but the separation of the isoenzymes is very inferior to that in starch gel.

J. S. ALABASTER: Have you studied lactate dehydrogenase in sea trout (*Salmo trutta*)?

N. P. WILKINS: No.

CHAPTER 7

BLOOD GROUPS AND HISTOCOMPATIBILITY IN MOUSE, RAT AND COYPU

Preliminary Studies on Histocompatibility Antigens in Wild Mice *(Mus musculus)*

P. IVÁNYI and P. DÉMANT

Institute of Experimental Biology and Genetics, Czechoslovak Academy of Sciences, Prague, Czechoslovakia

The number of histocompatibility loci (H loci) at which two inbred mouse strains differ was estimated at 13–30, representing only a small proportion of the total number of more than 430 histocompatibility sites in the mouse (for references see BAILEY, 1967). The alleles at the H loci in wild mouse populations can be subject to selective forces either directly or as a result of their linkage with other selectively non neutral loci, probably in supergene complexes. The concept of H-2-T region as a functional unit proposed by SNELL (1968) is one of the possible examples of this situation.

In the course of a more detailed study of H antigens in wild mice (described hereafter as W) an experiment was undertaken to estimate the number of histocompatibility loci in the heterozygous state in individual W mice. Wild males were mated with females of the C57B1/10 (referred to as B10) inbred strain and skin grafting was performed between the offspring. It can be shown that the proportion of surviving grafts equals $\left(\frac{1}{2}\right)$ where *n* is the number of H loci in heterozygous state in the male under study (presuming that B10 mice have different alleles at this locus).

Five wild males (Nos. 23–27) were captured in an area of 10×5 m in the Prague Zoological Garden and two males (Nos. 31 and 32) were captured in another place of the same area. Male No. 35 was captured in a country house near Prague. Skin grafts between the progeny of the mating of W males to B10 females (F_1 hybrids) were exchanged following a circular scheme: F_1 mouse No. 1 was the skin donor for its littermates No. 2 and 3, F_1 mouse No. 2 was skin graft donor for F_1 mice Nos. 3 and 4, etc. The results are given in the Table.

It should be pointed out that the *n* values in this and similar studies are operational. Their calculation is based on certain premises (the conventional 200 days as an observation period, and the presumption that the alleles determine antigens strong enough to elicit graft rejections). The value of *n* lies in its comparative aspects. Therefore the number of heterozygous loci determined in randomly captured W mice

Survival of skin grafts exchanged between $(W \times B10)F_1$ hybrids

Wild male No.	Total number of grafts	Number of rejected grafts on days after grafting																			Number of H loci (*n*)** heterozygous in the wild male
		10	15	20	25	30	40	50	60	70	80	90	100	120	140	160	180	200	250	250	
23–2	27	0	8	9	2	0	0	0	1	5*	0	0	0	1	0	0	0	0	0	1*	5
24	20	1	4	5	4+1*	5*															
25–3	19	2	7	5	2	0	2	0	0	0	0	0	0	0	0	0	0	0	0	1*	4–5
26–4	40	2	6	7	4	2	1	1	0	0	1	0	0	5	2	3	0	6*			3
27–1	40	1	13	9+1*	1	1	3*	5	2	1	0	0	0	0	0	2	0	1*			6
30–5	10	0	0	4	1	0	1	0	0	0	0	0	1	3*							2
31–7	16	0	1	6	2	1	4	0	0	0	0	0	0	0	2*						3
35–6	30	0	1	3	5	3	12	2	0	0	0	0	0	1*	3*						4

* Surviving grafts not further evaluated.

** Proportion of surviving grafts = $\left(\frac{1}{2}\right)^n$ (calculated for the given period).

and those of F_1 hybrids of two inbred lines in mice and rats are directly compared in the Figure.

The results obtained strongly indicate that the W males studied in this work were homozygous at a great number of their histocompatibility loci. This reflects the possibility that W mice may be "partially inbred" probably because they live in closed communities subjected to periodical reductions of the effective population

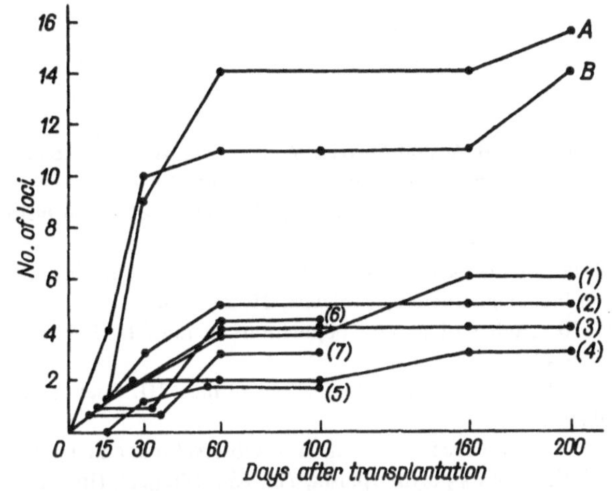

Comparison of an estimate of the number of histocompatibility loci in the heterozygous state in F_1 hybrids of two inbred lines of mice and rats and in randomly captured Wild male mice. The number of loci was calculated from parental skin grafts survival in F_2 hybrids (Exp. A and B; BARNES and KROHN, 1957; BILLINGHAM *et al.*, 1962) and from the survival of skin grafts exchanged between (Wild × C57Bl/10)F_1 hybrids (Exp. 1–7). The data indicate that of the 14 histocompatibility systems in mice (this number represents the lower limit of the estimation) not more than 6 are in the heterozygous state in the Wild males tested.

size. The maintenance of the polymorphism under these circumstances might be dependent on the action of strong selective forces. Partial inbreeding was observed by PETRAS (1967 a, b, c) in W mice and by PALM *et al.* (1967) in wild hamsters. The advantage of experiments the type presented is the high number of loci studied in one experiment.

The direct testing of H-2 alleles of wild mice by serotyping and skin grafting from B10 congenic lines to (W×B10)F_1 hybrids is in progress. In two (W×B10.A)F_1 hybrids derived from the same W male, B10 skin grafts survived for more than 20 and 70 days respectively indicating that at least one H-2 allele of this male was identical or very similar to H-2^b allele. It is interesting that in almost all studies in which the H-2^b allele has been involved, this allele seems to indicate resistance to oncogenic viruses (for review see SNELL, 1968). The H-2 alleles a, d, f, g, h and k were not detected in the other three wild males sufficiently tested.

Sterility of (W×B10)F_1 hybrid males was observed in the course of further study, while females were completely fertile. Preliminary experiments with Bc1 individuals showed that the male sterility is in linkage with H-2 indicating the involvement of some gene(s) in the IX linkage group.

REFERENCES

Bailey, D. W., 1967.* Advances in Transplantation, Munksgaard, Copenhagen, 317–324.

Barnes, A. D. and Krohn, P. L., 1957. The estimation of the number of histocompatibility genes controlling the successful transplantation of normal skin in mice, *Proc. Roy. Soc. B.*, **146**, 505–526.

Billingham, R. E., Hodge, B. A. and Silvers, W. K., 1962. An estimate of the number of histocompatibility loci in the rat, *Proc. Nat. Acad. Sci.*, **48**, 138–147.

Palm, J., Silvers, W. K. and Billingham, R. E., 1967. The problem of histocompatibility in wild hamsters, *J. Heredity*, **58**, 40–44.

Petras, M. L., 1967a. Studies of natural populations of mus. I. Biochemical polymorphisms and their bearing on breeding structure, *Evolution*, **21**, 259–274.

Petras, M. L., 1967b. Studies of natural populations of mus. II. Polymorphism at the T locus, *Evolution*, **21**, 466–478.

Petras, M. L., 1967c. Studies of natural populations of mus. III. Coat color polymorphisms, *Can. J. Genet. Cytol.*, **9**, 287–296.

Snell, G. D., 1968. The H-2 locus of the mouse: Observations and speculations concerning its comparative genetics and its polymorphism, *Fol. biol.* (Prague), (In press).

DISCUSSION

R. D. Owen: Do the partially sterile males give any evidence of being heterozygous for a translocation — resorbing embryos in the uterus of females bred to them?

P. Iványi: We did not make karyological studies yet. But the completely sterile hybrids have a complete stop of spermatogenesis. I don't presume, that the sterility is produced by translocation.

R. D. Owen: Have you checked for normal antibodies in your wild mice comparable to the system reported for wild mice in the U.S.A. by Foster?

P. Iványi: We did not test normal antibodies.

* The vastness and organization of the murine histocompatibility gene system as inferred from mutational data.

XITH EUROPEAN CONFERENCE ON ANIMAL BLOOD GROUPS AND BIOCHEMICAL POLYMORPHISM
Warsaw, July 2nd–6th, 1968

Further Analysis of the Histocompatibility-1 System of the Rat

O. ŠTARK, V. KŘEN and B. FRENZL

Department of Biology, Medical Faculty, Charles University, Prague, Czechoslovakia

SUMMARY. Antigens determined by the RtH-1 system were studied by means of allogeneic transplantation and by serological methods in 21 inbred strains of rats. Nine different alleles of this system were identified and their structure studied. Each allele determines a typical combination of some individual antigens whose total number has now reached 21. Eight antigens only are specific for different alleles, while the others are shared in common by complex products of two or more alleles. From the point of view of structure the RtH-1 system is analogous to other complex immunogenetic systems analysed in different species of animals.

INTRODUCTION

Immunogenetics of the rat has made great advances during the last two years. Knowledge has accumulated of the number of histocompatibility loci participating in the incompatibility barrier between different strains (see the right side of Table 2) and of the main histocompatibility system of the rat designated RtH-1 (ŠTARK *et al.*, 1966) or Ag-B (ELKINS and PALM, 1966).

The present paper presents information about the composition of the antigenic products determined by 9 different RtH-1 alleles identified so far by serological studies of 21 isohistogeneic strains of rat.

METHODS AND MATERIALS

Rat strains have been described earlier (ŠTARK *et al.*, 1968a) and will be reviewed here only with regard to their origin: AVN (Czech. Acad. Sci., Praha), BP, LEP, WP (Med. Fac., Praha), WR (Rosice), BN, DA, LEW (Wistar Inst., Philadelphia), BD II, BD V, BD VII, BD X (Max-Planck Inst., Freiburg), AS, AS2, BS, HS (Otago Univ. Med. Sch., Dunedin), AGA, VM, Y59 (Fac. Nat. Sci., Zagreb), CAP (Pol. Acad. Sci., Cracow), CDF (Med. Fac., Kiel). A standard panel of inbred strains used for comparative testing includes the AVN, BP, BN, BD V, BD VII, CAP, LEW and WP strains differing at the RtH-1 locus. Production of congeneic strains by successive back-cross mating of RtH-1 different animals to the LEW background was started as described by IVÁNYI (1966).

RtH-1 antigens were detected by dextran haemagglutination and cytotoxicity tests (ŠTARK *et al.*, 1967). Immunization was performed by repeated skin grafting, tumour inoculation and lymphoid cell injections. The alloimmune system used in most cases was prepared in LEW recipients from congeneic or allogeneic donors. In this way 7 LEW antisera to different RtH-1 products and AVN anti-LEW serum were obtained, all of them showing marked polyvalent characters. These sera were absorbed by cross-reacting cells and 7 monospecific antisera produced. This set of polyvalent and monospecific reagents was used for testing all newly examined strains (ŠTARK *et al.*, 1968b, c). The new strain was then used as donor for immunization of LEW recipients, and LEW antisera analysed by absorption. In cases of suspected identity with a standard strain at the RtH-1 locus reciprocal immunization was carried out and antibodies searched for. Absence of antibody formation, and identical results in absorption analysis, were considered as relatively reliable proof of identity at the RtH-1 locus.

RESULTS

Absorption analysis and immunization experiments verified by immunogenetic tests for allelism provided a basis for gradual development of a working hypothesis on composition of RtH-1 antigenic products, illustrated in Table 1. Products of different alleles are composed of 4–10 individual antigens, some of them being detectable directly by monospecific antisera (antigens 1, 3, 5, 6, 10, 11, 12, 19). The remaining antigens, common for two or more allelic products, were detected indirectly by cross-

TABLE 1

Composition of antigenic products of nine RtH-1 alleles and their occurence in 21 isohistogenic strains of rats

RtH-1 Allele	Individual antigens composing the allelic products: 1	2	3	4	5	6	7	8	9	10	11	12	13	14	15	16	17	18	19	20	21	Strains of rats
-1^{a}	**1**	2	–	4	–	–	7	–	–	–	–	–	13	14	–	–	17	–	–	20	21	**AVN**, DA
-1^{b}	–	2	**3**	–	–	–	–	8	9	–	–	–	13	–	–	–	–	18	–	–	–	**B P**
-1^{c}	–	–	–	–	–	–	–	–	9	**10**	–	–	13	–	15	–	17	–	–	–	–	**CAP**, Y 59
-1^{d}	–	–	–	4	–	–	–	8	9	–	–	**12**	13	14	–	16	–	18	–	20	21	**BD V**, BD X
-1^{e}	–	2	–	4	–	–	–	–	–	–	–	–	13	14	–	–	17	–	–	–	21	**BD VII**
-1^{f}	–	–	–	4	–	–	–	8	–	–	–	–	13	–	–	–	–	18	**19**	20	21	**AS 2**
-1^{l}	–	–	–	–	**5**	–	7	8	–	–	–	–	–	–	–	16	17	–	–	–	–	**LEW**, AGA, CDF, AS, BS, HS
-1^{n}	–	–	–	–	–	–	–	–	9	–	**11**	–	13	–	15	–	–	–	–	–	–	**B N**
-1^{w}	–	–	–	4	–	**6**	–	–	9	–	–	–	–	–	–	–	17	–	–	–	–	**W P**, WR, VM LEP, BD II

Bold numbers — antigens detected by monospecific antisera.
Bold symbols of strains — members of the standard panel of strains.

reacting sera and their absorption. Evidence of genetic determination of these antigenic complexes by a single chromosome locus was obtained from testing back-cross hybrid populations by corresponding sets of cross-reacting antisera. Antigens 1, 2, 4, 13, 14 were tested in hybrids LEW × (LEW × AVN), and no exception to the rule of common inheritance was found among 130 RtH-1^{a} positive animals.

TABLE 2

A review of results obtained by immunogenetic studies in different laboratories

Strain	Allele of the main histocompatibility system			Interstrain relationship: number of segregating H-loci		
AVN	RtH-1^{a}			AVN-WP	: 2–3	ŠTARK and KŘEN, 1967a,b
BP	RtH-1^{b}			BP-WP	: 10–12	
WP	RtH-1^{w}					
LEW	RtH-1^{l}	Ag-B^{1}	ELKINS and PALM, 1966	LEW-BN	: 14–16	BILLINGHAM *et al.*, 1962
BN	RtH-1^{n}	Ag-B^{3}				
DA	RtH-1^{a}	Ag-B^{4}		LEW-DA	: 6–7	RAMSEIER and PALM, 1967
AS2	RtH-1^{f}	Ag-B	difference HESLOP, 1968b	AS2-AS	: 4–5	HESLOP, 1968b
AS	RtH-1^{l}			AS-BS	: 9–10	ZEISS, 1967
BS	RtH-1^{l}	Ag-B	identity	AS-HS	: 12–14	
HS	RtH-1^{l}			BS-HS	: 8–11	

Immunogenicity of individual antigens within the same complex seems to differ considerably as does that of the complex products of different alleles. On the evidence of the stimulation of antibody formation, products of the RtH-1^{a}, RtH-1^{d} and RtH-1^{f} alleles, and individual antigens 1, 2, 5, 6, 12, 14, 19 act as the "strongest" ones. The number of different alleles now recognized is much lower than the number of strains tested. Together with the newly supposed RtH-1^{f} allele of the AS2 strain, the evidence for which arises from serological study but has not so far been verified by a test for allelism, only nine different alleles are present in 21 strains. Those occurring most frequently are the RtH-1^{l} and RtH-1^{w} alleles — in 6 and 5 strains respectively.

DISCUSSION

Our collection of 21 strains includes 7 strains from other three laboratories dealing with rat immunogenetics, which enables us to compare antigens detected by different methods (see the left part of Table 2). A good correlation was found not only with detection of identical strains (AS, BS and HS rats), but also with the described strength of interstrain difference evaluated according to antibody production (RAMSEIER and PALM, 1967; HESLOP, 1968a) or to local graft-versus-host reaction (ELKINS and PALM, 1966; HESLOP, 1968b).

Present formulation of the hypothesis of the composition of antigenic products of the RtH-1 system is limited by the unilateral detection system using mainly the LEW antisera. This problem may be solved by the use of homozygous congeneic strains whose reciprocal immunization will undoubtedly extend and add precision to our present ideas about RtH-1 antigens. A further limitation of our recently acquired knowledge arises out of the variable immunogenicity of some RtH-1 antigens. But this situation creates great difficulties in the study of all known complex immunogenetic systems.

The high frequency of occurrence of the $RtH\text{-}1^1$ and $RtH\text{-}1^w$ alleles in our collection of inbred strains might be explained by the common origin of these strains from a relatively homogeneous starting colony of albino Wistar rats. But the identity of strains evidently derived from crosses with "wild" types (AGA, BS, HS and LEP rats) suggest a possible selectional advantage of the $RtH\text{-}1^1$ and $RtH\text{-}1^w$ alleles in homozygous state for inbreeding, while other homozygous combinations may be less advantageous. This would be in a good agreement with an "antigen-associated selective factor" assumed to be present Wistar rats by MICHIE and ANDERSON (1966).

ACKNOWLEDGEMENTS

Authors are indebted to Mr. L. VOJČÍK for valuable technical assistance.

REFERENCES

BILLINGHAM, R. E., HODGE, B. A. and SILVERS, W. K., 1962. An estimate of the number of histocompatibility loci in the rat, *Proc. Nat. Acad. Sci.*, **48**, 138–147.

ELKINS, W. L. and PALM J., 1966. Identification of a strong histocompatibility locus in the rat by normal spleen-cell transfer, *Ann. N. Y. Acad. Sci.*, **129**, 573–580

HESLOP B. F., 1968a. The production of haemagglutinins after skin allografts in the rat, *Proc. Univ. Otago Med. Sch.*, **46**, 6–7.

HESLOP, B. F., 1968b. Histocompatibility antigens in the rat, the AS2 strain in relation to the AS, BS and HS strains, *Aust. J. Exp. Biol. Med. Sci.* (In press).

IVÁNYI, P., 1966. Serological production of congenic lines in rats. In: Polymorphismes biochimiques des animaux, Paris, p. 507–510.

MICHIE, D. and ANDERSON, N. F., 1966. A strong selective effect associated with a histocompatibility gene in the rat, *Ann. N.Y. Acad. Sci.*, **129**, 88–93.

RAMSEIER, H. and PALM J., 1967. Further studies of histocompatibility loci in rats, *Transplantation*, **5**, 721–729.

ŠTARK, O., FRENZL, B. and KŘEN, V., 1968b. Erythrocyte and transplantation antigens in inbred strains of rats. VII. H-1 alleles of the LEP, CAP, BN, BD V, BD VII and BD X strains, *Fol. biol.* (Praha), **14**, 169–175.

ŠTARK, O. and KŘEN, V., 1967a. Erythrocyte and transplantation antigens in inbred strains of rats. II. Antigens of the AVN strain, *Fol. biol.* (Praha), **13**, 93–99.

ŠTARK, O. and KŘEN, V., 1967b. Erythrocyte and transplantation antigens in inbred strains of rats. III. Antigens of the BP strain, *Fol. biol.* (Praha), **13**, 299–305.

ŠTARK, O., KŘEN, V. and FRENZL, B., 1966. Histocompatibility locus in the rat. In: Polymorphismes biochimiques des animaux, Paris, p. 501–506.

ŠTARK, O., KŘEN, V. and FRENZL, B., 1967. Erythrocyte and transplantation antigens in inbred strains of rats. I. Serological analysis of strain-specific antigens, *Fol. biol.* (Praha), **13**, 85–92.

ŠTARK, O., KŘEN, V., FRENZL, B. and BRILIČKA, R., 1968a. The main histocompatibility system of the rat. In: Advance in transplantation, Copenhagen, p. 331–339.

ŠTARK, O., KŘEN, V., FRENZL B. and KŘENOVÁ, D., 1968c. Histocompatibility-1 alleles in inbred rats of the BD II, WR, AGA, CDF, AS, BS, HS and DA strains, *Fol. biol.* (Praha). (In press).

ZEISS, J. M., 1967. Antigenic interrelationships and number of independently segregating histocompatibility loci in three isohistogenic strains of rats, *Transplantation*, **5**, 1393–1399.

DISCUSSION

H. BALNER: What is the best method to produce rather specific typing sera?

O. ŠTARK: The immunization course consisted of subsequent skin grafting, tumour inoculation and buffy coat or Cymphoid cell injections in 3 week intervals. The highest titres of haemagglutinins were reached against donor cells, but the maximum of cross reactions were irregular e.g. anti-4 antibodies in the LEW anti H-1^a serum appeared after the third immunization and, on the contrary, anti-9 antibodies in the LEW anti H-1^w serum disappeared.

H. BALNER: Does hyper immunization cause a rapid loss of specificity?

O. ŠTARK: The antisera obtained after the first immunization were never strain-specific but showed marked cross-reactions with red cells of some other strains. After the repeated immunization there appeared some further weaker cross-reactions depending on the RtH-1 difference used. Thus it seems that the immunogenicity of individual antigens composing the RtH-1 allelic products is expressed with different strength in different antigenic relationships.

H. BALNER: What are the skin graft survival times between the H_1 compatible strains?

O. ŠTARK: MST of skin graft exchanged between RtH-1 different strains is not longer than 12 days and in RtH-1 identical interstrain combinations the longest survival time found was 16 days (LEW-AS) which indicates the participation of other non H-1 systems.

R. OWEN: Have you checked blast transformation to see whether all of the H-1 alleles in your strains are equally effective in that test system?

O. ŠTARK: The blast transformation test was not used because of technical difficulties, but another similar cellular test is now under development.

XITH EUROPEAN CONFERENCE ON ANIMAL BLOOD GROUPS AND BIOCHEMICAL POLYMORPHISM
Warsaw, July 2nd–6th, 1968

H-1 Antigenicity of Rat Tumours after *In Vivo* Passage and *In Vitro* Cultivation

Drahomira KŘENOVÁ, V. KŘEN, and O. ŠTARK

Faculty of General Medicine, Charles University, Department of General Biology, Prague, Czechoslovakia

SUMMARY. Tumour host compatibility of H-1 alleles was found to be most important for tumour growth, indicating the predominant role of the H-1 system in the rat strains tested. The explanted tumour cells preserved their H-1 antigenicity during long term cultivation *in vitro* as proved by immunization and absorption techniques.

INTRODUCTION

The use of a tumour inoculum has proved to be suitable for histocompatibility system analysis in mice. Our paper summarizes results of four rat tumour implantations into inbred rat strains and their hybrids, thus demonstrating the importance of RtH-1 allele antigenic products in determining the fate of the tumour. Three of these tumours were explanted *in vitro* and the antigenic properties were followed during long term cultivation in order to confirm possible changes of antigenicity reported in literature. The presence of H-1 antigens was tested for by means of isoimmunization and absorption techniques.

MATERIALS AND METHODS

The rat strains used and their RtH-1 alleles were described in a previous report (ŠTARK *et al.*, 1967). Tumours (spindle cell sarcomas) were induced by iron-dextran complexes (KŘEN *et al.*, 1968) in the WP, LEW and AVN strains and denoted FDX-WP, FDX-LEW and FDX-AVN. CaM-LEW is a spontaneous mammary carcinoma of LEW rats. Transplantation and cultivation techniques, dextran haemagglutination methods and immunization and absorption techniques, are all described in the paper of KŘENOVÁ *et al.* (1968). Two groups of LEW rats were simultaneously immunized by the same doses of tumour cells cultivated *in vitro* and from a freshly-trypsinized *in vivo*-passaged FDX-AVN tumour. The same procedure was performed in LEW and WP rats using the FDX-AVN tumour and also for BN and

LEW-BN rats using the CaM-LEW tumour. The antisera obtained were evaluated by testing them with RBC of different rat strains and the cross reactions of the antisera to both *in vivo* and *in vitro* cell inocula were compared. The H-1 antigenicity of *in vitro* and *in vivo* cultivated CaM-LEW cells was studied by the absorption of anti H-1^1 sera at room temperature using increasing doses of both cell types.

RESULTS

(1) After trocar tumour implantations into rats of different strains and inter-strain hybrids had been performed, all four tumours were shown to preserve the specificity of the strain, in which they were induced. The tumours killed 100% of the original strain recipients and H-1 compatible hybrids, while in H-1 incompatible recipients the tumours did not grow and inoculum destruction was accompanied by anti H-1 haemagglutinin production. The main condition of the tumour growth is thus tumour-host compatibility in H-1 allele. The only exception was destruction of FDX-LEW inocula in AGA and CDF rats compatible for the H-1 allele, but in spite of repeated implantations no antibodies were detected. The explanation of this lies probably in the participation of other H-systems. On the contrary, the CaM-LEW tumour grew in some BN and BP rats, incompatible for the H-1 allele, most probably as a consequence of tumour malignancy overcoming the immune barrier, as in such cases anti H-1^1 haemagglutinins were detected.

(2) Repeated immunization of LEW rats by the same doses of FDX-WP cells from 4, 23 and 42 passages *in vitro* stimulated anti H-1^w haemagglutinin production at the same titre level. By using cross reactions of these antisera with erythro-

TABLE 1

The antigenicity comparison of FEDEX-AVN tumour cells from 30th passage in vitro and from in vivo passage. Each column represents cross reactions of the serum, obtained from one animal immunized by tumour cells

Cross reactions with RBC	Serum LEW anti FEDEX-AVN passaged		Antigens involved	Serum WP anti FEDEX-AVN passaged		Antigens involved
	in vivo	*in vitro*		*in vivo*	*in vitro*	
AVN	+ +	+ +	1, 2, 4, 13, 14	+ +	+ +	1, 2, 7, 13, 14 B1
BDV	+ +	+ +	4, 13, 14	+ +	+ +	13, 14 B1
BD VII	+ +	+ +	2, 4, 13, 14	+ +	+ +	2, 13, 14 B1
BN	+ −	+ −	13	+ +	+ +	13 B1
BP	+ +	+ +	2 13	+ +	+ +	2 13 B1
CAP	− −	− −	13	− −	− −	13 B1
LEW	− −	− −	—	+ +	+ +	7 B1
WP	− −	− −	4	− −	− −	—
WR	− −	− −	4	+ +	+ +	— B1

cytes CAP and AVN strains, the antibodies against specificities 6 and 9 of the H-1^w allele were detected at the same titre. The absorption of LEW anti H-1^w antiserum revealed the identical absorbing capacity of FDX-WP cells from 4, 23 and 42 passages. On immunizing WP and LEW rats by cells of *in vivo* and *in vitro* passaged FDX-AVN tumour the same titre levels were obtained in rats injected both by *in vivo* and *in vitro* cells. The presence of antibodies against single specificities of H-1^a complex was proved by testing these antisera with erythrocytes of 9 strains (Table 1).

TABLE 2

The antigenicity comparison of CaM-LEW tumour cells from 60th passage in vitro and from in vivo passage

Cross reactions with RBC	Serum LEW-BN anti CaM-LEW passaged		Antigens involved	Serum Bn anti CaM-LEW passaged		Antigens involved
	in vivo	*in vitro*		*in vivo*	*in vitro*	
LEW	+ +	+ +	5, 7, 16, 17	+ +	+	5, 7, 16, 17 P-2
AVN	+ +	+ +	7 17	+ +	+	7 17
BDV	+ +	+ +	16	+ +	+	16
BD VII	+ −	+ −	17	+ +	+	17
BN	− −	− −	—	− −	−	—
CAP	− −	− −	17	+ +	+	17
LEW-BN	− −	− −	—	+ +	+	— P-2
WP	− −	+ −	17	+ +	+	17

Moreover, the cross reaction positivity of WR erythrocytes with WP anti H-1^a serum was evidence of B1 erythrocyte antigen (FRENZL *et al.*, 1965) preservation on cultivated FDX-AVN tumour cells. On immunizing BN and congeneic LEW-BN rats by the same doses of tumour cell suspensions from an animal and from 60th passage of the CaM-LEW tumour *in vitro* haemagglutinins were formed, whose titres were the same in both groups. The antibodies against single specificities of H-1^1 allele were demonstrated by cross reactions of these antisera with erythrocytes of 8 strains (Table 2). Cross reactions of BN anti CaM-LEW sera with LEW-BN erythrocytes also demonstrated an antibody against antigen 2 (PALM, 1962), which segregates independently of the H-1 system. The absorbing capacity of CaM-LEW cells is identical both during *in vivo* and *in vitro* passages (the titres of BN, AVN and LEW-BN anti FDX-LEW sera were similarly lowered by the same cell doses).

DISCUSSION

Apart from the assumed variability of tumour malignancy and incompatibility in further H-systems, which affects the fate of tumour inocula in the recipient in different ways, compatibility in the H-1 system was shown to play a predominant role in the rat strains tested. A possible further use of *in vitro* cultivated cells for studies on the H-1 system is connected with confirmation of their antigenic properties during long term cultivation. The preservation of H-1 antigenicity (and B1 and P-2 antigens, too) during long term cultivations of tumour cells both qualitatively and quantitatively was demonstrated by immunization and by absorption techniques. The discrepancy between these findings and data on the antigenic loss during cultivation of human cells from normal and tumour strains can be explained by the differences in material and techniques used. The isoimmunization method cannot be used for antigenicity testing of human strains, although it is more reliable than mixed haemagglutination, which depends for results on antigen presentation on the cell surface. Human cell lines most probably originate from heterozygous donors and an antigenic loss might occur by cell selection during cultivation. The tumour cells used in our experiments, however, originated from inbred animals, homozygous for H-1 allele antigens.

REFERENCES

FRENZL, B., BRDIČKA, R., KŘEN, V. and ŠTARK, O., 1965. Erythrocyte B1 antigen in inbred rat strains. In: Blood groups of animals, Prague, p. 197.

KŘEN, V., BRAUN, A. and KŘENOVÁ, D., 1968. Transplantability of tumour induced in rats by Ferridextran Spofa, *Neoplasma*, **15**, 29–38.

KŘENOVÁ, D., KŘEN, V., and ŠTARK, O., 1968. Properties of the rat tumour FEDEX-WP during long-term cultivation *in vitro*, *Neoplasma* (In press).

PALM, J., 1962. Current status of blood groups in rats, *Ann. N.Y. Acad. Sci.*, **97**, 57–68.

ŠTARK, O., KŘEN, V., FRENZL, B., and BRDIČKA, R., 1968. The main histocompatibility system of the rat. In: Advance in transplantation, Copenhagen, p. 331.

DISCUSSION

H. BALNER: Did you try a passive enhancement of the one tumour that grew across the H_1 barrier?

D. KŘENOVÁ: The possibility of CAM-LEW tumour enhancement by anti-$H\text{-}1^2$ antibodies is probable but we have not tested it.

The investigation of the antigenic character of tumours of heterozygous origin is the second step of our work. We consider the confirmation of antigen preservation in homozygous tumours *in vitro* to be a very important because of contradictory data in literature.

XITH EUROPEAN CONFERENCE ON ANIMAL BLOOD GROUPS AND BIOCHEMICAL POLYMORPHISM
Warsaw, July 2nd–6th, 1968

Tolerance Induction to H-1 Antigenic Differences in the Rat

V. KŘEN and O. ŠTARK
Department of Biology, Medical Faculty, Charles University, Prague, Czechoslovakia

SUMMARY. Long-term skin graft tolerance by spleen cells was induced in H-1 compatible strain combinations only indicating the major importance of RtH-1 among rat histocompatibility systems. Partial or complete suppression of antibody formation was the only tolerance manifestation during donor skin graft rejection in all H-1 incompatible strain combinations tested. Moreover, 19 S antibody preponderance was revealed in partly tolerant animals. Thus the dissociation of the cellular and antibody mechanisms of immune response and of two immunoglobulin types was demonstrated on an allogeneic partial tolerance system.

INTRODUCTION

In some combinations of rat strains spleen was successfully used for skin graft tolerance induction (WOODRUFF and SIMPSON, 1955; EGDAHL *et al.*, 1958; BILLINGHAM and SILVERS, 1959; GROZDANOVIČ, 1959). On the contrary, in the LEW-BN combination BILLINGHAM *et al.* (1962) ascertained small tolerance inducing capacity of different types of lymphoid cells, including splenic, in comparison with bone marrow cells. Having used spleen cells for tolerance induction in several strain combinations we found similar variability of results. The detection of the main histocompatibility system in the rat (RtH-1) and the finding of rat strains with identical H-1 alleles provided the basis for an attempt to interpret the incidence of the runt syndrome and the induction of tolerance in different interstrain combinations.

MATERIAL AND METHODS

Rat strains used were described in ŠTARK *et al.* (1968). The preparation of spleen cell suspension was as described in KŘEN *et al.* (1962). The double transfer of BP spleen cells was performed by the i.p. injection of spleen cell suspension from WP donors (6–20 days old) runted by previous neonatal administration of adult BP spleen cells into them as newborn recipients. Skin grafting (BRDIČKA *et al.*, 1962), serological methods, mercaptoethanol (ME) treatment and Sephadex G 200 fractionation (KŘEN *et al.*, 1968) have been described earlier.

RESULTS

The intraperitoneal administration of adult allogeneic spleen cells to newborn rats induced runt syndrome (r.s.) in some of the interstrain combinations only and its incidence was not fully explainable by the H-1 antigenic differences as shown in Table 1. The importance of the H-1 system became clear when the skin graft tolerance of the surviving rats was tested (Table 2). Short-term and even long-term

TABLE 1

Runt syndrome incidence after intraperitoneal administration of adult allogenic spleen cells (2–5 × 10^7) into newborn rats

Strain combination					
H-1 Incompatible	WP	3/42 →	BP	207/216 →	WP
	LEP	0/13 →	BP	15/19 →	LEP
	LEW	0/8 →	BP	30/33 →	LEW
	AVN	3/14 →	BP	15/15 →	AVN
H-1 Compatible	AVN	33/106 →	LEW	7/14 →	AVN
	WP	10/50 →	LEW		
	LEP	3/47 →	WP	11/45 →	LEP
			WP	0/12 →	BD II
			CDF	2/17 →	LEW

TABLE 2

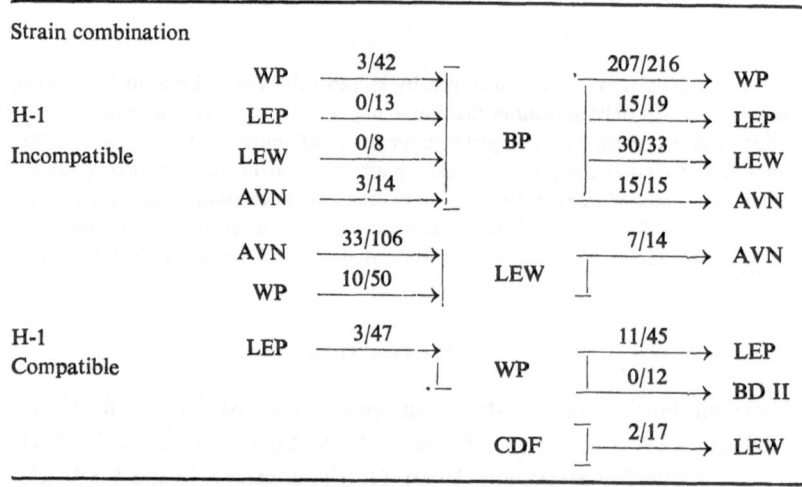

Skin graft tolerance after intraperitoneal spleen cell administration into newborn rats. The survivors were test grafted at 2–4 months of age. In BP-LEW combination 2 spontaneous survivors and 5 survivors after LEW anti-BP serum treatment started on day 10 after the spleen cell injection

Donor (Spleen → Skin)	Recipient	Number	Skin craft survival >15	>50	Across
AVN	BP	8	0	0	
AVN	LEW	13	0	0	
WP	LEW	18	0	0	H-1
BP	WP	4	0	0	
BP	LEW	2+5	0	0	
LEW	BP	8	0	0	
WP	LEP	15	14	10	
WP	BD II	12	11	11	NON
LEP	WP	8	7	5	H-1
CDF	LEW	13	13	5	

tolerance was induced only in donor–host combinations where the H-1 alleles concordance was detected.

The immune graft versus host interaction could hinder the development of the tolerance as indicated in the double transfer of BP spleen cells to newborn WP rats (Table 3). The r.s. incidence in secondary WP recipients was ten fold lower than that in primary recipients of spleen cells and less than a third of that in BP blood treated WP newborns. On the other hand, short-term but significant tolerance of BP skin graft was found.

TABLE 3

The double transfer of BP spleen cells into WP newborn rats and BP skin graft tolerance in the secondary recipients. The doses were 2–4×10⁷ of BP spleen cells for primary recipients and 5–16×10 of the spleen cells from runted primary recipients for the secondary recipients. BP heparinized blood was injected i.p. or i.v. into WP newborns in a quantity of 0.1 ml

			BP → WP				Skin graft survival		
			Number	Died	%	BP Skin grafted	> 15	> 20	FUR
	Intact	WP	—	—	—	25	0	0	0
BP	Blood ⟶	WP	90	27	30.0	17	0	0	0
BP	Spleen ⟶	WP	216	207	95.3	4	0	0	0
BP	Spleen ⟶ WP Spleen ⟶	WP	107	10	9.3	28	18	5	5

Depression or complete suppression of antibody formation was the only tolerance manifestation in r.s. survivors during and after skin graft rejection. Figure 1 presents this finding by comparing the curves of the average haemagglutinin titres after skin grafting of normal and spleen treated animals. In all 4 strain combinations there is a significant difference between the immune responses of normal and treated animals. The depression of antibody formation was directed specifically towards spleen donor antigens. LEW recipients injected at birth with WP spleen cells and at 2 months of age grafted simultaneously with WP and AVN skin produced anti-AVN agglutinins only during rejection of both grafts, the anti-WP antibody response being completely suppressed. Moreover, in those animals where the haemagglutinin production was only partly depressed 19 S antibodies were found to preponderate throughout the whole period of antibody response. Figure 2 demonstrates the preponderance of 19 S (ME-sensitive) antibody production in partly tolerant LEW rats during rejection of AVN skin grafts. The effect of ME on the normal LEW anti-AVN skin antibody response is given for comparison. The suppression of 7 S antibody formation found by ME-treatment was confirmed using Sephadex G 200 fractionation.

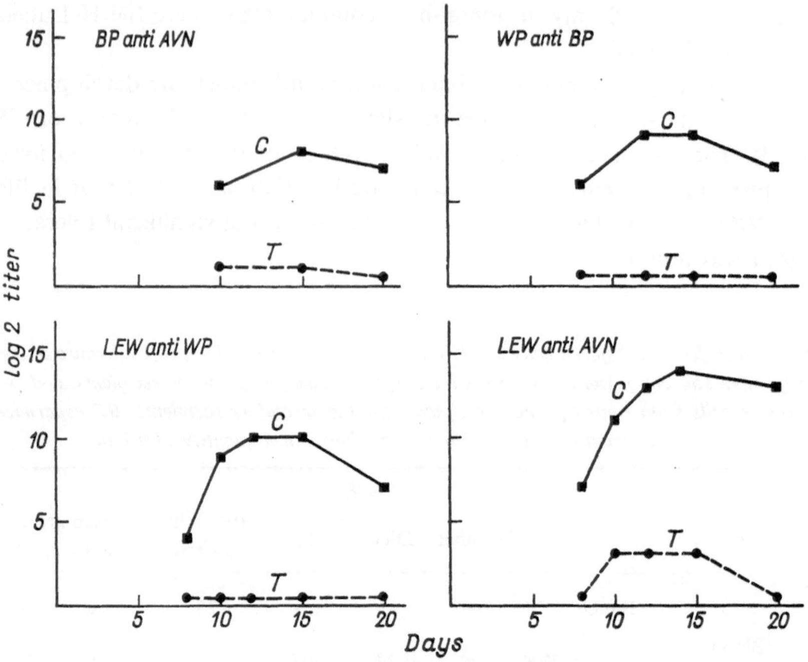

FIG. 1. Depression of the haemagglutinin formation in runt syndrome survivors after transplantation and rejection of the skin graft of the spleen cell donor. The curves of the average haemagglutinin titres of the control (C) and spleen treated animals (T). The haemagglutination reactions of the WP anti-BP sera were performed in saline, the others in the dextran medium.

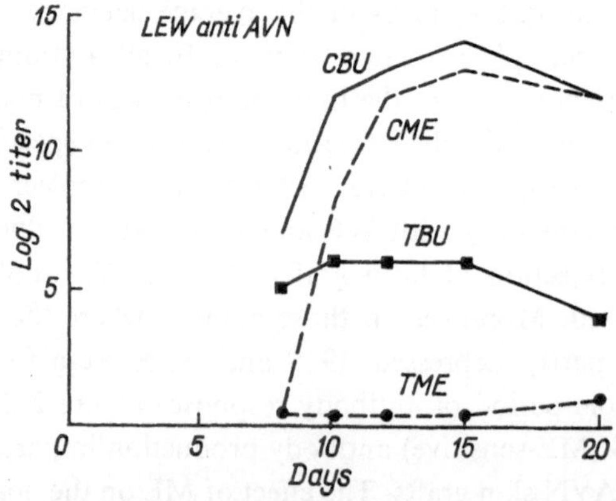

FIG. 2. The preponderance of 19 S haemagglutinins in AVN graft-rejecting partly-tolerant LEW animals. The curves of average haemagglutinin titres after incubation of the sera in buffered saline (BU) and in mercaptoethanol (ME), control (C) and spleen treated (T) animals, dextran medium.

DISCUSSION

The suppression of haemagglutinin formation and 19 S antibody preponderance in skin graft rejecting r.s. survivors could be caused by the residual chimaerism found in mice as well as in the rats (BILLINGHAM *et al.*, 1962). With regard to the simultaneous suppression of cytotoxic antibody production (as demonstrated in LEW-AVN and WP relationships) we consider our finding to be due to the dissociation of antibody and cellular mechanisms of the immune reaction. This dissociation could have some relationship to the immune deviation phenomenon (ASHERSON, 1967). The variability of the r.s. incidence after spleen cell administration is not completely attributable to the H-1 differences between donor and recipient and might be caused, in general, by strain different immune reactivity. In tolerance induction, however, the H-1 antigenic barrier was shown to have an effect greater than the other antigenic systems in strain combinations tested. Even spleen cells inducing long-term skin graft tolerance in the case of H-1 compatibility do not induce any tolerance in H-1 incompatible recipients. The explanation could lay in the damage to lymphoid immunologically competent cells caused by the immune interaction with strong H-1 antigens of the recipient. The quantity of cells preserved in the recipient (blood forming cells) is thus not sufficient for the induction of tolerance. In the case of non H-1 differences the immune reaction is not as strong and so the multiplication of donor lymphoid cells could be sufficient for the establishment of tolerance. The positive results in tolerance induction by spleen reported by the authors quoted above could be due to the possible H-1 compatibility of strains used especially when due regard is taken of the finding of several strains possessing identical H-1 alleles.

REFERENCES

ASHERSON, G. L., 1967. Antigen-mediated depression of delayed hypersensitivity, *Brit. Med. Bull.*, **23**, 24–29.

BILLINGHAM, R. E. and SILVERS, W. K., 1959. The induction of tolerance of skin homografts in rats with pooled cells from multiple donors, *J. Immunol.*, **83**, 667–679.

BILLINGHAM, R. E., DEFENDI, V., SILVERS, W. K. and STEINMULLER, D., 1962. Quantitative studies on the induction of tolerance of skin homografts and on runt disease in neonatal rats, *J. Nat. Cancer Inst.*, **28**, 365–435.

BRDIČKA, R., KŘEN, V., FRENZL B. and ŠTARK, O., 1962. Interlineal relationships in rats, *Fol. biol.* (Praha), **8**, 352–359.

EGDAHL, R. H., ROLLER, F. R., SWANSON, R. L. and VARCO, R. L., 1958. Acquired tolerance to homografts and heterografts in the rat, *Ann. N.Y. Acad. Sci.*, **73**, 842–847.

GROZDANOVIČ, J., 1959. Long-term passage of skin grafts in immunologically tolerant rats, *Fol. biol.* (Praha), **5**, 108–112.

KŘEN, V., BRAUN, A., ŠTARK, O., KRAUS, R., FRENZL, B. and BRDIČKA, R., 1962. The runting syndrome in rats and its inhibition by homologous sera, *Fol. biol.* (Praha), **8**, 341–351.

KŘEN, V., BRDIČKA, R. and ŠTARK, O., 1968. Erythrocyte and transplantation antigens in inbred strains of rats. VIII. The dynamics of 19S and 7S LEW anti-H-1 haemagglutinin production, *Fol. biol.* (Praha). (In press).

ŠTARK, O., KŘEN, V., FRENZL, B. and BRDIČKA, R., 1968. The main histocompatibility system of the rat. In: Advance in transplantation, Copenhagen, p. 331–339.

WOODRUFF, M. F. A. and SIMPSON, L. O., 1955. Induction of tolerance to skin homografts in rats by injection of cells from the prospective donor soon after birth, *Brit. J. Exp. Path.*, **36**, 494–499.

DISCUSSION

H. BALNER: Were the partially tolerant animals (in the H-1 compatible combinations) chimeras? Were you able to show that?

V. KŘEN: The cell chimaerismus was not tested in all cases but in 4 LEW rats, injected intravenously with AVN bone marrow at birth the chromosomal marker of AVN origin was found by our coworker Dr. Židek.

R. OWEN: In the marrow-injected rats, H-1 compatible, could you use the C/D marker to check for rat-cell chimaerism?

V. KŘEN: Unfortunately, LEW and AVN strains were identical in C antigen and therefore his antigenic marker could not be used for cell chimaerism testing in this relationship.

P. IVÁNYI: Were both haemagglutinating and cytotoxic antibodies suppressed?

V. KŘEN: In all the strain combinations tested (BP-WP, WP-LEW and AVN-LEW) haemagglutinins and cytotoxins were simultaneously suppressed.

XITH EUROPEAN CONFERENCE ON ANIMAL BLOOD GROUPS AND BIOCHEMICAL POLYMORPHISM
Warsaw, July 2nd–6th, 1968

Studies on Antigenic Differentation of Blood in the Coypu *(Myocastor coypus molina 1792)*

Ewa SZYNKIEWICZ

The Department of Genetics, Warsaw Agricultural University (SGGW), Warsaw, Poland

SUMMARY. The work presented, describes the methodology, the production and examination of the heteroimmune and isoimmune sera, and the examination of about 700 normal sera in the "cross tests", using agglutination and lysis.

INTRODUCTION

The aim of this research begun in 1964, was to examine the antigenic properties of the blood cells, as an additional factor in the research work, on the genetics and biology of the species.

Because this was the first work in this field, I used the methodology already widely used in research work on blood groups of the cattle (NEIMANN–SØRENSEN, 1958), and I was also guided in my research work by the publications of a number of authors, dealing with the groups of blood in rats (BOGDEN and APTEKMAN, 1962; FRENZL *et al.*, 1960; OWEN, 1962; PALM, 1962).

METHODS AND MATERIAL

Methodology widely accepted in research on blood groups, was used, but necessary modifications were introduced. To produce heteroimmune sera, rabbits were given, intravenously or intraperitoneally a 50 per cent suspension of haemolyzed or intact coypu blood cells and also the stroma of coypu blood cells.

Similar methods were used to produce isoimmune sera, i.e. intact blood cells, haemolyzed cells and the stroma of blood cells.

The stroma was prepared according to the methods of CALVIN *et al.* (1946), MOSKOWITZ *et al.* (1950), KOŚCIELAK (1963), GASPARSKI (1965).

Stroma was given to the recipients emulsified in Freund's adjuvant. The coypus were reimmunized after 6 months, the recipients were given centrifuged blood cells together with 25 and 50 per cent homogenates of coypu spleen.

To produce isoimmune sera, I was guided by the principle of choosing closely related donor–recipient pairs. The sera of immunized animals were subjected to three fold absorption at a ratio of 1: 1 and 0.75 : 1, as well as to agglutination tests.

To evaluate reaction of the isoimmune sera, different media for suspension of the blood cells were used: 2 and 6 per cent dextrane (according to GRUBB, 1949), and suspension of trypsinated blood cells in normal sera (according to MORTON and PICKLES, 1956). Also different times and incubation temperatures, were applied to the agglutination tests, these varied from 4–24 hr in temperatures from +4 to about 20°C. In investigating normal sera, "cross tests", using both agglutination and lysis were performed. Blood samples were taken from the heart and from the dorsal vein of the ankle into Alsevers solution.

Animals used in the production of sera were rabbits of White-Popielnian breed and coypus of the standard variety and sapphire variety from the experimental farm.

RESULTS

The effect of immunizing rabbits was determined by the titres of sera obtained. They oscillated from 1: 256 up to 1: 512 and from 1: 4096 up to 1: 65536. No differences between the methods of immunizing were noticed.

From 43 heteroimmune sera, two were isolated and marked CO_3 and CO_4, detecting two different antigens. 516 coypus were examined with these sera, the coypus originating from various breeding centres in Poland. The frequency of positive reaction in agglutination tests with serum CO_3 was 26 coypus reacting positively, and for serum CO_4 — 17 positive reactors. The positive reactions were obtained only with blood cells of animals which were not related to the material from the breeding farm. Any blood cells of coypus related to the donors and used for absorption, removed all antibody from serum Co_3 and Co_4, and investigation of coypus from the breeding farm, did not indicate any positive reactions. I suggest therefore that in the whole group being investigated and originating from the experimental farm, there is one and only one antigen. Similarly, OWEN investigating 5000 rats in 1948, noticed only one antigen, A, in them.

After examining normal sera from 710 individual animals only four sera were found agglutinating the blood cells of unrelated coypus.

By the use of successive absorptions of the sera, two kinds of antibody were detected, CO_1 and CO_2, the first to be detected in the species (SZYNKIEWICZ and GASPARSKI, 1966).

So far it has proved impossible to stimulate isoantibody production in related material. After the reimmunizations, made after an interval of 6 months, in 22 isoimmunized coypus, the sera of one animal did responded positively in an agglutination test with the unrelated blood cells. However this serum did not react simultaneously with the donor and it is therefore possible that positive reactions were the result of the natural antibody.

All the tests made on the sera of that series of immunizations, were read only after 24 hr of incubation at a temperature +4°C.

In order to make sure that the coypus used in the isoimmunizations were immunologically competent 10 were immunized with the blood cells of cows, of known antigenic composition. In the resultant sera, the anti-cattle antibody content was tested, and then, each serum was absorbed with the cattle blood cells used for the immunization. Next, a haemolytic test was performed, using the test sera as controls. It appeared that the coypus in gestation produce specific antibody to the cattle blood cells.

In a number of separate experiments, efforts were made to identify blood cells antigens of coypus, using the testing sera of the other species. Lytic and agglutination tests of coypu blood cells were carried out with human, bison, cattle, and horse test sera. With the exception of the bison (GASPARSKI, 1965), no identification of antigens of coypu blood cells by means of other sera was possible.

DISCUSSION AND CONCLUSIONS

On the basis of the results obtained, it was observed that:

(1) The coypus derived from the interbreeding of related animals do not possess the normal isoagglutinin. This observation was confirmed by the results obtained by FRENZL *et al.* (1960). These authors noticed that rats belonging to the same sibship do not possess normal isoagglutinins.

(2) The natural isoagglutinins were observed only in the unrelated animals. Similar conclusions were reached by SPITERI and EYQUEM (1964), after examination of 200 rats.

(3) The previous conclusions suggest that isoimmunization of related animals does not result in antibody formation.

(4) The normal sera have very low titres of short duration.

(5) Detection of antigens of blood cells in coypus is most easily performed with heteroimmune sera and the best antigen to produce such sera is the stroma of coypu blood cells.

(6) Identification of blood cells antigens of coypu using test sera of other species of animals was unsuccessful.

ACKNOWLEDGEMENTS

I am grateful to Dr. J. GASPARSKI for his very competent instruction in the methodology of research work, and his helpful discussions.

REFERENCES

BOGDEN, A. E. and APTEKMAN, P. M., 1962. *Ann. N.Y. Acad. Sci.*, **97**, 43–45.

CALVIN, P. M., EVANS, R. S., BEHRENDT, V. and GALVIN, G., 1946. *Proc. Soc. Exp. Biol. Med.*, **61**, 416–420.

FRENZL, B., KŘEN, V. and ŠTARK, O., 1960. *Folia Biol.*, Prague, **6**, 121–126.

GASPARSKI, J., 1965. Personal communication.

GRUBB, R., 1949. *Journ. Clin. Path.*, **2**, 223.

KOŚCIELAK, J., 1963. The Grouy Glicolypides of Human Erythrocytes of the Activity A, Hematology Institute, Warsaw.

MOSKOWITZ, M., DANDLIKER, W., CALVIN, M. and EVANS, R., 1950. *Journ. Immun.*, **65**, 385.

MORTON, I. and PICKLES, J., 1956. *Nature*, **158**, 880.

NEIMANN-SØRENSEN, A., 1958. Blood groups of cattle, A/S Carl Fr. Mortensen, Denmark, Copenhagen.

OWEN, R. D., 1948. *Genetics*, **33**, 623.

OWEN, R. D., 1962. *Ann. N.Y. Acad. Sci.*, **97**, 35–42.

PALM, J., 1962. *Ann. N.Y. Acad. Sci.*, **97**, 57–82.

SPITERI, M. and EYQUEM, A., 1964. Proc. of the 9th European Anim. Blood Group Conf. Prague, 205–209.

SZYNKIEWICZ, E. and GASPARSKI, J., 1966. Bull. of the Conference P.T.N.W., Lublin, 45–46.

CHAPTER 8

CORRELATIONS BETWEEN BLOOD GROUPS AND OTHER TRAITS

Maternal-fetal Incompatibility and Heterozygote Advantage in Mouse Transferrins

G. C. ASHTON*

Department of Genetics, University of Hawaii, Honolulu, Hawaii, U.S.A.

SUMMARY. A mouse colony segregating for serum transferrins was established by crossing two inbred strains with different transferrin alleles, and randomly breeding subsequent generations. The progeny from the F3, F4, F5 and F6 generations were examined for transferrin type. There was significant deviation ($P < 0.025$) from the expected 1 : 1 ratio in the compatible matings a♂ × ab♀ and b♂ × ab♀, with 46.9% homozygotes and 53.1% heterozygotes. There was no segregation disturbance in the incompatible matings, equal numbers of homozygotes and heterozygotes being produced.

The segregation from ab × ab matings did not differ from expectation, although there were significantly more Trf^b than Trf^a genes in the progeny ($P < 0.05$).

It is suggested that the advantage of the heterozygote in compatible matings may be one component of the balancing mechanism for this polymorphism.

INTRODUCTION

In 1958 ASHTON reported aberrant segregation ratios in certain transferrin matings in cattle. Subsequent data lead to a proposed mechanism for balancing transferrin polymorphism in this species (ASHTON, 1965). Two of the factors involved in the proposed mechanism are (1) an advantage of heterozygous over homozygous fetuses when the mother is genetically compatible with all fetuses she carries, and (2) a disadvantage of heterozygotes when the mother is homozygous and therefore genetically incompatible with any heterozygous fetus she carries. The results of six investigations, summarized by ASHTON (1965), plus the more recent results of BOUQUET (1966), gave 846 homozygotes and 987 heterozygotes from compatible matings ($P < 0.005$, without significant heterogeneity), and 633 homozygotes and 559 heterozygotes from incompatible matings ($P < 0.05$, without significant heterogeneity).

Superficially similar data are available for serum albumins in cattle (ASHTON, 1964); compatible matings gave 36 homozygotes and 56 heterozygotes, incompatible matings 28 homozygotes and 18 heterozygotes ($P < 0.025$). KRISTJANSSON (1966) has presented data for serum albumins in pigs; compatible matings gave 77 homozy-

* Supported by USPHS grant No. 1R01 HD 01831.

gotes and 114 heterozygotes, incompatible matings 139 homozygotes and 102 heterozygotes ($P < 0.005$).

These results prompted investigation of transferrin segregation in the laboratory mouse, and data obtained over the past two years are presented here.

MATERIALS AND METHODS

Mouse transferrin polymorphism was described independently by COHEN (1960), SHREFFLER (1960), and ASHTON and BRADEN (1961). The procedure used for identifying the three transferrin phenotypes was that described in the latter publication. The modified Trf-b phenotype (ASHTON and BRADEN, 1961; KLEIN *et al.*, 1966) was not seen during the present investigation, and no ambiguous phenotypes were found. The two phenotypes Trf-a and Trf-b each show three zones in starch gels, and represent the genotypes Trf^a/Trf^a, and Trf^b/Trf^b (nomenclature of COHEN and SHREFFLER, 1961). The heterozygous phenotype has the same appearance as that derived from a mixture of the two homozygous types.

Transferrin polymorphism in mice is unusual insofar as it has not yet been reported in wild mice, these animals carrying only Trf^b. All strains of laboratory mice so far examined, except CBA and strains derived from it, carry Trf^b, while CBA carries Trf^a.

A laboratory mouse population was established by reciprocal crossing from 5 males and 15 females of each of two strains, CBA/J and C3HeB/FeJ (Jackson Laboratory, Bar Harbor, Maine, U.S.A.). Subsequent generations were produced from this Fl by single pair matings within each generation, no female producing more than one litter. (Space precludes detailed description of the breeding system; this will be presented elsewhere.) To date the progeny from the F6 have been analyzed, and the colony will be continued.

RESULTS

The Table shows the segregation in the progeny from the F3, F4, F5, and F6 generations. Two main results were obtained:

(1) From the genetically compatible matings a♂ × ab♀ and b♂ × ab♀, there were 663 homozygotes (46.9%) and 752 heterozygotes (53.1%), which differs significantly ($\chi^2 = 5.72$, $P < 0.025$; SNEDECOR and COCHRAN, 1967) from the expectation of equality, without evidence of heterogeneity between the eight results ($\chi^2_7 = 10.12$; $P < 0.1$).

The genetically incompatible reciprocal matings on the other hand produced 697 homozygotes and 682 heterozygotes, not differing from the expectation of equality, nor evidencing heterogeneity.

Showing the segregation obtained from transferrin matings in mice

Mating	Distribution of progeny types from given generation									
	F3		F4		F5		F6		Total	
♂ ♀	homos.	hets.	homos.	hets.	homos.	hets.	homos.	hets.	homos.	hets.
a×ab	17	31	65	85	169	151	132	156	383	423
ab×a	25	28	92	90	128	121	98	120	343	359
b×ab	18	17	69	84	101	133	92	95	280	329
ab×b	26	22	66	66	162	141	100	94	354	323
ab×ab	81 a	171	77 a	169	124 a	263	136 a	246	418 a	849
	93 b		115 b		134 b		140 b		482 b	

(2) The ab×ab matings produced 418a, 849ab, and 482b progeny, not differing significantly from expectation. However in these progeny there were 1685 Trf^a genes, and 1813 Trf^b genes, differing significantly ($\chi^2 = 4.76$, $P < 0.05$) from expectation. There was no heterogeneity between generations ($\chi_3^2 = 4.58$; $P < 0.1$).

There was no difference in average litter size between the different mating groups, nor any effect on sex ratio.

DISCUSSION AND CONCLUSIONS

Even a slight advantage of a heterozygote is sufficient to maintain a balanced polymorphism. The data presented here support the hypothesis previously advanced for cattle transferrins that a heterozygous fetus is at an advantage in a genetically compatible mother, and that this is one component of the balancing mechanism. In cattle there is a deficiency of heterozygotes from incompatible mothers, but this was not evident in mice. As yet we have insufficient information to assess whether the third effect found in cattle occurs in mice, that is superior fertility of matings between like homozygotes compared with those between unlike homozygotes.

The significant excess of Trf^b alleles from heterozygous matings may reflect linkage of the transferrin locus with a locus affecting fertility. If this is so then it might be anticipated that this effect would disappear in subsequent generations, after crossing over has randomized the genes at the two loci, unless linkage disequilibrium is occurring. However, the results may reflect a genuine superiority of the Trf^b allele, and account for the fact that so far Trf^a has not been reported from wild populations of mice.

REFERENCES

Ashton, G. C., 1958. β-globulin polymorphism and early foetal mortality in cattle, *Nature*, **183**, 404–405.

Ashton, G. C., 1964. Serum albumin polymorphism in cattle, *Genetics*, **50**, 1421–1426.

ASHTON, G. C., 1965. Cattle serum transferrins: a balanced polymorphism, *Genetics*, **52**, 983–997.

ASHTON, G. C. and BRADEN, A. W. H., 1961. Serum β-globulin polymorphism in mice, *Aust. J. Biol. Sci.*, **14**, 248–253.

BOUQUET, Y. H., 1966. Bloedgroepenonderzoek op Belgische rundveepopulaties, *Mededelingen der Veeartsenijschool van de Rijksuniversiteit van Gent*, **10**, 1, 1–204.

COHEN, B. L., 1960. Genetics of plasma transferrins in the mouse, *Genet. Res.*, **1**, 431–438.

COHEN, B. L. and D. C. SHREFFLER, 1961. A revised nomenclature for the mouse transferrin locus, *Genet. Res.*, **2**, 306–308.

KLEIN, P. A., ROOP, B. L. and ROOP, W. E., 1966. Starch gel electrophoretic patterns of murine transferrin, *Nature*, **212**, 1, 376–377.

KRISTJANSSON, F. K., 1966. Fractionation of serum albumin and genetic control of two albumin fractions in pigs, *Genetics*, **53**, 675–679.

SHREFFLER, D. C., 1960. Genetic control of serum transferrin type in mice, *Proc. Nat. Acad. Sci.*, (Wash.), **46**, 1, 378–384.

SNEDECOR, G. W. and COCHRAN, W. G., 1967. Statistical methods, 6th ed., Iowa State University Press, Ames, Iowa, p. 248.

DISCUSSION

A. JAMIESON: A paper by BOYD *et al.* is in the current volume of the *Journal of Reproduction and Fertility.* It reports numbers of typed cows mated to typed bulls. The cows were slaughtered soon after mating. The Fallopian tubes in the slaughtered cows were examined for the presence of embryos. The minute embryos were not typed. This precise information about early embryonic survival provided no statistical evidence connecting blood transferrin types with fertility or survival.

XITH EUROPEAN CONFERENCE ON ANIMAL BLOOD GROUPS AND BIOCHEMICAL POLYMORPHISM
Warsaw, July 2nd–6th, 1968

On the Association between Egg-white Mucin-globulins and the Resistance of Offspring in the Perinatal Period

Ü. PAVEL, O. PÕDER, K. PETERSON and A. MÖÖL

Tartu State University, Estonian Academy of Agriculture, and Estonian Research Institute of Animal Breeding and Veterinary Science, Tartu, *U.S.S.R.*

The aim of the present investigation was to study the effect of the egg-white mucin-globulins on the viability of the embryo and on chick resistance to *Salm. gallinarum* during the first 3 weeks of life. The hens studied were provided by the Kurtna Poultry Breeding Experimental Station. The study of the effect of the hen was carried out on the four most frequent types of mucin-globulins (ABDD, BBDD, ABCD, and AADD). The chicks were experimentally infected with live cultures of *Salm. gallinarum* at a dosage rate of $1\text{–}3.1 \times 10^7$ bacterial cells by injecting the bacterial suspension subcutaneously between the first and the third day after hatching.

Altogether five experiments were carried out with 4073 eggs from 753 hens of New Hampshire breed. The ratios obtained were transformed to the arc $\sin\sqrt{\%}$ scale and were subjected to dispersion analysis. A statistically significant effect of the hen on embryonic mortality was found in one experiment only. In one case it was observed that the effect of the dam on the resistance of hatched chicks during the first 9 days after experimental infection approached probability.

In order to sum the results of all five experiments the influences of each experiment was reduced by subtracting the value of each hen from the mean value of the experiment. The deviations obtained were subjected to dispersion analysis.

Calculated deviations of the characteristics (the stages of ontogenesis) studied are shown in Table 1. As regards the effect of the dam, statistically significant differences in the phenotypes of the hen were discovered only in the viability of the embryo. It can, however, be seen from the Table that hens with the phenotypes ABDD and ABCD showed a fertilization rate which in the majority of experiments was higher than the mean values of the experiments.

The offspring of hens with the phenotype BBDD were characterized by the lowest embryonic mortality. Embryonic mortality was highest in the offspring of hens with the phenotype ABCD, although not in the all experiments.

The offspring of the hens with the phenotype AADD proved to be the most resistant to *Salm. gallinarum.* The offspring of the hens with the phenotype ABCD

TABLE 1

The offspring mortality of hens with different phenotypes

Stage	No. of experiment	Mother's phenotype			
		ABDD	BBDD	ABCD	AADD
Fertilization	1	+0.6363	−3.1174	+1.0477	+0.9591
	2	−0.3343	+0.8826	+0.2142	−0.1724
	3	+2.2423	−1.3604	−1.1085	−3.7254
	4	+1.2940	+3.4138	+4.8179	−7.3671
	5	+1.2163	−6.3759	+5.8095	+0.1683
	S	+5.0546	−6.5573	+10.7808	−10.1375
Embryonic mortality	1	−0.1842	−6.4415	−1.2192	+7.3105
	2	−0.5321	−1.0995	+7.9083	−2.0410
	3	−0.4332	+1.0788	+8.0268	+4.4286
	4	−0.6360	−5.0840	+17.9692	−2.9642
	5	+0.7443	−5.2757	−0.2907	+1.9441
	S	−1.0412	−15.8219	+32.3944	+8.6780
Post-embryonic mortality in 9 days after infection	1	+1.7878	−2.8383	+3.5090	−3.4575
	2	+1.4971	−1.9067	+3.4745	−3.3176
	3	+0.3229	−1.2706	+4.7380	−2.1620
	4	−2.1824	+6.9538	+4.5957	−1.7120
	5	+1.0712	−0.9045	−7.6546	+1.2422
	S	+2.4826	+0.0337	+8.6626	−9.4069
Post-embryonic mortality in 21 days after infection	1	+1.1940	−0.6378	+0.2681	−3.9625
	2	+1.2327	−0.9590	+0.0545	−1.9071
	3	+1.6168	−1.4494	−0.7991	−2.9870
	4	−0.9546	+5.7450	+0.8514	−2.5135
	5	+1.0086	−1.2309	−4.6577	+0.4870
	S	+4.0975	+1.4679	−4.2828	−10.8831

Note: The fenotypes ABDD, BBDD, ABCD and AADD in our notation correspond to Lush's symbols $II^A\,II^B\,III^B\,III^B$, $II^B\,II^B\,III^B\,III^B$, $II^A\,II^B\,III^A\,III^B$ and $II^A\,II^A\,III^B\,III^B$.

showed high mortality in the first nine days after experimental infection which, however, dropped steeply within the subsequent twelve days (the respective values were +8.6626 and −4.2828). Appreciable chick mortality was also found in the hens with the phenotype ABDD.

It is evident from the data presented in the Table that low fertilization is compensated for by low embryonic mortality (the BBDD hens) or by low post-embryonic mortality (the AADD hens). High fertilization is compensated for either by higher embryonic mortality (the ABCD hens) or by post-embryonic mortality (the ABDD hens).

TABLE 2

The offspring mortality of hens with different phenotypes

Stage	Mother's phenotype			
	ABDD	BBDD	ABCD	AADD
Fertilization:				
Hens	371	131	76	175
Eggs	2081	657	404	931
Fertilized eggs	1585	479	324	651
Fertilization	0.7617	0.7291	0.8020	0.6992
Embryonic mortality:				
Fertilized eggs	1585	479	324	651
Dead embryos	586	162	158	283
Mortality	0.3697	0.3382	0.4877	0.4347
Post-embryonic mortality in 9 days after infection:				
Hatched chicks	999	317	166	368
Dead chicks	740	242	129	276
Mortality	0.7407	0.7634	0.7771	0.7500
Post-embryonic mortality in 21 days after infection:				
Dead chicks	898	286	149	328
Mortality	0.8989	0.9022	0.8976	0.8913
Perinatal mortality up to 9th post-infection day:				
Fertilized eggs	1585	479	324	651
Deaths	1326	404	287	559
Mortality	0.8366	0.8434	0.8858	0.8587
Perinatal mortality up to 21st post-infection day:				
Deaths	1484	448	307	611
Mortality	0.9363	0.9353	0.9475	0.9386

The adjustment of levels of survival of the offspring in subsequent stages of development is also evident from the summary of all data. Thus, it appears from Table 2 that the greatest differences effecting the mortality of offspring of different dams were revealed in the embryonic period. Mortality was lowest in the offspring of the hens with the phenotype BBDD (33.82%) and highest in the offspring of the hens with the phenotype ABCD (48.77%). In the post-embryonic period, however, differences in dam effect decreased to the minimum. While the difference in offspring mortality between the extreme phenotypes (ABDD and ABCD) in the nine successive days after experimental infection was still approximately 3%, the difference over the whole experimental period of 21 days diminished still further. When studying the general perinatal mortality including both embryonic and post-embryonic mortality, one can also notice a decrease in the difference of dam effect. Thus a selective factor such as pullorum disease after hatching levelled off differences resulting from dam's phenotypes in the embryonic period.

XITH EUROPEAN CONFERENCE ON ANIMAL BLOOD GROUPS AND BIOCHEMICAL POLYMORPHISM
Warsaw, July 2nd–6th, 1968

Embryonic Mortality and Egg-white Protein Polymorphism in Chicken

D. O. SCHMID and P. THEIN

Institute of Animal Blood Group and Resistance Research, Livestock Breeding Research Organization Munich, G.F.R.

SUMMARY. We have studied egg-white protein polymorphism in 4450 incubated and unincubated eggs from White Leghorn chickens.

We found the B-allele of Lush locus II to predispose to embryonic mortality. In two other populations we found the A-allele of locus II to predispose to embryonic mortality. It is therefore concluded, that embryonic mortality is not controlled by one factor alone.

The distribution of genotypes within a polymorphism will be balanced by fitness factors. Fitness factors are differences between genotypes in fertility, mortality and adaptation to environment. These characters are of considerable interest and potential practical importance to livestock breeders. The investigation of relationship between genetically determined blood characters and fitness in animals has been influenced by the findings in humans of relationship between blood groups and disease, blood groups and fertility and hemoglobins and disease.

In contrast to the situation in humans, the use of artifical insemination in animals makes detection of relationships between fertility and biochemical polymorphisms relatively easy. In dairy cattle there is a possible effect of the blood group system J. in heifers possessing the natural serum antibody anti-J mated to bulls with the blood group antigen J^{cs} or J^{s} we find a lower conception rate (41%) in comparison to a control group in which the heifers were negative for anti-J (53%) (JAMIESON, 1960).

Studying serum transferrins in cattle ASHTON and FALLON (1962) have found, that the mating between homozygotes are more fertile than the mating involving heterozygotes. OGDEN and WOOLFE (1963) confirmed these findings. In sheep matings between transferrin homozygotes are more fertile than matings involving heterozygotes (ASHTON *et al.*, 1963).

Last but not least KHATTAB *et al.* (1964) by the examination of 282 matings of Welsh mountain sheep have confirmed, that the effect of the transferrin C gene maternal-foetal incompatibility results from disturbed segregation rather than selective fertilization.

* This work was supported by Grant Schm208/6 of the Deutsche Forschungsgemeinschaft.

During the last year we dealt with related problems in the chicken. The egg-white of the chicken (*Gallus domesticus*) is a mixture of different proteins. In four of these proteins a genetical determinated polymorphism has been established by starch-gel electrophoresis, the ovalbumin of locus I (LUSH, 1961, 1963; BAKER and MANWELL, 1962; LAW, 1963), the globulin loci II and III (LUSH, 1961; ANNAU and COCHRANE, 1962; BAKER and MANWELL, 1962) and also conalbumin, which is identical with serum transferrin polymorphism. These loci are autosomal and normally control the two alternate alleles A and B.

We have examined 2750 eggs of White Leghorn chickens in looking for a relationship between embryonic mortality and genetically determined protein structure. Egg-white samples were collected from unincubated eggs and also from incubated eggs in all stages of development. Examination was carried out using horizontal starch-gel electrophoresis with the tris-boric buffer system recommended by KRISTJANSSON and modificated by STRATIL (1966).

It was remarkable, that the detection of locus III was more difficult in incubated eggs than in unincubated eggs. Locus III was absent in the most cases. In locus II we found significant differences in gene frequencies between 451 unincubated eggs and 590 incubated dead germs of all stages of embryonic development. Allele A had a gene-frequency of 0.906 and allele B a gene-frequency of 0.094 in unincubated eggs. On the other hand we found an A frequency of 0.790 and a B frequency of 0.210 in incubated eggs with dead embryos up to 6 cm in length. In the group of dead embryos of length from 6 to 11 cm we found frequencies of 0.777 for A and 0.223 for B. This shows a significant excess of the B-allele in dead embryos.

MORTON *et al.* (1965) and PAVEL and MÖÖL (1967) had similar results. Our investigation shows, that the animals with the B-allele are predisposed to embryonic death (SCHMID and THEIN, 1967).

On account of these results we tested 1700 unincubated eggs from another Leghorn stock including all mating combinations with exception of the mating BB×BB, which we did not find. This examination also showed significant excess of the allele A.

In contrast to these results in two other White Leghorn populations we found excess of the allele A in incubated eggs with dead embryos. These results were initially a surprise but we interpret them to mean, that in flock gene A and on another gene B indicates predisposition to embryonic mortality. We think therefore we have detected two different factors responsible for embryonic death (SCHMID and THEIN, 1968; THEIN and SCHMID, 1968).

DISCUSSION

Performance and the production characteristics in farm animals are conditioned by multiple factors and therefore genetical analysis of them is difficult, and therefore we sought a more simple model. We believe, that the chicken is appropriate for such experiments. Our experiments have dealt with the question of how far embryonic mortality in the chicken can be influenced by the genotype of maternal egg-white proteins. As markers we used the egg-white proteins, which show clear polymorphism in chicken. According to our observations certain lines seem to have a predisposition to embryonic mortality. Our results are confirmed by the experiments of SHERIDAN (1964) and SOMES and SMITH (1967), who state that there is often linkage between embryonic mortality and genetically determined features in certain sire-lines and who observed, that this embryonic mortality is limited biphasically to the development of the first and last week.

We think, that, breeding programmes using the egg-white proteins as markers for embryonic mortality can only be applied within genetically closed breeding groups with well-known gene-frequencies. On these premises these results are repeatable and can be influenced in a selective way by means of breeding experiments.

REFERENCES

ANNAU, E. and COCHRANE, D., 1962. Comparative starch gel electrophoretic studies of fowl egg-white and plasma, *Nature*, **193**, 879.

ASHTON, G. C., EVANS, J. V. and TURNER, H. N., 1963. Personal communication.

ASHTON, G. C. and FALLON, G. R., 1962. Beta-globulin types, fertility and embryonic mortality in cattle, *J. Reprod. Fertil.*, **3**, 93.

BAKER, C. M. A. and MANWELL, C., 1962. Molecular genetics of avian proteins I. The egg-white proteins of the domestic fowl, *Brit. Poultry Sci.*, **3**, 161.

JAMIESON, A., 1960. J-phenotypes in relation to conception rates, *Immunogenetica Edinburgensis*, 1960–156 d.

KHATTAB, A. G. H., WATSON, J. H. and AXFORD, R. F. E., 1964. 140th Meeting of Genetical Society, London 9–10, 11, 1962. *Animal Prod.*, **6**, 207.

LAW, G. R. J., 1963. Independence of genetic variants of egg-white proteins of domestic fowl, Proc. XI Intern. Congr, *Genetics*, **I**, 193.

LUSH, I. E., 1961. Genetic polymorphisms in the egg albumen proteins of domestic fowl, *Nature*, **189**, 981.

LUSH I. E., 1963. Polymorphisms in the egg albumen proteins of the domestic fowl, Proc. XI Intern. Congr, *Genetics*, I, 153.

MORTON, J. B., GILMOUR, D. G., MCDERMID, E. M. and OGDEN, A. L., 1965. Association of blood-group and protein polymorphisms with embryonic mortality in the chicken, *Genetics*, **51**, 97.

OGDEN, A. L., MORTON, J. R., GILMOUR, D. G. and MCDERMID, E. M., 1962. Inherited variants in the transferrins and conalbumins of the chicken, *Nature*, **195**, 1026.

OGDEN, A. L. and WOOLFE, B., 1963. Personal communication.

PAVEL, U. and MÖÖL, A., 1967. Concerning the effect of the genotype of the motherbird on the fertilization and resistance of the embryo, International Symposium on Immunology of Spermatozoa and Fertlization, Sept. 27–29, 1967 Varna, Bulgaria.

SCHMID, D. O. and THEIN, P., 1967. Embryonic mortality and egg-white protein polymorphism in chicken, International Symposium on Immunology of Spermatozoa and Fertilization, Sept. 27–29, 1967, Varna, Bulgaria.

SCHMID, D. O. and THEIN, P., 1968. Embryonaler Fruchttod und Eiklarprotein Polymorphismus beim Huhn Zuchthygiene. (In print).

SHERIDAN, A. K., 1964. A sex-linked mutation causing low hatchability in the broiler chicken, *Proc. Austral. Poultry Sci. Conv.*, **1**, 87.

SOMES, R. G. and SMITH, J. R., 1967. Prenatal, a sex-linked lethal mutation of the fowl, *J. Hered.*, **58**, 25.

STRATIL, A., 1966. A contribution towards the study of transferrins and conalbumins of the domestic fowl, Proc. X Europ. Conf. Animal Blood Groups, Paris, 241.

THEIN, P. and SCHMID, D. O., 1968. Embryonic mortality in chickens and its diagnosis by examination of egg-white protein polymorphism, VI Congrès de Reproduction et Insemination Artificielle, Paris.

XITH EUROPEAN CONFERENCE ON ANIMAL BLOOD GROUPS AND BIOCHEMICAL POLYMORPHISM
Warsaw, July 2nd–6th, 1968

Naturally Occurring and Immune Antibodies as a Possible Cause of Hemolytic Disease in the Domestic Mink

J. RAPACZ, R. M. SHACKELFORD and J. HASLER

University of Wisconsin, Department of Genetics and Department of Meat and Animal Science, Madison, Wisconsin, U.S.A.

SUMMARY. The sera of more than 1500 adult mink from three sources were examined for presence of agglutinins. Antibodies against six red cell antigens; A, A_2, B, B_2, C and E (RAPACZ and SCHACKELFORD, 1966) were detected by the use of saline agglutination tests. Numerous sera with unidentified antibodies, usually of very low titre, were found. Normal antibodies were found in several mink sera following vaccination for virus enteritis: with this exception antibodies were found in the sera of females only after the pregnancy period and usually against the blood antigens of the male to which they had been mated.

INTRODUCTION

Naturally occurring isoantibodies against red cells were discovered more than sixty years ago by LANDSTEINER (1901), but the question as to whether they arise "without provocation" in the appropriate genetic background, or, as a result of immunization is still to be settled. Naturally occurring antibodies have been found in the sera of the human and such domestic species as cattle, swine and sheep. Within the first year of a two year investigation of the immunogenetics of the domestic mink (*M. vison*) such antibodies were detected in the sera of occasional individuals. The results of a systematic search for naturally occurring and artificially produced antibodies as a possible cause of hemolytic disease in new born mink is the subject of this paper.

OBSERVATIONS AND RESULTS

The sera of more than 1500 adult mink have been examined for naturally occurring antibodies by the use of agglutination, hemolytic or antigammaglobulin tests. All sera were examined by agglutination test, more than half by the hemolytic test and a selected few by the antiglobulin tests. Except in two cases, where nonspecific hemolysins were found, all specific antibodies functioned as agglutinins. From the first series of 450 tested sera, a single mink was found with anti-B_2. The sera of another group of 340 mink from the research herd and two commercial ranches were tested and eight additional individuals with isoagglutinins were found; three sera with anti-A and five with anti-B_2. All mink with normal antibodies came from the commercial ranches, where routine vaccination against virus enteritis is practiced and it is possible that antibodies may have been produced artificially as a result of exposing individuals to the tissues of the mink used in making the vaccine. The University herd was not vaccinated.

It seemed appropriate to investigate the possibility that hemolytic disease resulting from incompatible matings might contribute to the high mortality of mink kits within a few days of birth and the large number of "misses" especially with females at their first pregnancy, a condition sometimes observed on commercial ranches. Eight minks were immunized once, immediately after the young had been weaned in 1962. Two which had successfully raised large litters were included in the group: No. 1 (blood type BC) was immunized against the prospective mate No. 352 (blood type A) and No. 90 (blood type A) against her mate No. 357 (blood type C) the former producing anti-A and the latter anti-B_2. As indicated in Table 1 the antibody titre in the sera of these two females fell from first testing to late February or early March, increased during presumed pregnancy (No. 1 gave birth to seven, which died within 3 days, and No. 90 failed to litter) and began to drop following the gestation period; the highest titre we have observed was in sera of No. 90. By contrast, antibodies in the other six females disappeared within two months after immunization. The behavior observed in No. 1 and 90 led to a check of all mated females in the research herd which had failed to litter, or which lost a significant proportion of their kits soon after birth. Sera from 126 females and 54 males were tested as shown in Table 2: the females were divided into one group of 80 with surviving litters of which 17 had normal antibodies and another group of 46 that failed to litter or lost some of their young, of which 22 had normal antibodies. The testing was done within 2 to 6 weeks after the "kittening" period (1–15 May). This group was tested again and in three sera only were antibodies detected (one anti-A and two anti-B_2). The males mated to these females were tested the same time as the first check none had antibodies except for a single individual (369) with autoantibodies.

DISCUSSION

Naturally and artificially produced antibodies have been found against all except D and G of the seven blood antigens discovered in these studies. Many sera were found with unidentified antibodies either for factors within the known systems but at sufficiently low titres as to escape identification, or against blood factors yet to be studied. These observations seem unusual as compared to the situation in other species where the variety and frequency of normal antibodies is much lower than

Table 1

Changes in level of antibodies in two sera during the period of a year

Red cells		Serum No. 1 (anti-A)								Serum No. 90 (anti-B$_2$)							
mink No.	pheno-type	Date of testing: 10/8 Titre: 1/8	6/1 1/4	7/1 1/4	27/2 1/2	28/3 1/3	14/5 1/16	3/7 1/16	13/8 1/2	10/8 1/16	10/11 1/32	18/12 1/32	7/3 1/32	20/5 1/512	15/6 1/128	3/7 1/64	13/8 1/8
1	BC	.	.	.	.	.	.	.	.	++++	+++	++++	++	+++	+++	+++	++++
90	A	++++	+++	+++	+++	++++	+++	++++	+++	.	.	.	.	.	.	.	.
176	C	.	.	.	.	.	.	.	.	++++	++++	++++	+	++++	+++	+++	+++
8-8	B	.	.	.	.	.	.	.	.	++++	+++	+++	++	+++	+++	+++	+++
147	AB	++++	+++	++	+++	++++	+++	++++	+++	+++	++	+++	+	+++	+++	++	+++
58	B	.	.	.	.	.	.	.	.	++++	+++	+++	++	+++	+++	+++	+++
139	B	.	.	.	.	N.T	N.T*	N.T	N.T	++++	+++	++++	++	N.T	N.T	N.T	N.T
70	AC	++++	++++	++	+++	N.T	N.T	N.T	N.T	+++	+++	+++	+	N.T	N.T	N.T	N.T

* N.T — Non-tested.

TABLE 2

The frequency and type of "normal" antibodies in mink of the research herd

No. and sex of tested mink	No. of mink with normal antibodies	Type of antibodies in sera							
		anti-A	anti-B	anti-B_2	anti-C	anti-AB	anti-AC	Uni-denti-fied	Un-spe-cific
54 ♂♂	1	0	0	0	0	0	0	0	1
126 ♀♀ → Litter survived 80	17	1	0	7	0	0	0	9	0
126 ♀♀ → "Missed" or litter partially lost 46	22	0	3	4	0	0	0	15	0

in the mink groups composing this study. This species does not appear to be comparable to humans as (1) more than the appropriate genotype seems to be required for antibody production, or (2) the antibodies, if produced, are as such low levels the techniques employed here were not sufficient to detect them. More than 1000 individuals from the research herd have been examined; only after pregnancy were normal antibodies found, and then against blood types of the males to which the females involved had been mated. This could be interpreted to mean that in some way the unborn kits transfer to their mother antigenic substances inherited from their father, or that some other mechanism causes the rise of antibody titre during this period.

It seems probable that there are at least two sources of stimulation to antibody production in this species: pregnancy and inadvertent immunization as a side effect of virus enteritis vaccination. GOODWIN *et al.* (1955) concluded that hemolytic disease of newborn pigs is an artificial condition induced by injection of crystal violet hog cholera vaccine. Hemolytic disease resulting from incompatible matings may account for some of the high incidence of mortality in very young mink—a condition well known in all species studies except ruminants. A small amount of evidence in support of this suggestion is available; of the six females with the highest titre of antibodies, two failed to litter, and four gave birth but lost all young within 3 days. The mechanism and time of transfer of antibodies from mother to foetus is unknown, if such occurs. Antigenic substances may be transferred from unborn young to their mother directly through the placenta, or as a result of foetal-maternal blood vessel anastomosis. Unborn kits may receive antibodies by these same channels, or newborn kits through their mothers milk. Subsequent to our study, PORTER (1965) has demonstrated by the use of proteins trace-labelled with ^{131}I that immunoglobulins are transferred via the colostrum from mothers to their kits. Since normal antibodies against factors A and B_2 were found to be of most frequent occurrence and in highest titre, incompatible matings within system A would seem the most likely cause of the early deaths in cases of hemolytic disease. However, anti-B and -C, without anti-B_2 or -A, were found only in those "normal" sera collected during pregnancy, never from isoimmune sera which may mean that selectivity is involved. If hemolytic disease occurs in the mink, additional mechanisms may be at work since 17 females with antibodies from research herd gave birth and did not loose their young. The situation in mink may be comparable to that in humans where more than the Rh+(D) factor is involved as discussed by LEVIN (1962).

REFERENCES

GOODWIN, R. F. W., SAISON, R. and COOMBS, R. R. A., 1955. The blood groups in pig. II. Red cell isoantibodies in the sera of pigs injected with crystal violet swine fever vaccine, *J. Comp. Path. Therap.*, **65**, 79.

LANDSTEINER, K., 1901. Ueber Agglutinationserscheinungen normales menschlichen Blutes, *Wien. klin. Wschr.*, **14**, 1132–1134.

LEVIN, P. 1962. ABO and RH incompatibility as effecting fertility, Proc. of Conference on Immuno-Reproduction, La Jolla, California. September 9–11, 1962, 133–141.

PORTER, D. D., 1965. Transfer of gamma globulin from mother to offspring in mink, *Proc. Soc. Exp. Biol. and Med.*, **119**, 131–133.

RAPACZ, L. and SCHACKELFORD, R. M., 1966. The inheritance of seven erythrocyte antigens in the domestic mink, *Genetics*, **54**, 917–922.

DISCUSSION

C. STORMONT: Did you observe any increase of the titre of antibodies in females which were older or which were pregnant several times?

J. RAPACZ: The best example of that would be these 18 females, all incompatible in the system A, from which 16 did not litter last year. This year we changed males. Recently I have received information that the females littered above average in 1968. The studies are still on the way.

XITH EUROPEAN CONFERENCE ON ANIMAL BLOOD GROUPS AND BIOCHEMICAL POLYMORPHISM
Warsaw, July 2nd–6th, 1968

Transferrin Type and Milk Yield in Dairy Cattle

G. C. ASHTON* and R. W. HEWETSON

Department of Genetics, University of Hawaii, Honolulu, Hawaii, U.S.A.
C.S.I.R.O., Division of Animal Genetics, Ryde, N.S.W., Australia

SUMMARY. Production data for 932 lactations from approximately 460 cows from eight herds in Northern N.S.W., Australia for the years 1963 and 1964 were examined in relation to serum transferrin type. The mean milk yield of A/D2 cows was 151±72 lb ($P < 0.05$) more than that of the A/A cows, and they lactated 7.4±3.7 days ($P < 0.05$) longer. The mean yield of D2/D2 cows was 521±152 lb ($P < 0.001$) more than the A/A cows.

The best estimate of the effect of replacing a Tf^{A} gene with a Tf^{D2} gene was to increase yield by 170±56 lb ($P < 0.005$) and lactation length by 6.5±2.8 days ($P < 0.025$). Replacement by a Tf^{D1} gene had no effect.

It is suggested that there may be a breed difference in the relationship between transferrin type and milk yield.

INTRODUCTION

An association between milk production and serum transferrin genotype was first demonstrated by ASHTON (1960) from an analysis of the progeny tests of 130 bulls used in A.I. in England and Wales. It was found that the progeny of A/A bulls gave on average 260 lb of milk less than those of "D/D" bulls. JAMIESON and ROBERTSON (1967) have extended this investigation to 879 bulls and obtained a difference of 110 lb. A direct demonstration of the effect in cows was reported by ASHTON *et al.* (1964) who obtained differences in yield per lactation of 435 lb for Jersey cows and 463 lb for Australian Illawarra Shorthorn (AIS) cows.

Other workers have either had insufficient data to obtain a significant effect (LARSEN, 1961; DATTA *et al.*, 1965; VASENIUS, 1965) or have not found any difference (MEYER, 1967). Other workers have found A/A cows superior (YOUNG and HUNTER, 1966; RAUSCH *et al.*, 1968).

This paper reports data for milk production in eight herds of cows in northern N.S.W., Australia. There were two objectives (1) to examine the relationship between milk yield and transferrins typed for D1 and D2 (2) to reassess the relationship between milk yield and lactation length previously reported (ASHTON *et al.*, 1964).

* Supported by USPHS grant No. 1R01 HD 01831.

MATERIALS AND METHODS

The production records for eight commercial herds of dairy cows were analyzed in relation to transferrin type (and other polymorphism not reported here). Six herds were composed of Jersey, or mainly Jersey cows, one was mainly Guernseys, and one was mainly Ayrshires. All lactations up to the fifth were considered, not just the first. Data for 932 lactations from approximately 460 cows were available, for 1963 and 1964 (Table).

Distribution of cows in herds sampled

Herd	Breed	1963	1964
1	Mostly Jersey, some Guernsey	106	106
2	Mostly Jersey, some Guernsey	67	82
3	Jersey	78	77
4	Mostly Jersey	32	34
5	Mostly Ayrshire, some Jersey	43	42
6	Jersey	37	41
7	Jersey	52	53
8	Mostly Guernsey	39	36
Total		454	471

The records were analysed by multiple linear regression analysis within herds, years and lactations. The variables milk yield, fat yield, fat percentage and lactation lenght were regressed on transferrin type, with genotype A/A or gene A as baseline and the results expressed as deviations from this. Only the data for milk yield and lactation length are reported as fat yield followed milk yield, and no effect on fat percentage was found. In these herds milking is seasonal, yield regressing linearly on season expressed as the number of days after June 1 that lactation commenced. Accordingly in the comparisons reported below variation due to season has been removed before computing the variation between genotypes. Lactation ends when yield falls to 4 lb/day.

RESULTS

Genotype comparisons

The deviation of mean milk yield and mean lactation lenght for each genotype from the mean for A/A cows was as follows:

Tf	Milk yield	lb	Lactation lenght	Days
A/Dl	97 ± 127	N.S.	5.2 ± 6.5	N.S.
A/D2	151 ± 72	$P < 0.05$	7.4 ± 3.7	$P < 0.05$
Dl/Dl	-370 ± 304	N.S.	-15.5 ± 15.5	N.S.
Dl/D2	-327 ± 199	N.S.	0.2 ± 10.2	N.S.
D2/D2	521 ± 152	$P < 0.001$	15.3 ± 7.8	N.S.

Comparisons between genes

The best estimate of replacing a Tr^A gene by another is given below:

Tf	Milk yield	lb	Lactation lenght	Days
D1	-88 ± 92	N.S.	-2.8 ± 4.7	N.S.
D2	170 ± 56	$P < 0.005$	6.5 ± 2.8	$P < 0.025$

DISCUSSION

Our data suggest that in Jerseys and Guernseys Tf^{D2} has a significant effect on milk yield and lactation length, while Tf^{D1} does not. If there is a real difference between D1 and D2 this may help to explain some of the discrepancies reported in the literature, because "D/D" animals have not been subtyped except by JAMIESON and ROBERTSON (1967). Their Table 3 shows Tf^{D2} to have greater effect than Tf^{D1} in Jerseys, Guernseys and Shorthorns while the reverse effect is shown in Friesians and Ayrshires, although the differences are not significant.

Another reason for inconsistency of results between investigators may be due to the nature of the production data used. Part of the transferrin effect is due to increased lactation length, although removing variation in yield due to this still leaves a significant difference between A/A and D2/D2 cows of 318 ± 112 lb ($P < 0.01$). Consequently, where milk yield is expressed in terms of fixed lactation length through the use of arbitrary correction factors, real differences between genotypes may be obscured or reversed.

OSTERHOFF (1964) has shown that the frequency of Tf^D increased with advancing age group in a survey of dairy cows in S. Africa. In our herds we have found a similar effect by regressing gene frequency on lactation number within herds and years. Tf^A showed a significant negative regression with increasing lactation number ($P < 0.025$) and Tf^{D2} a significant positive regression ($P < 0.01$). This presumably is due to culling, and reflects the superiority of Tf^{D2} cows in an indirect but alternative way.

REFERENCES

ASHTON, G. C., 1960. β-globulin polymorphism and economic factors in dairy cattle, *J. Agric. Sci.*, **54**, 321–328.

ASHTON, G. C., FALLON, G. R. and SUTHERLAND, D. N., 1964. Transferrin type and milk and butterfat production in dairy cows, *J. Agric. Sci.*, **62**, 27–34.

DATTA, S. P., STONE, W. H., TYLER, W. J. and IRWIN, M. R., 1965. Cattle transferrins and their relation to fertility and milk production, *J. Anim. Sci.*, **24**, 313–318.

LARSEN, B., 1961. Serum-haemoglobin — og maelketypers mulige indflydelse pa den Kvantitative og Kvalitative maelkeproduktion hos Kvaeg, *Aarsberetn. Instit. f. Steritetsfoyskn.*, 125–134

JAMIESON, A. and ROBERTSON, A. 1967. Cattle transferrins and milk production, *Animal Production*, **9**, 491–500.

MEYER, H., 1967. Untersuchungen zum Transferrin-Polymorphisms beim Rind, *Zbl. Vet. Med.*, A, **14**, 335–247.

OSTERHOFF, D. R., 1964, Recent research on biochemical polymorphism in livestock, *J. S. Afr. Vet. Med. Ass.*, **35**, 363–380.

RAUSCH, W. H., BRUM, E. W. and LUDWICK, T. M., 1968. Preliminary report on the relationship between blood type and predicted differences in production of Guernsey sires in A.I., *Immunogen. Letter*, **5**, 153–157.

VASENIUS, L., 1965. Transferrin polymorphism in Finnish Ayrshire Cattle, *Ann. Acad. Sci. Fenn.*, A, IV, 98.

YOUNG, C. W., and HUNTER, A. G., 1966. Transferrin polymorphism studies in Holstein cattle. *J. Dairy Sci.*, **49**, 735.

DISCUSSION

J. RENDEL: Dr. Ashton said in his interesting lecture that the effect of the transferrins on milk yield was operating through an effect on lactation length. Numerous previous studies have shown that the variations in the milk yield at the end of lactation period is largely non-genetic, i.e. the variations is caused by the stage of pregnancy. I find it a little strange that a mainly non-genetic trait (tail end lactation yield) should be influenced by strictly gene-controlled characteristics such as the transferrins.

G. C. ASHTON: The lactation lengths in these herds, which are under Australian conditions were usually of the order of 240 or 250 days, although some reached 360 days. Lactations are not terminated until yield falls to 4 lb/day. I don't find it the least bit strange that the tail end is under environmental rather than genetic control. But the main bulk of the lactation that produces 90% of the milk does have a reasonably high heritability.

R. L. SPOONER: Is any danger of sampling error. Some of our bulls with the best contemporary comparison are homozygous T_g^a/T_g^a.

G. C. ASHTON: Bull data being very difficult to interpret, one should use cows. Bulls may be selected for other factors explaining why the best bulls are T_g^a/T_g^a.

W. MICHALAK: My question is in two parts. Do you not think it strange that Tf^{D_1} and Tf^{D_2} should act in different ways in different breeds? Secondly, was there any environmental effect on milk yield and did this differ between breeds and between herds?

G. C. ASHTON: There are other instances of the same gene acting in different ways in different genetic backgrounds. Until we know how the transferrin genes affect milk production, the differential effect of D_1 and D_2 will remain "strange".

In reply to your second question, the different herds had different mean milk yields. However, the differences between A/A and D_2/D_2 cows for the eight herds were not heterogeneous and were pooled for the overall estimate of the effect. The individual data will be published elsewhere. The analyses were carried out the effect of season taken out, because there was a significant effect of season on milking. And this was done for each herd and breed. So I'm hopeful that the environmental effects have been suitably dealt with.

The only way that one can really cope with effects other than the genetic effect is to repeat the observations in large numbers of herds, large numbers of breeds, in many different parts of the world. I think that all one can hope for is that other people will do this and do it in a similar manner and eventually we'll be able to answer questions such as the one you've raised. The effect of environment can be held to a minimum by taking large herds in which the number of bulls used is reasonably large. I think if you start looking at bulls through their progeny that one is going to get into difficulties.

C. C. Oosterlee: Do you have data on young bulls? I ask this question, because in Dutch cattle breeds we found for one B locus allele a positive and for another a negative relation with the fat percentage of the milk. By studying a large number of daughters of one bull, heterozygous for these two B locus alleles, there was an excess of the number of daughters which inherited the allele with the positive effect. When we studied these changes in the B locus alleles, we saw an opposite effect. It is quite possible that a gene has a positive relation with one character and can have a negative one with another character related for instance to the selection in young bulls. So it is possible that a positive correlation will not be very useful in selection. Do you agree?

G. C. Ashton: Certainly one has to be careful. An inverse relationship between milk yield and butterfat percentage is well known. Data for bulls have been published by Jamieson and Robertson (see references).

M. Żurkowski: I want to ask you to explain, how you can interpret that effect that cows which had gene D_2 they had longer lactation. We know that length of lactation is a law of inheritance and that it is mainly dependent on many environmental factors.

G. C. Ashton: Thank you for giving me the chance to talk about lactation length. The important thing about these results is that lactation length is increased for about half the time necessary to account for the extra production. Therefore, if one tries to partition the effect of lactation length on increased milk yield, one finds that it accounts for about half the extra milk. So there is also an effect other than that due to extra lactation length. In the D_2/D_2 cows, if the effect of lactation length is removed, there is still a significant difference between the A/A and the D_2/D_2 cows.

Now I think that in data in which lactation length has been automatically adjusted, for example to 305 days, half of the effect is going to disappear and this may be one reason why one is not able always to detect an effect.

I think the two effects, increased lactation length and increased production per day, are related physiologically.

J. Rendel: I woud lika to comment on the lactation length. Usually one cuts the lactation length of 305 days because one has found that the lactation yield in the length of the lactation is mainly influenced by other factors than heredity. It is influenced by the stage of pregnancy. And I must say I found therefore a little strange that genetic factors like transferrin D_2 would influence lactation length and thereby lactation yield, as previons studies had in general found that lactation length and the tailend of production yield is mainly non-genetic.

XITH EUROPEAN CONFERENCE ON ANIMAL BLOOD GROUPS AND BIOCHEMICAL POLYMORPHISM
Warsaw, July 2nd–6th, 1968

Immunogenetical Studies on Cows of High, and Low Butterfat Production

H. BALBIERZ, A. KACZMAREK, Maria NIKOŁAJCZUK, M. ŚWITEK and Z. DORYNEK

College of Agriculture in Poznań, Department of Animal Husbandry, Poznań, Poland
College of Agriculture in Wrocław, Department of Obstetrics and Gynecology, Laboratory of Immunopathology, Wrocław, Poland

SUMMARY. Using the data available on the production levels of 2105 cows in their first lactation, we selected 109 cows of low and 108 cows of high butterfat production for further study. In the second part of the investigations, out of total number of 453 cows, we selected 98 animals of low and 94 of high butterfat production.

The blood samples taken from all experimental cows were tested for determination of blood groups and of transferrins in the serum. In the cows with high butterfat per cent the B-allele $BO_3Y_2A'E_3'G'P'$ was present in high frequency. The allele GY_2E_1' was more frequent in cows with higher butterfat production whilst in those with low fat yield (in kg and %) the B-allele BO_1Y_2D' was more frequent.

The gene M was more frequent in cows of low butterfat production. Also the frequency of the gene determining transferrin D was higher in animals of low butterfat production.

INTRODUCTION

Studies on immunogenetical characteristics of animals have already been carried out by many investigators (ASHTON *et al.*, 1964; CONNEALY and STONE, 1965; KRAAY, 1964; MITSCHERLICH *et al.*, 1959; NEIMANN-SØRENSEN, 1959; NEIMANN-SØRENSEN and ROBERTSON, 1961; OOSTERLEE, 1964; RENDEL, 1959; RENDEL, 1961; TOLLE and MITSCHERLICH, 1959; and WHITE *et al.*, 1967) during the last two years. Results of some of these studies suggest that some blood group factors and some serum protein fractions may be connected with production characters. In a search for such a relationship we tried to characterize the blood groups and transferrins of cow of low and high butterfat production (in kg and in % content in milk).

MATERIALS AND METHODS

Investigations were carried out in two parts. The first part used cows of the Lowland Black and White breed in their first lactation periods in 1966 and 1967.

They belonged to 24 herds of total number 2105 cows, the progeny of 81 bulls. On the basis of the weight of butterfat produced by each cow during 305 days of lactation, we chose 108 cows of high and 109 cows of low butterfat yield (10.45% of the herds). The selected cows originated from 49 bulls and represented the upper and lower limits of tolerance, i.e. those animals of extremely high or extremely low butterfat production. The blood cell antigens and transferrin types of each group were determined.

The second part of investigation was carried out on cows in their first lactation period in 1967, belonging to 3 herds of total number 453 cows, the progeny of 36 bulls. Calculation of mean per cent of butterfat in milk and its standard deviation was used as a basis for chosing 2 groups of cows. The first group included animals whose butterfat level was the arithmetic mean plus a standard deviation, and the other, those of arithmetic mean minus standard deviation in butterfat per cent in milk. In such a way we obtained a group of 94 cows of high and the other group of 98 cows of low butterfat per cent in milk. They originated from 23 bulls. The blood samples from all these animals were tested with 47 test sera. Transferrin types were determined by starch gel technique. Frequencies of genes in B blood group system were calculated following the method described by BRAEND (1963).

RESULTS AND DISCUSSION

Table 1 shows the differences in frequency of only those among 61 alleles of B system which were the more frequent in the tested groups of cows.

The allele BO_1Y_2D' had a considerably higher frequency in the group of low weight of butterfat produced, while the allele $GY_2E'_1$ appeared to be more frequent in cows of high weight of butterfat. The remaining B-alleles: $BO_3Y_2A'E'_3G'P'$, $BGKO_xY_2A'O'$, I′, GO_1, and E'_3 showed little tendency at different levels of frequency between the groups. The gene frequencies in the remaining blood group systems, as seen in Table 2, are similar, except for gene M which was more frequent in the group of cows of low butterfat production.

The transferrin study showed a higher frequency of gene D in the group of cows of low weight of butterfat produced (Table 3).

The lower parts of Tables 1, 2 and 3 illustrate the results of the second part of our investigations. They show a higher frequency of the gene $BO_3Y_2A'E'_3G'P'$ in the group of cows of high butterfat per cent in milk, observed in the first part to have only a small effect on butterfat production. The allele BO_1Y_2D—as in the first part—was more frequent in cc ws of low fat content in milk.

It is difficult to draw any general conclusion concerning the remaining B-alleles though in some cases there were significant differences between groups.

We realize that the weight of butterfat produced depends on the amount of milk and on per cent butterfat content in it. In the results of some publications, the presence of B-allele BO_1Y_2D' coincides with high butterfat per cent in milk, but

TABLE 1

Differences in frequencies of B-allelles in cows of low and high butterfat yield (in kg and %)

B-allele	Number of B-allelles	Frequency	Number of B-allelles	Frequency	χ^2
	Cows of low butterfat production in kg		Cows of high butterfat production in kg		
$BO_3Y_2A'E'_3G'P'$	18	0.0833	26	0.1229	1.454
$BGKO_xY_2A'O'$	31	0.1499	23	0.1077	1.185
I'	13	0.0595	22	0.1028	2.3134
GO_1	5	0.0224	1	0.0044	2.6663
E'_3	5	0.0224	12	0.0546	2.8821
$GY_2E'_1$	12	0.0546	26	0.1229	**5.1574**
BO_1Y_2D'	14	0.0642	1	0.0044	**11.2627**
	Cows of low butterfat % in milk		Cows of high butterfat % in milk		
$BO_3Y_2A'E'_3G'P'$	9	0.0450	28	0.1554	**9.759**
$BGKO_xY_2A'O'$	23	0.1213	31	0.1743	1.185
I'	23	0.1213	16	0.0850	1.275
GO_1	—	—	4	0.0206	**4.000**
E'_3	5	0.0247	12	0.0632	2.882
$GY_2E'_1$	34	0.1851	21	0.1138	3.073
BO_1Y_2D'	8	0.0399	2	0.0103	3.600

TABLE 2

Gene frequencies in blood group systems A, J, L, M, Z, FV in cows of low and high butterfat yield (in kg and %)

Gene frequency in %			
Cows of low butterfat production in kg		Cows of high butterfat production in kg	
A = 13.79	a = 86.21	A = 18.92	a = 81.08
J = 15.40	j = 84.60	J = 14.48	j = 85.52
L = 21.60	l = 78.40	L = 22.42	l = 77.58
M = **15.40**	m = 72.02	M = **9.23**	m = 90.77
Z = 27.98	z = 72.02	Z = 24.85	z = 75.15
F = 86.70	v = 13.30	F = 85.65	v = 14.35
Cows of low butterfat % in milk		Cows of high butterfat % in milk	
A = 9.09	a = 90.91	A = 13.10	a = 86.90
J = 14.28	j = 85.72	J = 12.49	j = 87.51
L = 23.07	l = 76.93	L = 27.07	l = 72.93
M = 14.29	m = 85.71	M = 14.33	m = 85.67
Z = 6.33	z = 93.67	Z = 8.33	z = 81.67
F = 89.80	v = 10.20	F = 89.89	v = 10.11

TABLE 3

Frequencies of genes determining transferrins in cows of low and high butterfat yield (in kg and %)

Transferrin type	Number of cows	Frequency	Number of cows	Frequency
	Cows of low butterfat production in kg		Cows of high butterfat production in kg	
A	15		24	
AD	43	$q^A = 0.378$	39	$q^A = 0.479$
AE	5	$q^D = 0.568$	1	$q^D = 0.510$
D	35	$q^E = 0.053$	27	$q^E = 0.040$
DE	4		7	
E	1		—	
	Cows of low butterfat % in milk		Cows of high butterfat % in milk	
A	14		18	
AD	44	$q^A = 0.406$	35	$q^A = 0.426$
AE	6	$q^D = 0.532$	4	$q^D = 0.540$
D	26	$q^E = 0.062$	29	$q^E = 0.034$
DE	6		2	
E	—		—	

has a negative influence on milk production (RENDEL, 1961). We can not exclude that in the tested population the milk yield decreased simultaneously with little deviations in butterfat per cent. For this reason many more animals possessing this B-allele were placed in the group showing low weight of butterfat produced. In addition, higher frequency of gene M in the group of low butterfat production could also influence the lowering of the weight of butterfat produced, as according to RENDEL (1965) the presence of this gene decreased milk yield. In the second part of our study we did not observe any differences in the remaining blood group systems. In contrast to the first part, the investigations on transferrin types did not show significant differences in frequencies of individual genes in the groups being compared, as can be seen in Table 3.

We believe, however, that the results obtained are sufficiently encouraging as to suggest a relationship between butterfat content and immunogenetical characteristics of individual animals and we intend to continue the studies.

REFERENCES

ASHTON, G. C., FALLON, G. R. and SUTHERLAND, D. N., 1964. Transferrin (β-globulin) type and milk and butterfat production in dairy cows, *Journal of Agricultural Science*, **62**, 27.

BRAEND, M., 1963. Estimation of gene frequencies in the B system of cattle, *Immunogenetic Letter*, **3**, 2.

Conneally, P. M. and Stone, W. H., 1965. Association between a blood group and butterfat production in dairy cattle, *Nature*, **3**, 115.

Kraay, G. J., 1964. Aspects of relationships between genetically determined characters in cattle, Proceedings of the 9th European Animal Blood Group Conference, August 18–22, 1964, Prague 1965, 87.

Mitscherlich, E., Tolle, A. and Walter, E., 1959. Untersuchungen über das Bestehen von Beziehungen zwischen Blutgruppen und Milchleistung des Rindes, *Z. Tierz. Zucht. Biol.*, **72**, 289.

Neimann-Sørensen, A., 1959. Bloodgroups and production characters of cattle, Bericht u.d. VI. Int. Blutgr. Congres, München, 25.

Neimann-Sørensen, A. and Robertson, A., 1961. The association between blood groups and several production characteristics in three Danish cattle breeds. *Acta scand.* **XI**, 2, 163.

Oosterlee, C. G., 1964. Some aspects of studies on relationship between blood groups and economic characteristics in farm animals, The World's Poultry Science Association, Sezione Italiana, 1964, 457.

Rendel, J., 1959. A study on relationships between blood groups and production characters in cattle, Bericht u. d. VI. Int. Blutgr. Congres, München, 8.

Rendel, J., 1961. Relationships between blood groups and the fat percentage of milk in cattle, *Nature*, **189**, 408.

Rendel, J., 1961. Recent studies on relationships between blood groups and production characters in farm animals, *Z. Tierz. Zucht. Biol.*, **75**, 2, 97.

Rendel, J., 1965. Zusammenhänge zwischen Blutgruppen und Produktionseigenschaften, *Der Tierzüchter*, **17**.

Tolle, A. and Mitscherlich, E., 1959. Grundlagen und Untersuchungsergebnisse von Beziehungen Blutgruppenfaktoren und Färsenlaktationen, Bericht u. d. VI. Int. Blutgr. Congres, München, 40.

White, M. B. and Banfield, J. C., 1967. The distribution of serum transferrin types in dairy cattle and their relationship to milk and butterfat production, *Austr. J. of Exp. Agric. and An. Husb.*, 7, 28, 396.

XITH EUROPEAN CONFERENCE ON ANIMAL BLOOD GROUPS AND BIOCHEMICAL POLYMORPHISM
Warsaw, July 2nd–6th, 1968

A Study on Correlation between Blood Groups and Carcass Quality in the Pig*

I. WIATROSZAK and S. ALEXANDROWICZ

College of Agriculture in Poznań, Department of Animal Husbandry, Laboratory of Research in Animal Blood Groups, Poznań, Poland

SUMMARY. Investigations were carried out on 2325 bacon pigs evaluated as to their carcass characteristics at the Pig Progeny Testing Stations. The following features were taken into consideration: (1) weight of lean in primary cuts, (2) weight of lean in ham, (3) back-fat thickness, mean from 5 measurements, (4) mean daily gain in weight (during fattenning from 50 to 80 kg live weight), (5) length of carcass, (6) slaughter age, (7) area of loin "eye".

Blood samples from all animals were tested with at least 36 test sera belonging to 12 blood group systems and the alleles in individual systems were determined. Statistical correlations were carried out for 5 systems: E, A, F, G, K. Special attention was paid to the E system.

It was found that in the E blood group system homozygote animals of genotype bdg/bdg had less lean in primary cuts and in ham than had homozygote animals edh/edh.

The presence of the allele F^a was connected with less lean in primary cuts.

Barrows possessing the allele G^b had less lean in ham. In gilts this difference was not statistically significant.

In the remaining 5 carcass characters the differences between animals were insignificant.

INTRODUCTION

In the Laboratory of Research in Animal Blood Groups, Department of Animal Husbandry of the College of Agriculture in Poznań, we have been trying to find a relationship between blood group alleles and some carcass characteristics in pigs. We know of only three studies devoted to searching for correlation between antigens and quality of pig carcass (BALTZER, 1963; SCHRAPE, 1966; TIKHONOV, 1966). In this reported study, however, we investigated alleles and not antigens, thus we cannot compare our results with those of the other authors.

* This research has been financed in part by a grant made by the United States Department of Agriculture under P. L. 480.

MATERIAL AND METHODS

The experimental material included 2325 bacon pigs (1178 barrows and 1147 gilts). The tested animals belonged to 3 breeds: Polish Large White (1441 pigs), Polish Landrace (768 pigs) and White Zlotnicka (116 pigs). In our study however we did not distinguish between the breeds as we were interested generally in whether we could find any connections between blood group alleles and features of economic importance in pigs. In our further work we shall examine the problem of such relationship within individual breeds. The experimental animals were fattened in uniform conditions at the Pig Progeny Testing Stations, then slaughtered and dissected on the spot.

The following carcass characteristics, with rather high coefficient of heritability were taken into consideration:

(1) Weight of lean in primary cuts.
(2) Weight of lean in ham.
(3) Backfat thickness, mean from 5 measurements.
(4) Mean daily gain in weight (during fattening from 50 to 80 kg live weight).
(5) Lenght of carcass.
(6) Slaughter age.
(7) Area of loin "eye".

TABLE 1

Blood groups considered in statistical calculations

Blood group system	Allele	Genotypes group I	Genotypes group II	Notes
E	E^{bdg} E^{edh}	bdg/bdg bdg/edh edh/edh		E^{edh} includes individuals E^{efdh} and E^{edgh}
A	A^{A}	—/—	A/—	a^0 allele was not considered
F	F^{a}	—/—	a/—	
G	G^{b}	a/a	a/b b/b	
K	K^{ac}	b/b b/— —/—	ac/b ac/—	

The blood samples from all experimental animals and from their parents were tested with at least 36 test sera belonging to 12 blood group systems, but statistical correlations were carried out for the alleles of 5 systems only, as seen in Table 1.

In the study reported we investigated the significance of differences in mean values for carcass characters between individuals of various genotypes in E blood group system, taking into account the presence or absence of the alleles A^A, F^a, G^b and K^{ac}, successively, because preliminary calculations showed that in this system the largest differences could be expected.

The statistical calculations were carried out according to the method of analysis of variance of designs with two non-orthogonal classiffications when interactions are present (FEDERER and ZELEN, 1966).

RESULTS

Highly significant differences were found between individuals possessing the allele E^{bdg} and the allele E^{edh} in weight of lean in primary cuts and in ham. As seen in Tables 2, 3, 4 and 5, homozygous animals of both sexes of E system genotype bdg/bdg have less lean in primary cuts and less lean in ham than the homozygous animals edh/edh.

TABLE 2

Mean weight of lean in primary cuts in barrows

Genotypes in E system	Animals F^a negative		Animals F^a positive	
	number	mean kg	number	mean kg
bdg/bdg	149	15.95	12	15.48
bdg/edh	373	16.13	28	16.10
edh/edh	462	16.27	27	15.77
Difference between edh/edh and bdg/bdg		0.32		0.29
t — test		3.144**		0.765

F — test for the significance between F^a neg. and F^a pos.: 5.79.*
F — test for the significance of interaction E×F: non-significant.
* Significant at 5% level.
**Significant at 1% level.

In addition the presence of allele F^a appeared to be connected with less lean in primary cuts in carcasses of both barrows and gilts (Tables 2, 3).

We also found less lean in ham from barrows possessing the allele G^b (Table 4). In gilts this difference was not statistically significant (Table 5).

The differences found are not of importance from a practical point of view, although they show high statistical significance.

For the remaining 5 carcass characteristics, the differences found were insignificant.

In this short report we have presented only the preliminary results of our investigations and we are not going to draw any conclusions. It may be that we discovered

TABLE 3

Mean weight of lean in primary cuts in gilts

Genotypes in E system	Animals F^a negative		Animals F^a positive	
	number	mean kg	number	mean kg
bdg/bdg	129	16.93	7	16.21
bdg/edh	383	17.17	33	16.90
edh/edh	460	17.26	24	16.63
Difference between edh/edh and bdg/bdg		0.33		0.42
t — test		2.915**		0.856

F — test for the significance between F^a neg. and F^a pos.: 5.28.*
F — test for the significance of interaction E×F: non-significant.

TABLE 4

Mean weight of lean in ham in barrows

Genotypes in E system	Animals G^b negative		Animals G^b positive	
	number	mean kg	number	mean kg
bdg/bdg	20	4.658	141	4.515
bdg/edh	40	4.763	361	4.567
edh/edh	49	5.078	440	4.563
Difference between edh/edh and bdg/bdg		0.420		0.048
t — test		3.185**		1.004

F — test for the significance between G^b neg. and G^b pos.: 12.12.**
F — test for the significance of interaction E×G: 5.33.*

TABLE 5

Mean weight of lean in ham in gilts

Genotypes in E system	Animals G^b negative		Animals G^b positive	
	number	mean kg	number	mean kg
bdg/bdg	18	5.140	118	4.729
bdg/edh	34	5.233	382	4.852
edh/edh	42	5.330	442	4.871
Difference between edh/edh and bdg/bdg		0.189		0.142
t — test		1.305		2.657**

F — test for the significance between G^b neg. and G^b pos.: 12.37.**
F — test for the significance of interaction E×G: 10.48.*
* Significant at 5% level.
** Significant at 1% level.

merely a coincidental connection between some blood group alleles and carcass quality, as the experimental material was not a large sample and the differences found were not significant in many comparisons. Nevertheless we will continue research on this subject.

REFERENCES

BALTZER, J., 1963. Untersuchungen über das Bestehen von Beziehungen zwischen Blutgruppenfaktoren und Daten des Schlachtkörperwertes und Mastleistung des Schweines, Dissertation, Göttingen.

FEDERER, W. T. and ZELEN, M., 1966. Analysis of multifactor classifications with unequal numbers of observations, *Biometrics*, **22**, 3, 525–552.

SCHRAPE, H., 1966. Untersuchungen über Beziehungen zwischen Blutgruppenfaktoren und Leistungseigenschaften beim Schwein, Dissertation, Göttingen.

TIKHONOV, W. N., 1966. Immunogenetic analysis of polymorphism in blood groups in connection with some problems of selection, Thesis, Novosibirsk.